Robert N. Bateson, P.E.
Anoka–Ramsey Community College

INTRODUCTION TO CONTROL SYSTEM TECHNOLOGY

Fourth Edition

Merrill, an imprint of
Macmillan Publishing Company
New York

Maxwell Macmillan Canada
Toronto

Maxwell Macmillan International
New York   Oxford   Singapore   Sydney

Editor: Stephen Helba
Developmental Editor: Monica Ohlinger
Production Editor: Louise N. Sette
Art Coordinator: Lorraine Woost
Text Designer: Anne Flanagan
Cover Designer: Cathleen Norz
Production Buyer: Pamela D. Bennett

This book was set in Times Roman by Syntax International and was printed and bound by R. R. Donnelley & Sons Company. The cover was printed by Phoenix Color Corp.

Macmillan Publishing Company
866 Third Avenue
New York, NY 10022
Macmillan Publishing Company is part of the
Maxwell Communication Group of Companies.

Maxwell Macmillan Canada, Inc.
1200 Eglinton Avenue East, Suite 200
Don Mills, Ontario M3C 3N1

Library of Congress Cataloging-in-Publication Data

Bateson, Robert.
    Introduction to control system technology/Robert N. Bateson.—
    4th ed.
        p.   cm.
    Includes bibliographical references and index.
    ISBN 0-02-306463-3
    1. Automatic control. I. Title.
TJ213.B33   1992
629.8—dc20   92-6961
                                                                CIP

Printing: 1 2 3 4 5 6 7 8 9   Year: 3 4 5 6 7

Preface

The goal of this book is to provide both a textbook on control system technology and a reference that engineers and technicians can include in their personal library. The text covers the terminology, concepts, principles, procedures, and computations used by engineers and technicians to analyze, select, specify, design, and maintain all parts of a control system. Emphasis is on the application of established methodology with the aid of examples, calculators, and computer programs.

Calculus is not a prerequisite for this book, but neither is it avoided. Terms such as *derivative control mode*, *integral control mode*, and *integral process* are part of the language. A student of control must be reasonably comfortable using these terms in discussions about control systems. Helping the student master the language of control is one of the overall objectives of this text. Derivatives and integrals are introduced and explained as they are required. Emphasis is on developing an intuitive grasp of how derivatives and integrals relate to physical systems.

Control is by nature an interdisciplinary subject. That is what makes it such an interesting field of study. The model of a single control loop may include electrical, thermal, mechanical, fluid flow, and chemical elements. A notable feature of this book is the use of analogies to develop common elements for modeling and analyzing electrical, fluid flow, thermal, and mechanical components. These analogies help students translate knowledge they have of one type of component to components of other types.

The fourth edition includes a lot of "fine tuning" and several notable additions. A section on data sampling and conversion has been added to Chapter 7, and a section on control system compensation has been added to Chapter 17. In Chapter 16, the coverage of Nyquist diagrams and root locus has been expanded. High resolution graphics has been added to the design program, providing an excellent visual aid that greatly improves the usefulness of this program for both teaching and design.

The new program "DESIGN" produces Bode plots that clearly show how each control mode changes the shape of the open-loop frequency response of the system. Design decision data is also displayed on the screen. The designer uses the design decision data to determine the PID control mode values. A major feature of program

"DESIGN" is the ease with which the designer can go back and change any control mode. This enables the designer to use a "what-if" analysis in a "design by trial" procedure to search for the best possible control system design. Program "DESIGN" is the culmination of a dream the author has pursued for a number of years. Listings of the program are included in Appendix F, and executable programs are available on a disk provided with the Instructor's Resource Manual. Copies of the disk can also be obtained by writing to the author in care of Macmillan Publishing Company, P.O. Box 29761, Brooklyn Center, MN 55429 ($5 to cover postage and handling would be appreciated).

The book consists of five parts. Part One is an introduction to the terminology, concepts, and methods used to describe control systems. Parts Two, Three, and Four cover the three operations of control: measurement, manipulation, and control. Part Five is concerned with the analysis and design of control systems. Each chapter begins with a set of learning objectives and ends with a glossary of terms. There is sufficient material for a two-semester course, and there are a number of possible sequences of selected chapters for a one-semester course. The following are some suggested sequences for one- and two-semester courses.

Suggested One-Semester Sequences

A. Process Control Analysis and Design:
   Chapters 1, 2, 4, 5, 6, 7 (7.1–7.3), 11 (11.4, 11.5), 14 (14.1–14.3), 15, 16 (16.1–16.9), 17 (17.1–17.6)

B. Servo Control Analysis and Design:
   Chapters 1, 2, 4, 5, 6, 7 (7.1–7.3), 11 (11.3), 12, 14, 15, 16, 17 (17.1, 17.5–17.7)

C. Sequential and PID Control:
   Chapters 1, 2, 3, 6, 11, 12 (12.4), 13, 14, 17 (17.1–17.4)

D. Data Acquisition and Control:
   Chapters 1, 2, 6, 7, 8, 9, 10, 14, 17 (17.1–17.4)

Suggested Two-Semester Sequence

Semester 1: Data Acquisition: Chapters 1–10
Semester 2: Control: Chapters 11–17

ACKNOWLEDGMENTS

I would like to acknowledge and thank the many people who supported me in the preparation of this book. My wife, Betty, has been wonderfully patient and understanding throughout the project. My children, Mark, Karen, and Paul, were always patient, understanding, and supportive.

Special recognition must go to Professor Samuel Kraemer, Oklahoma State University; Dr. Lee Rosenthal, Fairleigh Dickinson University; and Keith D. Graham, Honeywell, Inc. Professor Kraemer provided the impetus and the ideas for the sections on root locus and data sampling, and much of the "fine tuning" is due to his helpful suggestions. Dr Rosenthal forced many rewrites with his insistence on clear, concise statements and sound pedagogy. Mr. Graham provided the insight of a control engineer and also contributed to the "fine tuning."

Don Craighead, CEO of Power/mation, provided a wealth of technical information and was a wellhead of inspiration. Ed Lawrence, a friend and supporter for over eighteen years, contributed both ideas and encouragement.

Thanks also to the team at Macmillan Publishing: Stephen Helba, Executive Editor; Monica Ohlinger, Developmental Editor; Louise Sette, Production Editor; Jeff Smith, Software Editor; and Lorraine Woost, Art Coordinator.
I would also like to thank the following reviewers for their helpful comments and suggestions: A. O. Brown, III, Pittsburg State University; Rex Klopfenstein, Jr, Bowling Green State University; Steven D. Rice, Missoula Vocational Technical Center; Sammy Shina, University of Lowell; and Paul F. Weingartner, Cincinnati Technical College.

Robert N. Bateson

Contents

PART ONE

Introduction

CHAPTER 1

Basic Concepts and Terminology

OBJECTIVES

Every profession, every subject, even every hobby has its own language: a set of concepts, symbols, and words that people in that field use to express ideas.

The purpose of this chapter is to introduce you to the fundamental language of control system technology. After completing this chapter, you will be able to

1. Define a control system
2. Name two relationships established by the transfer function of a component
3. Determine the gain, the phase difference, and the transfer function of a linear component at a specified frequency given values of the input and output signals
4. Differentiate between open-loop control and closed-loop control and give an example of each
5. Sketch the block diagram of a closed-loop control system and name each component and each signal in the diagram
6. Describe the function of each component and signal in Figure 1.7
7. Describe the operations performed by a feedback control system
8. Describe the control actions of the P, I, and D control modes and name the four combinations commonly used in control systems
9. Read control system drawings that use the ISA instrumentation symbols and identification code
10. List six general benefits of automatic control
11. Discuss load and load changes in a control system
12. List and explain three criteria of good control
13. Use the terms listed at the end of this chapter in discussions about control systems

1.1 INTRODUCTION

Control systems are everywhere around us and within us.* Many complex control systems are included among the functions of the human body. An elaborate control system centered in the hypothalamus of the brain maintains body temperature at 37 degrees Celsius (°C) despite changes in physical activity and external ambience. In one control system—the eye—the diameter of the pupil automatically adjusts to control the amount of light that reaches the retina. Another control system maintains the level of sodium ion concentration in the fluid that surrounds the individual cells.

Threading a needle and driving an automobile are two ways in which the human body functions as a complex controller. The eyes are the sensor that detects the position of the needle and thread, or of the automobile and the center of the road. A complex controller, the brain, compares the two positions and determines which actions must be performed to accomplish the desired result. The body implements the control action by moving the thread or turning the steering wheel; an experienced driver will anticipate all types of disturbances to the system, such as a rough section of pavement or a slow-moving vehicle ahead. It would be very difficult to reproduce in an automatic controller the many judgments that an average person makes daily and unconsciously.

Control systems regulate temperature in homes, schools, and buildings of all types. They also affect the production of goods and services by ensuring the purity and uniformity of the food we eat and by maintaining the quality of products from paper mills, steel mills, chemical plants, refineries, and other types of manufacturing plants. Control systems help protect our environment by minimizing waste material that must be discarded, thus reducing manufacturing costs and minimizing the waste disposal problem. Sewage and waste treatment also requires the use of automatic control systems.

A *control system* is any group of components that maintains a desired result or value. From the previous examples it is clear that a great variety of components may be a part of a single control system, whether they are electrical, electronic, mechanical, hydraulic, pneumatic, human, or any combination of these. The desired result is a value of some variable in the system, for example, the direction of an automobile, the temperature of a room, the level of liquid in a tank, or the pressure in a pipe. The variable whose value is controlled is called the *controlled variable*.

To achieve control, there must be another variable in the system that can influence the controlled variable. Most systems have several such variables. The control system maintains the desired result by manipulating the value of one of these influential variables. The variable that is manipulated is called the *manipulated variable*. The steering wheel of an automobile is an example of a manipulated variable.

---

* An excellent idea of the scope of control systems is given in an Instrument Society of America film, "Principles of Frequency Response," 1958.

> *Definition of a Control System*
>
> A control system is a group of components that maintains a desired result by manipulating the value of another variable in the system.

1.2 BLOCK DIAGRAMS AND TRANSFER FUNCTIONS

Although it is not unusual to find several kinds of components in a single control system, or two systems with completely different kinds of components, any control system can be described by a set of mathematical equations that define the characteristics of each component. A wide range of control problems—including processes, machine tools, servomechanisms, space vehicles, traffic, and even economics—can be analyzed by the same mathematical methods. The important feature of each component is the effect it has on the system. The *block diagram* is a method of representing a control system that retains only this important feature of each component. *Signal lines* indicate the input and output signals of the component, as shown in Figure 1.1.

Each component receives an input signal from some part of the system and produces an output signal for another part of the system. The signals can be electric current, voltage, air pressure, liquid flow rate, liquid pressure, temperature, speed, acceleration, position, direction, or others. The signal paths can be electric wires, pneumatic tubes, hydraulic lines, mechanical linkages, or anything that transfers a signal from one component to another. The component may use some source of energy to increase the power of the output signal. Figure 1.2 illustrates block representations of various components.

Block Diagrams

A block diagram consists of a block representing each component in a control system connected by lines that represent the signal paths. Figure 1.3 shows a very simple block diagram of a person driving an automobile. The driver's sense of sight provides the two input signals: the position of the automobile and the position of the center of the road. The driver compares the two positions and determines the position of the steering wheel that will maintain the proper position of the automobile. To implement the decision, the driver's hands and arms move the steering wheel to the new position.

Figure 1.1   Block representation of a component. The energy source is not shown on most block diagrams. However, many components do have an external energy source that makes amplification of the input signal possible.

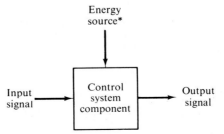

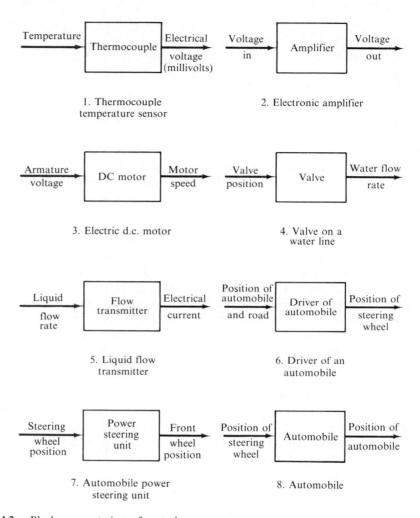

Figure 1.2   Block representations of control system components.

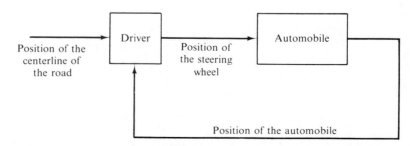

Figure 1.3   Simplified block diagram of a person driving an automobile.

The automobile responds to the change in steering wheel position with a corresponding change in direction. After a short time has elasped, the new direction moves the automobile to a new position. Thus, there is a time delay between a change in position of the steering wheel and the position of the automobile. This time delay is included in the mathematical equation of the block representing the automobile.

The loop in the block diagram indicates a fundamental concept of control. The actual position of the automobile is used to determine the correction necessary to maintain the desired position. This concept is called *feedback*, and control systems with feedback are called *closed-loop control systems*. Control systems that do not have feedback are called *open-loop control systems* because their block diagram does not have a loop.

Transfer Functions

The most important characteristic of a component is the relationship between the input signal and the output signal. This relationship is expressed by the *transfer function* of the component, which is defined as the ratio of the output signal divided by the input signal. (Mostly, it is the Laplace transform of the output signal divided by the Laplace transform of the input signal—further details are covered in Chapter 5.) Refer to the block diagram of a thermocouple shown in Figure 1.2, item 1. If we represent the input temperature by T, the output voltage by V, and the transfer function by H, then $H = V/T$ and $V = HT$. Thus, if we know the input signal and the transfer function, then we can compute the output signal by multiplying the input by the transfer function.

The transfer function consists of two parts. One part is the *size* relationship between the input and the output. The other part is the *timing* between the input and output. For example, the size relationship may be such that the output is twice (or half) as large as the input, and the timing relationship may be such that there is a delay of 2 seconds between a change in the input and the corresponding change in the output.

If the component is linear and the input signal is a sinusoidal signal, the size relationship is measured by *gain* and the timing is measured by *phase difference*. Figure 1.4 illustrates a linear component with a sinusoidal input signal. The gain of the component is the ratio of the amplitude of the output signal divided by the amplitude of the input signal.

$$\text{Gain} = \frac{\text{amplitude of the output signal (output units)}}{\text{amplitude of the input signal (input units)}}$$

The phase difference of the component is the phase angle of the output signal minus the phase angle of the input signal.

$$\text{Phase difference} = \text{output phase angle} - \text{input phase angle}$$

Complex numbers (in polar form) are most conveniently used to represent values of the input, the output, and the transfer function. In Figure 1.4, the input is represented

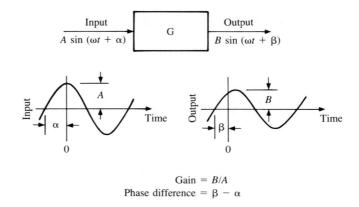

$$\text{Gain} = B/A$$
$$\text{Phase difference} = \beta - \alpha$$

Figure 1.4 Gain and phase difference of a linear component.

by the complex number $A\underline{/\alpha}$ and the output by the complex number $B\underline{/\beta}$. The transfer function, G, is represented by the complex number obtained by dividing the output, $B\underline{/\beta}$, by the input, $A\underline{/\alpha}$:

$$G = B\underline{/\beta}/A\underline{/\alpha} = (B/A)\underline{/\beta - \alpha}$$

Thus the transfer function, G, is represented by the complex number whose magnitude is the gain of the component, B/A, and whose angle is the phase of the output minus the phase of the input.

The gain of a component is often expressed as the ratio of the change in the amplitude of the output divided by the corresponding change in the amplitude of the input.

$$\text{Gain} = \frac{\text{change in output amplitude (output units)}}{\text{change in input amplitude (input units)}}$$

The gain of a component has the dimension of output units over input units. Thus an amplifier that produces a 10-volt (V) change in output for each 1-V change in input has a gain of 10 V per volt. A direct-current (dc) motor that produces a change in speed of 1000 revolutions per minute (rpm) for each 1-V change in input has a gain of 1000 rpm per volt. A thermocouple that produces an output change of 0.06 millivolt (mV) for each 1°C change in temperature has a gain of 0.06 mV/°C.

The gain and phase difference of a component for a given frequency are referred to as the *frequency response* of the component at that frequency. As an example, at a frequency of 1 Hz, a certain control system component has a gain of 0.995 and a phase difference of 5.71°. At a frequency of 10 Hz, the gain is 0.707 and the phase difference is 45°. At a frequency of 100 Hz, the gain is 0.0995 and the phase difference is 84.29°. These figures are given here only as an illustration. A more complete discussion is reserved for later chapters.

> The transfer function of a component describes the size and timing relationship between the output signal and the input signal.

Example 1.1

The input to a linear control system component is a 0.5-Hz sinusoidal signal with a peak amplitude of 5.3 V and a phase angle of 30°. The output of the component has a peak amplitude of 14 milliamperes (mA) and a phase angle of 25°. Determine the gain, the phase difference, and the transfer function for these conditions.

Solution

$$\text{Gain} = 14 \text{ mA}/5.3 \text{ V}$$
$$= 2.64 \text{ mA}/\text{V}$$
$$\text{Phase difference} = 25 - 30$$
$$= -5°$$
$$\text{Transfer function} = 2.64 \underline{/-5°} \text{ mA}/\text{V}$$

1.3   OPEN-LOOP CONTROL

An open-loop control system does not compare the actual result with the desired result to determine the control action. Instead, a calibrated setting—previously determined by some sort of calibration procedure or calculation—is used to obtain the desired result.

The needle valve with a calibrated dial shown in Figure 1.5 is an example of an open-loop control system. The calibration curve is usually obtained by measuring the flow rate for several dial settings. As the calibration curve indicates, different calibration lines are obtained for different pressure drops. Assume that a flow rate of F_2 is desired and a setting of S is used. As long as the pressure drop across the valve remains equal to P_2, the flow rate will remain F_2. If the pressure drop changes to P_1, the flow rate will change to F_1. The open-loop control cannot correct for unexpected changes in the pressure drop.

The firing of a rifle bullet is another example of an open-loop control system. The desired result is to direct the bullet to the bull's-eye. The actual result is the direction of the bullet after the gun has been fired. The open-loop control occurs when the rifle is aimed at the bull's-eye and the trigger is pulled. Once the bullet leaves the barrel, it is on its own: If a sudden gust of wind comes up, the direction will change and no correction will be possible.

The primary advantage of open-loop control is that it is less expensive than closed-loop control: It is not necessary to measure the actual result. In addition, the controller is much simpler because corrective action based on the error is not required. The disadvantage of open-loop control is that errors caused by unexpected

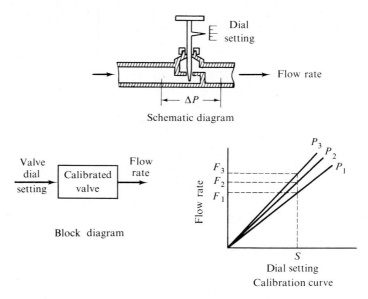

Figure 1.5 A calibrated needle valve is an example of an open-loop control system.

disturbances are not corrected. Often a human operator must correct slowly changing disturbances by manual adjustment. In this case, the operator is actually closing the loop by providing the feedback signal.

1.4 CLOSED-LOOP CONTROL: FEEDBACK

Feedback is the action of measuring the difference between the actual result and the desired result, and using that difference to drive the actual result toward the desired result. The term *feedback* comes from the direction in which the measured value signal travels in the block diagram. The signal begins at the output of the controlled system and ends at the input to the controller. The output of the controller is the input to the controlled system. Thus the measured value signal is fed back from the output of the controlled system to the input. The term *closed loop* refers to the loop created by the feedback path.

Block diagrams of closed-loop control systems are shown in Figures 1.6 and 1.7. Figure 1.6 is used in the design of servomechanisms, while Figure 1.7 is used in the design of process control systems. The names of the components and variables in Figure 1.7 are used throughout this book. However, we will develop the closed-loop transfer function of both the servo control system (Figure 1.6) and the process control system (Figure 1.7). You should be thoroughly familiar with these terms and the following operations, which form the basis of a feedback control system.

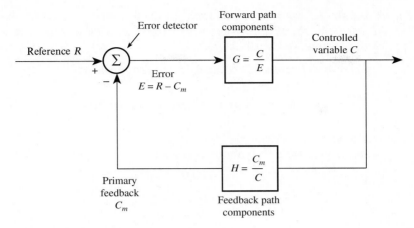

Figure 1.6   Block diagram of a closed-loop servo control system.

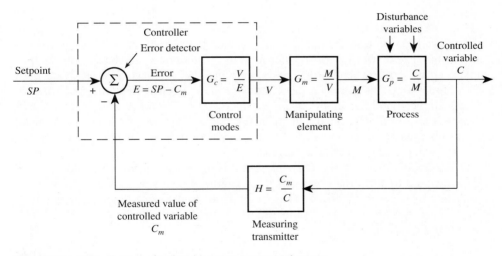

Figure 1.7   Block diagram of a closed-loop process control system.

> *Operations Performed by a Feedback Control System*
>
>   *Measurement:* measure the value of the controlled variable
>
>   *Decision:* compute the error (desired value minus measured value) and use the error to form a control action
>
>   *Manipulation:* use the control action to manipulate some variables in the process in a way that will tend to reduce the error

In Figure 1.6, the reference (R) is the input to the servo control system and the controlled variable (C) is the output. We now proceed to derive the transfer function, C/R, of the closed-loop servo control system.

$$\text{Error} = \text{reference} - \text{controlled variable, measured}$$
$$E = R - C_m$$
$$\text{Controlled variable} = \text{error} \times \text{system transfer function}$$
$$C = EG$$

Controlled variable, measured
$$= \text{controlled variable} \times \text{measuring transmitter transfer function}$$

$$C_m = CH$$
$$C = (R - C_m)G$$
$$C = (R - CH)G$$
$$C + CGH = RG$$
$$C(1 + GH) = RG$$
$$\frac{C}{R} = \frac{G}{1 + GH} \qquad \text{(1.1)}$$

Equation (1.1) is the transfer function of a closed-loop servo control system (servomechanism). The forward path transfer function (G) contains all the system components, such as motors, generators, gears, amplifiers, and so on. The feedback path transfer function (H) is usually a passive device that converts the controlled variable into a suitable signal for input to the error detector. The letter B is often used to represent the primary feedback signal in the block diagram of a servomechanism. The author prefers the notation C_m for this signal to indicate its relationship as the measured value of the controlled variable.

Occasionally, the primary feedback signal is inverted and must be added to the reference signal to form the error signal. In Figure 1.6, this positive feedback is accomplished by changing the sign of the lower input to the error detector from minus to plus. It is left as an exercise to show that the transfer function for positive feedback is given by Equation (1.2):

$$\frac{C}{R} = \frac{G}{1 - GH} \qquad \text{(1.2)}$$

In Figure 1.7, the setpoint (SP) is the input to the process control system, and the controlled variable (C) is the output. The feedback path consists of one component, the measuring transmitter with transfer function H. The forward path consists of three components (the control modes, the manipulating element, and the process) with transfer functions G_c, G_m, and G_p, respectively. The overall forward transfer function (G) is the product of the three component transfer functions:

$$G = G_c G_m G_p$$

The performance of a control system is usually based on a comparison between the setpoint (SP) and the measured value of the controlled variable (C_m). The reason

C_m is used instead of C is that C_m is measurable and available, but C is not. We now proceed to derive the transfer function, C_m/SP, of the closed-loop process control system:

$$E = SP - C_m$$
$$C = EG$$
$$C_m = CH$$
$$C_m = EGH$$
$$C_m = (SP - C_m)GH$$
$$C_m + C_mGH = (SP)GH$$
$$C_m(1 + GH) = (SP)GH$$
$$\frac{C_m}{SP} = \frac{GH}{1 + GH} \tag{1.3}$$

Equation (1.3) is the transfer function of a closed-loop process control system. The following is a description of each component in Figure 1.7.

Process

The *process block* in Figure 1.7 represents everything performed in and by the equipment in which a variable is controlled. The process includes everything that affects the controlled or *process variable* except the controller and the final control element.

Measuring Transmitter

The *measuring transmitter* or *sensor* senses the value of the controlled variable and converts it into a usable signal. Although the measuring transmitter is considered as one block, it usually consists of a primary sensing element and a signal transducer (or signal converter). The term *measuring transmitter* is a general term to cover all types of signals. In specific cases, the word *measuring* is replaced by the name of the measured signal (e.g., temperature transmitter, flow transmitter, pressure transmitter, etc.).

Figure 1.8 shows the input/output curve of a typical temperature transmitter. The primary element could be a thermocouple, a resistance element, a thermistor,

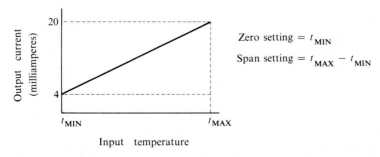

Figure 1.8   Input-output graph of a temperature-measuring transmitter.

or a filled thermal element. The signal transducer receives the output of the primary element and produces an electric current signal. For example, a thermocouple converts temperature into a millivolt signal, and the thermocouple transducer converts the millivolt signal into an electric current in the range 4 to 20 mA. A resistance element converts temperature into a resistance value, and the resistance transducer converts the resistance value into an electric current signal. Other primary elements are handled in a similar manner.

Controller

The controller includes the error detector and a unit that implements the control modes. The *error detector* computes the difference between the measured value of the controlled variable and the desired value (or setpoint). The difference is called the *error* and is computed according to the following equation:

$$\text{Error} = \text{setpoint} - \text{measured value of controlled variable}$$

or

$$E = SP - C_m$$

The *control modes* convert the error into a control action or *controller output* that will tend to reduce the error. The three most common control modes are the proportional mode (P), the integral mode (I), and the derivative mode (D). The three modes are defined mathematically in Chapter 14. In this chapter you need only to know the names of the three modes and have an intuitive understanding of how they work. The following discussion will show that the names of the modes suggest the types of control action that is formed.

The *proportional mode* (P) is the simplest of the three modes. It produces a control action that is proportional to the error. If the error is small, the proportional mode produces a small control action. If the error is large, the proportional mode produces a large control action. The proportional mode is accomplished by simply multiplying the error by a gain constant, K.

The *integral mode* (I) produces a control action that continues to increase its corrective effect as long as the error persists. If the error is small, the integral mode increases the correction slowly. If the error is large, the integral action increases the correction more rapidly. In fact, the rate that the correction increases is proportional to the error signal. Mathematically, the integral control action is accomplished by forming the integral of the error signal.

The *derivative mode* (D) produces a control action that is proportional to the rate at which the error is changing. For example, if the error is increasing rapidly, it will not be long before there is a large error. The derivative mode attempts to prevent this future error by producing a corrective action proportional to how fast the error is changing. The derivative mode is an attempt to anticipate a large error and head it off with a corrective action based on how quickly the error is changing. Mathematically, the derivative control action is accomplished by forming the derivative of the error signal.

The proportional mode may be used alone or in combination with either or both of the other two modes. The integral mode can be used alone, but it almost never is. The derivative mode cannot be used alone. Thus the common control mode combinations are: P, PI, PD, and PID.

Manipulating Element

The *manipulating element* uses the controller output to regulate the manipulated variable and usually consists of two parts. The first part is called an *actuator,* and the second part is called the *final controlling element.* The actuator translates the controller output into an action on the final controlling element, and the final controlling element directly changes the value of the manipulated variable. Valves, dampers, fans, pumps, and heating elements are examples of manipulating elements. The valve that controls the fuel flow in a home heating system is another example of a manipulating element.

A pneumatic control valve is often used as the manipulating element in processes (see Figure 1.9). The actuator consists of an air-loaded diaphragm acting against a spring. As the air pressure on the diaphragm goes from 3 psi to 15 psi, the stem of the valve will move from open to closed (air to close) or from closed to open (air to open).

Variable Names

The *controlled variable (C)* is the process output variable that is to be controlled. In a process control system, the controlled variable is usually an output variable that is

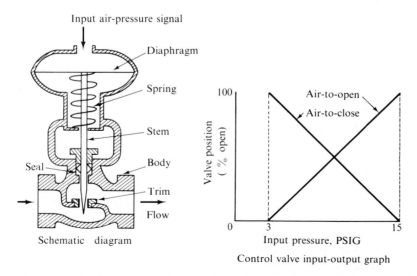

Figure 1.9 A pneumatic control valve has two possible actions: air-to-close and air-to-open. In an air-to-close valve, the valve stem moves from open to closed as the air pressure goes from 3 to 15 psi. An air-to-open valve moves from closed to open with the same change in air pressure.

a good measure of the quality of the product. The most common controlled variables are position, velocity, temperature, pressure, level, and flow rate.

The *setpoint (SP)* is the desired value of the controlled variable.

The *measured variable* (C_m) is the measured value of the controlled variable. It is the output of the measuring means and usually differs from the actual value of the controlled variable by a small amount.

The *error (E)* is the difference between the setpoint and the measured value of the controlled variable. It is computed according to the equation $E = SP - C_m$.

The *controller output (V)* is the control action intended to drive the measured value of the controlled variable toward the setpoint value. The control action depends on the error signal *(E)* and on the control modes used in the controller.

The *manipulated variable (M)* is the variable regulated by the final controlling element to achieve the desired value of the controlled variable. Obviously, the manipulated variable must be capable of effecting a change in the controlled variable. The manipulated variable is one of the input variables of the process. Changes in the *load* on the process necessitate changes in the manipulated variable to maintain a balanced condition. For this reason, the value of the manipulated variable is used as a measure of the load on the process.

The *disturbance variables (D)* are process input variables that affect the controlled variable but are not controlled by the control system. Disturbance variables are capable of changing the load on the process and are the main reason for using a closed-loop control system.

The primary advantage of closed-loop control is the potential for more accurate control of the process. There are two disadvantages of closed-loop control: (1) closed-loop control is more expensive than open-loop control, and (2) the feedback feature of a closed-loop control system makes it possible for the system to become unstable. An unstable system produces an oscillation of the controlled variable, often with a very large amplitude. (Stability will be studied in detail later.)

> A closed-loop (or feedback) control system measures the difference between the actual value of the controlled variable and the desired value (or setpoint) and uses the difference to drive the actual value toward the desired value.

1.5 CONTROL SYSTEM DRAWINGS

The Instrument Society of America has prepared a standard, "Instrumentation Symbols and Identifications," ANSI/ISA-S5.1-1984, to "establish a uniform means of designating instruments and instrument systems used for measurement and control." This standard presents a designation system that includes symbols and an identification code that is "suitable for use whenever any reference to an instrument is required." Applications include flow diagrams, instrumentation drawings, specifications, construction drawings, technical papers, tagging of instruments, and other uses. Appendix D includes material abstracted from ANSI/ISA-S5.1-1984.

A circular symbol called a *balloon* is the general instrument symbol. The instrument is identified by the code placed inside its balloon. The identification code consists of a *functional identification* in the top half of the balloon and a *loop identification* in the bottom half. The first letter in the functional identification defines the measured or initiating variable of the control loop (e.g., flow, level, pressure, temperature). Up to three additional letters may be used to name functions of the individual instrument (e.g., indicator, recorder, controller, valve). The standard also defines symbols for instrument lines, control valve bodies, actuators, primary elements, various functions, and other devices.

Figure 1.10 illustrates the use of the standard symbols and identification code in a process control drawing. The process blends and heats a mixture of water and

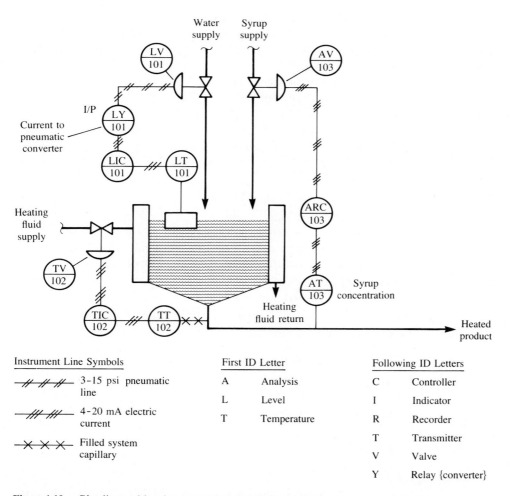

Instrument Line Symbols		First ID Letter		Following ID Letters	
┼┼┼	3–15 psi pneumatic line	A	Analysis	C	Controller
┼┼┼	4–20 mA electric current	L	Level	I	Indicator
✕✕✕	Filled system capillary	T	Temperature	R	Recorder
				T	Transmitter
				V	Valve
				Y	Relay {converter}

Figure 1.10   Blending and heating system instrumentation drawing.

syrup. This system has three control loops, with loop identification numbers of 101, 102, and 103. The first two digits designate the area in the plant where this system is located. The third digit identifies a particular control loop. Loop 101 is a level control loop, as indicated by the first letter in the function code of each instrument in the loop. The meaning of each code is as follows:

LT-101	Level transmitter Uses a float to sense the level of the liquid in the tank and transduces the signal into an electric current in the range 4 to 20 mA.
LIC-101	Level-indicating controller Uses the milliampere signal from the level transmitter to produce a control signal in the range 4 to 20 mA.
LY-101	Level current to pneumatic converter Converts the milliampere output from the controller into a pneumatic signal in the range 3 to 15 pounds per square inch (psi).
LV-101	Level control valve Uses the pneumatic signal from the I/P converter to position the stem of the level control valve.
TT-102	Temperature transmitter Uses a filled bulb to sense the temperature of the product leaving the blending tank and transduces the signal into an electric current in the range 4 to 20 mA.
TIC-102	Temperature-indicating controller Uses the milliampere signal from the temperature transmitter to produce a control signal in the range 4 to 20 mA.
TV-102	Temperature control valve Uses the milliampere signal from the temperature controller to position the stem of the temperature control valve.
AT-103	Analysis transmitter Senses the concentration of syrup in the product and transduces the signal into an electric current in the range 4 to 20 mA.
ARC-103	Analysis recording controller Uses the milliampere signal from the analysis transmitter to produce a control signal in the range 4 to 20 mA.
AV-103	Analysis control valve Uses the milliampere signal from the analysis controller to position the stem of the analysis control valve.

1.6 NONLINEARITIES

Most control system analysis and design is done with the assumption that all components in the system are linear. Actually, there are several different forms of nonlinearity

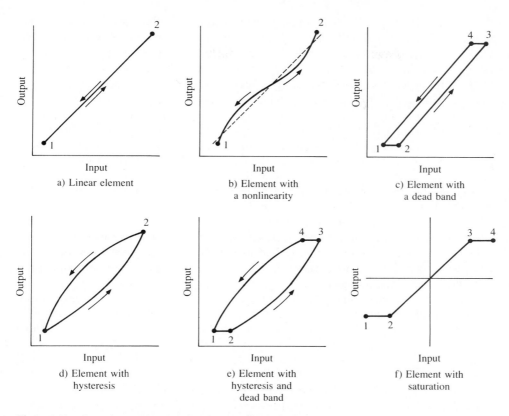

Figure 1.11 Input/output graphs of various nonlinear elements.

that occur in components. The intent of this section is to give you an intuitive understanding of linearity, nonlinearity, hysteresis, dead band, and saturation.*

Linearity means that the input/output (I/O) graph of the component is a perfectly straight line, as illustrated in Figure 1.11a. The term *linearity* also refers to how nearly the I/O graph approximates a straight line (see Figure 1.11b). We express linearity as the maximum deviation between an average I/O graph and a straight line so positioned that it minimizes the maximum deviation. The average I/O graph is constructed by averaging the values obtained from at least two full-range traverses in each direction. Thus we measure nonlinearity and express it as linearity.

Dead band is the range of values through which an input can be varied without producing an observable change in the output. Backlash in gears is one example of dead band and the term *backlash* is sometimes used in place of *dead band*. Figure 1.11c

---

* Much of the material in this section is based on *Standards and Practices for Instrumentation*, 6th ed. (Research Triangle Park, N.C.: Instrument Society of America, 1980).

shows the I/O graph of a component with a dead band. The locus of input versus output (I/O) points can fall on line (2–3), line (4–1), or anyplace between those two lines. When the input is increasing, the locus of I/O points first moves horizontally until it is on line (2–3). As the input continues to increase, the locus moves up line (2–3). If the input reverses direction and begins to decrease, the locus of I/O points first must move horizontally across to line (4–1). As the input continues to decrease, the locus moves down line (4–1). Line (2–3) is the increasing I/O line and line (4–1) is the decreasing I/O line. Anytime the input reverses direction, the locus of I/O points must first move horizontally from one of these lines to the other. The dead band is the horizontal distance between the increasing and decreasing I/O lines. Dead band is measured by slowly increasing (or decreasing) the input until a change in the output is observed. Record the input value when the change in output was first detected. Then slowly change the input in the opposite direction until a change in the output is again observed. Record the input value when the second change in output was first detected. The dead band is the difference between the two recorded output values. Dead band can also be expressed as the difference between the upscale and downscale output for a single test cycle. This is usually done when a component has both dead band and hysteresis.

Hysteresis is the nonlinearity that causes the value of the output for a given input to depend on the history of previous inputs. The I/O graph of a component with hysteresis forms a loop when the input is changed from one value to a second value and then back to the first value (see Figure 1.11d). Hysteresis is expressed as the maximum difference between the upscale output and the downscale output for a single test cycle.

Figure 1.11e shows the I/O graph of a component with both hysteresis and dead band. *Hysteresis plus dead band* is expressed as the maximum difference between the upscale output and the downscale output for a single test cycle.

Saturation refers to the limitations on the range of values for the output of a component (see Figure 1.11f). All real components reach a saturation limit when the input is increased (or decreased) beyond its limiting value. For example, a control valve may go from closed to open as the pressure in the actuator increases from 3 to 15 psi. The valve remains closed when the pressure is decreased below 3 psi. The valve remains open when the pressure is increased above 15 psi. We say that the valve reaches saturation when the pressure is below 3 psi or above 15 psi.

1.7 BENEFITS OF AUTOMATIC CONTROL

Control systems are becoming steadily more important in our society. We depend on them to such an extent that life would be unimaginable without them. Automatic control has increased the productivity of each worker by releasing skilled operators from routine tasks and by increasing the amount of work done by each worker. Control systems improve the quality and uniformity of manufactured goods and services; many of the products we enjoy would be impossible to produce without automatic controls. Servo systems place tremendous power at our disposal, enabling us to control large equipment such as jet airplanes and ocean ships.

Control systems increase efficiency by reducing waste of materials and energy, an increasing advantage as we seek ways to preserve our environment. Safety is yet another benefit of automatic control. Finally, control systems such as the household heating system and the automatic transmission provide us with increased comfort and convenience.

In summary, the benefits of automatic control fall into the following six broad categories.

1. Increased productivity
2. Improved quality and uniformity
3. Increased efficiency
4. Power assistance
5. Safety
6. Comfort and convenience

1.8 LOAD CHANGES

A control system must balance the material or energy gained by the process against the material or energy lost by the process, to maintain the desired value of the controlled variable. Usually, the material or energy loss is the load on the process, and the manipulated variable must supply the balancing material or energy gain. However, sometimes the opposite condition exists and the manipulated variable must provide the material or energy loss.

To maintain the desired inside temperature, a home heating system must balance the heat supplied by the furnace against the heat lost by the house. The heat lost is the load on the control system, and the energy supplied to the furnace is regulated by the manipulated variable.

To maintain the level at the desired value, a liquid-level control system must balance the input flow rate against the output flow rate. The output flow rate is the load on the system and the input flow rate is the manipulated variable.

To maintain the desired pump speed, the control of a variable-speed motor driving a pump must balance the input power to the motor against the power delivered to the pump. The power delivered to the pump is the load on the system, and the power input to the motor is regulated by the manipulated variable.

To maintain the desired room temperature, an air-conditioning system must balance the heat removed by the air conditioner against the heat gained by the room. The heat gained by the room is the load on the system, and the heat removed by the air conditioner is regulated by the manipulated variable.

The load on a process is always reflected in the manipulated variable. Therefore, the value of the manipulated variable is a measure of the load on the process. Every load change results in a corresponding change in the manipulated variable and, consequently, a corresponding change in the setting of the final controlling element. Consider a sudden increase in the load for the pump control system described above. The increase in load tends to reduce the motor speed. The controller senses the reduced motor speed and produces a control action that increases the power input to the motor. In an ideal situation, the control action will cause the manipulated variable to match the increased load and the pump speed will remain at the desired value. Within the control loop, the only variable that reflects the load change is the manipulated variable. For this reason, it makes sense to define the load on the control system in terms of the manipulated variable.

> The load on a control system is measured by the value of the manipulated variable required by the process at any one time in order to maintain a balanced condition.

The load on a control system does not remain constant. Any uncontrolled variable that affects the controlled variable is capable of causing a load change. Each load change necessitates a corresponding change in the manipulated variable in order to maintain the controlled variable at the desired value. A closed-loop control system automatically makes the necessary change in the manipulated variable; an open-loop control system does not make the necessary change. Thus a closed-loop control system is necessary if automatic adjustment to load changes is desired.

There are usually several uncontrolled conditions in a process that are capable of causing a load change. Some examples of load changes are:

1. A *change in demand* by the controlled medium. For example, opening the door of a house in winter necessitates more heat to keep the inside temperature at the desired value. Closing the door requires less heat. Both are load changes. In a manufacturing process, a change in production rate almost always results in a load change. In a heat exchanger, for example, a flowing liquid is continuously heated with steam. A change in the liquid flow rate is a load change because more heat is required.

2. A *change in the quality* of the manipulated variable. For example, a change in the heat content of the fuel supplied to a burner requires a change in the rate at which the fuel is supplied to the burner. In a neutralizing process, a solution of sodium bicarbonate is used to neutralize a fiber ribbon. A decrease in the concentration of sodium bicarbonate is a load change because more neutralizing solution is required.

3. A *change in ambient conditions.* For example, if the outside temperature drops, more heat is required to maintain the desired temperature in a house.

4. A *change in the amount of energy absorbed or supplied* within the process. For example, using the range to prepare supper supplies a house with a large quantity of heat. Thus less heat is required from the furnace to maintain the desired temperature. Chemical reactions often generate or absorb heat as part of the reaction; these are load changes because, as the process generates or absorbs heat, less or more heat is required from the manipulated variable.

1.9 DAMPING AND INSTABILITY

The gain of the controller determines a very important characteristic of a control system's response: the type of damping or instability that the system displays in response to a disturbance. The five general conditions are illustrated in Figure 1.12. As the gain of the controller is increased, the response changes in the following order: overdamped, critically damped, underdamped, unstable with constant amplitude, and unstable with increasing amplitude. Obviously, neither the unstable response nor the overdamped response satisfies the objective of minimizing the error. Typically, the optimum response is either critically damped or slightly underdamped. Exactly how much damping is optimum depends on the requirements of the process.

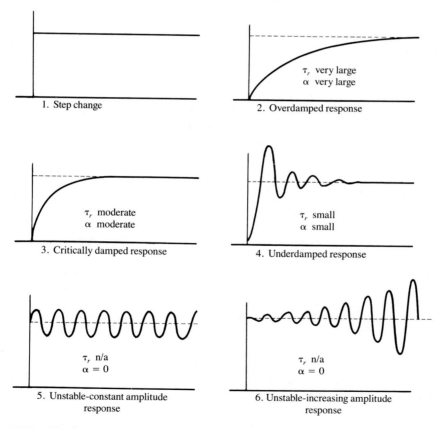

Figure 1.12   The five types of response to a step change in load or setpoint are characterized by the rise time, τ_r, and the damping constant, α. Rise time is the time it takes a signal to go from 10% to 90% of the total change in response to the step change. Damping constant is a measure of the amount of damping in the system.

Further insight about damping can be obtained by considering a familiar oscillating system—a child bouncing a ball. The ball will continue to bounce as long as the child pushes down when the ball is moving down (i.e., the force is in the same direction as the motion of the ball). The bouncing will die down quickly if the child pushes down when the ball is moving up (i.e., the force is in opposition to the motion of the ball). The oscillations of the ball are damped out by a force in opposition to the motion. Extending this concept to control systems, *damping* is a force or signal that opposes the motion (or rate of change) of the controlled variable.

Several stabilizing techniques are used to increase the damping in a system and thereby to allow a higher gain in the controller. The general idea is to find a force or signal that will oppose changes in the controlled variable. One such signal is the rate of change of the controlled variable. In mathematics, the derivative of a variable is equal to its rate of change, and this signal is referred to as the *derivative* of the controlled variable. Damping is increased if the derivative of the controlled variable is subtracted from the error signal before it goes to the controller. This technique is sometimes called *output derivative damping*.

Another stabilizing signal is the derivative of the error signal. If the setpoint is constant, this signal is equal to the negative of the derivative of the controlled variable. Damping is increased if the derivative of the error is added to the error signal before it goes to the controller. This technique is usually called the *derivative control mode*.

Viscous damping is a stabilizing technique sometimes used in position control systems. It operates on the fact that frictional forces always oppose motion. A simple brake, a fluid brake, or an eddy-current brake may be used to apply the damping force.

1.10 OBJECTIVES OF A CONTROL SYSTEM

At first glance, the objective of a control system seems quite simple—to maintain the controlled variable exactly equal to the setpoint at all times, regardless of load changes or setpoint changes. To do this, the control system must respond to a change before the error occurs; unfortunately, feedback is never perfect because it does not act until an error occurs. First, a load change must change the controlled variable; this produces an error. Then the controller acts on the error to produce a change in the manipulated variable. Finally, the change in the manipulated variable drives the controlled variable back toward the setpoint.

It is more realistic for us to expect a control system to obtain as nearly perfect operation as possible. Since the errors in a control system occur after load changes and setpoint changes, it seems natural to define the objectives in terms of the response to such changes. Figure 1.13 shows a typical response of the controlled variable to a step change in load.

One obvious objective is to minimize the maximum value of the error signal. Some control systems (with an integral mode) will eventually reduce the error to zero, whereas others require a residual error to compensate for a load change. In either case, the control system should eventually return the error to a steady, nonchanging value. The time required to accomplish this is called the *settling time*. A second ob-

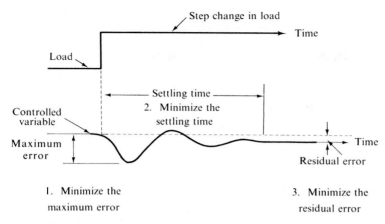

Figure 1.13 Three objectives of a closed-loop control system.

jective of a control system is to minimize the settling time. A third objective is to minimize the *residual error* after settling out.

Unfortunately, these three objectives tend to be incompatible. For instance, the problem of reducing the residual error can be solved by increasing the gain of the controller so that a smaller residual error is required to produce the necessary corrective control action. However, an increase in gain tends to increase the settling time and may increase the maximum value of the error as well. The optimum response is always achieved through some sort of compromise.

Control Objectives

After a load change or setpoint change, the control system should

1. Minimize the maximum value of the error
2. Minimize the settling time
3. Minimize the residual error

1.11 CRITERIA OF GOOD CONTROL

To evaluate a control system effectively, two decisions must be made: (1) the test must be specified, and (2) the criteria of good control must be selected. A step change in setpoint or load is the most common test. A typical step response test is illustrated in Figure 1.13. The three most common criteria of good control are: quarter amplitude decay, critical damping, and integral of absolute error. A discussion of each criterion follows.

1. *Quarter amplitude decay.* This criterion specifies a damped oscillation in which each successive positive peak value is one-fourth of the preceding positive

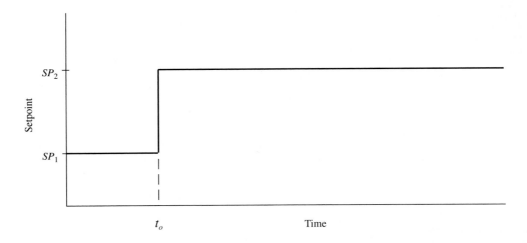

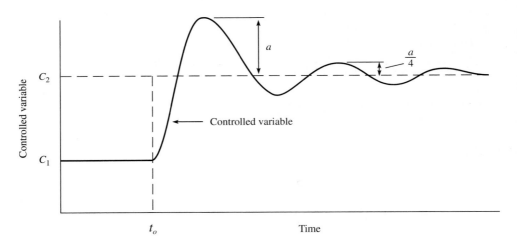

Figure 1.14 Quarter amplitude decay response to a step change in the setpoint.

peak value. Quarter amplitude decay is a popular criterion because it is easy to apply in the field and provides a nearly optimum compromise of the three control objectives. Figure 1.14 illustrates the quarter amplitude decay response.

A variation of quarter amplitude decay is peak percent overshoot (*PPO*). It is a measure of the peak overshoot of the controlled variable with respect to the size of the step change. A *PPO* of 50% is roughly equivalent to quarter amplitude decay. From Figure 1.14, the peak percent overshoot is given by the following equation:

$$PPO = 100\left(\frac{a}{C_2 - C_1}\right)$$

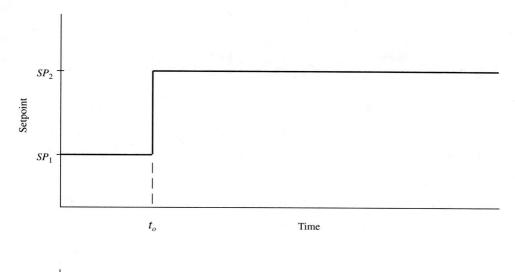

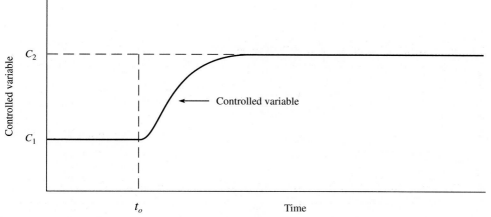

Figure 1.15   Critical damping response to a step change in the setpoint.

2. *Critical damping.* This criterion is used when overshoot above the setpoint is undesirable. Critical damping is the least amount of damping that will produce a response with no overshoot and no oscillation. Electrical instruments and some processes are critically damped. Figure 1.15 illustrates this critical damping criterion.

3. *Minimum integral of absolute error.* This criterion specifies that the total area under the error curve should be minimum. Figure 1.16 illustrates the minimum integral of absolute error criterion. The error is the distance between C_2 and the controlled variable curve. The integral of absolute error is the total shaded area on the curve. This criterion is easy to use when a mathematical model is used to evaluate a control system.

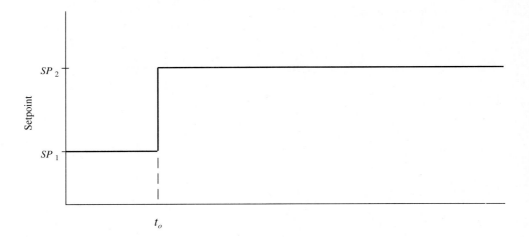

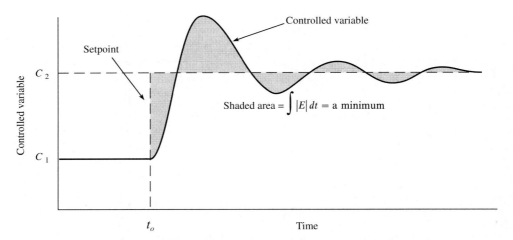

Figure 1.16   Minimum integral of absolute error response to a step change in the setpoint.

GLOSSARY

Actuator: An element that translates the controller output into an action on the final controlling element. (1.4*)

Closed-loop: The type of control system that uses feedback. (1.4)

Controlled variable: The process variable whose value is controlled by the control system. (1.4)

Controller: The component that computes the error signal and uses it to produce the control action. (1.4)

---

* Relevant section.

Controller output: The control action produced by the control modes. (1.4)

Control modes: Methods the controller uses to convert the error signal into a control action. (1.4)

Critical damping: A criterion of good control that permits no overshoot when the setpoint is changed. (1.11)

Damping: The progressive reduction or suppression of oscillation in a component. (1.9)

Dead band: The range of values through which an input can be varied without producing an observable change in the output. (1.6)

Derivative mode: A control mode that produces a control action that is proportional to the rate at which the error is changing. (1.4)

Disturbance variables: Process input variables that affect the controlled variable but are not controlled by the control system. (1.4)

Error: The signal in a controller that is obtained by subtracting the measured value of the controlled variable from the setpoint. (1.4)

Error detector: The element in a controller that computes the error signal. (1.4)

Feedback: The action of measuring the difference between the actual result and the desired result, and using that difference to drive the actual result toward the desired result. (1.4)

Gain: The ratio of the amplitude of the output signal of a component divided by the amplitude of the input signal. (1.2)

Hysteresis: The nonlinearity that causes the value of the output for a given input to depend on the history of previous inputs. (1.6)

Instability: An undesirable characteristic in which the error of a control system oscillates with constant or increasing amplitude. (1.9)

Integral mode: A control mode that produces a control action that is proportional to the accumulation of error over time. (1.4)

Load: The demand on the manipulated variable required to maintain the desired value of the controlled variable. (1.4)

Manipulated variable: The process variable that is acted on by the controller. (1.4)

Manipulating element: The component of a control system that uses the controller output to adjust the manipulated variable. (1.4)

Measured variable: A quantity or condition that is measured (e.g., temperature, flow rate, etc.). (1.4)

Measuring transmitter: The component of a control system that uses a sensing element to measure the controlled variable and converts the response into a usable signal. (1.4)

Minimum integral of absolute error: A criterion of good control that minimizes the accumulation of error over time. (1.11)

Open-loop: The type of control system that does not use feedback. (1.3)

Phase angle: An angular value that fixes the point on a sine wave where we start measuring time. It determines the value of the sinusoidal function when $t = 0$. (1.2)

Process: Everything performed in and by the equipment in which a variable is controlled. (1.4)

Process variable: Any variable in the process. Process controllers often refer to the controlled variable as the process variable. (1.4)

Proportional mode: A control mode that produces a control action that is proportional to the error. (1.4)

Quarter amplitude decay: A criterion of good control that progressively reduces the amplitude of oscillation by a factor of 4. (1.11)

Residual error: The error that remains after all transient responses have faded out. This is sometimes referred to as offset. (1.10)

Saturation: The characteristic that limits the range of the output of a component. (1.6)

Sensor: An element that responds to a parameter to be measured and converts the response into a more usable form. (1.4)

Setpoint: The controller signal that defines the desired value of the controlled variable. (1.4)

Settling time: The time, following a disturbance, that is required for the transient response to fade out. (1.10)

Transfer function: The mathematical expression that establishes the relationship between the input and the output of a component. (1.2)

EXERCISES

1.1 Write a sentence that describes how you act as a controller during a common activity such as taking a shower.

1.2 Draw a block diagram of a typical home heating system with the following components:

(1) A household thermostat where the input signal is the temperature in the living room and the output signal is either on or off.

(2) A solenoid valve where the input is the on or off signal from the thermostat and the output is the flow of gas to the furnace.

(3) A household heating furnace where the input is gas flow from the solenoid valve and the output is heat to the rooms in the house.

(4) The inside of a house where the input is heat from the furnace and the output is the temperature in the living room. *Note:* The output of the living room is also the input to the thermostat, so your diagram should form a closed loop.

1.3 Name the two parts of the relationship between the input and the output of a component, name the function that establishes this relationship, and give an example of each part from your own experience.

1.4 The input to a linear component is a sinusoidal signal with an amplitude of 3.25 psi, a frequency of 0.05 Hz, and a phase angle of 17°. The output is a sinusoidal signal with an amplitude of 1.6 gallons per minute, a frequency of 0.05 Hz, and a phase angle of 39°. Determine the gain, phase difference, and transfer function of the component at a frequency of 0.05 Hz.

1.5 Write a paragraph that describes an example of an open-loop control system taken from your own experiences.

1.6 Write a paragraph that describes an example of a closed-loop control system taken from your own experiences.

1.7 Assume that you are explaining feedback control to a friend who knows nothing about control systems. Explain the operations performed by a feedback control system. Name each component and each signal in the system and explain how the system works.

1.8 List the advantages and disadvantages of open-loop control.

1.9 List the advantages and disadvantages of closed-loop control.

1.10 A certain process consists of a kettle filled with liquid and heated by a gas flame. A thermocouple temperature transmitter measures the temperature of the liquid in the kettle. A control valve manipulates the flow of gas to the burner. The control system components are listed below. Name each component and sketch an instrumentation drawing for this system.

Component Code	Name	Input	Output
TT-201	_____	Temperature	4–20 mA
TRC-201	_____	4–20 mA	4–20 mA
TY-201	_____	4–20 mA	3–15 psi
TV-201	_____	3–15 psi	cfm[a]

[a] "Cfm" stands for gas flow rate in cubic feet per minute.

1.11 Test data from four components are given below. Each test consists of a complete traversal from an input value of 0 to an input value of 25. The data are listed in the order in which the traversal was made. Plot the data from each test on an input/output graph with the input on the horizontal axis and the output on the vertical axis. Use arrows to show the direction of traversal.

Component 1

Input	0	4	8	12	16	20	22	25	20	16	12	8	4	2	0
Output	20	20	69	118	167	216	240	240	240	191	142	93	44	20	20

Component 2

Input	0	5	10	15	20	25	20	15	10	5	0
Output	0	40	81	131	190	250	210	169	119	60	0

Component 3

Input	0	4	8	12	16	20	24	25	24	20	16	12	8	4	0
Output	0	23	60	115	170	215	242	250	242	215	170	115	60	23	0

Component 4

Input	0	5	9	13	17	21	25	19	15	11	7	3	0
Output	30	30	63	100	155	220	220	220	186	147	91	30	30

1.12 Identify all nonlinearities exhibited by each component in Exercise 1.11.

1.13 Determine the maximum output difference between the increasing line and the decreasing line of any component in Exercise 1.11 that has a loop in its input/output graph. To do this, find the input that has the greatest separation between the two lines. For this input, read the values of the output on the increasing and decreasing lines and take the difference between the two output values.

1.14 Select which of the following four types of load change is illustrated in each example.

Types of Load Change

(1) Change in demand by controlled medium

(2) Change in quality of manipulated variable

(3) Change in ambient conditions

(4) Change of energy supplied within the process

Examples

a. A chemical process in plant A mixes two ingredients that combine to form a compound. Heat is generated by the reaction, and a control system is used to control the temperature of the mixture. What type of load change is a change in the amount of heat generated by the reaction?

b. A process in plant B uses heated outside air to dry the product before packaging. A rainstorm raises the humidity of the outside air so that more heat is required to dry the product. What type of load change is this?

c. A food process in plant C uses a dryer to toast corn flakes. What type of load change is a change in production rate from 200 lb/hr to 300 lb/hr?

d. A food process in plant D uses a solution of sodium bicarbonate to neutralize synthetic meat fibers. The flow rate of sodium bicarbonate is the manipulated variable. What type of load change is a change in the concentration of sodium bicarbonate?

1.15 In a chemical process, two components are blended together in a large mixer. The temperature of the mixture must be maintained between 100 and 112°C. If the temperature exceeds 114°C, the finished product will not satisfy the specifications. Which of the three criteria of good control should be used for the temperature control system?

1.16 Derive Equation 1.2.

CHAPTER 2

Types of Control

OBJECTIVES

Control system technology has many facets, depending on what is controlled, how the control is accomplished, who produces the components, and who uses the control system.

The purpose of this chapter is to introduce you to various types of control systems and the language associated with each type. After completing this chapter, you will be able to

1. Explain the difference between analog signals and digital signals
2. Differentiate between regulator and follow-up control systems and give examples of each type
3. Describe process control and explain the operation of a typical process controller
4. Sketch a block diagram of a temperature control loop and name all components and signals
5. Describe servomechanisms and explain how hydraulic and dc motor position control systems work
6. Discuss sequential control and differentiate between event-sequenced processes and time-sequenced processes
7. Discuss ladder diagrams and timing diagrams
8. Discuss numerical control and robotics
9. Discuss microprocessor-based digital controllers and programmable logic controllers
10. Differentiate between centralized control and distributed control and give advantages and disadvantages of each
11. Identify each control system example presented in this chapter

2.1 INTRODUCTION

Control systems are classified in a number of different ways. They are classified as closed-loop or open-loop, depending on whether or not feedback is used. They are classified as analog or digital, depending on the nature of the signals—continuous or discrete. They are divided into regulator systems and follow-up systems, depending on whether the setpoint is constant or changing. They are grouped into process control systems or machine control systems, depending on the industry they are used in—processing or discrete-part manufacturing. *Processing* refers to industries that produce products such as food, petroleum, chemicals, and electric power. *Discrete-part manufacturing* refers to industries that make parts and assemble products such as automobiles, airplanes, appliances, and computers. They are classified as continuous or batch (or discrete), depending on the flow of product from the process—continuous or intermittent and periodic. Finally, they are classified as centralized or distributed, depending on where the controllers are located—in a central control room or near the sensors and actuators. Additional categories include servomechanisms, numerical control, robotics, batch control, sequential control, time-sequenced control, event-sequenced control, and programmable controllers. These general categories are summarized below.

Classifications of Control Systems
1. Feedback
   a. Not used—open-loop
   b. Used—closed-loop
2. Type of signal
   a. Continuous—analog
   b. Discrete—digital
3. Setpoint
   a. Seldom changed—regulator system
   b. Frequently changed—follow-up system
4. Industry
   a. Processing—process control
      (1) Continuous systems
      (2) Batch systems
   b. Discrete-part manufacturing—machine control
      (1) Numerical control systems
      (2) Robotic control systems
5. Location of the controllers
   a. Central control room—centralized control
   b. Near sensors and actuators—distributed control
6. Other categories
   a. Servomechanisms
   b. Sequential control
      (1) Event-sequenced control
      (2) Time-sequenced control
   c. Programmable controllers

As we discussed open- and closed-loop control systems in Chapter 1, we do not deal with them in this chapter but cover the other classifications.

2.2 ANALOG AND DIGITAL CONTROL

The signals in a control system are divided into two general categories: analog and digital. Graphs of an analog and a digital signal are shown in Figure 2.1.

An *analog signal* varies in a continuous manner and may take on any value between its limits. An example of an analog signal is a continuous recording of the outside air temperature. The recording is a continuous line (a characteristic of all analog signals). A *digital signal* varies in a discrete manner and may take only certain

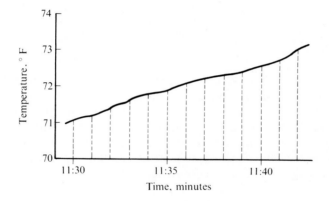

a) An analog signal of the outside air temperature

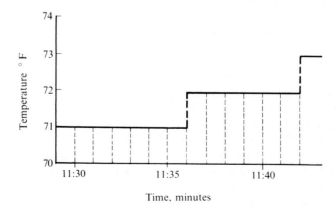

b) A digital signal of the outside air temperature

Figure 2.1   Examples of digital and analog signals of the same variable.

discrete values between its limits. An example of a digital signal is an outdoor sign that displays the outside air temperature to the nearest degree once each minute. A graph of the signal produced by the sign does not change during an interval, but it may jump to a new value for the next interval.

Analog control refers to control systems that use analog signals and *digital control* refers to control systems that use digital signals. The control system shown in Figure 2.3, for example, uses an analog controller; the control system in Figure 2.4 uses a digital controller.

2.3 REGULATOR AND FOLLOW-UP SYSTEMS

Control systems are classified as regulator systems or follow-up systems, depending on how they are used. A *regulator system* is a feedback control system in which the setpoint is seldom changed; its prime function is to maintain the controlled variable constant despite unwanted load changes. A home heating system, a pressure regulator, and a voltage regulator are common examples of regulator systems. Many process control systems are used to maintain constant processing conditions and hence are regulator systems.

A *follow-up system* is a feedback control system in which the setpoint is frequently changing. Its prime function is to keep the controlled variable in close correspondence with the setpoint as the setpoint changes. In follow-up systems, the setpoint is usually called the reference variable. A ratio control system, a strip chart recorder, and the antenna position control system on a radar tracking system are examples of follow-up systems. Many servomechanisms are used to maintain a position variable in close correspondence with an input reference signal and hence are follow-up systems.

2.4 PROCESS CONTROL

Process control involves the regulation of variables in a process. In this context, a *process* is any combination of materials and equipment that produces a desirable result through changes in energy, physical properties, or chemical properties. A continuous process produces an uninterrupted flow of product for extended periods of time. A batch process, in contrast, has an interrupted and periodic flow of product. Examples of a process include a dairy, a petroleum refinery, a fertilizer plant, a food-processing plant, a candy factory, an electric power plant, and a home heating system. The most common controlled variables in a process are temperature, pressure, flow rate, and level. Others include density, viscosity, composition, color, conductivity, pH, and hardness. Most process control systems maintain constant processing conditions and hence are regulator systems.

Process control systems may be either open-loop or closed-loop, but closed-loop systems are more common. The process control industry has developed standard, flexible, process controllers for closed-loop systems. Over the years these controllers have evolved from pneumatic analog controllers to electronic analog controllers to microprocessor-based digital controllers (microcontrollers). The driving force in this evolution has been increased capability and versatility, especially in microcontrollers, which tapped the power of the microprocessor.

Most process controllers share a number of common features. They show the value of the setpoint, the process variable, and the controller output in either analog or digital format. They allow the operator to adjust the setpoint and switch between automatic and manual control. When manual control is selected, they allow the operator to adjust the controller output to vary the manipulated variable in an open-loop control mode. They allow the operator to adjust the control mode settings to "tune the controller" for optimum response. Many controllers also provide for remote setting of the setpoint by an external signal, such as the output of another controller. A local/remote switch allows the operator to switch the setpoint between the local and remote settings. Figure 2.2 shows the front panel of a single station microcontroller.

Microcontrollers provide many additional features, some unique to one vendor and others common among a number of vendors. The following is a partial list of these features.

Choice of control modes: P, I, PI, PD, and PID (see discussion of Controllers in section 1.4)

Detects and annunciates alarms

Accepts several analog inputs (about four)

Accepts several digital inputs (three or four)

Provides more than one analog output (can be used to manipulate process variables)

Provides several digital outputs (can be used for ON-OFF control of heating elements, etc.)

Direct input from a thermocouple or RTD temperature sensor

Linearizes thermocouple inputs

Performs ratio, feedforward, or cascade control

Bumpless transfer between automatic and manual modes

Bumpless transfer between local and remote modes

Front-panel configuration of the controller

Adaptive gain: automatic adjustment of the proportional mode gain based on some combination of the process variable, the error, the controller output, and a remote input signal.

Self-tuning by process model: determination of the control mode parameters from a model that is formed from observations of the response to step changes of the setpoint. The step change and modeling process is repeated until the model matches the actual process.

Self-tuning by pattern recognition: automatic adjustment of the control mode parameters after a disturbance by scanning the recovery pattern and applying tuning rules that are stored in the controller's memory

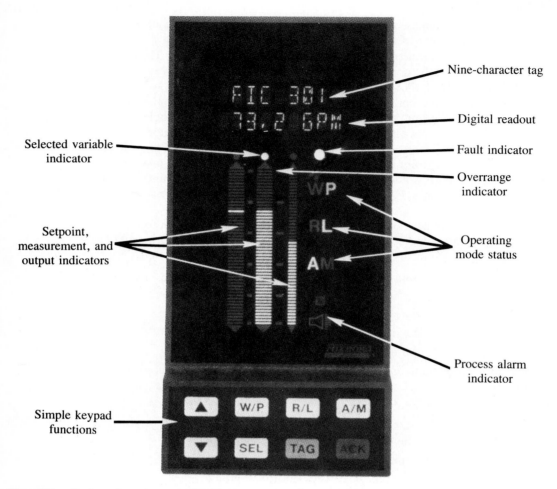

Figure 2.2 Single station microcontroller. The keypad provides easy adjustment of familiar controller operations and access to complete information about the process and the controller. Interactive prompting simplifies the setting of adjustable functions. The SEL key moves the selected variable indicator among the setpoint, measurement, and output indicators. The digital readout displays the value and engineering units of the setpoint, measurement, or output—whichever is chosen by the selected variable indicator. The two keys on the left side of the keypad are used to increase or decrease the setpoint or the output (depending on the position of the selected variable indicator and the operating mode status). The A/M key selects automatic or manual control. The R/L key selects remote or local setpoint. The W/P key chooses monitoring by a supervisory computer (workplace) or by an operator (panel). In the workplace setting, the controller is connected to a computer via a multidrop communication link. (Courtesy of The Foxboro Company, Foxboro, Mass.)

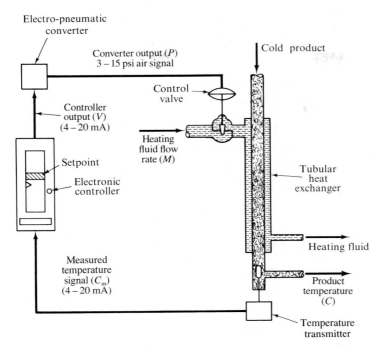

Figure 2.3 Schematic diagram of a temperature control system. An electronic analog controller regulates the temperature of a liquid product—for example, the pasteurization of milk. The heat exchanger consists of two concentric tubes; the product passes through the inner tube, which is surrounded by the heating fluid contained in the larger tube. Steam is the most common heating fluid, but hot water and hot oil are also used. The control valve manipulates the heating fluid flow rate, which determines the amount of heat transferred to the product. The temperature transmitter measures the temperature of the product as it leaves the heat exchanger. The controller compares the measured temperature with the setpoint and produces an output that manipulates the control valve to maintain the product temperature at the setpoint.

Self diagnostics: automatic detection and annunciation of certain types of failure

Mathematical operations, such as addition, subtraction, multiplication, division, and square root

Digital communication with a supervisory control computer

Figure 2.3 is a schematic diagram of an electronic analog controller in a routine process control system.

Figure 2.4 is an instrumentation diagram of a more involved process control system, a compensated mass flow control loop. The purpose of this system is to deliver a flow of gas in proportion by weight to the demand signal. The flow meter produces a signal that is proportional to the square of the *volume flow rate* of the gas. What is needed is a signal that is proportional to the *mass flow rate* of the gas. Equation (4.30) in section 4.4 gives the mass (m) of a volume (V) of gas as a function

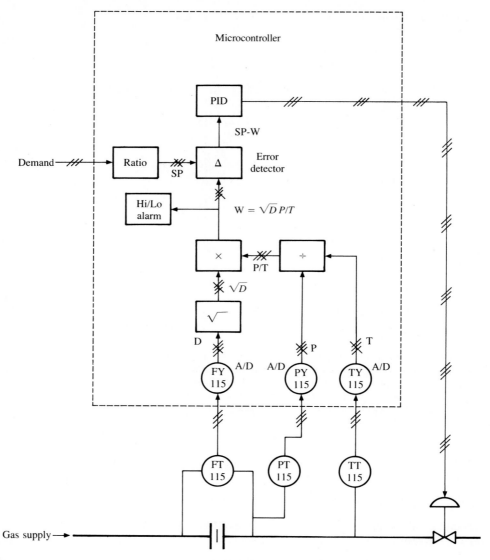

Figure 2.4 Instrumentation diagram of a compensated mass flow control loop. Before microprocessor-based digital controllers were available, this control loop required several more pieces of expensive hardware. [From *Bulletin C-404A* (Foxboro, Mass.: The Foxboro Company, November 1985), p. 4.]

of the absolute pressure (p), the absolute temperature (T), and the molecular weight (M) of the gas:

$$m = 1.2 \times 10^{-4} MV\left(\frac{p}{T}\right) \tag{4.30}$$

The equation to convert from volume flow rate (Q) to mass flow rate (w) is obtained by:

1. Dividing both sides of Equation (4.30) by time, t,

$$w = \frac{m}{t} = 1.2 \times 10^{-4} M\left[\frac{V}{t}\right]\left[\frac{p}{T}\right]$$

2. Substituting Q for V/t,

$$w = 1.2 \times 10^{-4} MQ\left[\frac{p}{T}\right] \tag{2.1}$$

where w = mass flow rate, kilogram/second

         Q = volume flow rate, cubic meter/second

         p = absolute pressure, pascal

         T = absolute temperature, kelvin

         M = the molecular weight of the gas (see "Properties of Gases" in Appendix A)

The volume flow rate, Q, is equal to the square root of the output of the flow meter, D, multiplied by the flow meter proportionality constant, k_f:

$$Q = k_f\sqrt{D} \tag{2.2}$$

Substitute (2.2) into (2.1) and replace $1.2 \times 10^{-4} Mk_f$ by k to get the following conversion equation:

$$w = k\sqrt{D}\left(\frac{p}{T}\right) \tag{2.3}$$

where k = mass flow proportionality constant

         D = flow meter differential pressure, pascal

         p = absolute pressure of the gas, pascal

         T = absolute temperature of the gas, kelvin

$$k = 1.2 \times 10^{-4} Mk_f \tag{2.4}$$

The controller converts the output of the flow meter (D) into a mass flow rate signal (w) by multiplying the square root of the flow meter signal by the quotient of pressure divided by temperature. The output of the multiplier is proportional to the mass flow rate and can be calibrated to the desired accuracy. This mass flow rate signal (w) is the measured variable input to the PID controller. The ratio unit multiplies the demand signal by a ratio value to form the setpoint input to the PID controller. The output of the controller is applied to the control valve to regulate the mass flow rate of the gas in a ratio to the demand signal.

2.5 SERVOMECHANISMS

Servomechanisms are feedback control systems in which the controlled variable is physical position or motion. Many servomechanisms are used to maintain an output position in close correspondence with an input reference signal and hence are follow-up systems. Servomechanisms are often part of another control system. Robotic control systems contain several servomechanisms, one for each joint in the robotic arm. Numerical control machines use servos to control the motion of the tool. Recorders use servos to position the recording pen. The driver and automobile control system in Figure 1.3 contains a power steering system, which is a servomechanism. If the car has cruise control, that is another servomechanism.

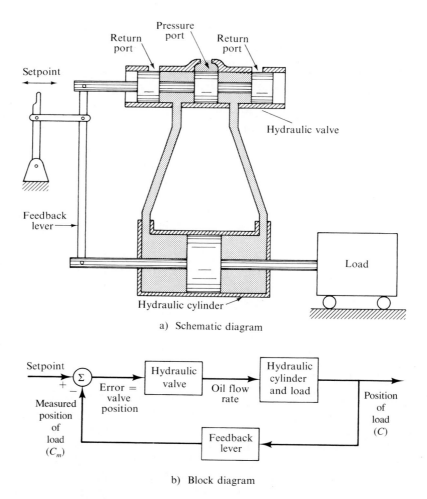

a) Schematic diagram

b) Block diagram

Figure 2.5   Hydraulic position control system.

There is no theoretical difference between a servomechanism and a closed-loop process control system; the same mathematical elements are used to describe each system, and the same methods of analysis apply to each. However, because servo control and process control were developed independently of one another, each has evolved different design methods and a different terminology. Servomechanisms usually involve relatively fast processes—the time constants may be considerably less than 1 second. Process control involves much slower processes—the time constants are measured in seconds, minutes, and even hours. The components in a servomechanism are usually well-defined mathematically, so the controller can be designed to meet the system specifications with little or no need for field adjustments. Processes are more difficult to define mathematically, so process control systems usually require field adjustments to obtain optimum response. Figures 2.5 and 2.6 provide examples of servomechanisms.

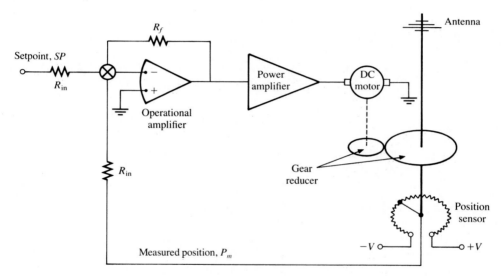

a) Schematic diagram

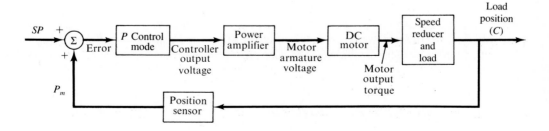

b) Block diagram

Figure 2.6   DC motor position control system.

The hydraulic position control system in Figure 2.5 uses a lever to provide a mechanical feedback signal. The hydraulic valve is shown in the neutral position. If the setpoint lever is moved to the right, the valve spool also moves to the right, thus connecting the left side of the hydraulic cylinder to the pressure port and the right side to the return port. Hydraulic fluid will flow into the left side of the cylinder, moving the piston and load to the right until the valve is back in the neutral position. If the setpoint handle were moved to the left, the load would be moved to the left. For each position of the setpoint handle, there is a corresponding position of the load that will place the valve in its neutral position. Not much force is required to move the lever and control the great force exerted by the hydraulic cylinder.

A dc motor position control system is illustrated in Figure 2.6. This system positions an antenna in response to a command voltage applied at the setpoint input. The position sensor is a 10-kilohm (kΩ) potentiometer with no stops and a 20° dead zone. The position sensor voltage output goes from $-V$ to $+V$ as the antenna rotates from its $+170°$ position to its $-170°$ position. The operational amplifier and the three resistors form a proportional (P) mode controller with a proportional gain of R_f/R_{in}.

The output of the controller is $-R_f/R_{in}$ times the algebraic sum of the setpoint voltage (SP) and the measured position voltage (P_m):

$$\text{controller output} = \frac{-R_f}{R_{in}}(SP + P_m)$$

The power amplifier inverts the controller output and increases the voltage by a factor of G_a, the gain of the power amplifier:

$$\text{power amplifier output} = \frac{G_a R_f}{R_{in}}(SP + P_m)$$

The power amplifier output is applied to the armature of the dc motor. The motor speed is proportional to the voltage applied to the armature and the direction is such that when the armature voltage is positive, the motor drives the position sensor toward $-V$, and when the armature voltage is negative, the motor drives the antenna toward $+V$. The result is that the motor drives the position sensor in the direction that will tend to make the sum of SP and P_m equal to zero. The summing junction of the op amp is the error detector, and the term $(SP + P_m)$ is the error signal $(SP - C_m)$ defined in section 1.4. Notice that $C_m = -P_m$ and the controller uses positive feedback. The negative feedback in this example is accomplished by making the sign of the position sensor voltage opposite the sign of the setpoint voltage.

2.6 SEQUENTIAL CONTROL

A *sequential control* system is one that performs a set of operations in a prescribed manner. The automatic washing machine is a familiar example of sequential control: The control system performs the operations of filling the tub, washing the clothes, draining the tub, rinsing the clothes, and spin drying the clothes. The automatic machining of castings for automobiles is another example of sequential control: A sequence of machining operations is performed on each casting to produce the finished part. Sequential control is covered in detail in Chapter 13; our objective in this chapter is to give an overview of sequential control systems.

The operations in a sequential control system can be categorized according to how they are initiated and terminated. One method is to initiate or terminate an operation when some event takes place. We use the term *event-driven* for this method. The other method is to initiate or terminate an operation at a certain time or after a certain time interval. We use the term *time-driven* for this method.

An automatic washing machine is an example of a time-driven sequential control system. The washing cycle starts out with one event-driven operation—the fill operation begins when someone presses the start button and it terminates when the tub is full. However, the remaining operations are all initiated and terminated by a timer. These include the wash operation, the drain operation, the rinse operation, and the spin-dry operation. Most batch process control systems are time-driven sequential systems. Time-driven systems are described by schematic diagrams and timing diagrams. Schematic diagrams show the physical configuration and timing diagrams define the sequential operations. The timing diagram of an automatic washing machine is shown in Figure 2.7.

A traffic counter is a simple example of an event-drive system. The counter is placed at the side of the road, and the sensor, which is a long rubber tube, is stretched across the road. Each time a vehicle axle passes over the tubular sensor, the counter increases its count by one. Thus an event (an axle passing over the sensor) drives the

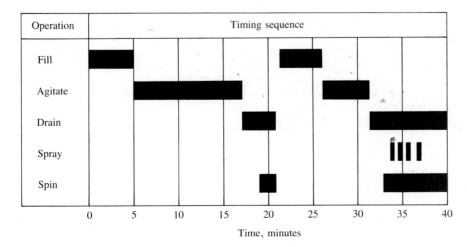

Figure 2.7   Timing diagram of an automatic washing machine.

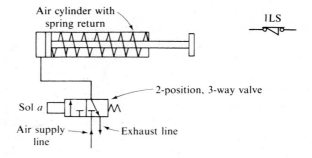

a) Schematic diagram

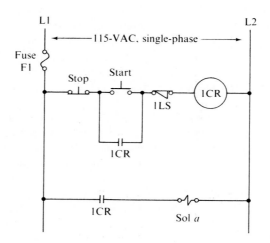

b) Electrical circuit diagram

Figure 2.8 Event-driven sequential control system for a pneumatic cylinder. The valve is shown in its deenergized position, which connects the air cylinder to the exhaust line. This allows the spring to force the piston to the retracted position shown in the diagram. The operator presses and releases the START switch to begin a cycle. This causes relay 1CR to energize, closing both contacts labeled 1CR. The 1CR contact that is connected in parallel with the START switch holds relay 1CR in the energized position. The other 1CR contact energizes solenoid (Sol a), which moves the valve to the right, connecting the cylinder to the air supply line. The air pressure forces the cylinder to the right until it reaches and opens limit switch 1LS. When 1LS opens, relay 1CR is deenergized, opening both 1CR contacts. The valve again connects the cylinder to the exhaust line, and the spring forces the cylinder back to the retracted position, ending the cycle.

counter. Manufacturing industries are principal users of event-driven sequential controllers. Before 1970, large relay panels were used to control event-driven operations. In 1968, the Hydramatic Division of General Motors Corporation specified the design criteria for the first programmable logic controller (PLC). The purpose was to replace inflexible relay panels with a computer-controlled solid-state system. The project succeeded beyond anyone's dreams. Programmable logic controllers have gone beyond replacement of relay panels to include PID modules for process control and communications interfaces that make it possible to link programmable controllers into an integrated manufacturing operation.

Event-driven systems are described by ladder diagrams and Boolean equations. The symbols of components used in ladder diagrams are included in Appendix D, and Boolean equations are covered in Chapter 3. The components most frequently used include switches, contacts, relays, contactors, motor starters, time-delay relays, pneumatic solenoid valves, pneumatic cylinders, hydraulic solenoid valves, and hydraulic cylinders. The pneumatic cylinder in Figure 2.8 is an example of an event-driven control system.

2.7   NUMERICAL CONTROL

Numerical control is a system that uses predetermined instructions to control a sequence of manufacturing operations. The instructions are coded numerical values stored on some type of input medium, such as punched paper tape, magnetic tape, or a common memory for program storage. The instructions specify such things as position, direction, velocity, and cutting speed. A *part program* contains all the instructions required to produce a desired part. A *machine program* contains all the instructions required to accomplish a desired process. Numerical control machines perform operations such as boring, drilling, grinding, milling, punching, routing, sawing, turning, winding (wire), flame cutting, knitting (garments), riveting, bending, welding, and wire processing.

Numerical control (NC) has been referred to as flexible automation because of the relative ease of changing the program compared with changing cams, jigs, and templates. The same machine may be used to produce any number of different parts by using different programs. The numerical control process is most justified when a number of different parts are to be produced on a particular machine; it is seldom used to produce a single part continually on the same machine. Numerical control is ideal when a part or process is defined mathematically. With the increasing use of computer-aided design (CAD), more and more processes and products are being defined mathematically. Drawings as we know them have become unnecessary—a part that is completely defined mathematically can be manufactured by computer-controlled machines. A closed-loop numerical control machine is shown in Figure 2.9.

The NC process begins with a specification (engineering drawing or mathematical definition) that completely defines the desired part or process. A programmer uses the specification to determine the sequence of operations necessary to produce the

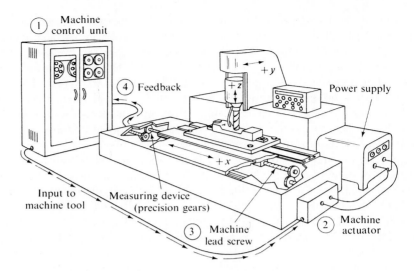

Figure 2.9 Numerical control machine that uses closed-loop systems to control x, y, and z positions. The x position controller moves the workpiece horizontally in the direction indicated by the $+x$ arrow. The y position controller moves the milling machine head horizontally in the direction indicated by the $+y$ arrow. The z position controller moves the cutting tool vertically as indicated by the $+z$ arrow. The following actions are involved in changing the x-axis position. (1) The control unit reads an instruction in the program that specifies a $+0.004$-in. change in the x position. (2) The control unit sends a pulse to the machine actuator. (3) The machine actuator rotates the lead screw and advances the x-axis position $+0.001$ in. (4) The position sensor measures the $+0.001$-in. change in x-axis position and sends this information to the control unit. (5) The control unit compares the $+0.004$ in. required motion with the $+0.001$-in. measured motion and sends another pulse. Steps (1) through (5) are repeated until the measured motion equals the desired $+0.004$ in. [From N. O. Olesten, *Numerical Control* (New York: John Wiley & Sons, Inc., 1970), p. 12.]

part or carry out the process. The programmer also specifies the tools to be used, the cutting speeds, and the feed rates. The programmer uses a special programming language to prepare a symbolic program. APT (Automatically Programmed Tools) is one language used for this purpose. A computer converts the symbolic program into the part program or the machine program. In the past, the part or machine program was stored on paper or magnetic tape. The numerical control machine operator fed the tape into the machine and monitored the operation. If a change was necessary, a new tape had to be made. Now, it is possible to store the program in a common database with provision for on-demand distribution to the numerical control machine. Graphic terminals at the machining center allow operators to review programs and make changes if necessary.

Computerized numerical control (CNC) was developed to utilize the storage and processing capabilities of a digital computer. CNC uses a dedicated computer to

accept the input of instructions and to perform the control functions required to produce the part. However, CNC was not designed to provide the information exchange
demanded by the recent trend toward *computer-integrated manufacturing* (CIM). The
idea of CIM is to "get the right information—to the right person—at the right time—
to make the right decision." "It links all aspects of the business—from quotation and
order entry through engineering, process planning, financial reporting, manufacturing,
and shipping—in an efficient chain of production."*

Direct numerical control (DNC) was developed to facilitate computer-integrated
manufacturing. DNC is a system in which a number of numerical control machines
are connected to a central computer for real-time access to a common database of
part programs and machine programs. General Electric used a central computer
connected to DNC machines through a communications network in the automation
of its steam turbine-generator operations (ST-GO). "A typical turbine-generator consists of more than 100,000 parts, some of which are manufactured in thousands of
different configurations to meet the specific needs of each custom-designed unit.
Through the CIM system, customers can specify a needed part and receive replacement components that suit the original configuration of more than 4000 operating
ST-GO installations. In some cases, the small-parts shop can now manufacture and
ship some emergency parts the same day the order is received."†

2.8 ROBOTICS

The *industrial robot* is a programmable manipulator designed to move material, parts,
tools, or other devices through a sequence of motions to accomplish a specific task.
Robots are used to move parts, load NC machines, operate die-casting machines, assemble products, weld, paint, debur castings, and package products. The most common robotic manipulator is an arm with from one to six axes of motion (or degrees
of freedom). The robotic arm shown in Figure 2.10 has six axes of movement:

1. Arm sweep (left or right at the waist)
2. Shoulder swivel (up or down at the shoulder)
3. Elbow extension (in or out at the elbow)
4. Pitch (up or down at the wrist)
5. Yaw (left or right at the wrist)
6. Roll (clockwise or counterclockwise at the wrist)

---

* Searle et al., "Computer-Integrated Manufacturing System Goes Beyond CAD/CAM," *Control Engineering*, February 1985, p. 50.

† Ibid., p. 51.

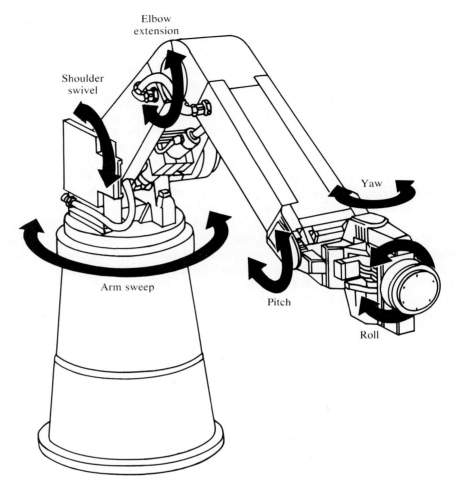

Figure 2.10 The Cincinnati Milacron T³ robot has six axes of movement, which duplicate the movements of a human arm. (From N. M. Morris, "Where Do Robots Fit in Industrial Control?" *Control Engineering*, February 1982, p. 59.)

Another type of robotic manipulator is a motorized cart that follows a programmed path to move parts from place to place in a factory.

Each axis of motion has its own actuator, connected to mechanical linkages that accomplish the motion of the joint. The actuator may be a pneumatic cylinder, a pneumatic motor, a hydraulic cylinder, a hydraulic motor, an electric servomotor, or a stepper motor. Pneumatic actuators are inexpensive, fast, and clean, but the compressibility of air limits their accuracy and ability to hold a load motionless. Hydraulic actuators can move heavy loads with precision and hold the load motionless, but they are expensive, noisy, relatively slow, and tend to leak hydraulic fluid. Electric actuators are fast, accurate, and quiet, but backlash in the gear train may limit their precision.

Industrial robots have three main parts: the controller, the manipulator, and the end effector. The *end effector* is a mechanical, vacuum, or magnetic device that is attached to the manipulator at the wrist and is used to grip parts or tools. The *manipulator* has already been described in some detail. The *controller* may be simple mechanical stops in a single-axis robot with open-loop control, or it may be a computer in a six-axis robot with closed-loop control. In any event, the controller stores the sequencing and positioning data in memory. It also initiates and stops each movement of the manipulator in the specified sequence of operations. If the controller is a computer, it may communicate with a host computer to download programs and provide management information. Each axis of motion is controlled by either an open-loop or a closed-loop control system. The open-loop control systems may be mechanical stops on a pneumatic cylinder, cam-actuated solenoid valves on a hydraulic motor, or an electric stepper motor. The closed-loop control systems are usually follow-up position control systems (servomechanisms). However, sight, tactile sensing, and voice recognition are also being used as inputs to the controller.

The simplest type of robot is the open-loop *pick-and-place* (PNP) robot. A PNP robot picks up an object and moves it to another location. The robot's movements are usually accomplished by pneumatic actuators controlled by limit switches, cam-actuated valves, or mechanical stops. The controller initiates movement along one axis at a time in an event-driven sequence. Each movement continues until a limit is tripped, stopping the motion. The controller then initiates movement on the next axis in the sequence. Typical applications include machine loading or unloading, palletizing, stacking, and general materials-handling tasks. Open-loop PNP robots are quite accurate, but they lack coordination of the various axes.

The second level of robots uses servo control on most axes and can be programmed to move from one point to another. If the path is not critical, the robot is called *point-to-point* (PTP). If the path is critical, the robot is called continuous path (CP). A PTP robot moves from point to point and performs a function at each point. Typical PTP functions include spot welding, gluing, drilling, and deburring. A CP robot moves from point to point on a specified path and performs an operation as it moves along the path. Typical CP applications include paint spraying, seam welding, cutting, and inspection. The second-level controllers are either programmable controllers or minicomputers. A teaching pendant is used to program the robot using a simple teach-by-doing method.

The third level of robots can also be programmed to move from point to point or in a continuous path. However, in addition to on-line programming using a teaching pendant, they can also be programmed off-line using a keyboard and CRT. These robots can communicate with a host computer. They use high-level languages and artificial intelligence to process information from a CAD/CAE database. They are capable of integration into computer-controlled workstations.

Robotic servo control systems use position and velocity feedback signals to control movements of the manipulator. The position signal can be either absolute or incremental. The robot's controller sends a setpoint signal to each servo to move along its axis to a given position (absolute position) or through a given distance (incremental position). Position and velocity are referred to as the internal feedback signals of the servo control loop. The robotic controller has other sensory inputs (external to the servo loop) that it can use to carry out its assigned task. These external inputs include vision, tactile sensing (touch), and voice recognition. The controller uses these external sensory inputs to detect the presence of an object, the dimensions of an object, or even the identity of an object. With a sense of vision or touch, a robot can calibrate its position sensors, search a defined area to locate a part, and identify any part it finds. Sensors are covered in detail in Chapters 9 and 10.

The industrial robot comes ready to do a job but does not know how. The user must program the robot to do its assigned task. First the user must determine the sequence of operations required to do the job. Then an operator must place this sequence of operations into the robot's memory. In simple PNP applications, the operator locates the mechanical stops or limit switches and establishes the logic of the event-driven sequence. In PTP and CP applications such as welding and spray painting, the operator uses a "teach-by-doing" method to program the robot. The operator puts the controller into the TEACH mode and moves the manipulator through the desired sequence of operations. The controller *learns* what it is taught by recording in its memory the various positions of the manipulator's joints. When finished, the operator puts the controller in the RUN mode and the robot follows the position data in its memory and exactly repeats the sequence of operations it was taught.

Robot manufacturers provide a *teaching pendant* to assist the operator in programming the robot. A typical pendant looks somewhat like a pocket calculator. It has an alphanumeric display and several pushbuttons. The programmer can move the manipulator by pressing the appropriate button on the pendant, or in some cases by actually moving the manipulator to the desired position. When each correct position is achieved, the programmer pushes another button to tell the computer to read and store the position of each joint in the robot's manipulator. When all positions have been recorded, the operator presses a REPEAT button to check the movements of the manipulator for any errors that might have occurred in the teaching process.

The "teach-by-doing" method works well for the simpler PTP and CP operations. However, when the operations become complex, the operator may have difficulty visualizing the program's structure. In these applications, off-line programming using a personal computer is more appropriate. One advantage of off-line programming is that valuable production time is not lost by on-line teaching of the robots. Another advantage is that the programmer can use the powerful software available for computer-aided design (CAD) and computer-aided engineering (CAE). A third advantage of off-line programming is that the robot programs can be prepared before the fixtures are built, shortening the time required to put a new product into production.

In the past decade, industry has made a major effort toward factory automation. One aspect of this is the "islands of automation" created by robotic work cells. The traditional method of grouping machine tools is to put all machines of a particular type in the same department and the same location. There would be a department of milling machines, a department of drill presses, a department of grinders, and so on. The problem with this arrangement is that it proved to be very inefficient. The Comptroller General, in a 1975 report to Congress, estimated that only 5% of the time consumed in producing a part is spent on the machine itself.* The other 95% of the time is expended in materials handling, record keeping, and so on. The robotic work cell is an arrangement that can significantly reduce the time spent on nonproductive activities.

A robotic work cell is a group of machine tools and robots arranged for the efficient production of a particular type of product. In this method, products that require the same machining operations are grouped together for production in a particular workstation. A computer controls the entire workstation, so the robots and machine tools work together in the most efficient manner. Figure 2.11 shows the top view of a robotic work cell for the production of forged parts.

2.9 THE EVOLUTION OF CONTROL SYSTEMS

Process control and machine control are going through an evolution that began with two separate and distinct systems and is approaching one integrated, distributed system of control for the entire manufacturing plant. Each step in the evolution was made possible by advancements in control system technology, and each step brought improvements to the control of manufacturing operations.

Process control began with sensing elements connected directly to recording controllers, which, in turn, were connected directly to the control valve. The control loop intelligence was distributed near the process it controlled. This distribution of loop intelligence produced good control of individual process variables; but operators could not adequately monitor all of the control loops, especially in spread-out processes such as oil refineries, paper mills, steel mills, and chemical plants. Control engineers could only dream of advanced control concepts, because there was no way to use inputs from several process variables to improve the control of critical variables in the process.

Then along came pneumatic transmitters, controllers, and valve actuators. Industry standardized on the 3- to 15-psi pneumatic signal, making it possible to obtain a measuring transmitter from one vendor, a controller from a second vendor, and a control valve from yet a third vendor. When the three components were assembled into a control loop, the system worked. Better still, the control engineer was able to collect all the controllers for a process into a single room, and the *central control*

---

* Morris et al., "Profitable Robotic Work Cells," *Control Engineering*, March 1985, p. 81.

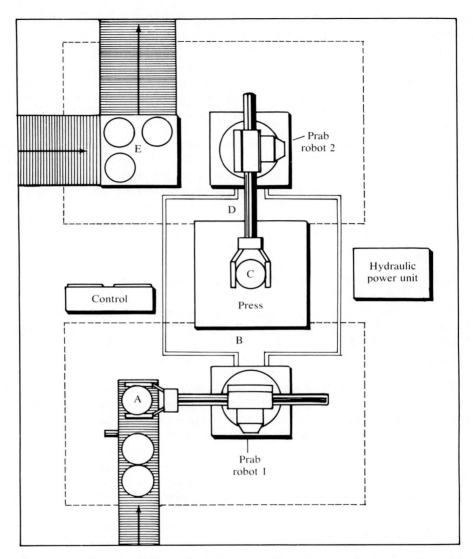

Figure 2.11 Top view of a robotic work cell for forging points. Robot 1 picks up a raw forging from (A) while robot 2 picks up a forged part at (C). Robot 1 then swings to (B) as robot 2 retracts out of the forge to (D). Robot 1 extends into the forge (C) and robot 2 swings to discharge pallet (E). Robot 1 sets the raw forging into the die (C) as robot 2 sets the forged part onto the pallet (E) in one of four positions. Robot 1 retracts from the forge and signals the forge's controller to cycle, and robot 2 swings to (D). Robot 1 then swings to the infeed conveyor (A) as robot 2 extends into the forge (C). The cycle then repeats. (From N. M. Morris, "Controlling Multiple Robot Arms," *Control Engineering*, November 1986, p. 146.)

room concept was born. Operators were now able to monitor the process in a way not possible before. The central control room was the evolutionary step brought on by the 3- to 15-psi pneumatic signal. This gathering of the control loop intelligence in the control room did have a couple of serious disadvantages. One was the cost of the pneumatic lines between the process and the control room. Each control loop required two signal lines: one from the measuring transmitter to the controller, the other from the controller to the control valve. In large plants, these lines measured hundreds, even thousands of feet long. Also, control was not quite as good, because of the transmission delay in the long pneumatic signal lines. Pneumatic signals travel at about 1000 ft per second. A process loop located 500 ft from the control room would suffer a 1-second delay in the control loop—a $\frac{1}{2}$-second delay from the measuring transmitter to the controller and another $\frac{1}{2}$-second delay from the controller to the control valve.

Next came electronic analog transmitters, controllers, and electropneumatic converters. The pneumatic signal lines were replaced by 4- to 20-mA electric signal lines. Since electric signals travel at nearly the speed of light, this eliminated the second disadvantage of the control room concept (i.e., signal delay was no longer a problem). Cost of the transmission lines was still significant, and a new disadvantage was introduced. The electric signal lines in a control loop make a good antenna, and the control loops picked up noise signals from the surrounding environment. Noise reduced the accuracy of the transmitted signal and caused problems as the controller carried out the control algorithm.

The digital computer entered the process control scene in 1959 when Texaco used a TRW-300 digital computer to control a polymerization unit. At last the control engineer had a tool to implement advanced control concepts. Input of all process variables into the computer memory meant that every process variable was available for use in a control algorithm. One method of using computers for control was to replace the analog controllers with a single large digital computer. The term *direct digital control* (DDC) was used to describe this type of system. Reliability was a problem, however. If the computer failed, the entire process was out of control. This happened with enough regularity that some form of backup was in order. One backup scheme used a standby computer that was ready to take over in case the primary computer failed. In extremely critical applications such as the space program, two backup computers were used. Another method used analog controllers for backup in case the DDC computer failed. Yet another method used analog controllers for loop control with the digital computer used in a supervisory role.

Machine control began with hard-wired panels of relay logic, motor starters, fluid actuators, and solenoid valves. Servo control systems used vaccum tubes, and variable-speed drives consisted of dc generators driving dc motors (the so-called rotating amplifiers). In the 1950s, numerical control machines appeared and machine control entered the world of digital control. These NC machines were programmed off-line and used punched paper tape for storage of the program. Programmable controllers appeared about 1970, replacing hard-wired relay logic with reprogrammable computer logic. Improvements in semiconductor switching elements and memory systems

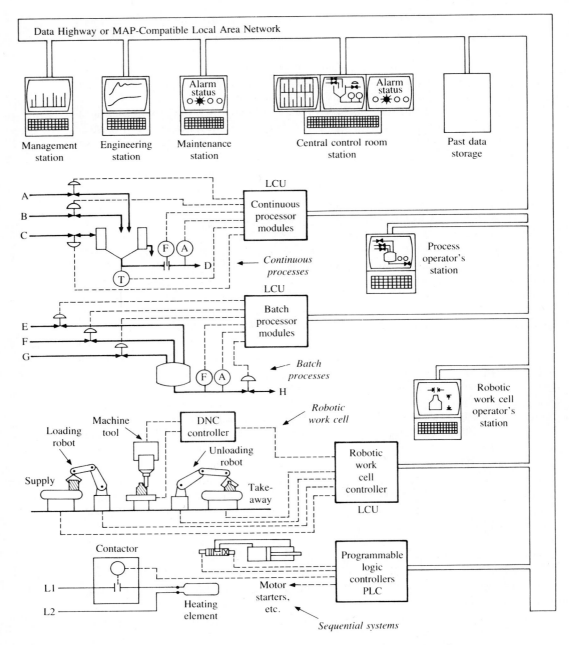

Figure 2.12 Integrated, distributed control system for a hypothetical manufacturing plant that contains both process and machine control systems.

set the stage for the microprocessor in the 1980s. The microprocessor generated the changes that are finally bringing process control and machine control under a single umbrella of unified control system technology.

The microprocessor gave us the means to move the control loop back to the plant floor and the ability to communicate with this distributed intelligence. We can have the "short loop" advantages of distributed control and the fully informed operator of a centralized control system. Figure 2.12 shows a fully integrated, distributed control system for a hypothetical manufacturing plant that has both process control and machine control systems. The path that crosses the top and runs down the left side represents the communication network. The terms *data highway* and *local area network* (LAN) are used for different approaches to the communication network. In the 1980s, industry began a major effort to solve the problem of communicating with all the various devices in a manufacturing plant. The purpose of this effort is to develop a set of standards called the *Manufacturer's Automation Protocol* (MAP). See Chapter 8 for further details about communication networks.

All of the control loops are closed on the plant floor. The local control units (LCUs) contain I/O modules that condition the input signals and controller modules that carry out the control algorithms. The control modules have access to all the inputs from the process, giving the control engineer complete freedom to apply advanced control techniques. The LCUs are designed to withstand the harsh environment of the plant floor, so the loop intelligence can be located close to the process it controls. The control loops are short, and the communication network is not bogged down with routine control signals. The measuring transmitters are intelligent—they contain their own microprocessor. These smart transmitters convert the analog signal into a digital signal, linearize the signal, eliminate noise, convert the signal to engineering units, store the ID tag of the transmitter, and store the date of the last calibration. They can even store values of past data for trend analysis. All modules in the system are addressed by their ID tag, and data can be obtained from any module in the system at any place in the system.

Local control units can handle continuous processes, batch processes, and robotic work cells. In addition, other units, such as programmable logic controllers (PLCs), can be interfaced with the communications network. Local operators, central control room operators, plant maintenance, plant engineering, and plant management all have access to all the information via color display and keyboards.

2.10 EXAMPLES OF CONTROL SYSTEMS

An example of the application of a microprocessor in a sequential control system is shown in Figure 2.13. This system monitors the voltage and current supplied to critical (or emergency) loads, such as a life support unit in a hospital. If a power failure occurs, the control system starts the emergency generator and switches the generator output to the critical load. This assures a continuous source of power in situations where loss of power would have serious consequences.

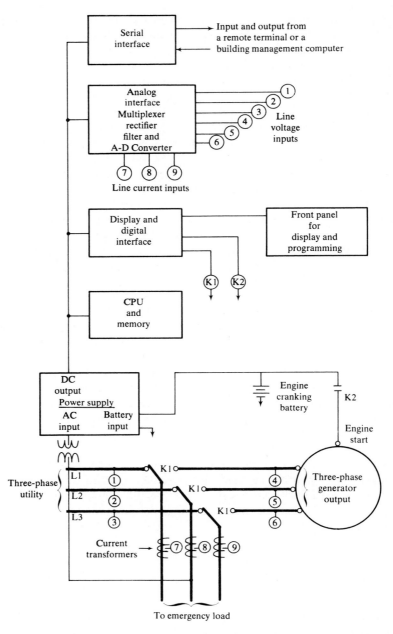

Figure 2.13 Microprocessor control system for an emergency generator. (Courtesy of ONAN Corporation.)

The microprocessor consists of six units: the power supply, the central processor (CPU) and memory unit, the display and digital interface, the front panel, the analog interface, and the serial interface. The power supply provides the dc voltage necessary to operate the microprocessor. The ac line is the normal source of power, but an input from the engine cranking battery is provided to ensure uninterrupted power to the microprocessor.

The CPU controls the interpretation and execution of instructions. It consists of an arithmetic section, working storage registers, logic circuits, timing and control circuits, decoders, and parallel buses for transfer of binary data and instructions. Memory includes random access memory (RAM) and read-only memory (ROM). The RAM memory has both read and write capability and is used primarily as scratch pad memory for storage and retrieval of temporary data and instructions. The ROM is used to store the main program and permanent data.

The display and digital interface contains the electronic circuits that match the CPU to the two control relays and the front panel. Relay K1 controls the emergency load switch. The actual switching action is performed by a linear motor (not shown in Figure 2.13), and the K1 relay operates the linear motor. The K2 relay is used to start and stop the generator engine.

The human interface provided by the front panel is one of the principal advantages of digital control. The front panel contains a function switch, a digital display, and a keyboard. The function switch selects the signal that will appear on the digital display. The keyboard is used for manual input of data or instructions into memory.

The analog interface matches the CPU to the nine analog input signals. It includes a multiplexer, a rectifier, a filter, and an analog-to-digital converter. The multiplexer selects the input signals one at a time for input to the rectifier. The rectifier converts the ac analog signals into dc signals and the filter removes the ripple in the rectifier output. The analog-to-digital converter changes the dc analog signal into a digital signal (usually an 8- to 16-bit binary number).

The serial interface matches the CPU with an external terminal or a building management computer. The ability to communicate with a remote computer or terminal is another principal advantage of digital control. This feature provides the building manager with immediate information about the status of the various systems in the building.

A variety of examples of control systems are illustrated in Figures 2.14 through 2.22. The examples include systems to control level, speed, liquid flow rate, gas pressure, engine speed, solid flow rate, sheet thickness, a hydraulic cylinder, and the composition of a liquid blend.

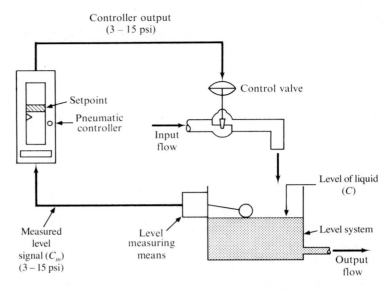

Figure 2.14 The level in the tank remains constant when the input flow rate equals the output flow rate. The level rises when the input flow rate is greater (or drops when the input flow rate is less) than the output flow rate. The controller uses the level signal to maintain a balance between the input and output flow rates.

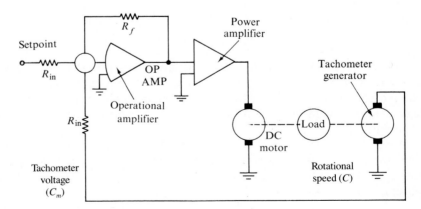

Figure 2.15 DC motor speed control system.

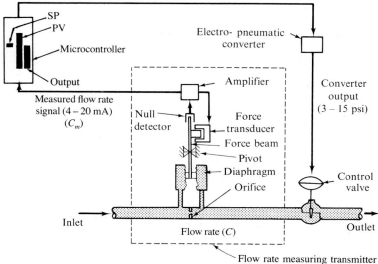

Figure 2.16   Liquid flow rate control system. This system consists of a flow rate measuring transmitter, a microcontroller (see Figure 2.2), an electro-pneumatic converter, and a control valve.

The measuring transmitter produces a 4- to 20-mA measured flow rate signal that is proportional to the square of the flow rate. The microcontroller converts the measured flow rate signal from analog to digital, computes the square root of the digital signal, converts the square root signal to flow rate in gallons per minute, and displays the flow rate on the controller's PV indicator. The microcontroller then computes the error signal, applies the control modes to compute the controller output, and converts the controller output from digital to an analog 4- to 20-mA controller output signal.

The electro-pneumatic converter converts the 4- to 20-mA controller output signal into a 3- to 15-psi air pressure signal. The control valve positions the valve stem according to the converter output signal and, indirectly, according to the controller output signal. The control valve manipulates the flow rate as directed by the microcontroller to maintain the desired flow rate as indicated by the setpoint.

The measuring transmitter is shown in some detail to show that it has its own feedback control system. The orifice is a thin plate with a small hole positioned so that all the flowing liquid must pass through the small hole. The flow of fluid through the orifice produces a pressure differential that is proportional to the square of the flow rate.

The measuring transmitter converts the pressure drop across the orifice into a 4- to 20-mA current signal as follows: The diaphragm arrangement converts the pressure difference across the orifice into a force on the lower end of the force beam. The force transducer at the other end of the beam produces the counterbalancing force. The null detector, amplifier, and force transducer make up a closed-loop control system that maintains the force beam in the null position. The null detector senses any displacement of the force beam, and the amplifier converts this displacement into an electric current. The amplifier current passes through the force transducer, generating the counterbalancing force, which is proportional to the current supplied by the amplifier. The amplifier current, which is proportional to the pressure differential, is also the 4- to 20-mA output signal from the measuring transmitter.

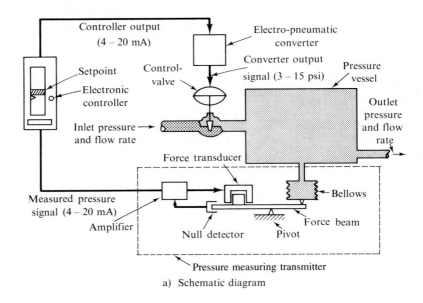

a) Schematic diagram

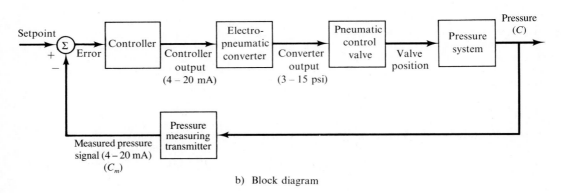

b) Block diagram

Figure 2.17   A pressure control system must maintain a balance between the input and output mass flow rates. The pressure measuring transmitter is very similar to the differential pressure transmitter in the liquid flow control system. It uses a force balance principle to convert the process pressure into a 4- to 20-mA current signal. The controller compares the measured pressure with the setpoint and manipulates the control valve to bring the two into correspondence.

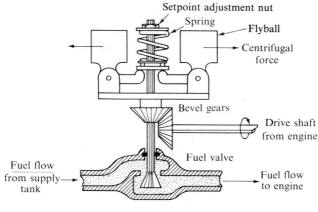

a) Schematic diagram

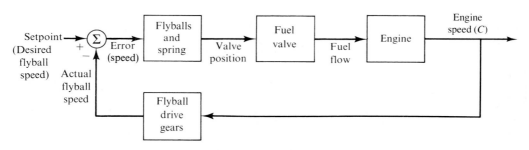

b) Block diagram

Figure 2.18 A mechanical speed control system or governor is used to control the speed of gasoline engines, gas turbines, and steam engines. A drive shaft from the engine rotates the flyball and spring assembly at a speed proportional to the engine speed. The rotation of the flyballs produces a centrifugal force that compresses the spring and raises the valve stem. Thus the valve stem position is proportional to the engine speed. As the engine speed increases, the valve decreases the fuel flow. As the engine speed decreases, the valve increases the fuel flow. The engine will settle out at a speed that results in just enough fuel flow to balance the load on the engine.

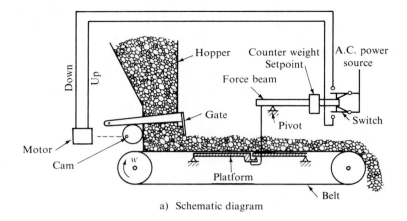

a) Schematic diagram

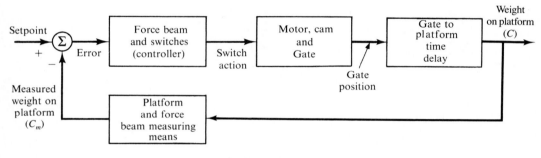

b) Block diagram

Figure 2.19 This solid flow rate control system uses a constant speed belt with a weigh platform. The solid flow rate is equal to the belt speed times the weight of material per unit length of belt. The weight of material on the platform applies a downward force on the left end of the force beam; the counterweight supplies the counterbalancing force on the right side. The force beam operates the two limit switches that control the cam drive motor. If the material on the belt is too heavy, the beam will close the upper limit switch, which drives the gate down; if the material is too light, the beam will close the lower limit switch, which drives the gate up.

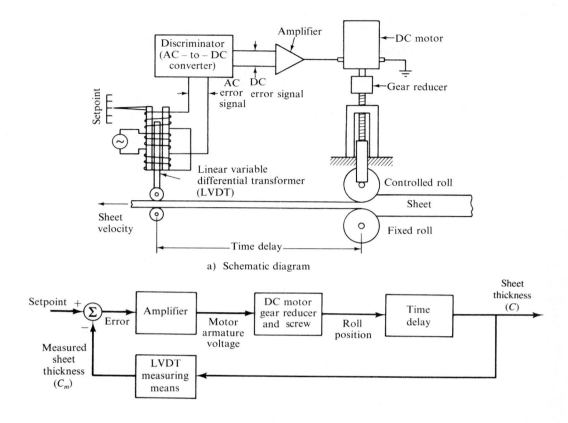

a) Schematic diagram

b) Block diagram

Figure 2.20 A sheet thickness control system uses a linear variable differential transformer as the thickness measuring means and error detector. The ac error signal from the LVDT is converted to a dc error signal by the discriminator (also called a phase-sensitive detector). A discriminator is an ac-to-dc converter whose output magnitude varies linearly with the input magnitude, and whose output sign depends on the relative phase of the input. The dc error signal is amplified and applied to the armature of the dc motor. The motor drives the upper roll, which determines the sheet thickness. After a change in the upper roll position, the sheet must travel to the sensor before the change in thickness is measured. This time delay is represented by a block in the block diagram.

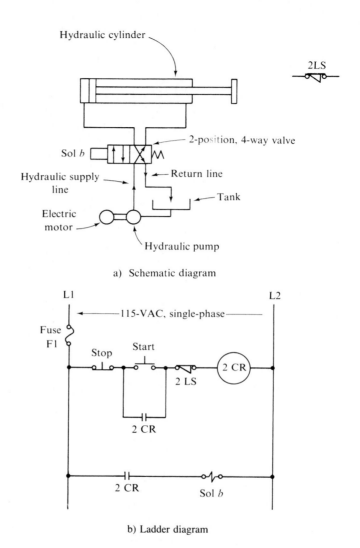

a) Schematic diagram

b) Ladder diagram

Figure 2.21 Control circuit for a hydraulic cylinder. The operator presses START to begin a cycle. Coil 2CR is energized, holding contact 2CR closed, "Sol b" is energized, the valve moves to the right, the pump delivers fluid to the left end of the cylinder, and the piston moves to the right. When the piston reaches switch 2LS and opens it, coil 2CR is deenergized, the valve moves to the left, and the piston returns to its original position.

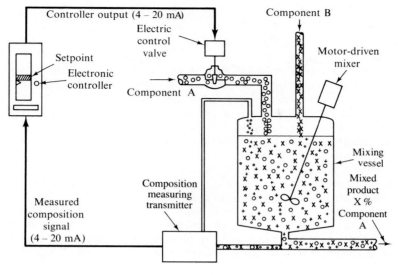

a) Schematic diagram

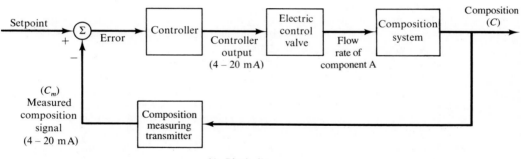

b) Block diagram

Figure 2.22   A composition control system maintains the desired mixture of components A and B by manipulating the input of component B. The measuring transmitter is some type of analyzer that measures the percentage of component A in the mixture. The mixing vessel blends the two components and smoothes out the fluctuations in the flow rates of both components.

GLOSSARY

Analog signal: A signal that varies in a continuous manner and may take on any value between its limits. (2.2)

Computerized numerical control (CNC): A numerical control system that uses a dedicated computer that accepts the input of instructions and performs the control functions required to produce a part. (2.7)

Digital signal: A signal that varies in a discrete manner and may take only certain discrete values between its limits. (2.2)

Direct numerical control (DNC): A system in which a number of numerical control machines are connected to a central computer for real-time access to a common database of part programs and machine programs. (2.7)

Event-driven operations: Operations in a sequential control system that are initiated and terminated when some event takes place. (2.6)

Follow-up system: A feedback control system in which the setpoint is frequently changing. Its primary function is to keep the controlled variable in close correspondence with the setpoint as the setpoint changes. (2.3)

Machine program: The set of instructions required to accomplish a desired process. (2.7)

Numerical control (NC): A system that uses predetermined instructions to control a sequence of manufacturing operations to produce a part. (2.7)

Part program: The set of instructions required to produce a desired part. (2.7)

Process control: The regulation of variables in a process. (2.4)

Regulator system: A feedback control system in which the setpoint is seldom changed. Its primary function is to maintain the controlled variable constant despite unwanted load changes. (2.3)

Robot: A programmable manipulator designed to move material, parts, tools, or other devices through a sequence of motions to accomplish a specific task. (2.8)

Sequential control: A control system that performs a set of operations in a prescribed manner. The automatic washing machine is a familiar example of a sequential control system. (2.6)

Servomechanism: A feedback control system in which the controlled variable is physical position or motion. (2.5)

Time-driven operations: Operations in a sequential control system that are initiated and terminated at a certain time or after a certain time interval. (2.6)

EXERCISES

2.1 The digital signal illustrated in Figure 2.1 is obtained by eliminating the decimal part of the analog temperature signal. This is referred to as truncating a signal. Thus the signals 72, 72.3, 72.56, and 72.999 are all converted to 72 by the analog-to-digital conversion. An alternative method of conversion would be to round off to the nearest integer. In this method, 72 and 72.3 would be converted to 72, and 72.56 and 72.999 would be converted to 73. Redraw Figure 2.1b using the nearest integer method of analog-to-digital conversion.

2.2 When an analog signal is converted to a digital signal, there is usually a difference between the digital and analog values. This difference is called the *quantization error*. Estimate the quantization error in Figure 2.1 at each minute from 11:30 to 11:42 (i.e., at the vertical dashed lines).

2.3 From your own experience, give an example of a regulator system and an example of a follow-up system. Identify the control system components in each example.

2.4 Draw a block diagram of the temperature control system in Figure 2.3 and identify each block in the diagram.

2.5 Draw an instrumentation diagram of the temperature control system in Figure 2.3. Use the loop number 203 in the identification tag for each instrument.

2.6 Draw a block diagram of the compensated mass flow control loop in Figure 2.4. Show the process as a block with one manipulated input (control valve position) and three outputs (volume flow rate, pressure, and temperature).

2.7 The mass flow control system in Figure 2.4 is to be used to control the flow rate of nitrogen gas in proportion to the demand signal. A test of the flow meter revealed that a nitrogen gas flow rate of 0.001 cubic meter per second produces a differential pressure of 2.5×10^4 pascal.

   a. Use Equation (2.2) to determine the flow meter proportionality constant, k_f.

   b. Determine the mass flow rate proportionality constant, k.

   c. Use Equation (2.3) to compute the mass flow rate given that $D = 2.5 \times 10^4$ pascal, $p = 2 \times 10^5$ pascal, and $T = 310$ kelvin.

2.8 Select an example of a servomechanism (perferably from your own experience), sketch a block diagram, and explain the operation of the system.

2.9 Name and explain the two types of sequential control based on how the operations are initiated and terminated.

2.10 Sketch a timing diagram of a four-operation time-driven sequential control system with the following specifications:

   (1) Operation A is on from 0 to 6 minutes and from 30 to 36 minutes.

   (2) Operation B is on from 5 to 15 minutes.

   (3) Operation C is on from 35 to 45 minutes.

   (4) Operation D is on from 30 to 32 minutes.

2.11 Construct a timing diagram of a series of time-driven sequential events from your own experience (e.g., a class day consists of a series of classes, which are time-driven events).

2.12 Redesign the pneumatic cylinder control system in Figure 2.8 so it will automatically cycle back and forth between the retracted position and the extended position. Draw a schematic diagram and an electric circuit diagram of your design. Write an explanation of the operation of your system similar to the caption of Figure 2.8. *Hint:* Replace the start switch with a normally open limit switch. Position the new limit switch so it will be closed when the cylinder is retracted and will open as soon as the piston moves from the retracted position.

2.13 Explain the following acronyms: NC, CNC, and DNC.

2.14 List the advantages and disadvantages of pneumatic actuators, hydraulic actuators, and electric actuators for robotic arms.

2.15 Describe the following types of robots: PNP, PTP, and CP.

2.16 Explain distributed control and centralized control, and list advantages and disadvantages of each type.

2.17 Place each control system in Figures 2.14 through 2.22 into one of the following categories: process control, servomechanism, sequential control.

2.18 Sketch block diagrams of the control systems in Figures 2.14, 2.15, and 2.16.

2.19 Label each example of a control system as one of the following types:

(1) a regulator system

(2) a followup system

(3) a time-driven sequential system

(4) an event-driven sequential system

Examples

a. Automobile steering system

b. Refrigerator temperature control system

c. Oven temperature control system

d. Automobile cruise control system

e. Automatic door opener

f. Automatic washing machine

g. Human body's temperature control system

h. Driver and automobile (Figure 1.3)

i. Position control system (Figure 2.6)

j. Hydraulic cylinder control system (Figure 2.21)

k. Sheet thickness control system (Figure 2.20)

CHAPTER 3

Digital Fundamentals

OBJECTIVES

Control has gone digital. The microprocessor is everywhere in control systems; it is the heart of process controllers and programmable controllers; it appears in "smart" sensors; it has made its way into final control elements. Control signals are transmitted in digital form on data highways and local area networks. A number of interface buses are available for custom-designed microprocessor control systems.

The purpose of this chapter is to review the digital fundamentals necessary to understand modern control systems. After completing this chapter, you will be able to

1. Convert numbers from one numbering system to another among the following numbering systems: binary, octal, decimal, or hexadecimal

2. Determine the one's-complement and the two's-complement representation of a negative binary number

3. Complete a truth table of any two-input logic symbol

4. Determine the logic level of the output of a logic circuit given the logic level of each input to the circuit

5. Design a decimal-to-binary encoder for a given code scheme

6. Design a binary-to-decimal decoder for a given code scheme

7. Write the Boolean function for the output of each gate in a logic diagram

8. Convert a Boolean equation into a truth table or Karnaugh map

9. Convert a truth table or Karnaugh map into a Boolean equation

10. Sketch the two-level logic circuit from a Boolean equation of the form $X = A \cdot B + C \cdot D$

11. Use loopings in a Karnaugh map to obtain the simplest form of a Boolean function

12. Describe the major parts of a microcomputer

13. Describe the process of preparing an assembly-coded program for a microcomputer

3.1 INTRODUCTION

The purpose of this chapter is to present the fundamental concepts and terminology related to digital systems. Considering the pervasiveness of digital operations in control, a knowledge of digital fundamentals can be very helpful.

Digital hardward works with binary numbers, but people prefer to work with decimal numbers. Therefore, number systems and conversion are a natural place to begin our study of digital fundamentals. Octal and hexadecimal numbers are included because they have a convenient relationship to binary, and people find them easier to work with than binary numbers.

Obviously, binary numbers in a digital system can represent numerical values in binary form. However, they can also represent decimal digits, alphanumeric characters, and various symbols, such as periods and commas. All we need is an agreed-upon code and a method of going between the binary representation and the symbol or character it represents. *Encoding* is the process of going from the symbol or character to the binary representation. *Decoding* is the reverse process. Thus encoders and decoders are another topic for our study.

Digital circuits consist of interconnections of elements that perform simple logical operations. The circuits combine these simple logic operations into a series of operations that perform a more complex logical function. Engineers use both mathematical and graphical methods to work with logic functions. Boolean algebra and Boolean equations are mathematical tools. Truth tables, Karnaugh maps, and timing diagrams are graphical tools. These tools are also a topic for this study. A discussion of the configuration and programming of a digital computer is a fitting conclusion for this chapter.

3.2 NUMBER SYSTEMS AND CONVERSION

Decimal Numbering System

The familiar decimal system is a good example to begin the study of number systems. Decimal refers to the fact that 10 symbols (0 through 9) are used to represent numbers. The number of symbols a numbering system uses is called the *base* of the numbering system. Thus the decimal numbering system is base 10.

Only one symbol is needed to represent values up to the base minus one (nine for the decimal system). To represent values greater than nine, additional symbols are used, and they are given weighted values according to their position. The positions are numbered from right to left, starting with 0 for the rightmost position. The positions are called *digits* and the position number is called the *order* of the digit. For example, in the number 345, the five is the zero-order digit, the 4 is the first-order digit, and the 3 is the second-order digit. The weighting value of a particular digit is equal to the base raised to the power of the position number (or order) of the digit. The zero-order digit has a weighted value of $10^0 = 1$, the first-order digit has a value of $10^1 = 10$, the second-order digit has a value of $10^2 = 100$, and so on. In the number 345, 5 represents five ones, 4 represents four tens, 3 represents three one-hundreds, and the value of the number is equal to the sum of the three weighted digits.

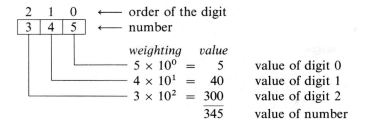

			weighting	value	
			$5 \times 10^0 =$	5	value of digit 0
			$4 \times 10^1 =$	40	value of digit 1
			$3 \times 10^2 =$	300	value of digit 2
				345	value of number

Binary Numbering System

The *binary numbering system* uses two symbols (0 and 1) and has a base of 2. The binary system works perfectly for describing the ON/OFF signals in computers and other digital systems. The OFF signal, which is usually 0 V, is assigned the binary value 0. The ON signal, which is usually 5 V, is assigned the binary value 1. Each position in the binary system is weighted by a power of 2: $2^0 = 1$, $2^1 = 2$, $2^2 = 4$, $2^3 = 8$, and so on (Table 3.1). The positions are called *bits* (*binary digits*), and they are named (from right to left) the zero-order bit, the first-order bit, the second-order bit, the third-order bit, and so on. The decimal value of a binary number is obtained by summing the weighted values of each bit.

Example 3.1

Determine the decimal equivalent of the binary number 100101.

Solution

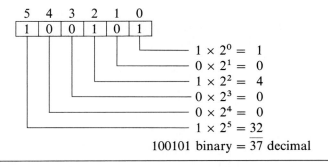

$$1 \times 2^0 = 1$$
$$0 \times 2^1 = 0$$
$$1 \times 2^2 = 4$$
$$0 \times 2^3 = 0$$
$$0 \times 2^4 = 0$$
$$1 \times 2^5 = 32$$

$$100101 \text{ binary} = \overline{37} \text{ decimal}$$

Table 3.1   Powers of 2

$2^0 = 1$	$2^9 = 512$
$2^1 = 2$	$2^{10} = 1{,}024$
$2^2 = 4$	$2^{11} = 2{,}048$
$2^3 = 8$	$2^{12} = 4{,}096$
$2^4 = 16$	$2^{13} = 8{,}192$
$2^5 = 32$	$2^{14} = 16{,}384$
$2^6 = 64$	$2^{15} = 32{,}768$
$2^7 = 128$	$2^{16} = 64{,}536$
$2^8 = 256$	

Octal Numbering System

The *octal numbering system* uses eight symbols (0 through 7) and has a base of 8. Each position in the octal system is weighted by a power of 8: $8^0 = 1$, $8^1 = 8$, $8^2 = 64$, $8^3 = 512$, and so on. The positions in an octal number are called *digits* and are named the same as decimal digits (zero-order digit, first-order digit, etc.). Octal numbers are sometimes used as a more convenient way to handle binary numbers. Each digit in an octal number is equivalent to three digits in a binary number as shown in Table 3.2.

Numbers can be easily converted from binary to octal or octal to binary, as demonstrated in Examples 3.2 and 3.3. Octal numbers are converted to decimal by summing the weighted values of each digit as shown in Example 3.4.

Example 3.2

Convert the octal number 3274 to binary.

Solution

$$
\begin{array}{lcccc}
\text{Octal:} & 3 & 2 & 7 & 4 \\
\text{Binary:} & 011 & 010 & 111 & 100
\end{array}
$$

3274 octal = 011 010 111 100 binary

Example 3.3

Convert the binary number 101 010 001 111 to octal.

Solution

Begin at the rightmost bit and convert each three binary digits into one octal digit. The three bit positions have the following values:

Right-hand bit has the value 1

Middle bit has the value 2

Left-hand bit has the value 4

$$
\begin{array}{lcccc}
\text{Binary:} & 101 & 010 & 001 & 111 \\
\text{Octal:} & 5 & 2 & 1 & 7
\end{array}
$$

101 010 001 111 binary = 5217 octal

Table 3.2   Octal and Binary Equivalents

Octal	Binary	Octal	Binary
0	000	4	100
1	001	5	101
2	010	6	110
3	011	7	111

Example 3.4

Convert the octal number 5217 to decimal.

Solution

Sum the values of the weighted digits.

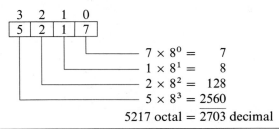

$$7 \times 8^0 = 7$$
$$1 \times 8^1 = 8$$
$$2 \times 8^2 = 128$$
$$5 \times 8^3 = 2560$$
$$5217 \text{ octal} = \overline{2703} \text{ decimal}$$

Hexadecimal Numbering System

The *hexadecimal numbering system* uses 16 symbols (0 through 9 and A through F) and has a base of 16. Each position in the hexadecimal system is weighted by a power of 16: $16^0 = 1$, $16^1 = 16$, $16^2 = 256$, $16^3 = 4096$, and so on. The positions in a hexadecimal number are also called *digits* and are named the same as decimal and octal digits. Hexadecimal numbers are often used as a more convenient way to handle binary numbers. Each hexadecimal number is equivalent to four digits in a binary number as shown in Table 3.3.

Numbers can easily be converted from binary to hexadecimal or hexadecimal to binary, as demonstrated in Examples 3.5 and 3.6. Hexadecimal numbers are converted to decimal by summing the weighted values of each digit as shown in Example 3.7.

Example 3.5

Convert the hexadecimal number A3B7 to binary.

Table 3.3   Decimal, Hexadecimal, and Binary Equivalents

Decimal	Hex	Binary	Decimal	Hex	Binary
0	0	0000	8	8	1000
1	1	0001	9	9	1001
2	2	0010	10	A	1010
3	3	0011	11	B	1011
4	4	0100	12	C	1100
5	5	0101	13	D	1101
6	6	0110	14	E	1110
7	7	0111	15	F	1111

Solution

Hexadecimal: A   3   B   7

Binary: 1010 0011 1011 0111

A3B7 hexadecimal = 1010 0011 1011 0111 binary

Example 3.6

Convert the binary number 0101 1110 0100 1100 to hexadecimal.

Solution

Begin at the rightmost bit and convert each four binary digits into one hex digit.

Binary: 0101 1110 0100 1100

Hexadecimal: 5   E   4   C

0101 1110 0100 1100 binary = 5E4C hexadecimal

Example 3.7

Convert the hexadecimal number A3B7 to decimal.

Solution

Sum the values of the weighted digits. Use the following decimal values for the symbols A . . . F:

$A = 10$ $B = 11$ $C = 12$ $D = 13$ $E = 14$ $F = 15$

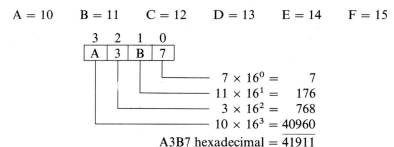

$$7 \times 16^0 = 7$$
$$11 \times 16^1 = 176$$
$$3 \times 16^2 = 768$$
$$10 \times 16^3 = 40960$$

A3B7 hexadecimal = 41911

Number Conversions

Conversions from any base to base 10 are a simple matter of summing the weighted values of the digits. Refer to Examples 3.1, 3.4, and 3.7 for further details in converting binary, octal, and hexadecimal numbers to the equivalent decimal number.

    Conversions from base 10 to any base can be done by the division algorithm for integer numbers and the multiplication algorithm for fractional numbers. (*Note:* Integer numbers have no digits to the right of the decimal point; fractional numbers have no digits except 0 to the left of the decimal point.)

Division Algorithm

1. Divide the integer decimal number by the base of the new numbering system. The *remainder* is the zero-order digit (converting to base 2, divide by 2; converting to base 8, divide by 8; converting to base 16, divide by 16).
2. Divide the quotient of the previous division by the base; the *remainder* is the next-higher-order digit (i.e., first-order digit, then second-order digit, etc.).
3. Continue step 2 until the quotient of the division is 0; the final remainder is the highest-order digit.

Example 3.8

Convert decimal 462 to binary.

Solution

$$462/2 = 231 \quad R \ 0$$
$$231/2 = 115 \quad R \ 1$$
$$115/2 = \ 57 \quad R \ 1$$
$$57/2 = \ 28 \quad R \ 1$$
$$28/2 = \ 14 \quad R \ 0$$
$$14/2 = \ \ 7 \quad R \ 0$$
$$7/2 = \ \ 3 \quad R \ 1$$
$$3/2 = \ \ 1 \quad R \ 1$$
$$1/2 = \ \ 0 \quad R \ 1$$

462 decimal = 1 1100 1110 binary

Check by summing the weighted digits of the binary number.

$$1 \times 256 + 1 \times 128 + 1 \times 64 + 1 \times 8 + 1 \times 4 + 1 \times 2 = 462 \text{ decimal}$$

Example 3.9

Convert decimal 607 to octal.

Solution

$$607/8 = 75 \quad R \ 7$$
$$75/8 = \ 9 \quad R \ 3$$
$$9/8 = \ 1 \quad R \ 1$$
$$1/8 = \ 0 \quad R \ 1$$

607 decimal = 1137 octal

Check by summing the weighted digits of the octal number.

$$1 \times 8^3 + 1 \times 8^2 + 3 \times 8^1 + 7 \times 8^0 = 607 \text{ decimal}$$

Example 3.10

Convert decimal 5676 to hexadecimal.

Solution

$$5676/16 = 354 \quad R \ 12 = C \ (hex)$$
$$354/16 = \ 22 \quad R \ 2$$
$$22/16 = \ \ 1 \quad R \ 6$$
$$1/16 = \ \ 0 \quad R \ 1$$
$$5676 \text{ decimal} = 162C \text{ hexadecimal}$$

Check: $1 \times 16^3 + 6 \times 16^2 + 2 \times 16^1 + 12 \times 16^0 = 5676$ decimal

Multiplication Algorithm

1. Multiply the fractional decimal number by the base of the new number system. The integer part of the product is the $-$first-order digit.
2. Multiply the fractional part of the previous product by the base; the integer part of the product is the next-lower-order digit (i.e., $-$first-order, then $-$second-order, then $-$third-order, etc.).
3. Continue step 2 until either the fractional part of the product is 0, or the desired degree of precision has been obtained.

Example 3.11

Convert decimal 0.625 to binary.

Solution

$$0.625 \times 2 = 1.250 \longleftarrow -\text{1st-order bit} = 1$$
$$0.250 \times 2 = 0.500 \longleftarrow -\text{2nd-order bit} = 0$$
$$0.500 \times 2 = 1.000 \longleftarrow -\text{3rd-order bit} = 1$$
$$0.625 \text{ decimal} = 0.101 \text{ binary}$$

Check: $1 \times 2^{-1} + 1 \times 2^{-3} = 0.625$ decimal.

Example 3.12

Convert decimal 0.428 to octal.

Solution

$$0.428 \times 8 = 3.424 \longleftarrow -\text{1st-order digit} = 3$$
$$0.424 \times 8 = 3.392 \longleftarrow -\text{2nd-order digit} = 3$$
$$0.392 \times 8 = 3.136 \longleftarrow -\text{3rd-order digit} = 3$$
$$0.136 \times 8 = 1.088 \longleftarrow -\text{4th-order digit} = 1$$
$$0.428 \text{ decimal} = \text{approximately } 0.3331 \text{ octal}$$

Check: $3 \times 8^{-1} + 3 \times 8^{-2} + 3 \times 8^{-3} + 1 \times 8^{-4} = 0.42797852$ decimal.

Example 3.13

Convert decimal 0.1 to hexadecimal.

Solution

$$0.1 \times 16 = 1.6 \longleftarrow -\text{1st-order digit} = 1$$
$$0.6 \times 16 = 9.6 \longleftarrow -\text{2nd-order digit} = 9$$
$$0.6 \times 16 = 9.6 \longleftarrow -\text{3rd-order digit} = 9$$
$$0.6 \times 16 = 9.6 \longleftarrow -\text{4th-order digit} = 9$$

Conclusion: All lower-order digits are 9.

$$0.1 \text{ decimal} = 0.1999 \ldots \text{ hexadecimal}$$

Check: $1 \times 16^{-1} + 9 \times 16^{-2} + 9 \times 16^{-3} + 9 \times 16^{-4} = 0.09999084 \ldots$ decimal.

Negative Numbers

Binary negative numbers can be represented in the following three ways:

1. Sign plus magnitude
2. One's complement
3. Two's complement

In each method, the *most significant bit* (MSB) of the number is used as the sign bit. A 1 in the MSB represents a negative number, and a 0 represents a positive number. The remainder of the number is where the difference occurs.

In the *sign-plus-magnitude method*, the remainder of the number is represented in its normal form, just as it is for positive numbers. For example, the 8-bit binary equivalent of decimal 10 is 0000 1010, and the sign-plus-magnitude representation of −10 is 1000 1010.

In the *one's-complement method*, the remainder of the number is represented in its one's-complement form. To form the one's complement of a number, every bit (except the sign bit) is inverted. The one's complement of −10 is 1111 0101.

In the *two's-complement method*, the remainder of the number is represented in its two's-complement form. To form the two's complement, move from right to left, and invert every bit *after* the first one is encountered. The two's complement of −10 is 1111 0110.

The one's- and two's-complement methods are related to the modulus of the digital system that is handling the binary numbers. The modulus is one more than the largest number the system can handle, and is equal to the number of unique values the system can represent. For example, in an 8-bit system, the binary numbers range from 0000 0000 to 1111 1111, which is 256 unique values going from 0 to 255 decimal. Thus the modulus of an 8-bit system is 256 decimal or 1 0000 0000 binary, and the modulus minus 1 is 255 decimal or 1111 1111 binary.

The one's complement of a binary number is obtained by subtracting the number from the modulus minus 1. Thus a second method of obtaining the one's-complement

Table 3.4   One's and Two's Complements

Decimal Number	Two's Complement		One's Complement	
	Binary	Hex	Binary	Hex
−1	1111 1111	FF	1111 1110	FE
−2	1111 1110	FE	1111 1101	FD
−3	1111 1101	FD	1111 1100	FC
−4	1111 1100	FC	1111 1011	FB
−5	1111 1011	FB	1111 1010	FA
−6	1111 1010	FA	1111 1001	F9
−7	1111 1001	F9	1111 1000	F8
−8	1111 1000	F8	1111 0111	F7
−9	1111 0111	F7	1111 0110	F6
−10	1111 0110	F6	1111 0101	F5
−11	1111 0101	F5	1111 0100	F4
−12	1111 0100	F4	1111 0011	F3

form of −10 is by subtracting 0000 1010 from 1111 1111 as follows: one's complement of 0000 1010 = 1111 1111 − 0000 1010 = 1111 1010.

The two's complement of a binary number is obtained by subtracting the number from the modulus and is always one more than the one's complement. We can obtain the two's complement of −10 by subtracting 0000 1010 from 1 0000 0000. We can also obtain the two's complement of −10 by adding 1 to the one's complement. Some one's and two's complements are shown in Table 3.4.

3.3 LOGIC ELEMENTS

Logic signals (or lines) and logic elements provide a convenient way to describe the operations performed in digital circuits. Signals in a digital circuit can have only two possible values, a high voltage or a low voltage. Consequently, only two numerical values are required to represent the two voltage levels. A high voltage is usually represented by the number 1 and a low voltage by the number 0. This is called *positive logic*. Sometimes the representations are reversed. The term *negative logic* is used when a high voltage is represented by the number 0 and a low voltage is represented by the number 1. In this book we use positive logic.

Positive Logic   *Negative Logic*

1 = HI voltage = true condition   1 = LO voltage = true condition

0 = LO voltage = false condition   0 = HI voltage = false condition

Logic elements perform logic operations on one or more logic signals to produce another logic signal that is the result of the operation. The signals the elements operate on are called the *inputs* to the element; the result of the operation is called the *output* of the element. Four logic operations are used to define the logic operations performed by the various logic elements. The four operators are the NOT operator,

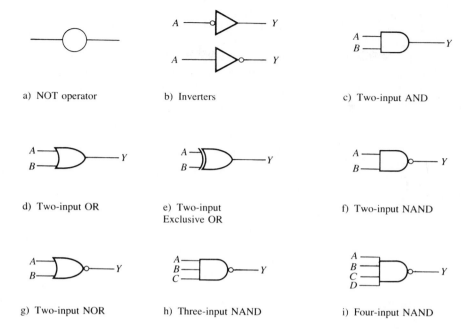

a) NOT operator

b) Inverters

c) Two-input AND

d) Two-input OR

e) Two-input
Exclusive OR

f) Two-input NAND

g) Two-input NOR

h) Three-input NAND

i) Four-input NAND

Figure 3.1   Symbols used to represent various logic elements in logic drawings. The letters A, B, C, and D are inputs; the letter Y is the output. The NOT operator is never used alone—it is always connected to the input or output of another logic element. An element that performs only the NOT operation is called an inverter. The symbol for an inverter is a triangle with the NOT circle at either the input or output.

the AND operator, the OR operator, and the Exclusive OR operator.* Two more operators, the NAND and the NOR, are formed by placing a NOT operator at the output of the AND and the OR operators (the term NAND is a contraction of NOT AND, and the term NOR is a contraction of NOT OR). The NOT operator has one input and one output; all the other operators have two or more inputs and one output. Figure 3.1 shows the symbols used to represent these logic operators in logic drawings.

The output of the NOT operator is always opposite the input. If the input to a NOT operator is 1, the output is 0; if the input is 0, the output is 1. The inverter works just like the NOT operator. The output of the AND operator is 1 if and only if all inputs are 1; otherwise, the output is 0. The output of the OR is 0 if and only if all inputs are 0; otherwise, the output is 1. The Exclusive OR is only defined for two inputs. The output of an Exclusive OR is 1 if and only if one input is 1 and the

* The Exclusive OR operator can be defined as a combination of the NOT, AND, and OR operators, but it is more convenient to define it as a separate operator.

other input is 0; otherwise, the output is 0. The logic operations can be defined by a table, called a *truth table*, that lists all possible combinations of the inputs and the resulting outputs. There really are not very many possibilities—with two inputs, there are only four possible combinations. With three inputs, there are eight possible combinations, and with four inputs, there are 16 possible combinations. Truth tables are usually confined to four or fewer inputs. In the following truth tables, the letters A, B, and Y match the letters in Figure 3.1.

Inverter

A	Y
0	1
1	0

Two-Input AND

A	B	Y
0	0	0
0	1	0
1	0	0
1	1	1

Two-Input OR

A	B	Y
0	0	0
0	1	1
1	0	1
1	1	1

Two-Input Exclusive OR

A	B	Y
0	0	0
0	1	1
1	0	1
1	1	0

Two-Input NAND

A	B	Y
0	0	1
0	1	1
1	0	1
1	1	0

Two-Input NOR

A	B	Y
0	0	1
0	1	0
1	0	0
1	1	0

A logic circuit consists of *logic elements* interconnected by lines that carry logic signals. Some of the inputs to the elements come from outside the logic circuit; these are called *external inputs*. The rest of the inputs to the elements come from other elements in the circuit; these are called *internal inputs*. The purpose of the logic circuit is to produce one or more outputs that are used externally to the logic circuit. These are called *external outputs* of the circuit. A trace of a logic circuit is a process in which a person starts with given values for the external inputs and works through the circuit to determine the logic values of the external outputs. When all inputs to a logic element are known, the truth table of the element is used to determine the output of the element. Figure 3.2 shows a typical trace of a logic circuit.

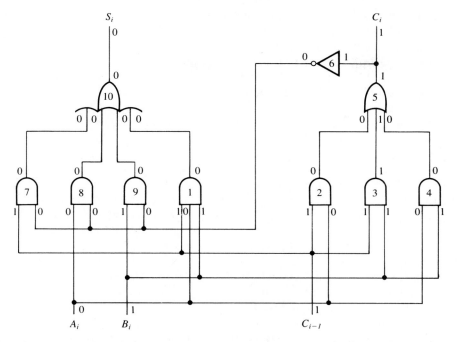

Figure 3.2 Logic diagram of one stage of a binary adder. Arbitrary logic values ($A_i = 0$, $B_i = 1$, $C_{i-1} = 1$) are assigned to the external inputs of the adder to illustrate a trace of logic levels in the circuit. A trace consists of writing the logic values of each input to an element and using the input values to determine the value of the output of the element. The logic elements are numbered in the order in which their outputs are determined. Notice that elements 5 through 10 have internal inputs that must be obtained before the output of the element can be determined.

3.4 CODES, ENCODERS, AND DECODERS

Codes

Binary numbers are only one type of data handled by digital systems—they just happen to be the type most suited to the system's internal workings. Other types of data include decimal numbers and text (i.e., letters, decimal digits, punctuation marks, and other symbols). To handle data in a digital system, a unique combination of 0's and 1's is defined for each letter, number, or symbol that must be represented. The complete set of symbols is called a *code*. There are three types of codes: those that represent numbers in a binary format, those that represent the 10 decimal digits, and those that represent text. The binary numbering system and the Gray code are the most common codes for representing numbers in binary format. The binary-coded decimal (BCD) code is the most popular code for decimal digits, and the American Standard Code for Information Interchange (ASCII) is the most popular code for text.

The binary numbering system was described in Section 3.2. It is the way numbers are usually represented inside a computer. However, the binary system has one problem when it is used for counting. If you have ever watched the odometer of a car, you may have noticed the source of the problem. Imagine that the odometer is just about to change from 59999 to 60000. Every digit must change at the same time, but they rarely do. A counting code that requires more than one digit to change when the count increases by 1 creates problems for digital systems. Digital systems do not look at the counter continuously, as we do. Rather, they take instantaneous readings called *samples*. The count obtained at a given sample is used until the next sample is taken some time later. Assume that the two right digits of the odometer have already changed from 9 to 0 but the other three digits have not yet changed. The odometer would read 59900, which is in error by 100 miles. The solution is to use a counting code that always changes only one digit each time the count increases by 1. The Gray code is just such a code. The Gray code is used in encoders and other digital counting devices. It is also used to number the rows and columns in Karnaugh maps, a topic covered in Section 3.6. Table 3.5 shows the Gray code for numbers from 0 to 15 (decimal).

The Gray code is a reflected code. Remove bit 3 from the Gray code numbers in Table 3.5 and notice that the remaining three digits of the Gray code in the right column are the mirror image of the remaining three digits of the Gray code in the left column (i.e., the right three digits of 8 and 7 are both 100, the right three digits of 9 and 6 are both 101, . . . , the right three digits of 15 and 0 are both 000). Notice also that the digit we removed, bit 3, is 0 in the left column and 1 in the right column. These observations can be used to obtain a Gray code for decimal values from 0 to 31. Start with the Gray code in Table 3.5, add an extra digit, x, on the left and follow this 5-bit code with its mirror image (i.e., start with $x1000$ for 16, $x1001$ for 17, $x1011$ for 18, . . . , $x0000$ for 31). Replace the x by 0 for the first half (0 to 15) and replace the x by 1 for the second half (16 to 31). The result is a Gray code for decimal values from 0 to 31. The same process can be used to expand the 5-bit Gray code into a 6-bit Gray code for values from 0 to 63, and so on.

It takes four binary digits (bits) to represent the 10 decimal digits (0 . . . 9). It seems impossible, but there are almost 30 billion unique ways to do this. Only a few of this multitude of possible codes have ever been used. Of these, the BCD code is

Table 3.5   The Gray Code

Decimal	Binary	Gray	Decimal	Binary	Gray
0	0000	0000	8	1000	1100
1	0001	0001	9	1001	1101
2	0010	0011	10	1010	1111
3	0011	0010	11	1011	1110
4	0100	0110	12	1100	1010
5	0101	0111	13	1101	1011
6	0110	0101	14	1110	1001
7	0111	0100	15	1111	1000

Table 3.6   Binary Codes for Decimal Digits

Decimal	BCD	Excess-3	Decimal	BCD	Excess-3
0	0000	0011	5	0101	1000
1	0001	0100	6	0110	1001
2	0010	0101	7	0111	1010
3	0011	0110	8	1000	1011
4	0100	0111	9	1001	1100

the most popular. Table 3.6 shows two such codes, the BCD code and the excess-3 code.

The excess-3 code has an interesting property. Take any two digits that add up to 9 and their codes complement each other. By complement, we mean that the 1's in one code are where the 0's are in the other code, and vice versa. Here are the five pairs of digits that add up to 9, together with their excess-3 codes.

$$0 \ (0011) \text{ and } 9 \ (1100) \qquad 1 \ (0100) \text{ and } 8 \ (1011)$$
$$2 \ (0101) \text{ and } 7 \ (1010) \qquad 3 \ (0110) \text{ and } 6 \ (1001)$$
$$4 \ (0111) \text{ and } 5 \ (1000)$$

A code with this property is called a self-complementing code. Self-complementing codes simplify some arithmetic operations.

The standard ASCII code is a 7-bit code with an eighth bit added for error checking. The 7 bits provide unique codes for 128 symbols, enough for upper- and lower-case letters, decimal digits, punctuation marks, control characters for controlling peripherals, and so on. The 7-bit ASCII code is shown in Table 3.7, with the ASCII code values in hexadecimal.

Example 3.14

Use the BCD code to convert the decimal number 49 to an 8-bit BCD representation, with the lower 4 bits used for the lower-order decimal digit and the upper 4 bits used for the higher-order decimal digit.

Solution

The BCD representation of 4 is 0100 and the BCD representation of 9 is 1001.

$$49 \ (\text{decimal}) = 0100 \ 1001 \ (\text{BCD})$$

Encoders

Encoding is the process of generating the code for any given symbol. A decimal-to-binary encoder is a logic circuit that receives an input on one of 10 decimal input lines and produces a unique 4-bit code on its four binary output lines. Encoding is

Table 3.7   Seven-Bit ASCII Code

ASCII Code (Hex)	Coded Symbol*	ASCII Code (Hex)	Coded Symbol	ASCII Code (Hex)	Coded Symbol	ASCII Code (Hex)	Coded Symbol
00	nul	20	space	40	@	60	'
01	soh	21	!	41	A	61	a
02	stx	22	"	42	B	62	b
03	etx	23	#	43	C	63	c
04	eot	24	$	44	D	64	d
05	enq	25	%	45	E	65	e
06	ack	26	&	46	F	66	f
07	bel	27	'	47	G	67	g
08	bs	28	(	48	H	68	h
09	ht	29	)	49	I	69	i
0A	lf	2A	*	4A	J	6A	j
0B	vt	2B	+	4B	K	6B	k
0C	ff	2C	,	4C	L	6C	l
0D	cr	2D	—	4D	M	6D	m
0E	so	2E	.	4E	N	6E	n
0F	si	2F	/	4F	O	6F	o
10	dle	30	0	50	P	70	p
11	dc1	31	1	51	Q	71	q
12	dc2	32	2	52	R	72	r
13	dc3	33	3	53	S	73	s
14	dc4	34	4	54	T	74	t
15	nak	35	5	55	U	75	u
16	syn	36	6	56	V	76	v
17	etb	37	7	57	W	77	w
18	can	38	8	58	X	78	x
19	em	39	9	59	Y	79	y
1A	sub	3A	:	5A	Z	7A	z
1B	esc	3B	;	5B	[	7B	{
1C	fs	3C	<	5C	\	7C		
1D	gs	3D	=	5D	]	7D	}
1E	rs	3E	>	5E	∧	7E	~
1F	us	3F	?	5F	–	7F	rub

* ASCII codes from 00 to 1F are control characters.

basically an OR operation. An encoder consists of four OR elements, one for each of the four output lines. Figure 3.3 shows the logic diagram for an excess-3 encoder.

Designing an encoder is a simple matter of determining the inputs to each of the four OR elements. To do this, just look down each column of the 4-bit binary code (i.e., bit 0 column, bit 1 column, bit 2 column, or bit 3 column). Record the decimal digit for each code that has a 1 in a particular column. The decimal values you record are the inputs to the OR element whose output has the same bit number as the column the 1 is in. For example, reading down the bit 0 column of the excess-3

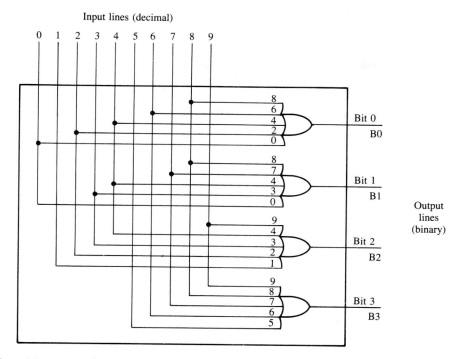

Figure 3.3 The excess-3 encoder receives a logic 1 on one of the 10 decimal input lines and produces the excess-3 code for that decimal digit on the four binary output lines. The remaining nine decimal input lines must all have logic 0 values.

code, we find a 1 in the following decimal rows: 0, 2, 4, 6, and 8. Therefore, the decimal inputs to the bit 0 OR element are 0, 2, 4, 6, and 8. Repeat this procedure for bits 1, 2, and 3 to obtain the inputs for the other three OR elements.

Decoders

Decoding is the process of determining the symbol for any given code. A binary-to-decimal decoder is a logic circuit that receives an input on four binary input lines and produces an output signal on one of the 10 decimal output lines. Decoding is basically an AND operation. A decoder consists of 10 four-input AND elements, one for each decimal output line. Figure 3.4 shows the logic diagram for an excess-3 decoder.

Designing a decoder is a simple matter of determining the inputs to each of the 10 AND elements. Each AND element has a bit 0 input, a bit 1 input, a bit 2 input, and a bit 3 input. Each input comes from one of two places, either directly from the binary input line or from an inversion of the binary input line. Inverted lines are indicated by placing a bar over the top of the bit designation. Notice that each of the four binary input lines passes through an inverter, and the normal and inverted values of

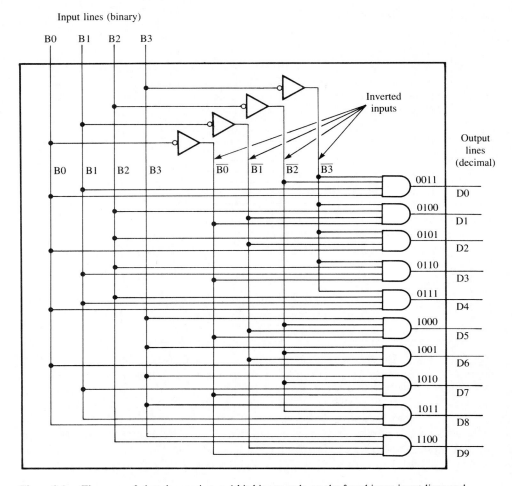

Figure 3.4   The excess-3 decoder receives a 4-bit binary code on the four binary input lines and produces a logic 1 on one of the 10 decimal output lines. The remaining nine decimal output lines will all have a logic 0 value.

each input are brought down for possible connection to one of the AND elements. The trick is to determine whether to use the normal or the inverted value for each input. To do this, look at the binary code for a particular decimal digit. For example, the excess-3 code for decimal 2 is 0101. Use the normal inputs for the two bits that have logic 1 values and the inverted inputs for the two that have logic 0 values. Thus the decimal 2 AND element has normal inputs from bits 0 and 2, and it has inverted inputs from bits 1 and 3. Repeat this process for the other nine AND elements, always using normal inputs for the bits that have a 1 value and inverted inputs for the bits that have a 0 value.

3.5 BOOLEAN ALGEBRA

In 1849, a mathematician named George Boole devised a form of algebra that can be used to represent logic. Of course, George Boole had no idea that his development would be used to help design digital logic circuits over 100 years later. Actually, Boole was attempting to devise a method of carrying out the logic of reasoning, an ancient form of philosophy. Although Boole's dream of reducing all reasoning to simple mathematical computations did not reach fruition, he did provide us with a mathematical method that applies to digital logic circuits. This mathematics of logic is named *Boolean algebra* in honor of George Boole.

Boolean algebra is more than a mathematical method for developing logic circuits. It is also a means of expressing complex logic circuits in terms of simple statements. Perhaps this is Boole's greatest contribution. Many of the design operations performed with Boolean algebra can be accomplished more easily with graphic techniques, such as looping in a Karnaugh map to reduce a logic circuit to a simpler form. However, the ability to express digital logic in concise statements is a very useful thing—it facilitates the communication of ideas between individuals. The objective of this section is to develop the ability to use Boolean algebra to express your ideas about digital logic and to give some understanding of the rules and operations that apply to Boolean expressions. Mastery of the mathematical methods of Boolean algebra is not a requirement in the remaining chapters of this book.

The variables in Boolean algebra are represented by a single letter or by a letter followed by more letters or numbers. *Boolean variables* have only two possible values, 0 or 1, which is perfect for representing the inputs and outputs of the elements in a logic circuit. Each external input and external output in a logic circuit is assigned a Boolean variable. Internal inputs may also be assigned a Boolean variable, but more often, they are represented by Boolean expressions of the external input variables.

Boolean algebra uses three logic operators (NOT, AND, and OR), and they have the same meaning as the logic operators with the same name (see Section 3.3). A fourth operator, the Exclusive OR, is also used with logic circuits. The AND operator symbol is a dot placed between the two operands (the variables the operation is to be performed on). The OR operator symbol is a plus sign placed between the two operators. The NOT operator symbol is a line drawn above the variable to which it is applied. The Exclusive OR operator symbol is a plus sign with a circle around it. A Boolean expression consists of Boolean variables separated by Boolean operators. The following Boolean expressions illustrate the use of the operator symbols.

Boolean Expressions

$$A \qquad A + B \qquad A \cdot B \qquad \bar{A} \cdot B + A \cdot \bar{B} \qquad \bar{A} + B$$

Boolean expressions can be used to express the output of a logic element in terms of the inputs to the element. For example, if the inputs to a two-input AND element are A and B, the output of the element may be expressed as $A \cdot B$; if the inputs to a three-input OR element are C, D, and E, the output of the element may be expressed as $C + D + E$. By working through a logic circuit element by element, it is possible

Figure 3.5 Logic circuit showing how Boolean variables and Boolean expressions can be used to represent the inputs and outputs of a logic circuit. The logic elements are numbered in the order in which the expressions for their outputs were determined.

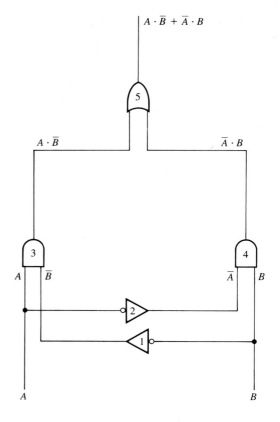

to obtain a Boolean expression for the output of a logic circuit in terms of the external inputs to the circuit. Figure 3.5 shows how Boolean expressions can be used to represent the outputs of the elements in a logic circuit and the output of the circuit.

The AND operator has precedence over the OR operator. Thus in the expression $A \cdot B + C \cdot D$, the two AND operations are performed before the OR operation. Parentheses may be used to change the order of operations. For example, if the previous expression were written as $A \cdot (B + C) \cdot D$, the OR operation would be performed first, followed by the two AND operations.

Equality between two Boolean expressions is indicated by the use of the equal sign in a *Boolean equation*. In Boolean algebra, the equal sign is both symmetric and transitive. By symmetric, we mean that $A = B$ implies that $B = A$. By transitive, we mean that $A = B$ and $B = C$ implies that $A = C$. Examples of Boolean equations are given below.

Boolean Equations

$$C = \bar{A} \cdot \bar{B} + A \cdot B$$
$$P = \bar{A} \cdot B + A \cdot \bar{B}$$
$$S = A \cdot B \cdot C + \bar{A} \cdot \bar{B} \cdot C + \bar{A} \cdot B \cdot \bar{C} + A \cdot \bar{B} \cdot \bar{C}$$

The following is a list of the rules that apply to Boolean algebra. They are similar to the rules of ordinary algebra, but there are a few differences, so be careful.

Rules of Boolean Algebra

Universal, Null, and Identity Rules

(1) $A + 1 = 1$ (2) $A + \bar{A} = 1$

(3) $A \cdot 0 = 0$ (4) $A \cdot \bar{A} = 0$

(5) $A \cdot A = A$ (6) $A + A = A$

(7) $A \cdot 1 = A$ (8) $A + 0 = A$

Commutative Law

(9) $A \cdot B = B \cdot A$ (10) $A + B = B + A$

Distributive Law

(11) $A \cdot (B + C) = (A \cdot B) + (A \cdot C)$

(12) $A + B \cdot C = (A + B) \cdot (A + C)$

Associative Law

(13) $(A \cdot B) \cdot C = A \cdot (B \cdot C)$

(14) $(A + B) + C = A + (B + C)$

DeMorgan's Theorem

(15) $\overline{A \cdot B} = \bar{A} + \bar{B}$ (16) $\overline{A + B} = \bar{A} \cdot \bar{B}$

Absorption

(17) $A + A \cdot B = A$ (18) $A \cdot (A + B) = A$

Double Inverse and Last Rule

(19) $\bar{\bar{A}} = A$ (20) $A + \bar{A} \cdot B = A + B$

The truth table provides us with a simple method to verify the foregoing rules. The idea is to develop a truth table for the expressions on either side of the equal sign. If the two expressions have the same truth table, the rule must be true. The following example illustrates this use of the truth table.

Example 3.15

Use a truth table to verify the first form of the distributive law, rule 11.

$$(11) \ A \cdot (B + C) = (A \cdot B) + (A \cdot C)$$

Solution

Construct the truth table with three input columns, two output columns for the left-hand expression in rule 11, and three output columns for the right-hand expression in rule 11. Label the three input columns A, B, and C, and fill in the input rows as shown in the table below. Label the output columns as shown below and proceed in the following order:

1. Use the values in columns labeled "B" and "C" to determine the values in the column labeled "B + C."
2. Use the values in columns "A" and "B + C" to determine the values in the column labeled "A · (B + C)." This is the left-hand expression in rule 11.
3. Use the values in the columns labeled "A" and "B" to determine the values in the column labeled "A · B."
4. Use the values in the columns labeled "A" and "C" to determine the values in the column labeled "A · C."
5. Use the results from steps 3 and 4 to determine the values in the column labeled "(A · B) + (A · C)." This is the right-hand expression in rule 11.
6. Compare the left-hand expression with the right-hand expression. Their truth tables are identical, which verifies rule 11.

A	B	C	B + C	A · (B + C)	A · B	A · C	(A · B) + (A · C)
0	0	0	0	0	0	0	0
0	0	1	1	0	0	0	0
0	1	0	1	0	0	0	0
0	1	1	1	0	0	0	0
1	0	0	0	0	0	0	0
1	0	1	1	1	0	1	1
1	1	0	1	1	1	0	1
1	1	1	1	1	1	1	1

↑
Left-hand side   ↑
Right-hand side

Example 3.16

Use Boolean algebra to reduce the following expression to a simpler form.

$$Z = \bar{A} \cdot B \cdot C + \bar{A} \cdot B \cdot \bar{C}$$

Solution

$$Z = \bar{A} \cdot B \cdot C + \bar{A} \cdot B \cdot \bar{C} \qquad \text{given}$$
$$Z = (\bar{A} \cdot B) \cdot C + (\bar{A} \cdot B) \cdot \bar{C} \qquad \text{rule 13}$$
$$Z = (\bar{A} \cdot B) \cdot (C + \bar{C}) \qquad \text{rule 11}$$
$$Z = (\bar{A} \cdot B) \cdot 1 \qquad \text{rule 2}$$
$$Z = \bar{A} \cdot B \qquad \text{rule 7}$$

3.6 ANALYSIS AND DESIGN OF LOGIC CIRCUITS

Analysis involves the determination of the outcomes of a logic circuit for each set of input conditions. Design involves the development of a logic circuit that satisfies a given set of output specifications. The steps in analysis and design of logic circuits are shown below.

Logic circuit

Analysis ↓   ↑ Design

Boolean equation

Analysis ↓   ↑ Design

Truth table or Karnaugh map

Analysis ↓   ↑ Design

Specification of outcomes

Analysis of logic circuits is facilitated by the ability to convert logic circuits into Boolean equations, Boolean equations into truth tables, and truth tables into a specification of outcomes. Design is facilitated by the ability to convert outcome specifications into truth tables, truth tables into Boolean equations, and Boolean equations into logic circuits. One objective of the design process is to obtain the simplest logic circuit that will do the job. In pursuit of the simplest circuit, the logic designer may use Boolean algebra or a Karnaugh map to reduce the Boolean equation to its simplest form. This section covers the conversions between logic circuits, Boolean equations, and truth tables; it also touches on the use of Boolean algebra and Karnaugh maps to obtain the simplest form of the Boolean equation.

From Logic Circuit to Boolean Equation

The conversion from a logic circuit to a Boolean equation consists of obtaining a Boolean equation for the output in terms of the external inputs. The equation has only the name of the output on the left-hand side and a Boolean expression of external inputs on the right-hand side. There will be one equation for each external output of the logic circuit. Conversion is largely a matter of determining the expression for the output of each circuit element in terms of the inputs to that element—starting with the elements at the circuit's input and ending with the element at the circuit's output. This process was first illustrated in Figure 3.5. Conversion of a more complex logic circuit is shown in Figure 3.6.

From Boolean Equation to Logic Circuit

The conversion from a Boolean equation to a logic circuit is just the reverse of the previous conversion. This discussion will be limited to Boolean equations that are of the form: $X = (\ \ ) + (\ \ ) \cdots + (\ \ )$, where the parentheses enclose expressions that consist of only variable names and AND operators. Here are some examples of this

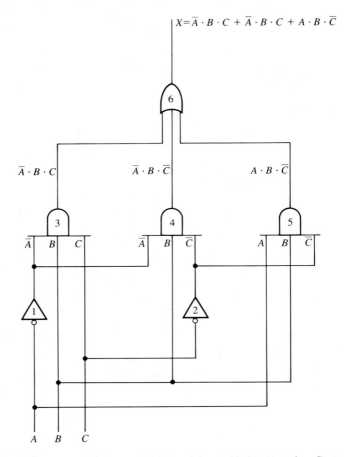

$X = \bar{A} \cdot B \cdot C + \bar{A} \cdot B \cdot C + A \cdot B \cdot \bar{C}$

Figure 3.6   Example of the conversion of a logic circuit into a Boolean equation. Conversion consists of determining the expression for the output of each circuit element in terms of the inputs to that element. The logic elements are numbered in the order in which the expression for each output was determined. The equation is obtained by equating the name of the output of the circuit (X) to the expression for the output from element 6.

form of equation:

$$X = A \cdot B + C \cdot D$$
$$Y = A \cdot \bar{B} \cdot C + \bar{A} \cdot B \cdot \bar{C} + \bar{A} \cdot \bar{B} \cdot \bar{C}$$
$$Z = A \cdot \bar{B} \cdot \bar{C} \cdot D + \bar{A} \cdot B \cdot C \cdot \bar{D} + A \cdot B \cdot C \cdot D$$

Equations of the form shown above can be implemented in a two-level logic structure consisting of one or more AND elements feeding an OR element. For this reason, we will refer to this as the AND–OR structure. The logic circuit will have one AND element for each of the terms in the right-hand side of the equation. The output of each AND element goes into a single OR element. Refer to Figure 3.6 again, but this time, assume that you are converting from the Boolean equation to the logic circuit. Start by drawing a row of AND elements, one for each term in the equation. Place a single OR element in the row above the AND elements. Connect the output of each AND element to an input to the OR element. You now have the general structure of the logic circuit. The final step is to connect the external inputs to the inputs of the AND elements, inverting external inputs where necessary. Often, both the true and complement values of the external inputs are available, and the inverters are not required.

From Boolean Equation to Truth Table

The conversion from a Boolean equation to a truth table consists of placing a 1 in each row of the truth table which has input values that, when substituted into the right-hand side of the equation, result in a value of 1. Since the terms are combined by the OR operator, the right-hand side will have a value of 1 if any one or more terms has a value of 1. This means that we can take the terms one at a time until every term has been considered. For example, the term $A \cdot B$ will have a 1 in every row of the truth table in which A and B both have a value of 1 [i.e., $(A = 1)$ AND $(B = 1)$]. A variable that is not included in the term can have either a 0 or a 1 value; it simply does not matter. In the following examples, the $\times$ means that the value of the variable can be either a 0 or a 1.

	Value of the Inputs for Which the Term Has a Value of 1			
Term	**A**	**B**	**C**	**D**
$\bar{A} \cdot B$	0	1	$\times$	$\times$
$\bar{A} \cdot \bar{B}$	0	0	$\times$	$\times$
$A \cdot \bar{B} \cdot C$	1	0	1	$\times$
$\bar{A} \cdot \bar{B} \cdot C$	0	0	1	$\times$
$A \cdot B \cdot \bar{C} \cdot \bar{D}$	1	1	0	0
$A \cdot \bar{B} \cdot \bar{C} \cdot D$	1	0	0	1

Example 3.17

Convert the following Boolean equation to a truth table.

$$Y = \bar{A} \cdot C + A \cdot B \cdot \bar{C} + A \cdot \bar{B} \cdot \bar{C}$$

Solution

The term $\bar{A} \cdot C$ will place a 1 in the truth table in every row for which A has a value of 0 and C has a value of 1. The term $A \cdot B \cdot \bar{C}$ will place a 1 in the truth table in the row for which the values of A, B, and C are 1, 1, and 0, respectively. The term $A \cdot \bar{B} \cdot \bar{C}$ will place a 1 in the truth table in the row for which the values of A, B, and C are 1, 0, and 0, respectively. The answer is shown in Table 3.8.

Table 3.8   Truth Table of
$Y = \bar{A} \cdot C + A \cdot B \cdot \bar{C} + A \cdot \bar{B} \cdot \bar{C}$

A	B	C	Y	
0	0	0	0	
0	0	1	1	⟵ $\bar{A} \cdot C$
0	1	0	0	
0	1	1	1	⟵ $\bar{A} \cdot C$
1	0	0	1	⟵ $A \cdot \bar{B} \cdot \bar{C}$
1	0	1	0	
1	1	0	1	⟵ $A \cdot B \cdot \bar{C}$
1	1	1	0	

From Truth Table to Boolean Equation

The conversion from a truth table to a Boolean equation consists of writing a "full" term for each 1 in the truth table. A "full" term (also called a *minterm*) is a term in which each input variable is present, either in its normal or inverted form. The minterms are combined by OR operators to form the right-hand side of the equation. The decision on which form of an input variable to use is based on the value the variable has for the object minterm. If the variable has a value of 0, the inverse form is used. If the variable has a value of 1, the normal form is used. Table 3.9 shows the minterm for each row in a three-variable truth table.

Example 3.18

An equation composed of minterms may not be the simplest form of the expression. Write the minterm form of Y in Table 3.8, and then use Boolean algebra to reduce the equation to a simpler form.

Solution

The minterm form of Y is

$$Y = \bar{A} \cdot \bar{B} \cdot C + \bar{A} \cdot B \cdot C + A \cdot \bar{B} \cdot \bar{C} + A \cdot B \cdot \bar{C}$$

The key to using Boolean algebra to reduce a logic circuit is to find two minterms that differ in only one variable. The variable that is different will be in its normal

Table 3.9   Minterm for Each Row
in a Three-Variable Truth Table

A	B	C	Y	
0	0	0	1	⟵ $\bar{A} \cdot \bar{B} \cdot \bar{C}$
0	0	1	1	⟵ $\bar{A} \cdot \bar{B} \cdot C$
0	1	0	1	⟵ $\bar{A} \cdot B \cdot \bar{C}$
0	1	1	1	⟵ $\bar{A} \cdot B \cdot C$
1	0	0	1	⟵ $A \cdot \bar{B} \cdot \bar{C}$
1	0	1	1	⟵ $A \cdot \bar{B} \cdot C$
1	1	0	1	⟵ $A \cdot B \cdot \bar{C}$
1	1	1	1	⟵ $A \cdot B \cdot C$

form in one minterm and in its inverse form in the other minterm. In the minterm equation for Y, the first two minterms are the same except for variable B. The rules of Boolean algebra can be applied to reduce these two terms to a single term, eliminating B in the process. Here is how it is done with Boolean algebra:

$$\bar{A} \cdot \bar{B} \cdot C + \bar{A} \cdot B \cdot C = \bar{A} \cdot C \cdot \bar{B} + \bar{A} \cdot C \cdot B \qquad \text{rule 9}$$

$$\bar{A} \cdot C \cdot \bar{B} + \bar{A} \cdot C \cdot B = \bar{A} \cdot C \cdot (\bar{B} + B) \qquad \text{rule 11}$$

$$\bar{A} \cdot C \cdot (\bar{B} + B) = \bar{A} \cdot C \cdot (B + \bar{B}) \qquad \text{rule 9}$$

$$\bar{A} \cdot C \cdot (B + \bar{B}) = \bar{A} \cdot C \cdot 1 \qquad \text{rule 2}$$

$$\bar{A} \cdot C \cdot 1 = \bar{A} \cdot C \qquad \text{rule 7}$$

Therefore,

$$Y = \bar{A} \cdot C + A \cdot B \cdot \bar{C} + A \cdot \bar{B} \cdot \bar{C} = \bar{A} \cdot C + A \cdot \bar{C}$$

Karnaugh Maps

The *Karnaugh map* contains the same information as the truth table arranged in a different format. In a Karnaugh map, the input variables are arranged in rows and columns that use the Gray code to arrange the variables by value (the truth table uses the binary number system to arrange the variables by value). A four-variable Karnaugh map is explained in Figure 3.7.

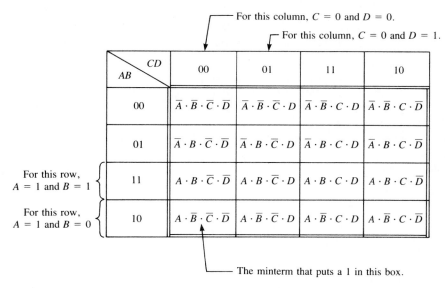

Figure 3.7   A four-variable Karnaugh map has four rows and four columns, which are numbered in Gray code. The rows and columns intersect to form boxes. Each box is associated with a single minterm, the one that is written inside the box.

Karnaugh map of X

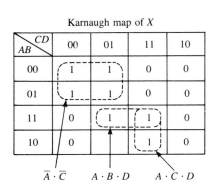

Karnaugh map of Y

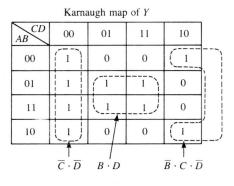

$$X = \overline{A} \cdot \overline{C} + A \cdot B \cdot D + A \cdot C \cdot D$$

$$Y = \overline{C} \cdot \overline{D} + B \cdot D + \overline{B} \cdot C \cdot \overline{D}$$

Figure 3.8   Examples of loopings in a Karnaugh map to reduce the Boolean equation to a simpler form. Loops may cover two, four, or eight boxes. There cannot be any 0's inside the loop. All ones in the map must be included in the final expression, either as an unlooped minterm or as a term reduced by looping.

The Gray code was chosen to arrange the input variables in the Karnaugh map for a very good reason. In the Gray code, successive numbers differ in only one digit. The significance of this in the Karnaugh map is that adjacent minterms differ in only one variable. We know from Example 3.18 that when two minterms differ in only one term, the two minterms can be combined into a single term with the elimination of the variable that is different. In the Karnaugh map, we circle any two adjacent 1's, determine which variable is different in the two minterms, and write a single term with the differing variable removed. In applying this looping technique, we assume that the Karnaugh map is like a flat map of the world. The double line on the left is actually the same place as the double line on the right. The same is true for the double lines on the top and the bottom of the map. Thus the minterms in the upper left corner and the lower left corner are adjacent and differ in only one term (the term that differs is *C*). It is also possible to loop four or eight adjacent 1's. A loop of four eliminates two variables, and a loop of eight eliminates three variables. Figure 3.8 illustrates some of the loops and the term produced by each loop. It is important to note that the reduced term for a loop covers all the 1's inside the loop, but a 1 can be used in two loops if that results in a simpler expression. The bottom line is to include all the 1's in the Karnaugh map with the simplest possible Boolean equation.

3.7 DIGITAL COMPUTERS

The evolution of digital electronics has come so rapidly that some have called it the microelectronic revolution. The microprocessor, in particular, has completely changed the way that digital computers are used in control systems. In the past, a digital computer was too expensive to devote to a single control loop. The computer had to be shared among many control loops to compete economically with conventional controllers. The low cost and reliability of the microprocessor have changed all that. Not only is it possible to use a microcomputer as a devoted controller for a single loop, but microprocessors are being distributed among almost all parts of the control loop. There are microprocessors in control valves, electric drives, signal conditioners, and measuring transmitters. The purpose of this section is to summarize the fundamental concepts and terminology of digital computers.

A digital computer consists of five major elements as shown in Figure 3.9. The *input unit* receives both analog and digital input signals and conditions the signals to conform to the computer's internal signal requirements. Since computers work only with binary signals, the analog signals are converted into binary numbers by a device called an *analog-to-digital converter* (ADC). The digital inputs are already in binary form, so the conditioning is simply a matter of producing the correct voltage levels for use in the computer. The *input unit* has memory called a *buffer* that holds the conditioned input data until the processor decides to accept them.

The *output unit* receives binary numbers from the memory unit and conditions them for use in external devices such as a digital indicator, a printer, a control valve, or some other final control element. If the external device requires an analog signal, the binary number from memory is converted by a *digital-to-analog converter* (DAC). If the external device requires a digital signal, the conditioning is a matter of producing the correct voltage and power for use by the external device.

The *memory unit* stores binary numbers. It can be visualized as a unit consisting of a large number of storage cells, each capable of storing one binary number. The memory unit in Figure 3.9 is much smaller than the memory unit of even the smallest computer—its purpose is to illustrate how computer memory is organized. Each

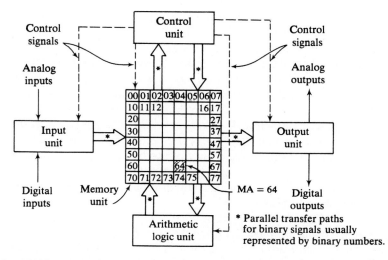

Figure 3.9   The five major elements of a computer are the input, output, memory, control, and arithmetic-logic units.

square in the memory unit in Figure 3.9 represents one storage cell in an 8-by-8 matrix with a total of 64 cells. Each cell is identified by the number of its row and column. For example, the crosshatched cell is in the sixth row and the fourth column, so it would be identified by the number 64. In this manner, each cell has a unique number, which is called its *memory address* (MA). Two important characteristics of a memory unit are its *word length* (the number of binary digits that can be stored in each memory cell), and the addressable memory capacity (the total number of memory cells that can be addressed).

As its name implies, the *arithmetic-logic unit* performs the arithmetic and logic operations. All operations are performed with binary numbers. However, some microcomputers use a binary code to represent decimal digits and are capable of performing decimal arithmetic using the binary-coded-decimal digits (BCD).

The *control unit* coordinates and sequences all operations. It fetches the program instructions one at a time from the memory unit, interprets each instruction, and then executes the operation specified by the instruction. The control unit can make decisions based on the outcome of an operation and then modify its operation based on the result of the decision. The program instructions are in the form of binary numbers.

Microprocessors and Microcomputers

A *microprocessor* is a single electronic component that contains the entire arithmetic and control section of a computer. It is a *large-scale integrated (LSI) circuit*, a term applied to an integrated circuit that contains more than 1000 transistors. Included in the microprocessor are several storage registers for holding binary numbers, an instruction decoder to interpret instructions, an arithmetic logic unit, and a control unit. The microprocessor controls the fetching and executing of the instructions in the program. A microprocessor is an electronic component whose function is determined by the sequence of instructions that make up the program (i.e., it is a programmable electronic component).

A *microcomputer* consists of a microprocessor, a section of read-only memory (ROM, PROM, or EPROM), a section of random-access memory (RAM), an input/output (I/O) interface, an address bus, and a data bus. Figure 3.10 is a simplified schematic of a microcomputer. The random-access memory is the working memory of the microcomputer. RAM can be written in or read out at any time. An operator can enter data or instructions into RAM from an input keyboard or terminal. The microprocessor can also write in RAM on command and then read out as required. The term *random access* refers to the ability of the microprocessor to go directly to any location in the memory unit for a READ or WRITE operation. This is in contrast to a nonrandom-access memory, such as magnetic tape, in which the contents are read out in sequence as they appear on the tape.

The type of RAM used in microcomputers has one disadvantage; it is volatile, that is, the contents of memory are lost when the power supply is interrupted or turned off. For this reason, a portion of the memory in the microprocessor is permanently set during the last stage of its manufacture. This type of memory is called read-only memory, or simply ROM. Once set, the contents of ROM cannot be changed, so it is used to store the permanent programs and data that will be used by the micro-

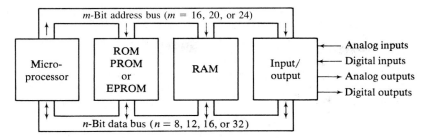

Figure 3.10   The major parts of a microcomputer are the microprocessor, the read-only memory (ROM), the random-access memory (RAM), the input/output unit, the address bus, and the data bus.

computer. A major disadvantage of ROM is that the user cannot alter the contents to make changes or correct errors in the permanently stored program.

Programmable read-only memory (PROM) is a type of memory that contains all ones or all zeros as manufactured. The user can create a permanent pattern of ones and zeros by applying a specified voltage that destroys a fusible link. Once the PROM is "burned in" in this manner, it is also permanent, but it does give the user more flexibility than ROM. Erasable PROM (EPROM) is also programmable electrically, but not permanently. The user can use an ultraviolet-light source to restore the EPROM so that a new program can be set into the memory. In this way, the EPROM can be reprogrammed many times.

A bus is a set of conductors intended to provide a transfer path between various components in the microcomputer. The number of conductors determines how many binary signals can be transferred at the same time (in parallel). Figure 3.10 shows two such transfer paths: the data bus and the address bus. As the name implies, the data bus is intended to transfer data (or instructions) from one component to another. The number of lines in the data bus coincides with the number of bits that can be transferred in parallel, and it is called the *word length* of the microcomputer. Standard word lengths of microcomputers are 4, 8, 12, 16, and 32 bits. The term *byte* is used to describe the 8-bit word length. The address bus is a set of conductors used to select one storage cell in the computer memory. The memory unit decodes the number on the address bus and selects the specified storage cell for a READ or WRITE operation. The number of conductors in the address bus determines the total number of memory cells that can be addressed. A 16-bit address bus can access 65,536 memory cells. With 20 bits, the addressable memory increases to 1,048,576; and with 24 bits, it is 16,777,216.

The input/output interface connects the microcomputer to the outside world. The most common means of input or output is through parallel I/O ports. One method of handling the input and output of data is to assign a memory address to each I/O port. To do this, the number of bits in the I/O port must be the same as the number of bits in a memory cell. It also means that the computer does not need special input/output instructions. The I/O ports are handled as if they were just another memory cell. This feature is called *memory-mapped I/O*. Another method of handling the input and output of data is with special I/O instructions. With this method, the I/O port can have fewer or more bits than a single memory cell.

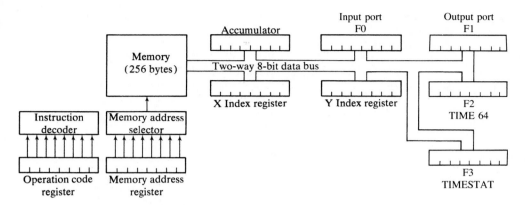

Figure 3.11 The ICST-1 is a hypothetical microcomputer used to illustrate assembly language programming for batch process control.

Programming a Microcomputer

A *program* is a sequence of instructions that defines the operations to be performed by the microcomputer. Programming is not a trivial task—it incurs a major portion of the development cost of a computer system. The purpose of this section is to illustrate the basic procedure for programming a microcomputer without getting excessively complicated. A hypothetical microcomputer called the ICST-1, shown in Figure 3.11, will be used for this purpose.

The ICST-1 is an 8-bit microcomputer that uses one or two bytes (8 or 16 bits) for each instruction. The first byte in each instruction specifies the operation to be performed. If the second byte is used, it specifies an address in memory used in the operation to be performed. The 8-bit address byte is enough to specify 256 addresses in memory. Most microcomputers have considerably more memory than the ICST-1, but 256 locations are enough to illustrate the programming procedure.

The arithmetic unit of the ICST-1 includes an 8-bit register called the *accumulator* (A) and two additional 8-bit registers called the *X index register* (X) and the *Y index register* (Y). The instructions used in the programming example are listed in Table 3.10. These instructions will be explained as the program is developed.

Writing a program for a microcomputer proceeds in five steps:

1. Define the program in a word statement that describes the desired result.
2. Construct a flowchart that shows the order of performing the major tasks necessary to achieve the desired result.
3. Determine the steps necessary to perform each major task and prepare a flowchart that shows the sequence in which these steps are to be performed.
4. Use the flowcharts to help prepare an assembly-coded program, using symbolic names for operation codes and addresses. This is called the *assembly language source code.*
5. Use the assembly language source code to prepare a *machine-coded program.* The machine-coded program is the only program the computer understands—

Table 3.10   Partial List of ICST-1 Instructions

Name	Binary Code	Hex Code	Number of Bytes	Operation Performed
DEX	0001 0000	10	1	Decrement X by 1
DEY	0001 0001	11	1	Decrement Y by 1
JMP	0010 0000	20	2	Jump to the memory location given in byte 2
JMS	0011 0000	30	2	Jump to the subroutine at the address given in byte 2
LDA	0100 0000	40	2	Load accumulator with the number stored in the memory location given in byte 2
RTN	1111 0000	F0	1	Return from subroutine to the instruction immediately after the JMS instruction that called the subroutine
SAP	0110 0011	63	1	Skip the next instruction if the accumulator is equal to or greater than 0
SIM	0110 0000	60	2	Skip the next instruction if the content of the address in byte 2 is negative (minus)
STA	1001 0000	90	2	Transfer the number in the accumulator to the address given in byte 2
STP	1111 1111	FF	1	Stop the computer
SXZ	0110 0001	61	1	Skip if index register X is storing the number 0
SYZ	0110 0010	62	1	Skip if index register Y is storing the number 0
TAX	0111 0000	70	1	Transfer the number in the accumulator to X index register
TAY	0111 0001	71	1	Transfer the number in the accumulator to Y index register

the operation codes and memory addresses must all be given in a binary form (i.e., binary, octal, or hexadecimal numbers). The machine-coded program is called the *assembly language object code*.

Example 3.19

The ICST-1 microcomputer is to be used as sequential controller for a batch process. The process consists of a blending tank with provisions to add water, syrup, and solids to the tank. A ready light indicates when the process has completed the sequence of operations and is ready to begin in a new sequence. A switch closure initiates the following sequence of operations.

1. Turn the tank drain valve OFF, turn the ready light ON, and wait for a switch closure to start sequence of operations.
2. Turn the ready light OFF, turn the water valve ON, and delay for 30 seconds with only the water valve ON.

3. Turn the syrup valve ON and delay for 85 seconds with both the water and the syrup valves ON.
4. Turn the syrup valve OFF, turn the solids feeder ON, and delay for 120 seconds with both the water valve and solids feeder ON.
5. Turn the water valve OFF, turn the solids feeder OFF, and delay for 175 seconds with everything OFF.
6. Turn the tank drain valve ON and delay 115 seconds to drain the tank.

Write a program for the ICST-1 microcomputer that will carry out the sequence of operations for the batch process.

Solution

Step 1: problem definition. First we must define the input and output connections. Connect bit 7 of the input port to the START switch for the batch process. In this way the SIM instruction can be used to test the condition of the switch (bit 7 of an 8-bit number is the sign bit; the number is negative if bit 7 is a 1 and positive if it is a 0). Connect the output port as follows: water valve to bit 0, syrup valve to bit 1, solids feeder to bit 2, tank drain valve to bit 3, and ready light to bit 4. All signals are arranged such that a 1 turns the device ON and a 0 turns the device OFF.

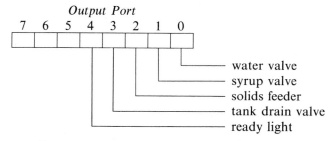

The delays can be obtained by the use of a subroutine that takes 1 second each time it is used. The 30-second delay can be obtained by repeating the delay subroutine 30 times, the 85-second delay by repeating the subroutine 85 times, and so on.

Step 2. The flowchart for step 2 is given in Figure 3.12. There are three tasks identified in the step 2 flowchart. Task 1 involves the operation of the various output devices (output with reference to the computer). Task 2 checks the status of the switch repeatedly until the switch is closed and the program advances to the next step. Task 3 accomplishes the various delays.

Step 3. The flowchart for step 3 is also given in Figure 3.12. Task 1 involves the loading of an 8-bit binary number into the I/O port to turn devices ON or OFF. The I/O port assignment in step 1 is used to determine the binary number required to accomplish the desired results. For example, the number 0000 0101 would turn ON the solids feeder and the water valve.

Task 2 is accomplished by a SKIP if the input port is negative (which means that the START switch is closed), followed by a jump back to the skip instruction. The computer will cycle back and forth between these two instructions until the switch is closed and the skip passes over the jump instruction.

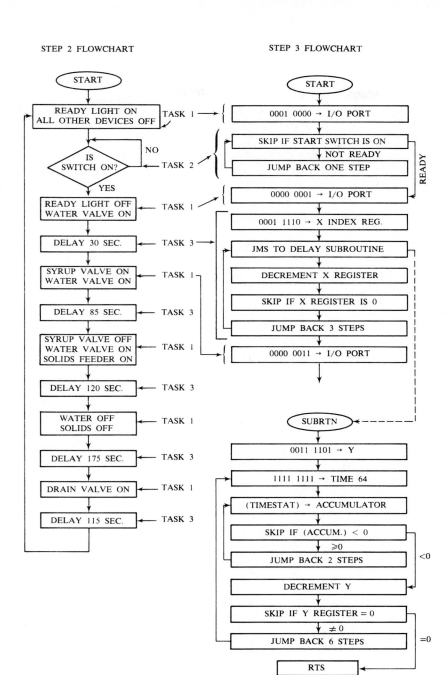

STEP 2 FLOWCHART

STEP 3 FLOWCHART

START

READY LIGHT ON
ALL OTHER DEVICES OFF — TASK 1

IS
SWITCH ON? ← TASK 2 — NO

YES

READY LIGHT OFF
WATER VALVE ON ← TASK 1

DELAY 30 SEC. ← TASK 3

SYRUP VALVE ON
WATER VALVE ON ← TASK 1

DELAY 85 SEC. ← TASK 3

SYRUP VALVE OFF
WATER VALVE ON
SOLIDS FEEDER ON ← TASK 1

DELAY 120 SEC. ← TASK 3

WATER OFF
SOLIDS OFF ← TASK 1

DELAY 175 SEC. ← TASK 3

DRAIN VALVE ON ← TASK 1

DELAY 115 SEC. ← TASK 3

START

0001 0000 → I/O PORT

SKIP IF START SWITCH IS ON — NOT READY
JUMP BACK ONE STEP

0000 0001 → I/O PORT

0001 1110 → X INDEX REG.

JMS TO DELAY SUBROUTINE

DECREMENT X REGISTER

SKIP IF X REGISTER IS 0

JUMP BACK 3 STEPS

0000 0011 → I/O PORT

READY

SUBRTN

0011 1101 → Y

1111 1111 → TIME 64

(TIMESTAT) → ACCUMULATOR

SKIP IF (ACCUM.) < 0 — ≥0
JUMP BACK 2 STEPS — <0

DECREMENT Y

SKIP IF Y REGISTER = 0 — ≠ 0
JUMP BACK 6 STEPS — =0

RTS

Figure 3.12   Flowcharts for Example 3.19.

105

Task 3 uses the X index register as a counter to count the desired number of seconds of delay. The skip instruction checks the counter after each decrement. If the counter has not reached zero, the program loops back and delays for an additional second. When the counter reaches zero, the skip moves to the next section of the program. The initial value in X is the binary equivalent of the number of seconds of delay that will result. Thus 0001 1110 (binary) = 30 (decimal) will result in a delay of 30 seconds.

The delay subroutine in task 3 makes use of an internal timer in the ICST-1 microcomputer. The timer is started by storing a number at the memory address with the label TIME64. The number in TIME64 is decremented once each 64 microseconds (μs) until it passes zero (i.e., when the content of TIME64 is decremented from 0 to -1). When the number in TIME64 passes zero, the number in the memory address labeled TIMESTAT is changed from a positive value to a negative value. If the number 1111 1111 (binary) = 255 (decimal) is loaded into TIME64, the timer will require 256 counts to pass zero. Each count takes 64 μs, so the total time required for the timer to run down is $64 \times 256 = 16{,}384$ μs. The program repeatedly checks the content of TIMESTAT until it becomes negative, indicating that the timer has run down.

Each time the timer runs down, the Y index register is decremented. The program loops back to run the timer again until the content of Y is decremented to zero. The Y register is used as a counter of the number of times the timer is used. If the number 0011 1101 (binary) = 61 (decimal) is loaded into the Y register, the timer will be used 61 times and the total delay will be $61 \times 16{,}384 = 999{,}424$ μs or 0.999424 s.

Step 4. Write the *assembly-coded program* using symbolic names for the operations from Table 3.10, the name INPUT for the input port, OUTPUT for the output port, TIME64 for the timer, and TIMESTAT for the timer status register. The completed assembly-coded program is shown on the left side of Table 3.11. The first step in writing the assembly-coded program is to assign names and addresses to the data that must be stored in memory for use by the program. The following is a list of the assigned values and addresses.

Name	Data Value	Data Address
OUT1	0001 0000	60
OUT2	0000 0001	61
OUT3	0000 0011	62
OUT4	0000 0101	63
OUT5	0000 0000	64
OUT6	0000 1000	65
DELAY1	0001 1110	66
DELAY2	0101 0101	67
DELAY3	0111 1000	68
DELAY4	1010 1111	69
DELAY5	0111 0011	6A
COUNT	0011 1101	6B
TIME	1111 1111	6C

Table 3.11   Assembly and Machine Language Programs

Assembly-Coded Program			Machine-Coded Program		
Name of Address	Operation Byte	Address Byte	Hex Address	Operation Byte	Address Byte
START	LDA	OUT1	00	40	60
	STA	OUTPUT	02	90	F1
LOC1	SIM	INPUT	04	60	F0
	JMP	LOC1	06	20	04
	LDA	OUT2	08	40	61
	STA	OUTPUT	0A	90	F1
	LDA	DELAY1	0C	40	66
	TAX		0E	70	
LOC2	JMS	SUBRTN	0F	30	4B
	DEX		11	10	
	SXZ		12	61	
	JMP	LOC2	13	20	0F
	LDA	OUT3	15	40	62
	STA	OUTPUT	17	90	F1
	LDA	DELAY2	19	40	67
	TAX		1B	70	
LOC3	JMS	SUBRTN	1C	30	4B
	DEX		1E	10	
	SXZ		1F	61	
	JMP	LOC3	20	20	1C
	LDA	OUT4	22	40	63
	STA	OUTPUT	24	90	F1
	LDA	DELAY3	26	40	68
	TAX		28	70	
LOC4	JMS	SUBRTN	29	30	4B
	DEX		2B	10	
	SXZ		2C	61	
	JMP	LOC4	2D	20	29
	LDA	OUT5	2F	40	64
	STA	OUTPUT	31	90	F1
	LDA	DELAY4	33	40	69
	TAX		35	70	
LOC5	JMS	SUBRTN	36	30	4B
	DEX		38	10	
	SXZ		39	61	
	JMP	LOC5	3A	20	36
	LDA	OUT6	3C	40	65
	STA	OUTPUT	3E	90	F1
	LDA	DELAY5	40	40	6A
	TAX		42	70	
LOC6	JMS	SUBRTN	43	30	4B
	DEX		45	10	
	SXZ		46	61	
	JMP	LOC6	47	20	43
	JMP	START	49	20	00

Table 3.11   *continued*

Assembly-Coded Program			Machine-Coded Program		
Name of Address	Operation Byte	Address Byte	Hex Address	Operation Byte	Address Byte
SUBRTN	LDA	COUNT	4B	40	6B
	TAY		4D	71	
LOC7	LDA	TIME	4E	40	6C
	STA	TIME64	50	90	F2
LOC8	LDA	TIMESTAT	52	40	F3
	SAP		54	63	
	JMP	LOC8	55	20	52
	DEY		57	11	
	SYZ		58	62	
	JMP	LOC7	59	20	4E
	RTN		5B	F0	

The first two instructions in the assembly code put the content of OUT1 into the output port to turn the ready light ON and everything else OFF:

```
START    LDA   OUT1
         STA   OUTPUT
```

The next two instructions examine the input port and skip the next instruction if the START switch is ON, or loop back to repeat if START is OFF:

```
LOC1     SIM   INPUT
         JMP   LOC1
```

The next two instructions put the content of OUT2 into the output port to turn the ready light OFF and the water valve ON:

```
         LDA   OUT2
         STA   OUTPUT
```

The next two instructions put the content of DELAY1 (decimal 30) into the X index register for counting down a 30-s delay:

```
         LDA   DELAY1
         TAX
```

The next four instructions call the delay subroutine 30 times for a total delay of 30 s:

```
LOC2     JMS   SUBRTN
         DEX
         SXZ
         JMP   LOC2
```

The next eight instructions put the content of OUT3 into the output port to turn the water and syrup valves on, put the content of DELAY2 into the X index register, and then call the delay subroutine 85 times for a total delay of 85 s:

```
        LDA   OUT3
        STA   OUTPUT
        LDA   DELAY2
        TAX
LOC3    JMS   SUBRTN
        DEX
        SXZ
        JMP   LOC3
```

After three more sections similar to the last one above, the program loops back to START to wait for the initiation of another batch.

Step 5. A computer program called an assembler is used to convert the assembly-coded program into the machine-coded program shown on the right-hand side of Table 3.11.

GLOSSARY

Base: The number of symbols used in a numbering system. (3.2)

Binary numbering system: A numbering system that uses two symbols (0 and 1) and has a base of 2. (3.2)

Boolean algebra: A mathematical method for developing logic circuits and a means of expressing complex logic circuits in terms of simple statements. (3.5)

Boolean equation: Two Boolean expressions separated by an equal sign, which means that the two expressions have the same value. (3.5)

Boolean expression: Boolean variables separated by Boolean operators. (3.5)

Code: A set of unique combinations of 0's and 1's that each define a letter, number, or symbol. There are three types of codes: those that represent numbers, those that represent the 10 decimal digits, and those that represent text. (3.4)

Decoding: The process of determining the symbol for any given code. (3.4)

Division algorithm: A procedure for converting the integer part of a number from decimal to any other base. (3.2)

Encoding: The process of generating the code for any given symbol. (3.4)

Hexadecimal numbering system: A numbering system that uses sixteen symbols (0 through 9 and A through F) and has a base of 16. (3.2)

Karnaugh map: A unique table that lists the outputs of a logic element for all possible combinations of the inputs. (3.6)

Logic circuit: A circuit made up of logic elements interconnected by lines that carry logic signals. (3.3)

Logic elements: Circuit elements that perform logic operations on one or more logic signals to form another logic signal that is the output of the element. (3.3)

Microcomputer: A computer made up of a microprocessor, a read-only memory unit, a random-access memory unit, an address bus, and a data bus. (3.7)

Microprocessor: A single electronic component that contains the entire arithmetic and control section of a computer. (3.7)

Multiplication algorithm: A procedure for converting the fractional part of a number from decimal to any other base. (3.2)

Octal numbering system: A numbering system that uses eight symbols (0 through 7) and has a base of 8. (3.2)

One's complement: A method of representing negative binary numbers in which the most significant bit is used to represent the sign of the number and the remainder of the bits represent the one's complement form of the number. (3.2)

Order: The position of a digit in a number when the digits are counted from right to left, starting with 0 for the rightmost digit. (3.2)

Program: A sequence of instructions that defines the operations to be performed by a computer. (3.7)

Sign-plus-magnitude: A method of representing negative binary numbers in which the most significant bit is used to represent the sign of the number and the remainder of the bits represent the magnitude of the number. (3.2)

Truth table: A table that lists the outputs of a logic element for all possible combinations of the inputs. (3.3)

Two's complement: A method of representing negative binary numbers in which the most significant bit is used to represent the sign of the number and the remainder of the bits represent the two's complement form of the number. (3.2)

EXERCISES

3.1 Convert the following binary numbers to the equivalent octal, decimal, and hexadecimal numbers.

    a. 1011   **b.** 0100 1001

    c. 1000 1010   **d.** 1100 0110

3.2 Convert the following octal numbers to the equivalent binary, decimal, and hexadecimal numbers.

    a. 17   **b.** 143

    c. 251   **d.** 4136

3.3 Convert the following decimal numbers to the equivalent binary, octal, and hexadecimal numbers.

    a. 47   **b.** 92

    c. 132   **d.** 29

3.4 Convert the following hexadecimal numbers to the equivalent binary, octal, and decimal numbers.

    a. 2A   **b.** B4

    c. 3A6   **d.** B3D0

3.5 Prepare a table of decimal numbers from 0 to 63 with the equivalent binary (6 bit), octal, and hexadecimal numbers.

3.6 An 8-bit binary number is often used to represent positive and negative values with bit 7 used as the sign bit. What is the largest positive binary number that this arrangement can represent, and what is the decimal equivalent of this number? What is the negative binary number with the largest magnitude that this arrangement can represent, and what is the decimal equivalent of this number?

3.7 Make a table of all negative numbers from -1 decimal to -128 decimal and the equivalent two's-complement hexadecimal numbers (see Table 3.4).

3.8 Construct truth tables of a three-input NAND element and a four-input NAND element.

3.9 Construct the truth table of a two-input negated-input OR element (an OR element with a NOT operator on each input). If A and B are the two inputs, the output is (NOT A) OR (NOT B). Compare your truth table with the truth table of the two-input NAND element whose output is NOT (A AND B). What do you conclude?

3.10 Determine the logic value of the output of each element in the binary adder in Figure 3.2 for the following input conditions: $A_i = 1$, $B_i = 1$, $C_{i-1} = 0$.

3.11 Prepare a Gray code for the numbers from 0 to 31 decimal.

3.12 Write the 16-bit binary number that represents the decimal number 4982 using the BCD code. Repeat using the excess-3 code.

3.13 Design an encoder for the following binary-to-decimal code:

Decimal	Binary	Decimal	Binary
0	0000	5	1000
1	0001	6	1001
2	0010	7	1010
3	0011	8	1011
4	0100	9	1100

3.14 Design a decoder for the code in Exercise 3.10.

3.15 Convert the binary adder circuit in Figure 3.2 into two Boolean equations for the two outputs S_i and C_i.

3.16 Convert the following truth table into a Boolean equation.

A	B	C	Z
0	0	0	0
0	0	1	1
0	1	0	0
0	1	1	1
1	0	0	0
1	0	1	0
1	1	0	1
1	1	1	0

3.17 Construct truth tables of the following two logic functions:

$$C = \bar{A} \cdot \bar{B} + A \cdot B$$
$$P = \bar{A} \cdot B + A \cdot \bar{B}$$

Comment on the differences between the two truth tables.

3.18 Use Boolean algebra to reduce the following Boolean equation to a simpler form.

$$X = \bar{A} \cdot \bar{B} \cdot C + \bar{A} \cdot \bar{B} \cdot \bar{C} + A \cdot B \cdot C + A \cdot B \cdot \bar{C}$$

3.19 Construct a Karnaugh map of the following Boolean equation and use loopings to reduce the equation to its simplest form.

$$Y = A \cdot B \cdot \bar{C} \cdot D + A \cdot B \cdot C \cdot D + A \cdot \bar{B} \cdot C \cdot D + \bar{A} \cdot B \cdot \bar{C} \cdot D$$
$$+ \bar{A} \cdot B \cdot C \cdot D + A \cdot B \cdot C \cdot \bar{D} + A \cdot \bar{B} \cdot C \cdot \bar{D}$$

Sketch a logic diagram of the original equation and another logic diagram of the simplified equation.

3.20 Write an assembly-coded program for the ICST-1 microcomputer to control the following batch blending process. The process consists of a blending tank that mixes five liquid ingredients. The ingredients are designated as *A*, *B*, *C*, *D*, and *E*. Each ingredient is controlled by an ON/OFF valve. The sequence is as follows:

(1) Turn the tank drain valve ON, turn the ready light ON, and wait for closure of the start switch to begin the cycle.

(2) When START is pressed, turn the tank drain valve OFF, turn the ready light OFF, turn the ingredient A valve ON, and delay for 15 s.

(3) Turn ingredient A OFF, turn C and D ON, and delay 40 s.

(4) Turn C and D OFF, turn B and E ON, and delay 65 s.

(5) Turn B OFF, turn E OFF, and delay 22 s.

(6) Turn the tank drain valve ON and delay 127 s.

CHAPTER 4

The Common Elements of System Components

OBJECTIVES

Control systems often consist of a variety of different types of components. In a temperature control loop, for example, the process and the primary sensor are thermal systems, the temperature transmitter and controller are electrical systems, and the final control element consists of fluid flow (pneumatic) and mechanical systems. This variety of components within a control system could complicate the analysis and design of the control loop. Fortunately, the behavior of these diverse systems is determined by the same four common elements: **resistance, capacitance, inertia** (or **inductance** in electrical systems), and **dead time.** The effects of these basic elements in one system are analogous to the effects of the same elements in any other system. Thus thermal resistance and capacitance have the same effect in a thermal system as electrical resistance and capacitance have in an electrical system.

The purpose of this chapter is to provide you with the means of determining the values of the four basic elements for the following types of systems: **electrical, thermal, liquid flow, gas flow,** and **mechanical.** After completing this chapter, you will be able to

1. Define electrical resistance, capacitance, inductance, and dead time in terms of voltage, current, charge, and time

2. Define liquid flow resistance, capacitance, inertance, and dead time in terms of pressure, volume flow rate, volume, and time

3. Define gas flow resistance and capacitance in terms of pressure, mass flow rate, mass, and time

4. Define thermal resistance and capacitance in terms of temperature, heat flow rate, quantity of heat, and time

5. Define mechanical resistance, capacitance, inertia, and dead time in terms of force, velocity, distance, and time

6. Determine electrical, liquid flow, gas flow, or mechanical resistance from a graph for both linear and nonlinear systems

7. Use a computer program to calculate liquid flow resistance (laminar or turbulent) given the density, viscosity, and flow rate of the fluid and the diameter and type of the pipe

8. Calculate the thermal resistance of a composite wall given the surface area, the thickness and thermal conductivity of each layer, and the two film coefficients

9. Compute thermal film coefficients for a variety of surface conditions

10. Calculate liquid flow, gas flow, thermal, or mechanical capacitance

11. Calculate the dead-time delay of a solid or liquid flow system

4.1 INTRODUCTION

Most control system components fit into one of the following types: electrical, mechanical, liquid flow, gas flow, or thermal. A particular control system may include two, three, or even five different types of components. This mix of component types could have made the analysis and design of control systems much more difficult, but it did not. The reason for this fortunate circumstance is that the behavior of components of one type is analogous to the behavior of components of any other type. This analogous behavior is determined by four elements that are common to the five types of components. The four elements are resistance, capacitance, inertia (or inductance), and dead-time delay. Some knowledge of these "basic" elements for the five component types will be quite helpful in understanding the behavior of control system components.

This chapter provides the equations and information necessary to calculate the elements of the five types of components. The equations for each element are included in a box to make them easy to locate. All terms are defined within the box, and the units are included for each term. Examples are included to show how to complete the calculations. Computer programs are included for computing liquid flow resistance and thermal resistance.

For each type of component, the four elements are defined in terms of three variables. *The first variable defines a quantity of material, energy, or distance. The second variable defines a driving force or potential that tends to move or change the quantity variable. The third variable is time.* For example, in a liquid flow component, the quantity variable is the volume of liquid that is moved, and the potential variable is the pressure drop that tends to cause the liquid to flow. Table 4.1 names the three

Table 4.1   The Three Variables Used to Define the Four Elements for Each Type of Component

Type of Component	Variable		
	Quantity	**Potential**	**Time**
Electrical	Charge	Voltage	Second
Liquid flow	Volume	Pressure	Second
Gas flow	Mass	Pressure	Second
Thermal	Heat energy	Temperature	Second
Mechanical	Distance	Force	Second

variables for each of the five types of components. Table 4.2 lists the symbols and units for the parameters used to define the four elements for the various systems.

Resistance is an opposition to the movement or flow of material or energy. It is measured by the amount of potential required to make a unit change in the quantity moved each second. Electrical resistance is the increase in the voltage across the terminals of a component required to move one more coulomb of charge through the

Table 4.2   Parameters Used to Define the Four Elements*

Symbol	Name	SI Units	English Units
a	Area (of pipes)	Square meter	Square foot
A	Area (of tanks)	Square meter	Square foot
d	Diameter (of pipes)	Meter	Foot
D	Diameter or distance	Meter	Foot
e	Potential	Volt	Volt
f	Friction factor	Dimensionless	Dimensionless
F	Force	Newton	Pound (force)
g	Gravitational acceleration	Meter per square second	Foot per square second
h	Thermal film coefficient	Watt/square meter kelvin	Btu/square foot Fahrenheit
H	Height	Meter	Foot
i	Current	Ampere	Ampere
l	Length	Meter	Foot
m	Mass	Kilogram	Pound (mass)
M	Molecular weight	Dimensionless	Dimensionless
P	Pressure	Pascal (newton per square meter)	Pound (force) per square foot
q	Charge	Coulomb	Coulomb
Q	Heat flow rate	Watt	Btu/hour
Q	Volume flow rate	Cubic meter per second	Cubic foot per second
S	Specific heat	Joule/kilogram kelvin	Btu/pound Fahrenheit
t	Time	Second	Second
T	Temperature	Celsius or Kelvin	Fahrenheit or Rankin
v	Velocity	Meter per second	Foot per second
V	Volume	Cubic meter	Cubic foot
W	Mass flow rate	Kilogram per second	Pound (mass) per second
μ	Viscosity (absolute)	Pascal second	Pound (mass) per foot second
ρ	Density	Kilogram per cubic meter	Pound (mass) per cubic foot

* See Appendix A for alternative units and conversions.

component each second (i.e., to increase the current by 1 ampere). Liquid flow resistance is the increase in the pressure drop between two points along a pipe required to increase the flow rate through the pipe by 1 cubic meter per second. Gas flow resistance is the increase in the pressure drop between two points along a pipe required to increase the flow rate through the pipe by 1 kilogram per second. Thermal resistance is the increase in temperature difference across a wall section required to increase the heat flow through the wall section by 1 joule per second. Mechanical resistance is the change in the force applied to an object required to increase the velocity of the object by 1 meter per second.

> Resistance is measured in terms of the amount of potential required to produce one unit of electric current, liquid flow rate, gas flow rate, heat flow rate, or velocity.

Capacitance is the amount of material, energy, or distance required to make a unit change in potential. It expresses the relationship between a change in quantity and the corresponding change in potential. Capacitance should not be confused with capacity, which is the total material- or energy-holding ability of a device. If we say a jug holds 1 liter, we are stating its capacity. If we say that 100 cubic meters of liquid must be added to a tank to increase the pressure at the bottom by 1 newton per square meter, we are stating the tank's capacitance.

Electrical capacitance is the coulombs of charge that must be stored in a capacitor to increase its voltage by 1 volt. Liquid capacitance is the cubic meters of liquid that must be added to a tank to increase the pressure by 1 pascal. Gas capacitance is the kilograms of gas that must be added to a tank to increase the pressure by 1 pascal. Thermal capacitance is the amount of heat energy that must be added to an object to increase its temperature by 1 degree Celsius. Mechanical capacitance is the amount of compression of a spring (in meters) required to increase the spring force by 1 newton.

> Capacitance is measured in terms of the amount of material, energy, or distance required to make a unit change in potential.

Inertia, inertance, or *inductance* is an opposition to a change in the state of motion. It is measured in terms of the amount of potential required to produce a unit change in the *rate* at which a quantity is moving. This element is usually important in electrical and mechanical systems, sometimes important in liquid and gas systems, and usually ignored in thermal systems. The term *inductance* is used with electrical systems, the term *inertance* is used with fluid systems, and the term *inertia* is used with mechanical systems.

Electrical inductance is the increase in voltage across an inductor required to increase the current by 1 ampere each second. Liquid flow inertance is the increase in the pressure drop between two points along a pipe required to accelerate the flow rate by 1 cubic meter per second per second. Mechanical inertia is the increase in force required to produce an acceleration of 1 meter per second per second.

> Inertia, inertance, or inductance is measured in terms of the amount of potential required to increase electric current, liquid flow rate, gas flow rate, or velocity by one unit per second.

Dead time is the time interval between the time a signal appears at the input of a component and the time the corresponding response appears at the output. A pure dead-time element does not change the magnitude of the signal; only the timing of the signal is changed. Dead time occurs whenever mass or energy is transported from one point to another. It is the time required for the mass or energy to travel from the input location to the output location. Dead time is also called transport time, pure delay, or distance–velocity lag. Control becomes increasingly more difficult as dead time increases.

If v is the velocity of the mass or energy and D is the distance traveled, the dead-time delay is equal to the distance divided by the velocity:

$$t_d = \frac{D}{v} \tag{4.1}$$

The effect of dead time is to delay the input signal by the dead-time delay(t_d). As an example, consider the system with a dead-time delay of 5 s. Let $f_i(t)$ represent the input signal and $f_o(t)$ represent the output signal. The (t) indicates that f_i and f_o have different values at different times. If $f_i(t)$ is 5 at a time $t = 0$ s, then $f_o(t)$ will be 5 at $t = 5$ s. If $f_i(t)$ is 7 at $t = 5$ s, $f_o(t)$ will be 7 at time $t = 10$ s. Several more time intervals are indicated below.

t	$f_i(t)$	$f_o(t)$	Relationship between f_i and f_o
0	5		
5	7	5	$f_o(5) = f_i(0)$
10	8	7	$f_o(10) = f_i(5)$
15	6	8	$f_o(15) = f_i(10)$
20		6	$f_o(20) = f_i(15)$

It is interesting to note that in each case, $f_o(t) = f_i(t - 5)$ or, more generally,

$$f_o(t) = f_i(t - t_d)$$

This equation is often used to represent a dead-time process. In simple terms, it expresses the concept that the output at any time t is the same as the input was t_d seconds before time t, that is, at time $t - t_d$.

4.2 ELECTRICAL ELEMENTS

Electrical resistance is that property of a material which impedes the flow of electric current. The unit of electric resistance is the *ohm*. Good conductors have a low resistance; insulators have a very high resistance. Electrical resistance (R) is expressed by *Ohm's law,* a statement of proportionality between the applied voltage (e) and the resulting current (i) in linear resistance elements. Equation (4.2) is the usual form of Ohm's law, while Equation (4.3) is more convenient as a definition of electrical resistance.

$$e = iR \tag{4.2}$$

$$R = \frac{e}{i} \tag{4.3}$$

Expressed as $e = iR$, Ohm's law is the equation of a straight line through the origin with a slope of R (see Figure 4.1a). The resistance can be determined by finding the slope between any two points on the graph, such as point a and point b in Figure 4.1a. The resistance is equal to the increase in voltage between points a and b, Δe, divided by the increase in current between the same two points, Δi. This gives us a slightly different definition of resistance, given as

$$R = \frac{\Delta e}{\Delta i} \tag{4.4}$$

When the resistive element is nonlinear, the determination of resistance becomes more involved, and Equations (4.3) and (4.4) give us different values for the resistance. Figure 4.1b shows a nonlinear resistive element. If we use Equation (4.3) to compute the resistance, we get a value of resistance that is equal to the slope of a straight line from the origin to the operating point. We call this the *static resistance* of the element. In a control system, the operation is usually confined to small variations around the operating point. In this situation, the operation is approximately along the tangent to the curve at the operating point rather than along the line from the origin to the operating point. The resistance obtained from the slope of the tangent line is called the *dynamic resistance* of the element at the operating point. The dynamic resistance of a nonlinear element is more accurate than the static resistance for control system analysis. In the remainder of this book, the term *resistance* will mean dynamic resistance.

Dynamic resistance can be determined graphically using Equation (4.4) as shown in Figure 4.1b. Begin with a tangent line at the operating point and pick a second point *on the tangent line*. Determine the increase in voltage, Δe, and the increase in current, Δi, between the two points. The dynamic resistance is equal to Δe divided by Δi, as given by Equation (4.4).

The mathematical determination of dynamic resistance makes use of the fact that the value of the derivative at a point is equal to the slope of the tangent line at that

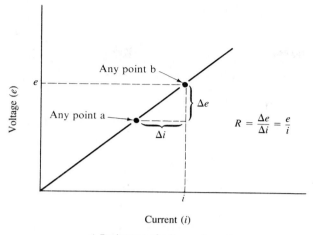

a) Resistance of a linear element

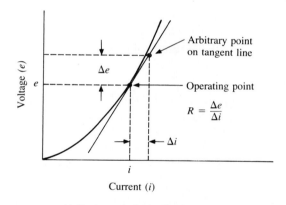

b) Resistance of a nonlinear element

Figure 4.1   Graphic determination of the resistance of linear and nonlinear electric components.

point, as given by the equation

$$R = \frac{de}{di} \qquad \textbf{(4.5)}$$

Equation (4.5) would be read to mean that dynamic resistance is equal to the rate of change of the voltage with respect to current.

Example 4.1

An electrical component is known to have a linear volt–ampere graph. A test with 24 V applied to the terminals of the component resulted in a measured current of 12 mA. Determine the resistance of the component.

Solution

The volt–ampere graph is a straight line, so Equation (4.3) applies:

$$R = \frac{e}{i} = \frac{24}{0.012} = 2000\ \Omega$$

Example 4.2

A light bulb is an example of an electric component with a nonlinear volt–ampere graph. The electrical resistance of a nonlinear component can be approximated using Equation (4.4) with a very small increment of voltage, Δe, and the resulting small increment of current, Δi. Determine the resistance of a light bulb at 6 V from the following information.

5.95 V results in 0.500 A

6.05 V results in 0.504 A

Solution

$$\Delta e = 6.05 - 5.95 = 0.10\ \text{V}$$
$$\Delta i = 0.504 - 0.500 = 0.004\ \text{A}$$
$$R = \frac{\Delta e}{\Delta i} = \frac{0.10}{0.004} = 25\ \Omega$$

Electrical capacitance is the quantity of electric charge (q coulombs) required to make a unit increase in the electrical potential (e volts). The unit of electrical capacitance is the *farad*.

$$\text{Capacitance} = C = \frac{\Delta q}{\Delta e} \tag{4.6}$$

A simple manipulation of Equation (4.6) results in the following relationship between the voltage and current in a capacitance:

$$\Delta q = C\,\Delta e$$
$$\frac{\Delta q}{\Delta t} = I = C\,\frac{\Delta e}{\Delta t}$$

where Δt = time in seconds required to make the Δq change in charge
   I = average current during the time interval Δt

If Δt is reduced until it approaches zero, $\Delta q/\Delta t$ becomes dq/dt, the instantaneous rate of change of charge; I becomes i, the instantaneous current; and $\Delta e/\Delta t$ becomes de/dt, the instantaneous rate of change of potential.

$$\frac{dq}{dt} = i = C\frac{de}{dt} \qquad (4.7)$$

Equation (4.7) would be read to mean that the current through a capacitor is equal to the capacitance (C) times the rate of change of the voltage across the capacitor with respect to time. Equation (4.7) is used in circuit analysis as one of the models of a capacitor. The following example shows how Equation (4.7) can be used to explain the 90° phase difference between the voltage and current in a capacitor.

Let $\qquad\qquad\qquad\qquad e = A \sin \omega t$

then $\qquad\qquad\qquad\qquad \dfrac{de}{dt} = A\omega \cos \omega t$

and $\qquad\qquad\qquad\qquad i = CA\omega \cos \omega t$

Thus the current i leads the voltage e by 90° in a capacitor.

Example 4.3

A current pulse with an amplitude of 0.1 mA and a duration of 0.1 s is applied to an electrical capacitor. The voltage across the capacitor is increased from 0 V to $+25$ V by the current pulse. Determine the capacitance (C) of the capacitor.

Solution

$$I = C\frac{\Delta e}{\Delta t}$$

$$0.1 \times 10^{-3} = C\left(\frac{25 - 0}{0.1}\right)$$

$$C = \frac{(0.1 \times 10^{-3})(0.1)}{25}$$

$$= 0.4\ \mu F$$

Electrical inductance is the voltage required to produce a unit increase in electric current each second. The unit of electrical inductance is the *henry*. Equations (4.8) and (4.9) define the voltage–current relationship for an inductor.

$$e = L\frac{\Delta i}{\Delta t} \qquad (4.8)$$

$$e = L\frac{di}{dt} \qquad (4.9)$$

Equation (4.9) would be read to mean that the voltage across an inductor is equal to the inductance (L) times the rate of change of the current through the inductor with respect to time. Equation (4.9) is used in circuit analysis as one of the models of an inductor. The following example shows how Equation (4.9) can be used to explain the 90° phase difference between the voltage and current in an inductor.

Let

$$i = A \sin \omega t$$

then

$$\frac{di}{dt} = A\omega \cos \omega t$$

and

$$e = LA\omega \cos \omega t$$

Thus the voltage e leads the current i by 90° in an inductor.

Example 4.4

A voltage pulse with an amplitude of 5 V and a duration of 0.02 s is applied to an inductor. The current through the inductor is increased from 1 A to 2.1 A by the voltage pulse. Assume that the resistance of the inductor is negligible, and determine the inductance (L).

Solution

Equation (4.8) may be used to determine L.

$$e = L\frac{\Delta i}{\Delta t}$$

$$5 = L\left(\frac{2.1 - 1}{0.02}\right)$$

$$L = (5)\left(\frac{0.02}{1.1}\right) = 0.0909$$

$$= 0.0909 \text{ H}$$

Electrical dead-time delay is the delay caused by the time it takes a signal to travel from the source to the destination (Figure 4.2). Although electrical signals travel at tremendous speeds (2×10^8 to 3×10^8 m/s), the transport time of an electrical signal constitutes a dead-time delay that has important consequences in some systems. In digital computer circuits, the delay on a transmission line is sometimes used to delay a signal deliberately to accomplish the desired logic function. In most control systems, however, the effect of electrical dead time is negligible because the delay is so small compared to the delay in other parts of the system.

The velocity of a signal on a transmission line is called the *velocity of propagation* (v_p). As mentioned before, v_p varies between 2×10^8 and 3×10^8 m/s. The dead-time

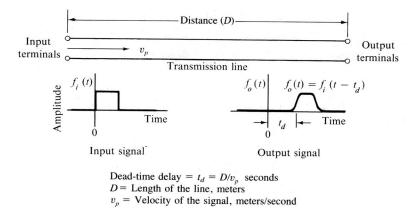

Figure 4.2 Electrical dead-time element: a transmission line.

delay of the line is equal to the distance the signal travels (D) divided by the velocity of propagation (v_p).

$$t_d = \frac{D}{v_p}$$ (4.1)

Example 4.5

a. Determine the dead-time delay of a 600-m-long transmission line if the velocity of propagation is 2.3×10^8 m/s.
b. Determine the dead-time delay of a signal from a space vehicle that is located 2000 km from the earth station receiving the signal. The signal travels at 3×10^8 m/s.

Solution

a. Equation (4.1) applies.

$$t_d = \frac{600}{2.3 \times 10^8} = 2.61 \times 10^{-6} \text{ s}$$

$$= 2.61 \ \mu\text{s}$$

b. Equation (4.1) applies.

$$t_d = \frac{2 \times 10^6}{3 \times 10^8} = 0.67 \times 10^{-2} \text{ s}$$

$$= 6.7 \text{ ms}$$

ELECTRICAL EQUATIONS

Resistance

$$R = \frac{e}{i} = \frac{\Delta e}{\Delta i} = \frac{de}{di} \qquad \text{(4.3 to 4.5)}$$

Capacitance

$$C = \frac{\Delta q}{\Delta e} \qquad \text{(4.6)}$$

$$i = C\frac{de}{dt} \qquad \text{(4.7)}$$

Inductance

$$e = L\frac{\Delta i}{\Delta t} \qquad \text{(4.8)}$$

$$e = L\frac{di}{dt} \qquad \text{(4.9)}$$

Dead-Time Delay

$$t_d = \frac{D}{v_p} \qquad \text{(4.1)}$$

where
C = capacitance, farad
D = distance between the input and the output, meter
e = applied voltage, volt
Δe = change in applied voltage, volt
i = instantaneous current, ampere
Δi = change in current, ampere
I = average current over interval Δt, ampere
L = inductance, henry
Δq = change in charge, coulomb
R = resistance, ohm
Δt = interval of time, second
t_d = dead-time delay, second
v_p = velocity of travel, meter/second

4.3 LIQUID FLOW ELEMENTS

Liquid flow resistance is that property of pipes, valves, or restrictions which impedes the flow of a liquid. It is measured in terms of the increase in pressure required to make a unit increase in flow rate. The SI unit for liquid flow resistance is "pascal second/cubic meter." The English unit is "psf/cfs," where psf means "pound per square foot" and cfs means "cubic foot per second." A more practical English unit is "psi/gpm," where psi means "pound per square inch" and gpm means "gallon per minute."

Liquid resistance is determined by the relationship between the pressure drop and the flow rate as expressed by a flow equation. There are two different types of flow: *laminar* and *turbulent*. Each has a different flow equation and hence a different liquid resistance. Laminar flow occurs when the fluid velocity is relatively low and the liquid flows in layers. A colored dye injected into the center of a laminar flow will move with the liquid and remain concentrated in the center. Turbulent flow occurs when the fluid velocity is relatively high and the liquid does not flow in layers. A colored dye injected into the center of turbulent flow is soon mixed throughout the flowing fluid.

The type of flow that occurs depends on four parameters: the density of the fluid, ρ; the inside diameter of the pipe, d; the dynamic viscosity of the fluid, μ; and the average velocity of the flowing fluid, v. The four parameters are arranged in a dimensionless grouping called the *Reynolds number*.

$$\text{Reynolds number} = \frac{\rho v d}{\mu} \qquad \text{(4.10)}$$

$$v = \frac{Q}{a} = \frac{4Q}{\pi d^2} \qquad \text{(4.11)}$$

Laminar flow occurs when the Reynolds number has a value less than 2000. Turbulent flow occurs when the Reynolds number has a value greater than 4000. A transition between laminar and turbulent flow occurs when the Reynolds number has a value between 2000 and 4000. Graphs of pressure versus flow rate for laminar and turbulent flow are shown in Figures 4.3 and 4.4.

Liquid flow resistance can be determined graphically in the same manner as electrical resistance. In Figures 4.3 and 4.4, the slope of the graph is the liquid flow resistance. The laminar flow graph is linear and the resistance has a constant value for all flow rates. The turbulent flow graph is nonlinear and the resistance increases as the flow rate is increased.

The flow equation for laminar flow in a round pipe is given by Equation (4.12) (called the *Hagen–Poiseuille law*). The laminar flow resistance, given by Equation (4.13), is a constant that depends on three parameters: the absolute viscosity, μ, the inside diameter of the pipe, d, and the length of the pipe, l.

$$p = R_L Q \qquad \text{(4.12)}$$

$$R_L = \frac{128\mu l}{\pi d^4} \qquad \text{(4.13)}$$

The flow equation for turbulent flow in a round pipe is given by Equation (4.14) (called the *Fanning equation*). The turbulent flow resistance, given by Equations (4.15) and (4.16), is considerably more complex than laminar flow resistance. Turbulent resistance depends on the flow rate, Q, and the turbulent flow coefficient, K_t. The turbulent flow coefficient depends on four parameters: the liquid density, ρ; the length of the pipe, l; the inside diameter of the pipe, d; and a factor called the friction factor,

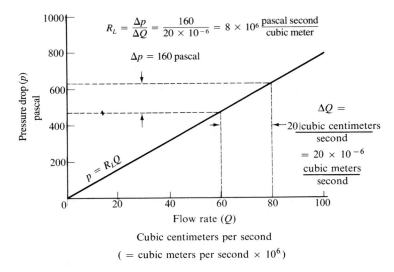

Figure 4.3 Graph of pressure versus flow rate for laminar flow is linear. Liquid flow resistance is equal to the slope of the graph and is constant for all values of flow rate.

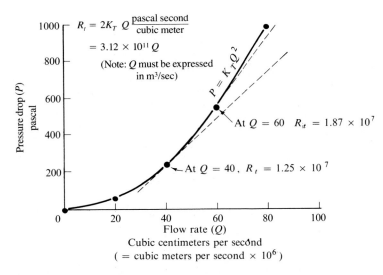

Figure 4.4 Graph of pressure versus flow rate for turbulent flow is nonlinear. Liquid flow resistance is equal to the slope of the tangent to the curve at any point. Turbulent flow resistance increases as the flow rate is increased.

Table 4.3   Values of the Friction Factor (f)

Type	Diameter (cm)	Reynolds Number*					
		4 × 10³	10⁴	10⁵	10⁶	10⁷	10⁸
Smooth	1–2	0.039	0.030	0.018	0.014	0.012	0.012
tubing	2–4	0.039	0.030	0.018	0.013	0.011	0.010
	4–8	0.039	0.030	0.018	0.012	0.010	0.009
	8–16	0.039	0.030	0.018	0.012	0.009	0.008
Commercial	1–2	0.041	0.035	0.028	0.026	0.026	0.026
steel pipe	2–4	0.040	0.033	0.024	0.023	0.023	0.023
	4–8	0.039	0.030	0.022	0.020	0.019	0.019
	8–16	0.039	0.030	0.020	0.018	0.017	0.017

Source: L. F. Moody, "Friction Factors for Pipe Flow," *ASME Transactions*, Vol. 66, No. 8, 1944, p. 671.
* Reynolds number = $\rho VD/\mu$.

f. The friction factor depends on the diameter of the pipe, the Reynolds number, and the smoothness of the inside of the pipe. A table of friction factors is included here as Table 4.3.

$$p = K_t Q^2 \qquad (4.14)$$

$$R_t = 2K_t Q \qquad (4.15)$$

$$K_t = \frac{8\rho f l}{\pi^2 d^5} \qquad (4.16)$$

The computation of liquid flow resistance and pressure drop begins with the determination of the Reynolds number. If the Reynolds number is less than 2000, the flow is laminar and Equations (4.12) and (4.13) are used. If the Reynolds number is greater than 4000, the flow is turbulent and Equations (4.14) to (4.16) are used along with the friction factor from Table 4.3. If the Reynolds number is between 2000 and 4000, the resistance and pressure drop will be somewhere between the values from the laminar flow equations and the values from the turbulent flow equations, so both sets of equations should be used. If the flow is turbulent, exact values of the friction factor (f) must be interpolated from Table 4.3. Conversion of units may also be required.

Computations may be done using a calculator, a spreadsheet, or a computer program. Example 4.6 illustrates the use of a calculator to compute laminar flow resistance and pressure drop. Example 4.7 does the same for turbulent flow resistance and pressure drop.

Program "LIQRESIS" (see listing in Appendix F) is a GW BASICA program for computing laminar or turbulent flow resistance and pressure drop. The program accepts practical inputs in either SI or English units. It then converts the units to the correct SI units, computes the Reynolds number, selects the correct equations, computes the resistance and pressure drop, and prints the results using the system of units selected for input. Examples 4.8 through 4.10 illustrate the use of the program.

Example 4.6

Oil at a temperature of 15°C flows through a horizontal tube which is 1 cm in diameter with a flow rate of 9.42 liters per minute (L/min). The line is 10 m long. Determine the Reynolds number, the resistance, and the pressure drop in the tube.

Solution

From "Properties of Liquids" in Appendix A,

$$\rho = 880 \text{ kilograms/cubic meter}$$
$$\mu = 0.160 \text{ pascal second}$$

Use Appendix B to convert d and Q to standard SI units:

$$d = 1 \text{ cm} = 0.01 \text{ meter}$$
$$Q = 9.42 \text{ L/min} = 9.42(1.6667\text{E}-05) = 1.57\text{E}-04 \text{ m}^3/\text{s}$$

Note: The E notation is used in problem solutions to represent numbers that are to be entered into a computer or a pocket calculator. Large and small numbers are written with a base number equal to or greater than 1 and less than 10. The base number is followed by the letter E (for exponent), a plus or minus symbol, and a two-digit power of 10. Thus 6.72×10^{-4} is written as $6.72\text{E}-04$.

Use Equation (4.11) to compute the average fluid velocity:

$$v = \frac{4Q}{\pi d^2} = \frac{4(1.57\text{E}-04)}{\pi(0.01)^2} = 2.00 \text{ m/s}$$

Use Equation (4.10) to compute the Reynolds number:

$$\text{Reynolds number} = \frac{\rho v d}{\mu} = \frac{(880)(2.00)(0.01)}{0.160} = 109.95$$

The flow is laminar, so we use Equations (4.12) and (4.13) to compute the resistance and pressure drop:

$$R_L = \frac{128\mu l}{\pi d^4} = \frac{(128)(0.160)(10)}{\pi(0.01)^4} = 6.519\text{E}+09 \text{ Pa}\cdot\text{s/m}^3$$
$$p = R_L Q = (6.52\text{E}+09)(1.57\text{E}-04) = 1.0235\text{E}+06 \text{ Pa}$$

Example 4.7

Water at 15°C flows through a commercial steel pipe that is 0.4 in. in diameter with a flow rate of 6 gal/min. The line is 50 ft long. Determine the Reynolds number, the resistance, and the pressure drop in the tube.

Solution

From "Properties of Liquids" in Appendix A,

$$\rho = 1000 \text{ kilograms/cubic meter}$$
$$\mu = 0.001 \text{ pascal second}$$

When working problems with English units, our approach will be to use SI units to complete the computations and then convert the answer to the desired English units. This avoids the confusing task of converting the equations to English units. Use Appendix B to convert d, Q, and l to standard SI units:

$$d = 0.4 \text{ inch} = (0.4)(0.0254) = 0.01016 \text{ meter}$$
$$Q = 6 \text{ gal/min} = (6)(6.3088\text{E}-05) = 3.7853\text{E}-04 \text{ m}^3/\text{s}$$
$$l = 50 \text{ feet} = (50)(0.3048) = 15.240 \text{ meters}$$

Use Equation (4.11) to compute the average fluid velocity:

$$v = \frac{4Q}{\pi d^2} = \frac{4(3.7853\text{E}-04)}{\pi(0.01016)^2} = 4.669 \text{ m/s}$$

Use Equation (4.10) to compute the Reynolds number:

$$\text{Reynolds number} = \frac{\rho v d}{\mu} = \frac{(1000)(4.669)(0.01016)}{0.001} = 47,440$$

The flow is turbulent, so we use Equations (4.14) to (4.16) to compute the resistance and pressure drop.

The first step is to obtain the friction factor by interpolation from Table 4.3, using the row for commercial steel pipe with a diameter of 1 to 2 inches. The following variables and values will be used in the interpolation:

$R = 47,440 = $ Reynolds number from Equation (4.10)
$R_a = 10,000 = $ largest Reynolds number column heading that is less than R
$R_b = 100,000 = $ smallest Reynolds number column heading that is greater than R
$f = $ friction factor for the Reynolds number R (47,440)
$f_a = 0.035 = $ friction factor in the R_a column
$f_b = 0.028 = $ friction factor in the R_b column

The interpolation formula is:

$$f = f_a + (f_b - f_a)\left(\frac{R - R_a}{R_b - R_a}\right)$$
$$f = 0.035 + (0.028 - 0.035)\left(\frac{4.744\text{E}4 - 1\text{E}4}{10\text{E}4 - 1\text{E}4}\right)$$
$$f = 0.035 - 0.007(3.744/9) = 0.032088$$

Use Equation (4.16) to compute K_t:

$$K_t = \frac{8\rho fl}{\pi^2 d^5} = \frac{(8)(1000)(0.032088)(15.24)}{\pi^2(0.01016)^5} = 3.6614E12$$

Use Equation (4.15) to compute R_t:

$$R_t = 2K_tQ = (2)(3.6614E12)(3.7853E-04) = 2.7719E09 \text{ Pa·s/m}^3$$

Use Appendix B to convert R_t to English units

$$R_t = (2.7719E09)(9.148E-09) = 25.357 \text{ psi/gpm}$$

Use Equation (4.14) to compute p:

$$p = K_tQ^2 = (3.66148E12)(3.785E-04)^2 = 5.246E05 \text{ Pa}$$

Use Appendix B to convert p to English units:

$$p = (5.246E05)(1.45E-04) = 76.07 \text{ psi}$$

Answer: $R_t = 25.4$ psi/gpm, $p = 76.1$ psi

Example 4.8

Use the GW BASICA program "LIQRESIS" to solve the problem presented in Example 4.6. Program "LIQRESIS" is available on disk (see Preface), and the program listing is in Appendix F.

Solution

The screen produced by a run of the program follows. The first eight lines are the title and input. The last three lines are the result of the run.

```
            LIQUID FLOW RESISTANCE AND PRESSURE DROP

Smooth tubing [T] or commercial pipe [P] ─────────────→  TUBE
SI units [S] or English units [E] ────────────────────→  SI
Flow rate in liters/minute ───────────────────────────→  9.42
Inside diameter of TUBE in centimeters ───────────────→  1
Length of TUBE in meters ─────────────────────────────→  10
Absolute viscosity in Pascal seconds ─────────────────→  0.160
Fluid density in kilograms/cubic meter ───────────────→  880

Reynolds number = 1.099E+02, Flow is laminar
Resistance = 6.519E+09 Pascal second/cubic meter
Pressure drop = 1.024E+06 Pascal
```

Example 4.9

Use the GW BASICA program "LIQRESIS" to solve the problem presented in Example 4.7.

Solution

```
            LIQUID FLOW RESISTANCE AND PRESSURE DROP

Smooth tubing [T] or commercial pipe [P] ──────────────→  PIPE
SI units [S] or English units [E] ─────────────────────→  ENGLISH
Flow rate in gallons/minute ───────────────────────────→  6
Inside diameter of PIPE in inches ─────────────────────→  0.4
Length of PIPE in feet ────────────────────────────────→  50
Absolute viscosity in pounds/foot second ──────────────→  6.72E-04
Fluid density in pounds/cubic foot ────────────────────→  62.43

Reynolds number = 4.743E+04, Flow is turbulent
Resistance = 2.536E+01 psi/gpm
Pressure drop = 7.607E+01 psi
```

Example 4.10

A fluid with a density of 880 kg/m³ and a viscosity of 0.0053 Pa·s is flowing through
a smooth tube with an inside diameter of 1 cm. The flow rate is 9.42 L/min. The line
is 10 m long. Determine the Reynolds number, the resistance, and the pressure drop
in the tube.

Solution

```
            LIQUID FLOW RESISTANCE AND PRESSURE DROP

Smooth tubing [T] or commercial pipe [P] ──────────────→  TUBE
SI units [S] or English units [E] ─────────────────────→  SI
Flow rate in liters/minute ────────────────────────────→  9.42
Inside diameter of TUBE in centimeters ────────────────→  1
Length of TUBE in meters ──────────────────────────────→  10
Absolute viscosity in Pascal seconds ──────────────────→  0.0053
Fluid density in kilograms/cubic meter ────────────────→  880

Reynolds number = 3.319E+03, Flow is in transition
Resistance = 2.159E+08 to 8.964E+08 Pascal second/cubic meter
Pressure drop = 3.390E+04 to 7.037E+04 Pascal
```

Liquid flow capacitance is defined in terms of the increase in volume of liquid in
a tank required to make a unit increase in pressure at the outlet of the tank.

$$C_L = \frac{\Delta V}{\Delta p} \qquad (4.17)$$

where C_L = capacitance, cubic meter/pascal

   ΔV = increase in volume, cubic meter

   Δp = corresponding increase in pressure, pascal

The increase in pressure in the tank depends on three things: the increase in level of the liquid (ΔH), the acceleration due to gravity (g), and the density of the liquid (ρ). The relationship is expressed by the following equation:

$$\Delta p = \rho g \, \Delta H \qquad \text{(4.18)}$$

The increase in level in the tank is equal to the increase in volume divided by the average surface area (A) of the liquid in the tank.

$$\Delta H = \frac{\Delta V}{A} \qquad \text{(4.19)}$$

By substitution of Equations (4.18) and (4.19) into Equation (4.17),

$$\Delta p = \frac{\rho g \, \Delta V}{A}$$

$$C_L = \frac{\Delta V}{\Delta p} = \frac{\Delta V A}{\rho g \, \Delta V}$$

$$= \frac{A}{\rho g} \qquad \text{(4.20)}$$

Example 4.11

A liquid tank has a diameter of 1.83 m and a height of 10 ft. Determine the capacitance of the tank for each of the following fluids:

   a. Water
   b. Oil
   c. Kerosene
   d. Gasoline

Solution

Equation (4.19) may be used to determine C_L.

$$C_L = \frac{A}{\rho g}$$

$$A = \frac{\pi D^2}{4} = \frac{\pi (1.83)^2}{4} = 2.63 \text{ m}^2$$

$$g = 9.81 \text{ m/s}^2$$

$$C_L = \frac{2.63}{9.81\rho} = \frac{0.268}{\rho}$$

Density is obtained from the "Properties of Liquids" in Appendix A.

   a. Water: $\rho = 1000$

$$C_L = \frac{0.268}{1000} = 2.68 \times 10^{-4} \text{ m}^3/\text{Pa}$$

b. Oil: $\rho = 880$

$$C_L = \frac{0.268}{880} = 3.05 \times 10^{-4} \text{ m}^3/\text{Pa}$$

c. Kerosene: $\rho = 800$

$$C_L = \frac{0.268}{800} = 3.35 \times 10^{-4} \text{ m}^3/\text{Pa}$$

d. Gasoline: $\rho = 740$

$$C_L = \frac{0.268}{740} = 3.62 \times 10^{-4} \text{ m}^3/\text{Pa}$$

Liquid flow inertance is measured in terms of the amount of pressure drop in a pipe required to increase the flow rate by 1 unit each second.

$$I_L = \frac{p}{\Delta Q/\Delta t} \qquad (4.21)$$

where I_L = inertance, pascal/(cubic meter/second^2)

   p = pressure drop in the pipe, pascal

   ΔQ = change in flow rate, cubic meter/second

   Δt = time interval, second

A more practical equation for inertance can be obtained as follows: The pressure drop acts on the cross-sectional area of the pipe (a) to produce a force (F) equal to the pressure drop (p) times the area (a):

$$F = p \times a$$

This force will accelerate the fluid in the pipe, according to Newton's law of motion:

$$F = pa = m \frac{\Delta v}{\Delta t} \qquad (4.22)$$

The mass of fluid in the pipe (m) is equal to the density of the fluid (ρ) times the volume of fluid in the pipe. The volume is equal to the cross-sectional area of the pipe (a) times the length of the pipe (l).

$$m = \rho a l \qquad (4.23)$$

The change in flow rate, ΔQ, is equal to the change in fluid velocity, Δv multiplied by the area, a:

$$\Delta Q = a \, \Delta v \qquad (4.24)$$

Combining Equations (4.22), (4.23), and (4.24) into Equation (4.21) results in the following equation for inertance:

$$I_L = \frac{\rho l}{a} \qquad (4.25)$$

Example 4.12

Determine the liquid flow inertance of water in a pipe that has a diameter of 2.1 cm and a length of 65 m.

Solution

Equation (4.25) applies:

$$I_L = \frac{\rho L}{a}$$

$$\rho = 1000 \text{ kg/m}^3 \quad \text{(Appendix A)}$$

$$a = \frac{\pi d^2}{4} = \frac{\pi(0.021)^2}{4} = 3.46\text{E}-04 \text{ m}^2$$

$$I_L = \frac{(1000)(65)}{3.46\text{E}-04} = 1.88\text{E}+08 \text{ Pa/m}^3/\text{s}^2$$

Dead time occurs whenever liquid is transported from one point to another in a pipeline. An example of liquid flow dead time is shown in Figure 4.5. The dead-time delay is the distance traveled (D) divided by the average velocity (v) of the fluid. Equations (4.11) and (4.1) are used to compute the average velocity (v) and the dead-time delay (t_d). Since $a = \pi d^2/4$,

$$v = \frac{Q}{a} = \frac{4Q}{\pi d^2} \tag{4.11}$$

$$t_d = \frac{D}{v} \tag{4.1}$$

Example 4.13

Liquid flows in a pipe that is 200 m long and has a diameter of 6 cm. The flow rate is 0.0113 m^3/s. Determine the dead-time delay.

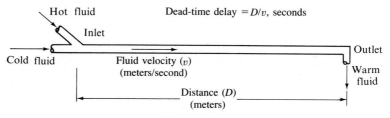

Figure 4.5   Liquid flow dead-time element. Hot and cold fluids are combined in a Y connection to produce a warm fluid which flows a distance D to the outlet of the pipe. The input to the system is the ratio of hot fluid to cold fluid. The output is the temperature of the fluid at the outlet end. The dead-time delay is the distance traveled (D) divided by the average velocity (v) of the fluid.

Solution

Equations (4.11) and (4.1) apply.

$$v = \frac{4Q}{\pi d^2} = \frac{(4)(0.0113)}{\pi(0.06^2)} = 4 \text{ m/s}$$

$$t_d = \frac{D}{v} = \frac{200}{4} = 50 \text{ s}$$

LIQUID FLOW EQUATIONS

Reynolds Number

$$\text{Reynolds number} = \frac{\rho v d}{\mu} \qquad (4.10)$$

Average Velocity

$$v = \frac{Q}{a} = \frac{4Q}{\pi d^2} \qquad (4.11)$$

Laminar Flow Resistance
(Reynolds Number Less Than 2000)

$$p = p_1 - p_2 = R_L Q \qquad (4.12)$$

$$R_L = \frac{128 \mu l}{\pi d^4} \qquad (4.13)$$

Turbulent Flow Resistance
(Reynolds Number More Than 4000)

$$p = p_1 - p_2 = K_t Q^2 \qquad (4.14)$$

$$R_t = 2 K_t Q \qquad (4.15)$$

$$K_t = \frac{8 \rho f l}{\pi^2 d^5} \qquad (4.16)$$

Capacitance

$$C_L = \frac{A}{\rho g} \qquad (4.20)$$

Inertance

$$I_L = \frac{\rho l}{a} \qquad (4.25)$$

Dead Time

$$t_d = \frac{D}{v} \qquad (4.1)$$

where a = cross-sectional area of pipe, meter^2

A = surface area of liquid in tank, meter^2

C_L = liquid flow capacitance, cubic meter/pascal

d = inside diameter of the pipe, meter

D = diameter of the tank or distance, meter

f = friction factor from Table 4.3

g = 9.81 meter/second^2 (gravitational acceleration)

I_L = inertance, pascal/(cubic meter/second^2)

K_t = turbulent flow coefficient

l = length of the pipe, meter

p = pressure drop from inlet to outlet, pascal

p_1 = pressure at inlet of pipe, pascal

p_2 = pressure at outlet of pipe, pascal

Q = flow rate, cubic meter/second

R_L = laminar flow resistance, pascal second/meter^3

R_t = turbulent flow resistance, pascal second/meter^3

v = average fluid velocity in pipe, meter/second

ρ = fluid density, kilogram/cubic meter*

μ = absolute viscosity of fluid, pascal second*

*See "Properties of Liquids" in Appendix A.

4.4 GAS FLOW ELEMENTS

Gas flow resistance is that property of pipes, valves, or restrictions which impedes the flow of a gas. It is measured in terms of the increase in pressure required to produce an increase in gas flow rate of 1 kg/s. The SI unit for gas flow resistance is "pascal second per kilogram." Gas flow in a pipe may be laminar or turbulent. In laminar flow, the pressure drop varies directly with the gas velocity. In turbulent flow, the pressure drop varies directly with the square of the gas velocity. In practice, gas flow is almost always turbulent, and the commonly used equations apply to turbulent flow.

If the pressure drop is less than 10% of the initial gas pressure, the equation for incompressible flow gives reasonable accuracy for gas flow. This is the equation we used for turbulent liquid flow with the volume flow rate (Q) replaced by mass flow rate (W).

$$p = p_1 - p_2 = K_g W^2 \tag{4.26}$$

$$R_g = 2K_g W \tag{4.27}$$

$$K_g = \frac{8fl}{\pi^2 d^5 \rho} \tag{4.28}$$

Equations (4.26), (4.27), and (4.28) do not account for the expansion of the gas as the pressure decreases, and the results are not accurate when the pressure drop is greater than 10% of the initial gas pressure. Also, the "Properties of Gases" table in

Appendix A gives the density and viscosity of gases at standard atmospheric conditions. To keep things from getting unnecessarily complicated, we will limit the gas flow examples to near-atmospheric conditions with pressure drops less than 10% of the inlet pressure. This will allow familiarization with the concept of gas flow resistance without worrying about the variations in viscosity and density caused by changes in pressure and temperature.

Example 4.14

Figure 4.6 is a graph of mass flow rate (W) versus pressure drop (p). Determine the gas flow resistance at the point where the mass flow rate is 0.6 kg/s.

Solution

Refer to Figure 4.6 for the solution.

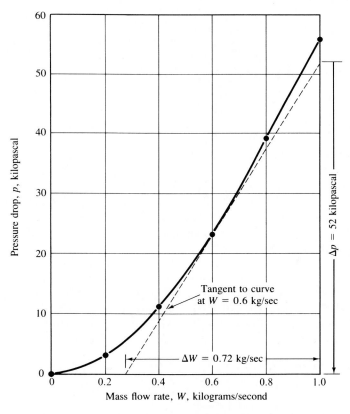

Figure 4.6   Graphical determination of gas flow rate at the point where $W = 0.6$ kg/s. First draw a line tangent to the curve at $W = 0.6$. The desired resistance is the slope of this tangent line. The rise of the tangent line is 52,000 Pa, and the run is 0.72 kg/s. The resistance is the rise divided by the run: $R = 52,000/0.72 = 72,222$ Pa·s/kg.

Example 4.15

A smooth tube is supplying 0.03 kg per second of air at a temperature of 15°C. The tube is 30 m long and has an inside diameter of 4 cm. Use Equations (4.26), (4.27), and (4.28) to find the pressure drop and the gas flow resistance. The outlet pressure (p_2) is 102 kPa. Find the inlet pressure (p_1).

Solution

The following alternative form of the Reynolds number equation is more convenient for this problem.

$$\text{Re} = \frac{4W}{\pi \mu d}$$

From Appendix A,

$$\mu = 1.81\text{E}-05 \text{ Pa·s}$$
$$\rho = 1.22 \text{ kg/m}^3$$
$$\text{Re} = \frac{(4)(0.03)}{\pi(1.81\text{E}-05)(0.04)} = 52{,}800$$

Interpolating from Table 4.3, we obtain friction factor f

$$f = 0.03 + (0.018 - 0.030)\left(\frac{5.28 - 1}{10 - 1}\right) = 0.023$$

$$K_g = \frac{8fl}{\pi^2 d^5 \rho} = \frac{(8)(0.023)(30)}{\pi^2(0.04)^5(1.22)}$$
$$= 4.469\text{E}+06$$
$$p = K_g W^2 = (4.469\text{E}+06)(0.03)^2 = 4.02 \text{ kPa}$$
$$p_1 = p + p_2 = 4.0 + 102 = 106.0 \text{ kPa}$$
$$R_g = 2K_g W = (2)(4.469\text{E}+06)(0.03)$$
$$= 268 \text{ kPa·s/kg}$$

Gas flow capacitance is defined in terms of the increase in the mass of gas in a pressure vessel required to produce a unit increase in pressure while the temperature remains constant. The SI unit of gas flow capacitance is kilogram per pascal. The perfect gas law may be used to determine the capacitance equation.

$$pV = \left(\frac{10^3 m}{M}\right)RT \tag{4.29}$$

where p = absolute pressure of the gas, pascal
           V = volume of the gas, cubic meter
           m = mass of the gas, kilogram

M = molecular weight of the gas (see "Properties of Gases" in Appendix A)

R = universal gas constant = 8.314 J/K·mol

T = absolute temperature, kelvin

Solving Equation (4.29) for m, the mass of the gas, yields

$$m = \left(\frac{1.2 \times 10^{-4} MV}{T}\right) p \qquad \textbf{(4.30)}$$

Equation (4.30) applies to a pressure vessel where V is the volume of the vessel, T the absolute temperature of the gas, and M the molecular weight of the gas. For a given pressure vessel and gas, V and M are constants. The temperature (T) must also be held constant to determine the gas capacitance. When M, V, and T are all constant, the term in parentheses is constant. Under these conditions, Equation (4.30) is linear, and the relationship between a change in mass (Δm) and the corresponding change in pressure (Δp) is given by

$$\Delta m = \left(\frac{1.2 \times 10^{-4} MV}{T}\right) \Delta p$$

$$\text{gas flow capacitance} = C_g = \frac{\Delta m}{\Delta p}$$

$$C_g = \frac{1.2 \times 10^{-4} MV}{T} \qquad \text{kg/Pa} \qquad \textbf{(4.31)}$$

Example 4.16

A pressure tank has a volume of 0.75 m³. Determine the capacitance of the tank if the gas is nitrogen at 20°C.

Solution

Equation (4.31) may be used to determine C_g.

$$C_g = \frac{1.2 \times 10^{-4} MV}{T}$$

M = 28.016 for nitrogen   (Appendix A)

V = 0.75 m³

T = 20°C = 293 K

$$C_g = \frac{(1.2\text{E}-04)(28.016)(0.75)}{293}$$

$$= 8.6 \times 10^{-6} \text{ kg/Pa}$$

GAS FLOW EQUATIONS

Reynolds Number

$$\text{Reynolds number} = \frac{4W}{\pi \mu d}$$

Resistance—Low-Pressure Drop

$$p = p_1 - p_2 = K_g W^2 \qquad \textbf{(4.26)}$$

$$R_g = 2K_g W \qquad \textbf{(4.27)}$$

$$K_g = \frac{8fl}{\pi^2 d^5 \rho} \qquad \textbf{(4.28)}$$

Capacitance

$$C_g = \frac{1.2 \times 10^{-4} MV}{T} \qquad \textbf{(4.31)}$$

where

C_g = gas flow capacitance, kilogram/pascal

d = inside diameter of the pipe, meter

K_g = turbulent flow coefficient

l = length of the pipe, meter

M = molecular weight of the gas (Appendix A)

p = pressure drop from inlet to outlet, pascal

p_1 = pressure at inlet of pipe, pascal

p_2 = pressure at outlet of pipe, pascal

R_g = gas flow resistance, pascal second/kilogram

T = temperature, kelvin

W = gas flow rate, kilogram/second

ρ = fluid density, kilogram/cubic meter*

μ = absolute viscosity of fluid, pascal second*

*See "Properties of Gases," Appendix A.

4.5 THERMAL ELEMENTS

Thermal resistance is that property of a substance which impedes the flow of heat. It is measured in terms of the difference in temperature required to produce a heat flow rate of 1 watt (joule per second). Normally, heat flow occurs through a wall that separates two fluids at different temperatures. The fluids may be either liquids or gases. Heat flows through the wall from the hotter fluid to the cooler fluid. If T_o is the temperature of the outside fluid and T_i is the temperature of the inside fluid, the temperature difference is $T_o - T_i$. The heat flow rate, Q, from the outside fluid to the inside fluid is equal to the temperature difference divided by the thermal resistance, R_T.

$$Q = \frac{T_o - T_i}{R_T} \qquad (4.32)$$

If the heat flow rate (Q) is negative because T_i is greater than T_o, it simply means that the heat is going from the inside fluid to the outside fluid.

The thermal resistance, R_T, in Equation (4.32) is the resistance of the entire wall separating the two fluids. It is often convenient to express the resistance of 1 square meter of the wall. We will use the term *unit resistance* and the symbol R_u for the resistance of 1 square meter of the wall separating the two fluids. The term *total resistance* and the symbol R_T will be used for the resistance of the entire wall. The total resistance is equal to the unit resistance divided by the surface area of the wall between the fluids.

$$R_T = \frac{R_u}{A} \qquad (4.33)$$

The heat flow rate, Q, is equal to the temperature difference multiplied by the surface area and divided by the unit resistance.

$$Q = \frac{(T_o - T_i)A}{R_u} \qquad (4.34)$$

The wall separating the two fluids is a composite, consisting of several layers, each contributing to the resistance of the wall. Figure 4.7 shows a composite wall separating two fluids which are at different temperatures. The two outside layers are thin films of stagnant fluid that form an insulating blanket around the wall. The two inner layers consist of different materials, such as wood, steel, or insulation.

The inside and outside films are always present, but their thickness varies depending on the motion of the main body of the fluid. A strong wind, for example, reduces the thickness of the air film surrounding our bodies, giving us the familiar "wind chill index." The resistance of the film depends on the thickness of the film and on the thermal resistance of the fluid. Empirical formulas have been developed to determine the resistance of fluid films under different conditions. A number of these formulas are summarized in Equations (4.40) to (4.47) at the end of this section. The formulas give the expression for the conductance of the film, represented by the letter h. Conductance is the reciprocal of the unit resistance. The film conductance, h, is referred to as the *film coefficient*. Thus h_o is the outside film coefficient and h_i is the inside film coefficient. The unit resistance of the film is equal to the reciprocal of the film coefficient.

$$R_u(\text{film}) = \frac{1}{h} \qquad (4.35)$$

The unit resistance of the inner layers in the composite wall depends on the thickness of the layer (x) and the thermal conductivity (k) of the material.

$$R_u(\text{inner layer}) = \frac{x}{k} \qquad (4.36)$$

Figure 4.7 Composite wall separating the outer fluid on the left from the inner fluid on the right. The two outside layers are thin films of stagnant fluid. The two inner layers consist of materials such as wood, steel, or insulation. The total thermal resistance of the wall is the sum of the resistances of each section.

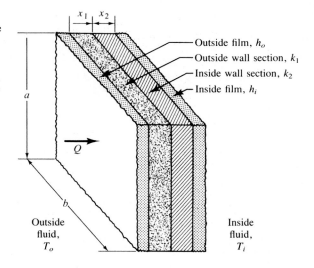

A = wall surface area = $a \times b$, square meters

R_T = thermal resistance = $\dfrac{1}{A}\left(\dfrac{1}{h_o} + \dfrac{x_1}{k_1} + \dfrac{x_2}{k_2} + \dfrac{1}{h_i}\right)\dfrac{\text{kelvin}}{\text{watt}}$

T_o = outside fluid temperature. ° Celsius

T_i = inside fluid temperature. ° Celsius

Q = heat flow rate = $(T_o - T_i)/R_T$ watts

The value of the thermal conductivity of a number of common substances is given in Appendix A, "Properties of Solids."

The unit resistance of the wall is the sum of the unit resistances of each layer. If we use the subscripts 1, 2, . . . for the inner layers, the unit resistance of a wall with n inner layers is expressed as

$$R_u(\text{wall}) = \frac{1}{h_o} + \frac{x_1}{k_1} + \frac{x_2}{k_2} + \cdots + \frac{x_n}{k_n} + \frac{1}{h_i} \tag{4.37}$$

The total resistance is equal to the unit resistance divided by the surface area, A.

$$R_T(\text{wall}) = \frac{1}{A}\left(\frac{1}{h_o} + \frac{x_1}{k_1} + \frac{x_2}{k_2} + \cdots + \frac{x_n}{k_n} + \frac{1}{h_i}\right) \tag{4.38}$$

The computation of thermal resistance may be done using a calculator, a spreadsheet, or a computer program. Example 4.17 illustrates the use of a calculator to compute the thermal resistance of a composite wall section. Example 4.18 shows the calculator computations of the thermal resistance of five different film conditions. Program "THERMRES" (see listing in Appendix F) is a GW BASICA program for computing the thermal resistance of a composite wall section.

Example 4.17

A wall section similar to Figure 4.7 has two inner layers: a steel plate 1 cm thick and insulation 2 cm thick. The inside fluid is still water. The difference between the water temperature and the surface temperature (T_d) is estimated to be 10°C. The outside fluid is air, which has a velocity of 6 m/s. The water temperature is 45°C, and the air temperature is 85°C. The wall dimensions are: $a = 2$ m, $b = 3$ m. Determine the unit resistance, the total resistance, the total heat flow, and the direction of the heat flow.

Solution

From "Properties of Solids" in Appendix A,

$$k_1 = 45 \text{ W/m} \cdot \text{K} \qquad \text{for steel}$$
$$k_2 = 0.036 \text{ W/m} \cdot \text{K} \qquad \text{for insulation}$$

The area of the wall section is:

$$A = 2 \times 3 = 6 \text{ m}^2$$

The inside film coefficient is given by Equation (4.43):

$$h_i = 2.26(T_w + 34.3)T_d^{0.5}$$
$$h_i = 2.26(45 + 34.3)10^{0.5}$$
$$h_i = 566.7 \text{ watt/meter}^2 \text{ kelvin}$$

The outside film coefficient is given by Equation (4.46):

$$h_o = 7.75v_a^{0.75}$$
$$h_o = 7.75(6)^{0.75}$$
$$h_o = 29.71 \text{ watt/meter}^2 \text{ kelvin}$$

The unit thermal resistance is given by Equation (4.37):

$$R_u = \frac{1}{h_o} + \frac{x_1}{k_1} + \frac{x_2}{k_2} + \frac{1}{h_i}$$
$$R_u = \frac{1}{29.71} + \frac{0.01}{45} + \frac{0.02}{0.036} + \frac{1}{566.7}$$
$$R_u = 0.003366 + 0.000222 + 0.5556 + 0.001765$$
$$R_u = 0.5912 \text{ kelvin meter}^2/\text{watt}$$

The total thermal resistance is given by Equation (4.33):

$$R_T = R_u/A = 0.09853 \text{ kelvin/watt}$$

The total heat flow is given by Equation (4.33):

$$Q = \frac{T_o - T_i}{R_T} = \frac{85 - 45}{0.09853} = 406 \text{ watts}$$

Example 4.18

Determine the unit thermal resistance of each of the following film conditions.

a. Natural convection in still air on a vertical surface where $T_d = 20°C$.
b. Natural convection in still water where $T_d = 30°C$ and $T_w = 20°C$.
c. Natural convection in oil where $T_d = 16°C$.
d. Forced convection in air with a velocity of 4 m/s.
e. Forced convection in water with a temperature (T_w) of 40°C, a pipe diameter of 5 cm, and a velocity of 4 m/s.

Solution

a. Equation (4.42) applies:

$$h = 1.78T_d^{0.25}$$
$$h = 1.78(20)^{0.25} = 3.764 \text{ watts/meter}^2 \text{ kelvin}$$
$$R_u = 1/h = 0.266 \text{ kelvin meter}^2/\text{watt}$$

b. Equation (4.43) applies:

$$h = 2.26(T_w + 34.3)T_d^{0.5}$$
$$h = 2.26(20 + 34.3)30^{0.5} = 672.2 \text{ watt/meter}^2 \text{ kelvin}$$
$$R_u = 1/h = 0.00149 \text{ kelvin meter}^2/\text{watt}$$

c. Equation (4.44) applies:

$$h = 7.0T_d^{0.25}/\mu^{0.4}$$
$$\mu = 0.160 \text{ pascal second (from Appendix A)}$$
$$h = 7.0(16)^{0.25}/0.160^{0.4}$$
$$h = 29.14 \text{ watt/meter}^2 \text{ kelvin}$$
$$R_u = 1/h = 0.03432 \text{ kelvin meter}^2/\text{watt}$$

d. Equation (4.45) applies:

$$h = 4.54 + 4.1v_a$$
$$h = 4.54 + 4.1(4)$$
$$h = 20.94 \text{ watt/meter}^2 \text{ kelvin}$$
$$R_u = 1/h = 0.0478 \text{ kelvin meter}^2/\text{watt}$$

e. Equation (4.47) applies:

$$h = \frac{20.93(68.3 + T_w)v_w^{0.8}}{d^{0.2}}$$

$$h = \frac{20.93(68.3 + 40)4^{0.8}}{0.05^{0.2}}$$

$$h = 12,510 \text{ watt/meter}^2 \text{ kelvin}$$
$$R_u = 7.994E-05 \text{ kelvin meter}^2/\text{watt}$$

Example 4.19

Use the GW BASICA program "THERMRES" to solve the problem presented in Example 4.17. Program "THERMRES" is available on disk (see Preface), and the program listing is in Appendix F.

Solution

A run of the program results in the following output.

```
        THERMAL RESISTANCE AND HEAT FLOW

   Surface area:          6.0 square meter
   Inside temperature:    45 degrees Celsius
   Outside temperature:   85 degrees Celsius
   Inside film:  Natural convection, water
     Td:  10 Celsius,   Tw:  45 Celsius
     hi:   5.67E+02 watt/square meter kelvin
   Outside film, Forced convection, air, smooth surface & inside pipe
     Fluid velocity:   6.0 meter/second
     ho:   2.97E+01 watt/square meter kelvin
   INNER LAYERS:
     x (cm)     1.0    2.0
     K (W/mk)  45.000  0.036

   Unit resistance:    5.91E-01 kelvin square meter/watt
   Total resistance:   9.85E-02 kelvin/watt
   Total heat flow:    4.06E+02 watts
   Heat flows from outside to inside.
```

Thermal capacitance is defined in terms of the increase in heat required to make a unit increase in temperature. The SI unit of thermal capacitance is "joule/kelvin." The heat capacity (or specific heat) of a substance is the amount of heat required to raise the temperature of 1 kilogram of the substance by 1 kelvin. Thus the thermal capacitance (C_T) of an object is simply the product of the mass (m) of the object times the heat capacity (S_h) of its substance.

$$C_T = mS_h \tag{4.39}$$

Example 4.20

Determine the thermal capacitance of 8.31 m³ of water.

Solution

Equation (4.39) may be used to determine C_T.

$$C_T = mS_h$$

The mass of water (m) is equal to the density (ρ) times the volume of water (8.31 m^3). From Appendix A, the density of water is 1000 kg/m^3, and the specific heat is 4190 J/kg·K.

$$C_T = (1000)(8.31)(4190)$$
$$= 3.48 \times 10^7 \text{ J/K}$$

THERMAL EQUATIONS

Resistance

$$R_u = \frac{1}{h_o} + \frac{x_1}{k_1} + \frac{x_2}{k_2} + \cdots + \frac{x_n}{k_n} + \frac{1}{h_i} \qquad (4.37)$$

$$R_T = \frac{R_u}{A} \qquad (4.33)$$

$$Q = \frac{T_o - T_i}{R_T} \qquad (4.32)$$

$$Q = \frac{(T_o - T_i)A}{R_u} \qquad (4.34)$$

Film Coefficients (see note 1 below)

1. Natural convection in still air:
    a. Horizontal surfaces facing up:

    $$h = 2.50T_d^{0.25} \qquad (4.40)$$

    b. Horizontal surfaces facing down:

    $$h = 1.32T_d^{0.25} \qquad (4.41)$$

    c. Vertical surfaces:

    $$h = 1.78T_d^{0.25} \qquad (4.42)$$

2. Natural convection in still water:

    $$h = 2.26(T_w + 34.3)T_d^{0.5} \qquad (4.43)$$

3. Natural convection in oil:

    $$h = \frac{7.0T_d^{0.25}}{\mu^{0.4}} \qquad (4.44)$$

4. Forced convection for air against smooth surfaces and inside straight pipes:
    a. Air velocity, $v_a \leq 4.6$ m/s:

    $$h = 4.54 + 4.1v_a \qquad (4.45)$$

    b. Air velocity, $v_a > 4.6$ m/s:

    $$h = 7.75v_a^{0.75} \qquad (4.46)$$

5. Forced convection for turbulent water flow in straight pipes:

$$h = \frac{20.93(68.3 + T_w)v_w^{0.8}}{d^{0.2}}$$ (4.47)

Capacitance

$$C_T = mS_h$$ (4.39)

where
A = area of the wall surface, meter²

C_T = thermal capacitance, joule/kelvin

d = inside diameter of pipe, meter

h = film coefficient, watt/meter² kelvin (h_o = outside film, h_i = inside film)

k = thermal conductivity, watt/meter kelvin

m = mass, kilogram

Q = heat flow rate, watt

R_u = unit thermal resistance, kelvin meter²/watt

R_T = total thermal resistance, kelvin/watt

S_h = heat capacity, joule/kilogram kelvin

T_d = temperature difference between the main body of the fluid and the wall surface, kelvin or Celsius (see note 2 below).

T_i = inside fluid temperature, Celsius

T_o = outside fluid temperature, Celsius

T_w = water temperature, Celsius

v_a = velocity of air, meter/second

v_w = velocity of water, meter/second

x = thickness of a layer, meter

μ = absolute viscosity, pascal second

Notes:

1. Equations (4.40) to (4.47) are based on empirical equations in J. Kenneth Salisbury, *Kent's Mechanical Engineers' Handbook: Power*, 12th edition (New York: John Wiley & Sons, 1950), pp. 3–17 to 3–20.

2. $T_o - T_i$ in kelvin is equal to $T_o - T_i$ in Celsius, so a temperature difference may be computed using either kelvin or Celsius units.

4.6 MECHANICAL ELEMENTS

Mechanical resistance (or friction) is that property of a mechanical system which impedes motion. It is measured in terms of the increase in force required to produce an increase in velocity of 1 m/s. The SI unit of mechanical resistance is the "newton second/meter."

An automobile shock absorber and a dashpot are examples of mechanical resistance devices. Figure 4.8 illustrates the operation of a dashpot. The cylinder is stationary and the piston rod is attached to the moving part (M). When part M moves, the fluid in the cylinder must move through the orifice around the piston from one side to the other. The flow rate of the fluid through the orifice is proportional to the velocity of part M, and a difference of pressure $(p_2 - p_1)$ is required to force the fluid

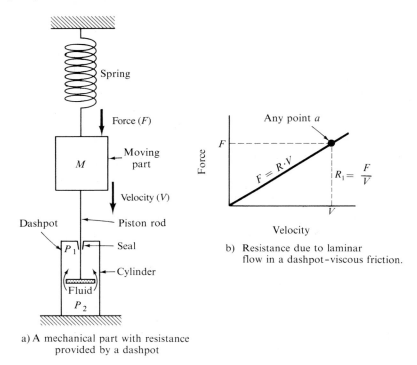

a) A mechanical part with resistance provided by a dashpot

b) Resistance due to laminar flow in a dashpot–viscous friction.

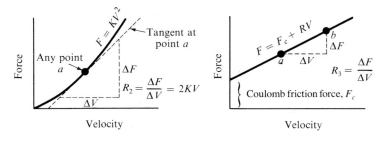

c) Resistance due to turbulent flow in a dashpot.

d) Resistance due to both viscous friction and coulomb friction.

Figure 4.8   Examples of mechanical resistance.

through the orifice. This difference in pressure produces a force that opposes the motion. This force is equal to the difference in pressure times the area of the piston $F = (p_2 - p_1)A$.

If the fluid flow rate through the orifice is small, the flow is laminar and the force is proportional to the velocity (Figure 4.8b). Mechanical resistance that produces a force proportional to the velocity is called *viscous friction*. If the flow rate is large, the flow is turbulent and the force is proportional to the square of the velocity (Figure 4.8c). The friction effects of the seal are neglected in the preceding examples. The seal produces a constant friction force that is independent of velocity. Mechanical resistance that produces a force independent of velocity is called *Coulomb friction*. The combination of Coulomb friction and viscous friction is illustrated in Figure 4.8d.

Pure viscous friction is the simplest to treat mathematically, and Figure 4.8b is usually used as a first approximation to mechanical resistance. The resistance is equal to the force (F) divided by the velocity (v) at any point.

$$R_m = \frac{F}{v} \tag{4.48}$$

In Figure 4.8c, the resistance is nonlinear, increasing as the velocity increases. At any point on the curve, the resistance can be determined by drawing a line tangent to the curve at that point. The resistance is equal to the slope of this tangent line. The slope is determined by locating two points on the tangent line that are far enough apart to give reasonable accuracy. The slope is equal to the change in force (ΔF) between the two points divided by the change in velocity (Δv) between the same two points.

$$R_m = \frac{\Delta F}{\Delta v} \tag{4.49}$$

In Figure 4.8d, the resistance is linear, but Coulomb friction prevents the line from passing through the origin. Equation (4.48) cannot be used to determine the resistance. However, Equation (4.49) can be used provided the two points are located approximately as shown in the figure.

Example 4.21

A dashpot is used to provide mechanical resistance in a packaging machine. The flow is laminar, so the viscous friction equation shown in Figure 4.8a applies. A test was conducted in which a force of 98 N produced a velocity of 24 m/s. Determine the mechanical resistance (R_m).

Solution

$$R_m = \frac{F}{v} = \frac{98}{24} = 4.08 \text{ N·s/m}$$

Example 4.22

A mechanical system consists of a sliding load (Coulomb friction) and a shock absorber (viscous friction). The force versus velocity curve is shown in Figure 4.8d. The following data were obtained from the system.

Run	Force, F (N)	Velocity, v (m/s)
a	7.1	10.5
b	9.6	15.75

Determine the resistance (R_m) and the Coulomb friction force (F_C). Write an equation for the applied force (F) in terms of the velocity (v).

Solution

$$R_m = \frac{\Delta F}{\Delta v} = \frac{F_b - F_a}{v_b - v_a} = \frac{9.6 - 7.1}{15.75 - 10.5} = \frac{2.5}{5.25} = 0.476$$

$$= 0.476 \ \text{N·s/m}$$

$$F = F_c + 0.476v \quad \text{(from Figure 4.8d)}$$

$$7.1 = F_c + (0.476)(10.5) \quad \text{(from run a)}$$

$$F_c = 7.1 - 5.0 = 2.1 \ \text{N}$$

The equation for the applied force is

$$F = 2.1 + 0.476v$$

where F = applied force, newton

v = velocity, meter/second

Mechanical capacitance is defined as the increase in the displacement of a spring required to make a unit increase in spring force. The SI unit of mechanical capacitance is the "newton/meter." The reciprocal of the capacitance is called the spring constant, K. Mechanical capacitance (C_m) is computed by dividing a change in spring displacement (Δx) by the corresponding change in spring force (ΔF).

$$C_m = \frac{\Delta x}{\Delta F} = \frac{1}{K} \tag{4.50}$$

Some useful conversion factors:

1 pound per inch = 175 newtons per meter

1 pound per foot = 14.6 newtons per meter

1 inch per pound = 0.0057 meter per newton

1 foot per pound = 0.0685 meter per newton

Example 4.23

A spring is used to provide mechanical capacitance in a system. A force of 100 N compresses the spring by 30 cm. Determine the mechanical capacitance.

Solution

Equation (4.50) may be used to determine C_m.

$$C_m = \frac{\Delta x}{\Delta F} = \frac{0.30}{100} = 0.003 \text{ m/N}$$

Mechanical inertia (mass) is measured in terms of the force required to produce a unit increase in acceleration. It is defined by Newton's law of motion, and the term *mass* is used for the inertia element. Equation (4.51) states Newton's law of motion in terms of the average force (F_{avg}), the mass (m), the change in velocity (Δv), and the interval of time during which the change takes place (Δt).

$$F_{avg} = m\frac{\Delta v}{\Delta t} \tag{4.51}$$

Equation (4.52) states Newton's law in terms of the instantaneous force (F), the mass (m), and the rate of change of velocity with respect to time (dv/dt).

$$F = m\frac{dv}{dt} \tag{4.52}$$

Example 4.24

Automobile A has a mass of 1500 kg. Determine the average force required to accelerate A from 0 m/s to 27.5 m/s in 6 s. Automobile B requires an average force of 8000 N to accelerate from 0 m/s to 27.5 m/s in 6 s. Determine the mass of B.

Solution

Equation (4.51) may be used for both problems.

$$F_{avg} = m\frac{\Delta v}{\Delta t}$$

Automobile A:

$$F_{avg} = (1500)\left(\frac{27.5}{6}\right) = 6875 \text{ N}$$

Automobile B:

$$8000 = m\left(\frac{27.5 - 0}{6}\right)$$

$$m = (8000)\left(\frac{6}{27.5}\right)$$

$$= 1745 \text{ kg}$$

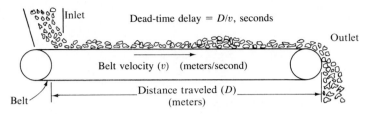

Figure 4.9 A belt conveyor is an example of a mechanical dead-time element.

Mechanical dead time is the time required to transport material from one place to another. A belt conveyor is frequently used to transport solid material in a process (see Figure 4.9). The dead-time delay is the time it takes for the belt to travel from the inlet end to the outlet end. Equation (4.1) may be used to compute the dead-time delay. The effect of the dead time is to delay the input flow rate by the dead-time delay (t_d).

MECHANICAL EQUATIONS

Resistance

$$R_m = \frac{F}{v} \qquad (4.48)$$

$$R_m = \frac{\Delta F}{\Delta v} \qquad (4.49)$$

Capacitance

$$C_m = \frac{\Delta x}{\Delta F} = \frac{1}{K} \qquad (4.50)$$

Inertia

$$F_{avg} = m\frac{\Delta v}{\Delta t} \qquad (4.51)$$

$$F = m\frac{dv}{dt} \qquad (4.52)$$

Dead Time

$$t_d = \frac{D}{v} \qquad (4.1)$$

where C_m = mechanical capacitance, meter/newton
D = distance, meter
F = force, newton
F_{avg} = average force, newton
ΔF = change in force, newton
K = spring constant, newton/meter
m = mass, kilogram
R_m = mechanical resistance, newton second/meter

t_d = dead time, second

Δt = increment of time, second

v = velocity, meter/second

Δv = change in velocity, meter/second

dv/dt = rate of change of velocity, meter/second^2

Δx = change in displacement, meter

Example 4.25

a. A belt conveyor is 30 m long and has a belt speed of 3 m/s. Determine the dead-time delay between the input and output ends of the belt.

b. Write the equation for the output mass flow rate $f_o(t)$ in terms of the input mass flow rate $f_i(t)$.

Solution

a. Equation (4.1) may be used.

$$t_d = \frac{D}{v} = \frac{30}{3} = 10 \text{ s}$$

b. The following equation from Section 4.1 may be used:

$$f_o(t) = f_i(t - t_d)$$
$$= f_i(t - 10)$$

GLOSSARY

Capacitance: The amount of material, energy, or distance required to make a unit increase in potential. (4.1)

Capacitance, electrical: The quantity of electric charge (q coulombs) required to make an increase in electrical potential of 1 V. (4.2)

Capacitance, gas flow: The increase in the mass of gas in a pressure vessel required to produce an increase in pressure of 1 pascal while the temperature remains constant. (4.4)

Capacitance, liquid: The increase in volume of liquid in a tank required to make an increase in pressure of 1 pascal at the outlet of the tank. (4.3)

Capacitance, mechanical: The increase in displacement of a spring required to make an increase in spring force of 1 newton. (4.6)

Capacitance, thermal: The amount of heat required to make an increase in temperature of 1 degree kelvin (or Celsius). (4.5)

Coulomb friction: Mechanical resistance that produces a force that is independent of the velocity. (4.6)

Dead time: The time interval between the time a signal appears at the input of a component and the time the corresponding response appears at the output. (4.1)

Dead time, electrical: The delay caused by the time it takes an electrical signal to travel from the source to the destination. (4.2)

Dead time, liquid flow: The time it takes for a liquid to flow from the input end of a pipe to the output end. (4.3)

Dead time, mechanical: The time required to transport material from one place to another place. (4.6)

Fanning equation: An equation that describes turbulent flow in a round pipe. (4.3)

Hagen-Poiseuille law: An equation that describes laminar flow in a round pipe. (4.3)

Inductance, electrical: The voltage required to increase electric current at a rate of 1 ampere per second. (4.1, 4.2)

Inertance, liquid flow: The amount of pressure drop in a pipe required to increase the flow rate at a rate of 1 cubic meter per second per second. (4.3)

Inertia: The force required to increase velocity at a rate of 1 meter per second per second (i.e., an acceleration of 1 meter per second^2). (4.6)

Laminar flow: An orderly type of flow that occurs when the Reynolds number is less than 2000 (low flow velocity). (4.3)

Resistance: An opposition to the movement or flow of material or energy. (4.1)

Resistance, electrical: The increase in voltage required to increase current by 1 ampere. (4.2)

Resistance, gas flow: The increase in pressure required to produce an increase in gas flow rate of 1 kilogram per second. (4.4)

Resistance, liquid flow: The increase in pressure required to make an increase in liquid flow rate of 1 cubic meter per second. (4.3)

Resistance, mechanical: The increase in force required to produce an increase in velocity of 1 meter per second. (4.6)

Resistance, thermal: The increase in temperature required to produce an increase in heat flow rate of 1 watt. (4.5)

Reynolds number: A dimensionless number used to predict the type of fluid flow (laminar or turbulent). See Equation (4.1). (4.3)

Turbulent flow: A disorderly type of flow that occurs when the Reynolds number is greater than 4000 (high flow velocity). (4.3)

Viscous friction: Mechanical resistance that produces a force that is proportional to the velocity. (4.6)

EXERCISES

4.1 An electrical component has a linear volt–ampere graph. An applied voltage of 65 V produces a current of 0.21 A. Determine the resistance of the component.

4.2 The following data were obtained in a test of a nonlinear electrical resistor:

Volts	0	5	10	15	20	25
Amperes	0	0.36	0.63	0.87	1.10	1.31

Plot a volt–ampere graph. Use the tangent line method to determine the resistance at 10 V and 20 V.

4.3 An electric current of 0.42 mA is applied to a capacitor for a duration of 0.2 s. The current pulse increased the voltage across the capacitor from 0 to 12 V. Determine the capacitance (C) of the capacitor.

4.4 A voltage pulse with an amplitude of 5.2 V and a duration of 0.025 is applied to an inductor. The current through the inductor is increased from 0 to 0.063 A by the voltage pulse. Assume that the resistance of the inductor is negligible and determine the inductance (L).

4.5 Determine the dead-time delay of a transmission line that is 20 km long. The velocity of propagation is 2.7×10^8 m/s.

4.6 Ethyl alcohol (at 15°C) flows through a 15-m-long commercial pipe with an inside diameter of 2.09 cm. The flow rate is 25 L/min. Determine the Reynolds number, the resistance, and the pressure drop in the pipe.

4.7 Glycerin (at 15°C) flows through a 25-ft-long smooth tube with an inside diameter of 1 in. The flow rate is 2 gal/min. Determine the Reynolds number, the resistance, and the pressure drop in the pipe.

4.8 Kerosene (at 15°C) flows through a 55-ft-long smooth tube with an inside diameter of 2 in. The flow rate is 50 gal/min. Determine the Reynolds number, the resistance, and the pressure drop in the pipe.

4.9 Gasoline (at 15°C) flows through a 35-m-long smooth tube with an inside diameter of 2.65 cm. The flow rate is 10.57 L/min. Determine the Reynolds number, the resistance, and the pressure drop in the pipe.

4.10 A liquid tank has a diameter of 1.6 m and a height of 3.5 m. Determine the capacitance of the tank for each of the following fluids.

    a. Water   **b.** Oil

    c. Kerosene   **d.** Gasoline

    e. Turpentine

4.11 Determine the liquid flow inertance of oil in a pipe that has a diameter of 1.2 cm and a length of 70 m.

4.12 Determine the dead-time delay in a pipe that is 25 m long and has a diameter of 2.2 cm. The flow rate is 12 L/min.

4.13 A smooth tube is supplying 0.042 kg/s³ of nitrogen gas at a temperature of 15°C. The tube is 25 m long and has an inside diameter of 4.5 cm. Use Equations (4.26), (4.27), and (4.28) to find the pressure drop and the gas flow resistance. The outlet pressure (p_2) is 103 kPa. Find the inlet pressure (p_1).

4.14 The following data were obtained for a gas flow system. Plot the data and graphically determine the resistance at flow rates of 0.4 and 0.8 kg/s.

$$W = \text{gas flow rate, kilogram/second}$$

$$p = \text{pressure drop, kilopascal}$$

W	0	0.2	0.4	0.6	0.8	1.0
p	0	6.2	25.2	55.8	99.6	155.3

4.15 A pressure tank has a volume of 1.4 m³. Determine the capacitance of the tank for each of the following gases.

a. Nitrogen at 40°C   **b.** Carbon dioxide at 30°C

c. Oxygen at 20°C   **d.** Carbon monoxide at 50°C

4.16 A composite wall section similar to Figure 4.7 has a surface area of 4.5 m² and three inner layers: a sheet of pine wood 1.9 cm thick, insulation 7.6 cm thick, and a sheet of oak panel 0.63 cm thick. The outside fluid is air with a velocity of 7 m/s. The inside fluid is still air (vertical surface). The estimated value of T_d for the inside film is 4°C. The fluid temperatures are −5°C for the outside and 21°C for the inside. Determine the unit thermal resistance, the total thermal resistance, the total heat flow, and the direction of the heat flow.

4.17 A composite wall section has an area of 5.2 m² and one inner layer: an aluminum plate 0.72 cm thick. The inside fluid is oil and the outside fluid is water. Both fluids are still. The value of T_d is estimated to be 1°C for the water film and 19°C for the oil film. The oil temperature is 95°C and the water temperature is 75°C. Determine the unit resistance, the total resistance, the total heat flow, and the direction of the heat flow.

4.18 Cold water (5°C) flows through a steel pipe that is exposed to warm air (20°C). Your job is to estimate how much heat flows through the pipe from the air to the water. The pipe is 20 m long with a mean diameter of 1.86 cm and a wall thickness of 0.277 cm. The total surface area of the pipe is $\pi d_m L$, where d_m is the mean diameter in meters and L is the length in meters. In your choice for the water film coefficient, consider the following: the water velocity is 1.8 m/s and the flow is turbulent. The "still air—vertical surface" is a good choice for the air film. Estimate the value of t_d for the air film and try several values for T_d between 1 and 15°C to see how sensitive your estimate is.

4.19 Determine the thermal capacitance of 1.8 m³ of ethyl alcohol.

4.20 A dashpot operates in the linear region and Coulomb friction is negligible. A test was conducted in which a velocity of 1.6 m/s produced a force of 135 N. Determine the mechanical resistance of the dashpot.

4.21 A mechanical system contains a sliding load (Coulomb friction) and a shock absorber (viscous friction). A force of 12 N produces a velocity of 2 m/s, and a force of 31.5 N produces a velocity of 15 m/s. Determine the mechanical resistance (R_m) and the Coulomb friction force (F_c). Write an equation for the applied force (F) in terms of the velocity (v).

4.22 A spring is used to provide mechanical capacitance. A force of 1500 N compresses the spring by 8.5 cm. Determine the mechanical capacitance.

4.23 A mechanism consists of a linear acceleration cam. The cam has two sections: A and B. Section A accelerates the load from 0 m/s to 0.95 m/s in 0.32 s. Section B decelerates the load from 0.95 m/s to 0 m/s in 0.22 s. Determine the inertial force in A and B if the load has a mass of 2.9 kg.

4.24 Combine equations (4.22), (4.23), and (4.24) into Equation (4.21) to derive Equation (4.25).

4.25 The braking distance from 60 miles per hour is one measure of the brakes of an automobile. In this test, the brakes are applied while the car is traveling at a steady 60 miles per hour. The braking distance is the distance from the point where the brakes were first applied to the point where the car comes to a complete stop. The average acceleration during braking is given by the following equation:

$$a = -\frac{V_o^2}{2D_b}$$

where V_o = initial velocity of the car

D_b = braking distance

Complete the following for a 2600 pound car that has a braking distance of 160 feet.

a. Convert the initial velocity (60 mph) to feet per second and then to meters per second. Convert the mass (2600 lbm) to kilograms, and the braking distance (160 ft) to meters. (The conversion factors are in Appendix B.)

b. Compute the average acceleration in meters/second^2, and then use $f = ma$ to compute the average braking force in newtons.

c. Convert the average braking force from newtons to pounds force (lbf).

d. Compute the average acceleration in feet/second^2. Use the engineering fps system units (see Appendix B) to compute the average braking force in pounds. The fps system uses pounds to measure force (lbf) and pounds to measure mass (lbm). In this system, the relationship between force, mass, and acceleration is given by the following equation:

$$f = ma/g_c$$

The factor g_c is the acceleration due to gravity. The nominal value of g_c is 32.2.

e. Compare the average braking force obtained in (d) with the converted result obtained in (c).

CHAPTER 5

Laplace Transforms and Transfer Functions

OBJECTIVES

Laplace transforms are used to (1) describe control systems conveniently, (2) facilitate the analysis of control systems, and (3) facilitate the design of control systems.

Control system components are described mathematically by an equation that establishes the time relationship between the input and output signals of the component. These equations are functions of time and often include derivative and/or integral terms. The Laplace transform converts these integral/differential equations into algebraic equations that are functions of frequency. These frequency-domain equations also determine a relationship between the input and output signals, namely the frequency response of the component. When the algebraic equation is solved for the ratio of the output signal over the input signal, the result is called the transfer function. With the transfer function, we can write a computer program to compute the frequency response of the component.

The purpose of this chapter is to give you the ability to use the Laplace transform to obtain the transfer function of a component and to use the transfer function to obtain the Bode diagram (frequency response) of the component. After completing this chapter, you will be able to

1. Use the Laplace transform tables to obtain the Laplace transform of a specified function of time

2. Use the Laplace transform tables to determine the transfer function of a component from the time-domain equation that defines the relationship between the component's input and its output

3. Use partial fraction expansion and transform pairs to obtain the inverse Laplace transform of a specified function of s

4. Use the transfer function of a component to obtain the output/input gain and phase shift of the component at a specified frequency

5. Explain three methods of obtaining the frequency response of a component

6. Use the computer-aided method to obtain frequency response data for a component and plot a Bode diagram from the data

5.1 INTRODUCTION

In this chapter you will study several important tools that will help you understand, analyze, and design control systems. The first tool is the Laplace transformation. The Laplace transform converts equations that involve derivatives and integrals (calculus) into equations that use only algebraic terms. We begin the chapter with the development of the equations that describe several simple components. We do this to show how derivative and integral terms are used to describe the behavior of a component. The derivative is a mathematical expression of the rate of change of a variable. For example, the speed of an automobile is the derivative (or rate of change) of the distance traveled. The integral is a mathematical expression for the accumulation of an amount (or quantity) of a variable. For example, the integral of the speed of an automobile is the distance traveled.

The Laplace transform is only one of a number of useful transforms. Logarithms and phasors are two examples of other useful transforms. We use the logarithmic transformation as a way of leading into the Laplace transformation. Our treatment of the Laplace transformation stresses use and application rather than theory and development. Tables of the most often used transforms are presented and methods of using the tables are explained. The inverse Laplace transformation is covered next. Although much of control system design avoids the inverse transformation, it is useful in explaining the Nyquist stability criteria (Chapter 16).

The next tool in our study is the transfer function that was introduced in Chapter 1 as a means of describing the relationship between the input and the output of a component. Now we can show how to use the Laplace transformation to convert the integrodifferential equation that describes the input/output relationship of the component into the transfer function of the component. The transfer function now becomes an algebraic expression for the ratio of the output over the input.

Our final tool is another method of describing a component called its *frequency response*. The frequency response of a component describes how the component responds to a sinusoidal input signal. More specifically, the frequency response describes how the gain and phase difference of the component vary with the frequency of the input signal. There are three methods of determining the frequency response of a component: experimentally, mathematically, and with the aid of a computer program. The last two methods use the transfer function as a starting point. When the s parameter in the transfer function of a component is replaced by $j\omega$, we obtain a mathematical definition of the frequency response of that component. In Chapter 1, we defined the transfer function as the ratio of the Laplace transform of the output to the Laplace transform of the input. Thus the Laplace transform is our means of obtaining frequency response by computation or a computer program. The chapter concludes with a computer program for determining the frequency response of a component, a very useful tool for analysis and design.

5.2 INPUT/OUTPUT RELATIONSHIPS

In this section we develop the mathematical equations that describe the relationship between the input and the output of five simple control system components. The five components are a self-regulating liquid tank, a nonregulating liquid tank, an electrical *RC* circuit, a liquid-filled thermometer, and a process control valve.

Self-Regulating Liquid Tank

Consider the self-regulating liquid tank in Figure 5.1. The liquid level in the tank remains constant when the inflow rate (q_{in}) is equal to the outflow rate (q_{out}). If the inflow rate is greater than the outflow rate, the liquid level will rise. If the inflow rate is less than the outflow rate, the level will fall. During a certain time interval (Δt), the amount of liquid in the tank will change by an amount (ΔV) equal to the average difference between the inflow rate and the outflow rate multiplied by Δt.

$$\Delta V = (q_{in} - q_{out})_{avg} \Delta t \qquad m^3 \qquad (5.1)$$

The change in the liquid level in the tank (Δh) is equal to the change in volume (ΔV) divided by the cross-sectional area of the tank (A).

$$\Delta h = \frac{\Delta V}{A} = \frac{(q_{in} - q_{out})_{avg} \Delta t}{A} \qquad \text{meters} \qquad (5.2)$$

The average rate of change of the level in the tank is equal to the change in level (Δh) divided by the time interval (Δt). For example, if the level changed 0.26 m during a time interval of 100 s, the average rate of change of level would be $0.26/100 = 0.0026$ m/s.

$$\text{Average rate of change of level} = \frac{\Delta h}{\Delta t} = \frac{(q_{in} - q_{out})_{avg}}{A} \qquad \text{m/s} \qquad (5.3)$$

Different time intervals may be used to determine the average rate of change. The average of 0.0026 m³/s could have resulted from a change of 0.026 m during a time

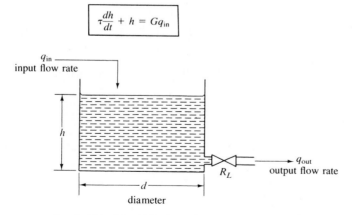

$$\tau \frac{dh}{dt} + h = G q_{in}$$

q_{in} — input flow rate

h

R_L

q_{out} — output flow rate

d — diameter

h = height of liquid in the tank, meters

Figure 5.1   A self-regulating liquid tank has a restriction with resistance R_L at the outlet of the tank. If the liquid level in the tank increases, the flow through the restriction also increases. If the level decreases, the flow through the restriction decreases. As long as the tank is not full, the level will automatically adjust until the outflow equals the inflow.

interval of 10 s or a change of 0.0026 m during a 1-s interval. When the time interval diminishes to 0 s, we call it an instant of time. As the time interval Δt diminishes to an instant of time, the average rate of change becomes the instantaneous rate of change. In mathematics, the instantaneous rate of change of liquid level is called the *derivative* of level (h) with respect to time (t) and is designated by the symbol dh/dt. If the time interval (Δt) in Equation (5.3) diminishes to an instant, the average rate of change of level becomes the instantaneous rate of change (dh/dt).

$$\text{As } \Delta t \to 0, \frac{\Delta h}{\Delta t} \to \frac{dh}{dt} = \frac{q_{\text{in}} - q_{\text{out}}}{A} \tag{5.4}$$

If the equation for flow out of the tank is linear (laminar flow), the outflow rate (q_{out}) is given by the following equation (see Chapter 4):

$$q_{\text{out}} = \frac{\rho g h}{R_L} \qquad \text{m}^3/\text{s} \tag{5.5}$$

where q_{out} = liquid flow rate, cubic meter/second

ρ = liquid density, kilogram/cubic meter

g = gravitational acceleration, meter/second^2

h = liquid level, meter

R_L = laminar flow resistance, pascal second/cubic meter

Substituting Equation (5.5) into Equation (5.4) gives us

$$\frac{dh}{dt} = \frac{q_{\text{in}} - \rho g h / R_L}{A}$$

or

$$R_L \left(\frac{A}{\rho g} \right) \frac{dh}{dt} = \left(\frac{R_L}{\rho g} \right) q_{\text{in}} - h$$

The term ($A/\rho g$) is the capacitance (C_L) of the liquid tank (see Chapter 4), and the entire term $R_L A/\rho g = R_L C_L$ is called the *time constant* (τ) of the liquid tank. The term $R_L/\rho g$ is the steady-state gain (G) of the system. Substituting τ and G into the preceding equation gives us the final form of the differential equation for the liquid tank.

$$\tau \frac{dh}{dt} + h = G q_{\text{in}} \tag{5.6}$$

where $$\tau = \frac{R_L A}{\rho g} \qquad \text{and} \qquad G = \frac{R_L}{\rho g}$$

Nonregulating Liquid Tank

A nonregulating liquid tank is shown in Figure 5.2. A positive-displacement pump provides a constant output flow rate, q_{out}. Equation (5.2), which was developed for the self-regulating liquid tank, also applies to the nonregulating tank.

$$\Delta h = \frac{\Delta V}{A} = \frac{(q_{\text{in}} - q_{\text{out}})_{\text{avg}} \Delta t}{A} \qquad \text{meters} \tag{5.2}$$

Figure 5.2 A nonregulating liquid tank has a positive-displacement pump at the outlet of the tank. The liquid level in the tank has no effect on the flow through the pump. If the pump flow rate (q_{out}) does not exactly equal the inflow rate (q_{in}), the level in the tank will change at a rate proportional to the difference, $q = q_{in} - q_{out}$. The level increases when q is positive and decreases when q is negative.

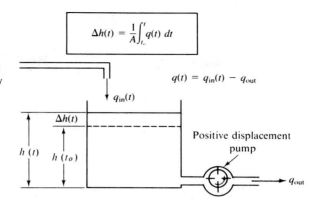

$$\Delta h(t) = \frac{1}{A}\int_{t_o}^{t} q(t)\,dt$$

$$q(t) = q_{in}(t) - q_{out}$$

$q_{in}(t)$

$\Delta h(t)$

Positive displacement pump

$h\,(t)$ $h\,(t_0)$

q_{out}

For convenience, we will define $q(t)$ as follows:

$$q(t) = q_{in} - q_{out}$$

In other words, $q(t)$ is the difference between the input flow rate and the output flow rate.

If the time interval, Δt, begins at time t_0 and ends at time t_1, then

$$\Delta t = t_1 - t_0$$

and

$$\Delta h(t) = h(t_1) - h(t_0)$$

The term $(q_{in} - q_{out})_{avg}\,\Delta t = q_{avg}(t)\,\Delta t$ represents the accumulation of liquid in the tank over the interval Δt. In calculus, this accumulation of liquid is given by the *integral* from t_0 to t_1 of $q(t)\,dt$.

$$\Delta h(t) = \frac{1}{A} \int_{t_0}^{t_1} q(t)\,dt \tag{5.7}$$

Electrical Circuit

Consider the electrical circuit of Figure 5.3. If the output terminals are connected to a high-impedance load, then the current through the resistor and the current through

Figure 5.3 The electrical *RC* circuit is sometimes used as a low-pass filter in a control system.

$$\tau\frac{de_{out}}{dt} + e_{out} = e_{in}$$

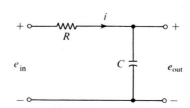

i

R

e_{in} C e_{out}

the capacitor are essentially equal. Let i represent the current that passes through the resistor and the capacitor. For the resistor, the current (i) is equal to the voltage difference ($e_{in} - e_{out}$) divided by the resistance (R).

$$i = \frac{e_{in} - e_{out}}{R} \tag{5.8}$$

For the capacitor, the current (i) is equal to the capacitance (C) times the instantaneous rate of change of the capacitor voltage.

$$i = C \frac{de_{out}}{dt} \tag{5.9}$$

Equating the right-hand side of Equations (5.8) and (5.9) and setting $RC = \tau$ results in the following differential equation for the electrical circuit:

$$\tau \frac{de_{out}}{dt} + e_{out} = e_{in} \tag{5.10}$$

where
$$\tau = RC$$

Liquid-Filled Thermometer

A liquid-filled thermometer is illustrated in Figure 5.4. The amount of heat (ΔQ) transferred from the fluid surrounding the bulb to the liquid inside the bulb depends on the thermal resistance (R_T) between the two fluids, the difference in temperature ($T_a - T_m$), and the time interval (Δt).

$$\Delta Q = \frac{(T_a - T_m)\,\Delta t}{R_T}$$

The change in temperature of the liquid in the bulb (ΔT_m) is equal to the amount of heat added (ΔQ) divided by the thermal capacitance (C_T) of the liquid inside the bulb.

$$\Delta T_m = \frac{\Delta Q}{C_T} = \frac{(T_a - T_m)\,\Delta t}{R_T C_T}$$

Dividing both sides by Δt gives us

$$\frac{\Delta T_m}{\Delta t} = \frac{T_a - T_m}{R_T C_T}$$

or, as $\Delta t \to 0$,

$$\tau \frac{dT_m}{dt} + T_m = T_a \tag{5.11}$$

where $\tau = R_T C_T$ = time constant, second

R_T = thermal resistance between the fluid in the container and the liquid inside the bulb, kelvin/watt

C_T = thermal capacitance of the liquid in the bulb, joule/kelvin

T_a = temperature of the liquid in the container, °C or K

T_m = temperature of the liquid in the bulb, °C or K

$$\tau \frac{dT_m}{dt} + T_m = T_a$$

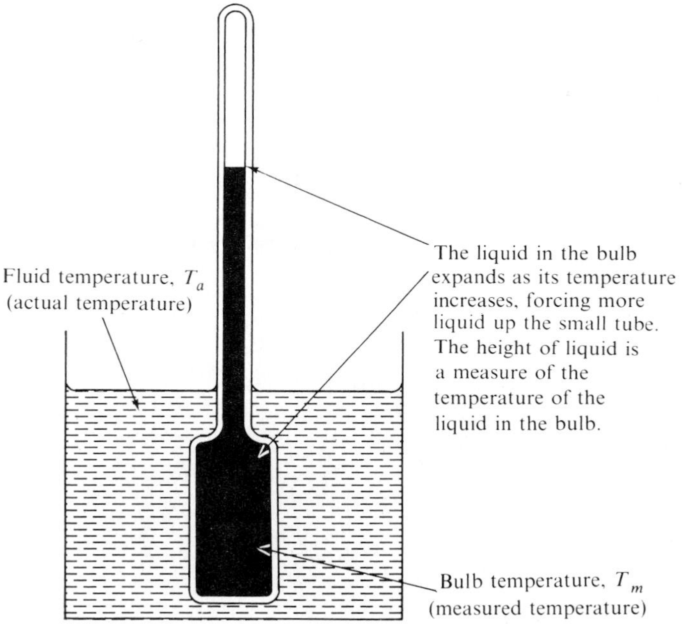

Fluid temperature, T_a
(actual temperature)

The liquid in the bulb
expands as its temperature
increases, forcing more
liquid up the small tube.
The height of liquid is
a measure of the
temperature of the
liquid in the bulb.

Bulb temperature, T_m
(measured temperature)

Figure 5.4   A liquid-filled thermometer is similar to a liquid-filled primary element for measuring temperature in a control system.

Process Control Valve

A process control valve is illustrated in Figure 5.5. The inlet signal is air pressure that enters the diaphragm chamber through the hole at the top of the valve. The air pressure applies a downward force on the diaphragm that is equal to the air pressure (p_{in}) multiplied by the area of the diaphragm (A):

$$\text{Applied force} = F_a = p_{in} A \tag{5.12}$$

This applied force is opposed by three reaction forces: the inertial force of the moving mass (F_I), the resistive force of the seal (F_R), and the compressive force of the spring (F_C):

$$F_a = F_I + F_R + F_C \tag{5.13}$$

The inertial force is equal to the mass of the moving parts (valve stem and diaphragm plate) multiplied by the acceleration of the moving mass. If x is the position of the moving mass, the acceleration is expressed mathematically by the second derivative of x with respect to time. The acceleration (or second derivative) and the inertial

$$p_{in}A = m\frac{d^2x}{dt^2} + R_m\frac{dx}{dt} + \frac{1}{C_m}x$$

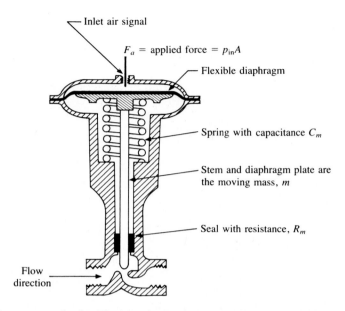

Inlet air signal

F_a = applied force = $p_{in}A$

Flexible diaphragm

Spring with capacitance C_m

Stem and diaphragm plate are the moving mass, m

Seal with resistance, R_m

Flow direction

Figure 5.5   Process control valve. The inlet air pressure acts on the top of the diaphragm to produce an applied force (F_a). This applied force is opposed by three reaction forces: the inertial force of the moving mass (F_I), the frictional force of the seal (F_R), and the compressive force of the spring (F_C); $F_a = F_I + F_R + F_C$.

force are written as follows:

$$\text{Acceleration} = \frac{d^2x}{dt^2}$$

$$\text{Inertial force} = F_I = m\frac{d^2x}{dt^2} \tag{5.14}$$

We will use the simple viscous friction model for the resistive force, F_R. In this model the friction force is equal to the resistance multiplied by the velocity of the moving mass.

$$F_R = R_m\frac{dx}{dt} \tag{5.15}$$

The spring force is equal to the compression of the spring divided by the mechanical capacitance value of the spring (C_m). If we measure the position of the moving valve stem (x) such that the spring force is 0 when x is 0, the equation for the spring force is

$$F_C = \frac{x}{C_m} \qquad (5.16)$$

Substituting Equations (5.12), (5.14), (5.15), and (5.16) into Equation (5.13) gives us the desired mathematical equation for the control valve:

$$P_{in}A = m\frac{d^2x}{dt^2} + R_m\frac{dx}{dt} + \frac{1}{C_m}x \qquad (5.17)$$

Equation (5.17) is called a second-order differential equation. Equations (5.6), (5.10), and (5.11) are called first-order differential equations. First- and second-order differential equations have applications in many types of engineering systems.

As the previous examples illustrate, many components, systems, and processes are modeled by some type of differential equation. At this point, a brief discussion of differential equation terminology will be helpful.

A *differential equation* is an equation that has one or more derivatives of the dependent variable with respect to the independent variable (e.g., dh/dt in Equation (5.6)). A *solution* of a differential equation is an expression for the dependent variable in terms of the independent variable that satisfies the differential equation. As an example, consider the following differential equation and its solution. The solution assumes that GK is a constant and $h = 0$ when $t = 0$ (the value of h at $t = 0$ is called an *initial condition* of the differential equation).

Differential equation:

$$\tau\frac{dh}{dt} + h = GK$$

Solution:

$$h = GK(1 - e^{-t/\tau})$$

The *order* of a differential equation is the order of the highest derivative that appears in the equation. Thus Equations (5.6), (5.10), and (5.11) are first-order differential equations, and Equation (5.17) is a second-order differential equation. Notice the remarkable similarity of the three first-order differential equations, Equations (5.6), (5.10), and (5.11), even though they describe three very different systems. This means these three dissimilar systems will have similar input-output relationships. By knowing the behavior of the *RC* circuit in Figure 5.3, we can predict the behavior of the thermometer in Figure 5.4 and the self-regulating liquid tank in Figure 5.1. This observation is very useful in the analysis of a control system.

A *linear* differential equation has only first degree terms in the dependent variable and its derivatives. Thus Equations (5.6), (5.10), and (5.17) are all linear differential equations. If any term in the equation is raised to a power other than 1, the equation is nonlinear.

A linear differential equation of second order can be written in the following standard form.

$$\frac{d^2x}{dt^2} + A(t)\frac{dx}{dt} + B(t)x = F(t)$$

If $F(t) = 0$, the equation is called *homogeneous*; otherwise, it is called *nonhomogeneous*. If $A(t)$ and $B(t)$ are constants, the equation is said to have constant coefficients. Many components of a control system are modeled by linear differential equations with constant coefficients. The Laplace transform is both a tool for solving these differential equations and for converting the equations into transfer functions. Once the conversion to transfer functions is made, the operations of calculus are replaced by algebraic operations.

Often the equation that models a component includes both derivative and integral terms. Equations that have at least one integral term and one derivative term are called *integro-differential equations*. The Laplace transform method also applies to integro-differential equations.

5.3 LOGARITHMS: A TRANSFORMATION

Logarithms are a familiar example of how a transformation makes problem solving easier. A brief review of the logarithmic transformation will make a good introduction to the Laplace transformation. Figure 5.6 illustrates the logarithmic transformation. The logarithmic and inverse logarithmic functions move numbers back and forth between two domains: the domain of positive real numbers and the domain of the logarithms of real numbers.

The domain of real numbers is the one we use in normal operations with numbers. We all know how to multiply and divide real numbers. We also know that it is usually easier to add and subtract numbers than it is to multiply and divide them. The logarithmic transformation allows us to replace the operations of multiplication

Figure 5.6 The logarithmic function transforms numbers from the domain of positive real numbers (N) to the domain of logarithms (L). The operations of multiplication and division in domain N are transformed into addition and subtraction in domain L. The operation of exponentiation in domain N is transformed into multiplication in domain L.

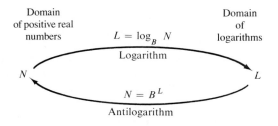

and division with the simpler operations of addition and subtraction. Most of us can square or cube a number, but what about raising a number to the 2.475 power? The logarithmic transformation allows us to replace the operation of exponentiation (raising to a power) with the simpler operation of multiplication.

If N, B, and L are three numbers such that

$$N = B^L$$

the exponent (L) is the logarithm of the number (N) to the base (B). In other words, the logarithm of a number (N) to a given base (B) is the exponent (L) to which the base must be raised to produce the number (N). The logarithm is usually written as follows:

$$L = \log_B N$$

We may think of the logarithm and antilogarithm as transformations between two regions or domains (see Figure 5.6). One domain contains all the positive real numbers (N) represented by decimal digits. The other domain contains all the logarithms (L) of the numbers (N) in the real-number domain. The logarithmic transformation transforms a number (N) into its corresponding logarithm (L) and is defined by the following relationship:

Logarithmic transformation: $L = \log_B N$

The antilogarithmic transformation transforms a logarithm (L) into its corresponding number (N). It is defined by the following relationship:

Antilogarithmic transformation: $N = B^L$

The three laws of exponents are the bases of the operations of multiplication, division, and exponentiation in the logarithmic domain. Consider two numbers, N_1 and N_2, with logarithms L_1 and L_2.

$$N_1 = B^{L_1} \qquad N_2 = B^{L_2}$$

Multiplication Law: $N_1 \cdot N_2 = (B^{L_1}) \cdot (B^{L_2}) = B^{L_1 + L_2}$

Division Law: $\dfrac{N_1}{N_2} = \dfrac{B^{L_1}}{B^{L_2}} = B^{L_1 - L_2}$

Exponentiation Law: $N^x = (B^L)^x = B^{xL}$

Operation in the Domain of Real Numbers	Corresponding Operation in the Domain of Logarithms
Multiplication, $N_1 \cdot N_2$	Addition, $L_1 + L_2$
Division, N_1/N_2	Subtraction, $L_1 - L_2$
Exponentiation, N^x	Multiplication, $x \cdot L$

Example 5.1

Compute (24.3)(18.2) using logarithms to the base 10.

Solution

Complete the following four steps to find (24.3)(18.2).

1. Transform 24.3 into the logarithmic domain. Transform 18.2 into the logarithmic domain.
2. Add the logarithms of the two numbers in the logarithmic domain to obtain the sum of 2.6457.
3. Inverse transform 2.6457 into the real-number domain to obtain 442.3, which is the product of 24.3 × 18.2.
4. Note that the answer could be obtained in the real-number domain by the more difficult operation of multiplication.

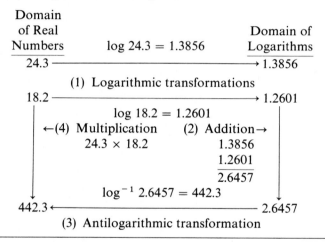

5.4 LAPLACE TRANSFORMS

In Section 5.2 we learned that even simple control system components are modeled by equations that include derivative and integral terms. The *Laplace transform* allows us to transform these integral/differential equations into simpler algebraic equations. By solving the algebraic equation for the ratio of the output over the input, we can obtain the transfer function of the component. With the transfer function, we can compute the frequency response of the component or program a computer to do the work for us. In this text, we will not derive Laplace transforms but will only apply them. Considering the widespread use of transfer functions in the control field, some knowledge of the Laplace transform is a decided asset.

In the preceding section, the logarithmic transformation took us from one domain to another in order to simplify mathematical operations. In a similar way, the Laplace

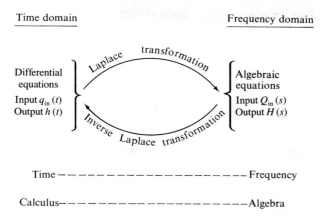

Time domain   Frequency domain

Differential equations
Input $q_{in}(t)$
Output $h(t)$

Algebraic equations
Input $Q_{in}(s)$
Output $H(s)$

Time — — — — — — — — — — — — — — Frequency

Calculus— — — — — — — — — — — — — — — —Algebra

Figure 5.7   The Laplace transform takes a function from the time domain to the frequency domain. The inverse Laplace transform takes a function from the frequency domain to the time domain. Equations in the time domain determine the size of a control signal at various times, t. Equations in the frequency domain determine the amplitude and phase angle of control signals at various frequencies, s. Operations on equations in the time domain involve calculus. Operations on equations in the frequency domain involve algebra.

transformation takes us from a domain called the *time domain* to another domain called the *frequency domain*. Figure 5.7 shows an overview of the Laplace transformation. In the time domain, the size of the signals is given (or determined) for each instant of time. We say that the input and output signals are functions of time and append a (t) to the variable name to indicate that time is the independent variable. In Figure 5.7, $q_{in}(t)$ represents the time domain input signal, and $h(t)$ represents the time domain output signal. The time-domain equation of a component defines the size of the output as a function of time and the input signal. Time-domain equations often include derivative and integral terms. The equations we developed in Section 5.2 are all time-domain equations.

In the frequency domain, the size and the phase angle of the signal are given (or determined) for each value of a complex frequency parameter designated by the letter s (s has units of second^{-1}). We say that the input and output signals are functions of frequency and append an (s) to the variable name to indicate that frequency is the independent variable. In Figure 5.7, $Q_{in}(s)$ is the frequency domain input signal, and $H(s)$ is the frequency domain output signal. The frequency domain equation of a component defines the size and phase angle of the output signal as a function of s and the input signal. We use the Laplace transform to obtain the frequency domain equation from the time domain equation.

Lowercase letters are used to represent signals in the time domain, and uppercase letters are used for signals in the frequency domain. In Figure 5.7, for example, the input signal is represented by $q_{in}(t)$ in the time domain and by $Q_{in}(s)$ in the frequency domain. Similarly, the output signal is represented by $h(t)$ in the time domain and by $H(s)$ in the frequency domain.

Figure 5.8 illustrates the solution of a differential equation using the Laplace transformation and the inverse Laplace transformation. The Laplace transform enables us to convert a differential equation from the time domain into an algebraic equation in the frequency domain. The algebraic equation is then solved to get a frequency-domain solution. An inverse transformation converts the frequency-domain solution into a time-domain solution. Of course, we can also obtain the time-domain solution directly using the methods of operational calculus.

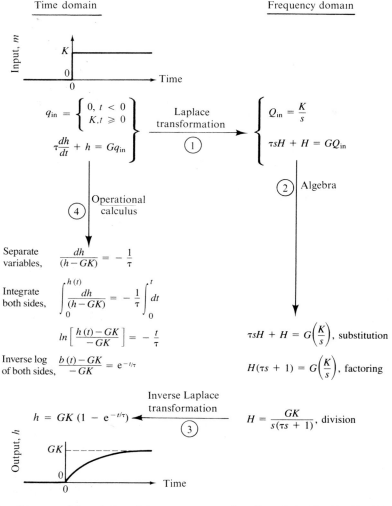

Figure 5.8 Diagram of the solution of the step response of a self-regulating liquid tank. The Laplace transform solution consists of the following three steps: (1) the Laplace transformation of $q_{in}(t)$ and $(\tau\, dh/dt + h = Gq_{in})$ to frequency domain, (2) the algebraic solution for $H(s)$, and (3) the inverse Laplace transformation of $H(s)$ to time domain $h(t)$. The calculus solution is shown as step 4.

The solution outlined in Figure 5.8 is for a step change in the inlet flow rate of the self-regulating liquid tank shown in Figure 5.1. Equation (5.6) is the time-domain equation for this component.

$$\tau \frac{dh}{dt} + h = Gq_{in} \tag{5.6}$$

Before time $t = 0$ s, the tank is empty and the input flow rate is zero (i.e., $h = q_{in} = 0$, for $t < 0$ s). At time $t = 0$, the input valve is opened and the input flow rate changes to K cubic meters/second (i.e., $q_{in} = K$, for $t > 0$ s). This type of change is called a *step change* in the input signal (q_{in}). A graph of input (q_{in}) versus time (t) is shown in Figure 5.8. The question is: What is the level of the tank after the step change in input? [That is, $h(t) = ?$ for $t > 0$.]

Solution

Step 1. Transform the input (q_{in}) and Equation (5.6) into the frequency domain. A table of Laplace transform pairs was used to make this transformation. The results of the transformation are listed below and in Figure 5.8.

Time Domain	*Frequency Domain*
$q_{in}(t) = K$	$Q_{in}(s) = \dfrac{K}{s}$
$\tau \dfrac{dh(t)}{dt} + h(t) = Gq_{in}$	$\tau s H(s) + H(s) = GQ_{in}(s)$

Step 2. Solve the two frequency-domain equations for H.

$$H(s) = \frac{GK}{s(\tau s + 1)} \tag{5.18}$$

Step 3. Use the table of Laplace transforms (Table 5.1) to transform Equation (5.18) into the time domain.

Time Domain	*Frequency Domain*
$h(t) = GK(1 - e^{-t/\tau})$	$H(s) = \dfrac{GK}{s(\tau s + 1)}$

The time-domain graph of $h(t)$ is shown in Figure 5.8.

Functional Laplace Transforms

Laplace transforms can be divided into two types: functional transforms and operational transforms. The first type is simply the Laplace transform of a particular function such as sin ωt or e^{-at}. The second involves the transform of the result of some operation, such as the sum of two functions, the derivative of a function, or the integral of a function. In this section we cover *functional transforms*. The notations

$\mathscr{L}\{f(t)\}$ and $F(s)$ are both used to indicate the Laplace transform of $f(t)$. The Laplace transform $F(s)$ of the function $f(t)$ is defined by the following relationship:

$$\mathscr{L}\{f(t)\} = F(s) = \int_0^\infty f(t)e^{-st}\, dt$$

As an illustration, we will use the defining equation to determine the Laplace transform of $f(t) = K$ where K is a constant.

$$F(s) = \int_0^\infty f(t)e^{-st}\, dt = \int_0^\infty Ke^{-st}\, dt$$

$$F(s) = K\,\frac{e^{-st}}{-s}\bigg|_0^\infty = -\frac{K}{s}(e^{-s\infty} - e^0)$$

$$F(s) = -\frac{K}{s}(0 - 1) = K/S$$

Table 5.1 gives the transform pairs for the functions most often encountered in the study of control systems. Using the table to find the Laplace transform of a time-domain function is simply a matter of looking up the function and its frequency-domain mate. The table works just as well in reverse to find the inverse Laplace transform of a frequency-domain function. Several examples follow.

Table 5.1 Functional Laplace Transform Pairs*

Time Domain, $f(t)$, $t > 0$	Frequency Domain, $F(s)$
1. K	$\dfrac{K}{s}$
2. Kt	$\dfrac{K}{s^2}$
3. Ke^{-at}	$\dfrac{K}{s+a}$
4. Kte^{-at}	$\dfrac{K}{(s+a)^2}$
5. $K \sin \omega t$	$\dfrac{K\omega}{s^2+\omega^2}$
6. $K \cos \omega t$	$\dfrac{Ks}{s^2+\omega^2}$
7. $Ke^{-at} \sin \omega t$	$\dfrac{K\omega}{(s+a)^2+\omega^2}$
8. $Ke^{-at} \cos \omega t$	$\dfrac{K(s+a)}{(s+a)^2+\omega^2}$
9. $K(1 - e^{-t/\tau})$	$\dfrac{K}{s(\tau s + 1)}$
10. $2Ke^{-at} \cos(bt + \theta)$	$\dfrac{K\underline{/\theta}}{s+a-jb} + \dfrac{K\underline{/-\theta}}{s+a+jb}$

* The letters a, K, and T represent any numerical constants.

Example 5.2

Use Table 5.1 to find the Laplace transform of each of the following functions.

    a. $f(t) = 12$

    b. $f(t) = 120 \cos 377t$

    c. $f(t) = 27t$

    d. $f(t) = 5e^{-2.5t}$

    e. $f(t) = 89e^{-2t} \sin 1000t$

Solution

    a. From entry 1 in Table 5.1,

$$F(s) = \frac{12}{s}$$

    b. From entry 6 in Table 5.1,

$$F(s) = \frac{120s}{s^2 + (377)^2}$$

    c. From entry 2 in Table 5.1,

$$F(s) = \frac{27}{s^2}$$

    d. From entry 3 in Table 5.1,

$$F(s) = \frac{5}{s + 2.5}$$

    e. From entry 7 in Table 5.1,

$$F(s) = \frac{89,000}{(s + 2)^2 + (1000)^2}$$

Example 5.3

Use Table 5.1 to find the inverse Laplace transform of each of the following functions.

    a. $F(s) = \dfrac{27.5}{s}$

    b. $F(s) = \dfrac{8}{s + 5}$

    c. $F(s) = \dfrac{19.6s}{s^2 + 2500}$

    d. $F(s) = \dfrac{17.4}{2s^2 + 32s + 128}$

e. $F(s) = \dfrac{350}{(s + 2)^2 + 10{,}000}$

Solution

a. From entry 1 in Table 5.1,

$f(t) = 27.5$

b. From entry 3 in Table 5.1,

$f(t) = 8e^{-5t}$

c. $F(s) = \dfrac{19.6s}{s^2 + (50)^2}$

From entry 6 in Table 5.1,

$f(t) = 19.6 \cos 50t$

d. $F(s) = \dfrac{8.7}{s^2 + 16s + 64} = \dfrac{8.7}{(s + 8)^2}$

From entry 4 in Table 5.1,

$f(t) = 8.7te^{-8t}$

e. $F(s) = \dfrac{(3.5)(100)}{(s + 2)^2 + (100)^2}$

From entry 7 in Table 5.1,

$f(t) = 3.5e^{-2t} \sin 100t$

Operational Laplace Transforms

Operational transforms provide the answer to questions such as: What is the Laplace transform of the sum of two or more functions when you know the transform of each function acting alone? What is the Laplace transform of the derivative or integral of a function? Table 5.2 lists the transform pairs for the most common operational transforms.

In Table 5.2, $f(t)$ can be any function of t, and $F(s)$ is the Laplace transform of $f(t)$. For example, if $f(t) = t$, then $F(s) = 1/s^2$. Entry 12 is of particular importance. It states that the Laplace transform of the sum of two or more functions is the sum of the transforms of each term taken alone.

Entries 13, 14, and 15 appear more complicated than they are in actual practice. The term $f(0)$, which appears in entries 13, 14, and 15, is the value of the function $f(t)$ when $t = 0$. In a mechanical system where $f(t)$ is the position of an object, $f(0)$ is the position at $t = 0$ (i.e., the initial position). The term $df(0)/dt$, which appears in entries 14 and 15, is the value of the derivative of $f(t)$ when $t = 0$. In the mechanical system, this would be the initial velocity of the object. In entry 15, the term $d^2 f(0)/dt^2$ is the initial acceleration of the object. In control systems, *the usual practice is to assume that all initial conditions are zero*. The transfer function is defined as the *s*-domain ratio of the output over the input with all initial conditions set to zero. When

Table 5.2   Operational Laplace
Transforms*

Time Domain, $f(t), t > 0$	Frequency Domain, $F(s)$
11. $Kf(t)$	$KF(s)$
12. $f_1(t) + f_2(t) - f_3(t)$	$F_1(s) + F_2(s) - F_3(s)$
13. $\dfrac{df(t)}{dt}$	$sF(s) - f(0)$
14. $\dfrac{d^2 f(t)}{dt^2}$	$s^2 F(s) - sf(0) - \dfrac{df(0)}{dt}$
15. $\dfrac{d^3 f(t)}{dt^3}$	$s^3 F(s) - s^2 f(0) - s\dfrac{df(0)}{dt} - \dfrac{d^2 f(0)}{dt^2}$
16. $\int f(t)\,dt$	$\dfrac{F(s)}{s}$
17. $f(t - a)$	$e^{-as}F(s)$

* K represents any numerical constant.

using Laplace transforms to get the transfer function of a component, we *use only the first term in entries 13, 14, and 15.*

Example 5.4

Use Tables 5.1 and 5.2 to obtain the Laplace transform of the following function:

$$f(t) = 12.3 + t + 5e^{-4t} + te^{-2t}$$

Solution

By entry 12, we know the Laplace transform of $f(t)$ will be the sum of the Laplace transforms of the four terms. Use entry 1 for the first term, entry 2 for the second term, entry 3 for the third term, and entry 4 for the fourth term.

$$F(s) = \frac{12.3}{s} + \frac{1}{s^2} + \frac{5}{s + 4} + \frac{1}{(s + 2)^2}$$

Example 5.5

Find the Laplace transform of the following function if all initial conditions are zero.

$$f(t) = 8\frac{d^2 x(t)}{dt^2} + 12\frac{dx(t)}{dt} + 7x(t)$$

Solution

First we note that the Laplace transform of $x(t)$ is $X(s)$. By entry 14, the Laplace transform of the first term is

$$\mathscr{L}\left\{8\frac{d^2 x(t)}{dt^2}\right\} = 8s^2 X(s)$$

By entry 13, the Laplace transform of the second term is

$$\mathcal{L}\left\{12\frac{dx(t)}{dt}\right\} = 12sX(s)$$

By entry 11, the Laplace transform of the third term is

$$\mathcal{L}\{7x(t)\} = 7X(s)$$

By entry 12, the Laplace transform of $f(t)$ is

$$F(s) = 8s^2X(s) + 12sX(s) + 7X(s)$$

Example 5.6

Repeat the problem in Example 5.5 with the following initial conditions:

$$x(0) = 6 \qquad \frac{dx(0)}{dt} = -4$$

Solution

By entry 14, the Laplace transform of the first term is

$$\mathcal{L}\left\{8\frac{d^2x(t)}{dt^2}\right\} = 8[s^2X(s) - s(6) - (-4)]$$
$$= 8s^2X(s) - 48s + 32$$

By entry 13, the Laplace transform of the second term is

$$\mathcal{L}\left\{12\frac{dx(t)}{dt}\right\} = 12sX(s) - 72$$

The Laplace transform of the third term is unchanged.

$$\mathcal{L}\{7x(t)\} = 7X(s)$$

By entry 12, the Laplace transform of $f(t)$ is

$$F(s) = 8s^2X(s) - 48s + 32 + 12sX(s) - 72 + 7sX(s)$$
$$F(s) = 8s^2X(s) + 12sX(s) + 7X(s) - 48s - 40$$

Example 5.7

Find the Laplace transform of the following function if all initial conditions are zero.

$$f(t) = 5.5\frac{dx(t)}{dt} + 24.2x(t) + 4.8\int x(t)\,dt$$

Solution

By entry 13, the Laplace transform of the first term is

$$\mathcal{L}\left\{5.5\frac{dx(t)}{dt^2}\right\} = 5.5sX(s)$$

By entry 11, the Laplace transform of the second term is

$$\mathcal{L}\{24.2x(t)\} = 24.2X(s)$$

By entry 16, the Laplace transform of the third term is

$$\mathcal{L}\{4.8 \int x(t) \, dt\} = \frac{4.8X(s)}{s}$$

By entry 12, the Laplace transform of the $f(t)$ is

$$F(s) = 5.5sX(s) + 24.2X(s) + \frac{4.8X(s)}{s}$$

Example 5.8

Entry 17 in Table 5.2 is used to obtain the Laplace transform of the dead time delay element. If $f_i(t)$ describes the input to a component that has a dead time delay of 5 seconds, then $f_o(t) = f_i(t - 5)$ will describe the output of that component. Entry 17 states that the Laplace transform of the output of a dead time element, $F_o(s)$ is equal to the Laplace transform of the input, $F_i(t)$, multiplied by the factor e^{-as}. In entry 17, the letter a represents the amount of dead time delay.

   Find the Laplace transform of the output of dead time delay elements with inputs and outputs as follows:

   a. $f_i(t) = 4t$ and $f_o(t) = 4(t - 6) = f_i(t - 6)$
   b. $f_i(t) = 6e^{-4t}$ and $f_o(t) = 6e^{-4(t-5)} = f_i(t - 5)$

Solution

   a. By entry 2, the Laplace transform of $f_i(t) = 4t$ is

$$F_i(s) = \frac{4}{s^2}$$

   By entry 17, the Laplace transform of $f_o(t) = f_i(t - 6)$ is

$$F_o(t) = e^{-6s}\left(\frac{4}{s^2}\right)$$

   b. By entry 3, the Laplace transform of $f_i(t) = 6e^{-4t}$ is

$$F_i(s) = \frac{6}{s + 4}$$

   By entry 17, the Laplace transform of $f_o(t) = f_i(t - 5)$ is

$$F_o(t) = e^{-5s}\left(\frac{6}{s + 4}\right)$$

5.5 INVERSE LAPLACE TRANSFORMS

The *inverse Laplace transform* converts a frequency-domain function of (s) into a time-domain function of (t). If the frequency-domain function is in the table of Laplace transform pairs, the inverse transform is the mating time-domain function in the table. This is how the inverse transform was obtained in the example illustrated in Figure 5.8. The frequency-domain equation in Figure 5.8 is given below, with (s) appended to H for clarity.

$$H(s) = \frac{GK}{s(\tau s + 1)} \tag{5.18}$$

The right-hand side of Equation (5.18) is essentially the same as the frequency-domain function of entry 9. The only difference is the arbitrary constant in the numerator, which is K in entry 9 and GK in Equation (5.18). Since K and GK both represent numeric values, we can simply replace K in the time-domain function of entry 9 by GK and we have the inverse transform of the right-hand side of Equation (5.18).

$$h(t) = GK(1 - e^{-t/\tau}) \tag{5.19}$$

Of course, the inverse Laplace transform is usually more difficult than a simple table conversion. The frequency-domain functions that occur in control system analysis are usually a ratio of two polynomials which have only integer powers of s. The right-hand side of Equation (5.20) is an example of such a function, with both polynomials in factored form.

$$X(s) = \frac{8(s + 3)(s + 8)}{s(s + 2)(s + 4)} \tag{5.20}$$

If we can break the right-hand side of Equation (5.20) into a sum of terms and each term is in a table of Laplace transform pairs, we can get the inverse transform of the equation. The method for breaking a ratio of polynomials into a sum of terms is called a *partial fraction expansion*. For Equation (5.20), the partial fraction expansion takes the following form, where K_1, K_2, and K_3 are undetermined numerical constants. Notice that each term on the right-hand side of Equation (5.21) is in Table 5.1, so we can determine the inverse transform. All that remains is to determine the values of K_1, K_2, and K_3.

$$X(s) = \frac{8(s + 3)(s + 8)}{s(s + 2)(s + 4)} = \frac{K_1}{s} + \frac{K_2}{s + 2} + \frac{K_3}{s + 4} \tag{5.21}$$

In general, there will be a term on the right-hand side for each root of the polynomial in the denominator of the left-hand side. Multiple roots for factors such as $(s + 2)^n$ will have a term for each power of the factor from 1 to n. Equation (5.22) illustrates the expansion of a factor with a multiple root of multiplicity 2.

$$Y(s) = \frac{8(s + 1)}{(s + 2)^2} = \frac{K_1}{s + 2} + \frac{K_2}{(s + 2)^2} \tag{5.22}$$

Complex roots are common, and they always occur in conjugate pairs. The two constants in the numerator of the complex conjugate terms are also complex conjugates. Equation (5.23) illustrates the expansion of a polynomial with complex-conjugate roots. In Equation (5.23), j represents the square root of -1, K is a complex constant, and K^* is the complex conjugate of K.

$$Z(s) = \frac{5.2}{s^2 + 2s + 5} = \frac{K}{(s + 1 - j2)} + \frac{K^*}{(s + 1 + j2)} \tag{5.23}$$

A two-step procedure is used to find each distinct (nonmultiple) root, real or complex. Assume that you wish to evaluate the constant in the second term on the right-hand side of Equation (5.21). The first step in evaluating the constant (K_2) is to multiply both sides of the equation by the factor in the denominator of the second term, $(s + 2)$. The second step is to replace s on both sides of the equation by the root of the factor by which you multiplied in step 1 [i.e., replace s by -2, the root of the term $(s + 2)$]. All other terms on the right-hand side will be zero, leaving just the single constant, K_2. The left-hand side reduces to a numerical value, the value of K_2. This process is repeated for each distinct term on the right-hand side. Example 5.9 completes the partial expansion and inverse transformation of Equation (5.21). The two-step procedure works for distinct complex roots, but the reduction of the left-hand side involves complex numbers, and the result is a complex number. Example 5.11 completes the partial expansion and inverse transformation of Equation (5.23).

The process is more involved for multiple roots, and we will limit the discussion to multiple roots with a multiplicity of 2. Example 5.9 completes the expansion and inverse transformation of Equation (5.22).

Example 5.9

Complete the partial fraction expansion and inverse Laplace transformation of Equation (5.21).

$$X(s) = \frac{8(s + 3)(s + 8)}{s(s + 2)(s + 4)} = \frac{K_1}{s} + \frac{K_2}{s + 2} + \frac{K_3}{s + 4} \tag{5.21}$$

Solution

1. $K_1 = \dfrac{8(s + 3)(s + 8)}{(s + 2)(s + 4)}\bigg|_{s=0} = \dfrac{8(0 + 3)(0 + 8)}{(0 + 2)(0 + 4)} = 24$

2. $K_2 = \dfrac{8(s + 3)(s + 8)}{s(s + 4)}\bigg|_{s=-2} = \dfrac{8(-2 + 3)(-2 + 8)}{-2(-2 + 4)} = -12$

3. $K_3 = \dfrac{8(s + 3)(s + 8)}{s(s + 2)}\bigg|_{s=-4} = \dfrac{8(-4 + 3)(-4 + 8)}{-4(-4 + 2)} = -4$

The partial fraction expansion of Equation (5.21) is

$$X(s) = \frac{24}{s} - \frac{12}{s + 2} - \frac{4}{s + 4}$$

The inverse Laplace transformation is easily seen from Table 5.1 to be

$$x(t) = 24 - 12e^{-2t} - 4e^{-4t}$$

Example 5.10

Complete the partial fraction expansion and inverse Laplace transformation of Equation (5.22).

$$Y(s) = \frac{8(s + 1)}{(s + 2)^2} = \frac{K_1}{s + 2} + \frac{K_2}{(s + 2)^2} \tag{5.22}$$

Solution

Evaluate K_2 as if it were a distinct root, and then add a third step to evaluate K_1.

Step 1. Multiply both sides of Equation (5.22) by the factor $(s + 2)^2$. Equation (5.24) is the result.

$$8(s + 1) = K_1(s + 2) + K_2 \tag{5.24}$$

Step 2. Replace s by the root of $(s + 2)$ (i.e., -2) and solve for K_2.

$$8(-2 + 1) = K_1(-2 + 2) + K_2$$
$$K_2 = -8$$

Step 3. Substitute K_2 into Equation (5.24) and solve for K_1. You can solve for K_1 algebraically, or you can replace s by any value except the multiple root (-2) and solve for K_1 arithmetically.

Algebraic Solution
$$8(s + 1) = K_1(s + 2) - 8$$
$$K_1 = \frac{8s + 16}{s + 2}$$
$$= 8$$

Arithmetic Solution
$$\text{let } s = 0$$
$$8(0 + 1) = K_1(0 + 2) - 8$$
$$2K_1 = 16$$
$$K_1 = 8$$

The partial expansion of Equation (5.22) is

$$Y(s) = \frac{8}{s + 2} - \frac{8}{(s + 2)^2}$$

The inverse Laplace transformation is obtained from entries 3 and 4 in Table 5.1, and entry 12 in Table 5.2.

$$y(t) = 8e^{-2t} - 8te^{-2t}$$

Example 5.11

Complete the partial fraction expansion and inverse Laplace transformation of Equation (5.23).

$$Z(s) = \frac{5.2}{s^2 + 2s + 5} = \frac{K}{s + 1 - j2} + \frac{K^*}{s + 1 + j2} \qquad (5.23)$$

Solution

Since K and K^* are complex conjugates, we will evaluate K and set K^* equal to the conjugate of K.

$$\text{Note: } s^2 + 2s + 5 = (s + 1 - j2)(s + 1 + j2)$$

Step 1:

$$\frac{5.2}{s^2 + 2s + 5}(s + 1 - j2) = \frac{5.2}{s + 1 + j2}$$

Step 2:

$$K = \frac{5.2}{s + 1 + j2}\bigg|_{s = -1 + j2} = \frac{5.2}{-1 + j2 + 1 + j2}$$

$$= \frac{5.2}{j4} = -j1.3 = 1.3\underline{/-90°}$$

$$K^* = 1.3\underline{/90°}$$

The partial fraction expansion is

$$Z(s) = \frac{1.3\underline{/-90°}}{s + 1 - j2} + \frac{1.3\underline{/90°}}{s + 1 + j2}$$

The inverse Laplace transformation is

$$z(t) = 2(1.3)e^{-t}\cos(2t - 90°)$$
$$= 2.6e^{-t}\cos(2t - 90°)$$

5.6   TRANSFER FUNCTION

A major use of the Laplace transformation in the study of control systems is to obtain the transfer function of a component. The *transfer function* is obtained by solving the frequency-domain algebraic equation for the ratio of the output signal over the input signal with all initial conditions set to zero. As an example, consider the liquid system described by Equation (5.6):

$$\tau \frac{dh}{dt} + h = Gq_{in} \qquad (5.6)$$

The frequency-domain algebraic equation is obtained by applying the Laplace transformation to each term in Equation (5.6) and solving for the ratio H/Q_{in}. *Note:* It is understood that h is a function of t and H is a function of s even though the (t) and (s) are not used. We will often drop the (t) and (s) for convenience in writing long expressions.

$$\tau s H + H = GQ_{in}$$

$$H(\tau s + 1) = GQ_{in}$$

$$H = \frac{G}{\tau s + 1} Q_{in}$$

$$\frac{H}{Q_{in}} = \frac{G}{1 + \tau s} = \text{transfer function} \tag{5.25}$$

The transfer function is defined by Equation (5.25). Notice that the output (H) can be obtained by multiplying the input (Q_{in}) by the transfer function (H/Q_{in}).

> The transfer function of a component is the ratio of the frequency-domain output over the input, with all initial conditions set to zero.
>
> The frequency-domain output of a component is equal to the product of the input times the transfer function.

Example 5.12

Determine the transfer function of the nonregulating liquid tank shown in Figure 5.2. The time-domain equation is as follows:

$$\Delta h(t) = \frac{1}{A} \int_{t_0}^{t_1} q(t)\, dt \tag{5.7}$$

Solution

Use entry 16 to transform the right-hand side of Equation (5.7). The term $\Delta h(t)$ will be considered to be the output variable of the component and will be transformed to $\Delta H(s)$.

$$\Delta H(s) = \frac{1}{As} Q(s)$$

$$\frac{\Delta H(s)}{Q(s)} = \frac{1}{As} = \text{transfer function} \tag{5.26}$$

Example 5.13

Determine the transfer function of the electrical circuit shown in Figure 5.3. The time-domain equation is as follows:

$$\tau \frac{de_{out}}{dt} + e_{out} = e_{in} \tag{5.10}$$

Solution

The Laplace transformation of Equation (5.10) is

$$\tau s E_{\text{out}} + E_{\text{out}} = E_{\text{in}}$$

$$\frac{E_{\text{out}}}{E_{\text{in}}} = \frac{1}{1 + \tau s} = \text{transfer function} \qquad (5.27)$$

Example 5.14

Determine the transfer function of the liquid-filled thermometer shown in Figure 5.4. The time-domain equation is as follows:*

$$\tau \frac{dT_m}{dt} + T_m = T_a \qquad (5.11)$$

Solution

The Laplace transformation of Equation (5.11) is

$$\tau s T_m + T_m = T_a$$

$$\frac{T_m}{T_a} = \frac{1}{1 + \tau s} = \text{transfer function} \qquad (5.28)$$

Notice that Equations (5.27) and (5.28) are the same, even though the first equation describes an electrical circuit and the second describes a thermometer.

Example 5.15

Determine the transfer function of the control valve shown in Figure 5.5. The time-domain equation is as follows:

$$P_{\text{in}} A = m \frac{d^2 x}{dt^2} + R_m \frac{dx}{dt} + \frac{1}{C_m} x \qquad (5.17)$$

Solution

The Laplace transformation of Equation (5.17) is

$$P_{\text{in}} A = m s^2 X + R_m s X + \frac{1}{C_m} X$$

$$\frac{X}{P_{\text{in}}} = \frac{A C_m}{1 + R_m C_m s + m C_m s^2}$$

---

* In Equation (5.11), an exception was made to the rule of using lowercase letters for time-domain variables. Since t is used to represent time, T is used to represent temperature in both the time and frequency domains.

5.7 FREQUENCY RESPONSE: BODE PLOTS

A feedback control system consists of several components connected by signal paths that form a closed loop. As the signal passes through each component, it is changed in ways that depend on the characteristics of the component. The effect that each component has on the signal is of major importance in the analysis and design of a control system. Indeed, the design of a controller consists of adding components that change the signal in a manner that provides the best possible control of the process.

Two types of input signals are used to describe the behavior of a component. The first type is a step change in the input to the component. Step changes are used to study the transient response of the component. The second type of input is the sinusoidal signal, which is used to study the steady-state response of the component. The response of a component to a sinusoidal input signal is called the frequency response of the component. A graph of the frequency response of a component is called a Bode diagram.

In Section 5.2, we developed the differential equations of five control system components. All five components are classified as linear components. The analysis of a control system is greatly simplified if all the components are linear. Fortunately, many of the systems we want to control can be modeled by linear components. A linear component has a very interesting and important characteristic: a sinusoidal input signal to a linear component will always produce a sinusoidal output signal that has the same frequency as the input. Only the amplitude and the phase angle of the signal will be changed by the component. Figure 5.9 shows typical input and output signals for a linear component.

Gain and Phase Angle

The change in amplitude of a sinusoidal signal is expressed by the ratio of the output amplitude divided by the input amplitude. This ratio is a dimensionless number called the *gain* of the component. Gain is often expressed in decibel units.

Figure 5.9 A linear component can change the amplitude and the phase of a sinusoidal signal, but the frequency of the output is the same as the input. In the diagram, the gain is A_h/A_q, and the phase angle is $-\theta$ (minus because the output lags the input).

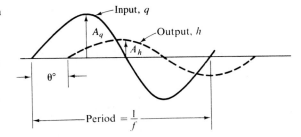

For $s = j\omega = j2\pi f$, the transfer function has a magnitude (m) and an angle $\theta°$ such that

$$m = \frac{A_h}{A_q} = \text{amplitude ratio (change in size)}$$

$$\theta° = \text{phase difference (change in timing)}$$

$$\text{Gain} = \frac{\text{output amplitude}}{\text{input amplitude}}$$

$$\text{Decibel gain} = 20 \log_{10}(\text{gain}) \qquad (\text{dB})$$

The change in timing of a sinusoidal signal is expressed by the number of degrees by which the phase of the output signal differs from the phase of the input signal. The difference is called the *phase angle* or *phase difference* of the component. The phase angle is usually measured in degrees. It is positive when the output leads the input, and negative when the output lags the input.

$$\text{Phase angle} = \text{output phase} - \text{input phase}$$

Bode Diagram

The *frequency response* of a component is the set of values of *gain* and *phase angle* that occur when a sinusoidal input signal is varied over a range of frequencies. For each frequency, there is a gain and a phase angle that give the characteristic response of the component *at that frequency*. The frequency response is plotted in a pair of graphs known as a *Bode diagram*. The two graphs share a common frequency scale on the *x*-axis. The *y*-axis of one Bode graph is the decibel gain; the other is the phase angle. Figure 5.10 shows a typical Bode diagram.

Determination of the Frequency Response

There are three ways to determine the frequency response of a component: direct measurement, hand computation, and computer computation. In the *direct measurement method*, a sinusoidal input signal is applied to the input of the component. The input and output signals are measured and the gain and phase angle are determined from the measurements. The frequency is then changed to a new value for another determination of gain and phase angle. The process is repeated until enough data are obtained to construct a Bode diagram.

In the *computation methods*, the frequency response is obtained from the transfer function by replacing *s* by *jω*. The *j* represents the square root of -1, and *ω* is the radian frequency of the sinusoidal input signal. For each value of *ω*, the transfer function reduces to a single complex number. The magnitude of this complex number is the gain of the component at that frequency. The angle of the complex number is the phase angle of the component at that frequency.

Complex numbers can be represented by a point on a two-dimensional plane called the *complex plane*. The *x*-axis is called the *real axis*, and the *y*-axis is called the *imaginary axis*. The graph of a complex number is shown in Figure 5.11. The complex number may be defined in rectangular form by its real coordinate (*a*) and its imaginary coordinate (*b*) (i.e., $N = a + jb$, where *j* identifies the imaginary coordinate). In this form, *a* is called the *real part* of the complex number, and *b* is called the *imaginary part*.

The complex number may also be defined in polar form by its distance from the origin (*m*) and the angle (*θ*). In the polar form, *m* is called the *magnitude* of the complex number, and *θ* is called the *angle* of the complex number (i.e., $N = m\underline{/\theta}$). A complex

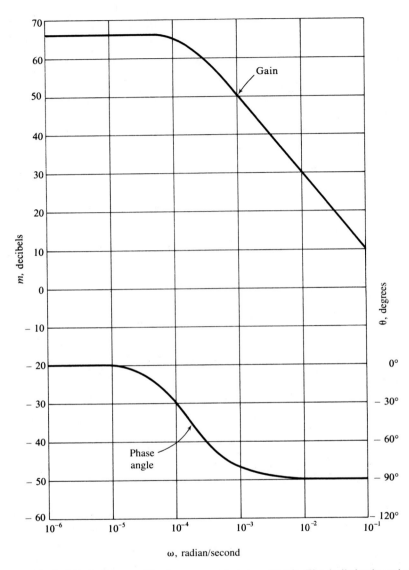

Figure 5.10 Typical Bode diagram. The frequency scale is logarithmic. The decibel gain and phase angle scales are linear.

number may be converted from one form to the other by using the appropriate conversion equation:

Rectangular-to-Polar Conversion   *Polar-to-Rectangular Conversion*

$$m = \sqrt{a^2 + b^2}$$ $$a = m \cos \theta$$

$$\theta = \arctan(b/a)$$ $$b = m \sin \theta$$

Figure 5.11 Graph of the complex number N. The number N can be represented by its rectangular coordinates as $a + jb$, or by its polar coordinates as $m\underline{/\theta}$.

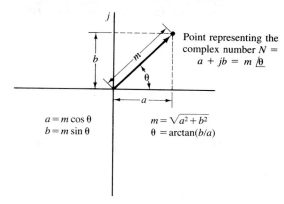

Point representing the complex number $N = a + jb = m\underline{/\theta}$

$a = m \cos \theta$
$b = m \sin \theta$

$m = \sqrt{a^2 + b^2}$
$\theta = \arctan(b/a)$

The frequency response of a component is obtained by substituting $j\omega$ for s in the transfer function. With s replaced by $j\omega$, the transfer function reduces to a complex number which can be converted to the polar form. The magnitude of the complex number is the gain of the component at the frequency ω. The angle of the complex number is the phase angle of the component at the frequency ω.

Example 5.16

A self-regulating liquid tank is described in the time domain by Equation (5.6) and in the frequency domain by its transfer function given by Equation (5.25). The input flow rate (q_{in}) varies sinusoidally about an average value with an amplitude of 0.0002 m³/s and a frequency (f) of 0.0001592 Hz. The time constant (τ) is 1590 s and the gain (G) is 2000 s/m². Determine the amplitude and phase of the output (h).

Solution

1. Determine the value of $j\omega$.

$$j\omega = j2\pi f = j2\pi 0.0001592$$
$$j\omega = j0.001 \text{ rad/s}$$

2. Replace s in the transfer function by $j\omega$ and reduce it to a complex number in polar form.

$$\frac{H}{Q_{in}} = \frac{G}{1 + \tau s} = \frac{2000}{1 + 1590(j0.001)}$$

$$= \frac{2000}{1 + j1.59} = \frac{2000\underline{/0°}}{1.879\underline{/57.8°}} = 1065\underline{/-57.8°}$$

Gain in decibels $= 20 \log 1065 = 60.5$

The output amplitude is equal to the input amplitude times the magnitude of the transfer function.

$$\text{Output amplitude } h = 1065 \times 0.0002 = 0.213 \text{ m}$$

The phase difference is $-45°$, so the output (h) lags the input (q_{in}) by $57.8°$.

Computer-Generated Frequency Response Data

The hand computation of frequency response data can become quite tedious. The large number of computations make this an ideal application for a computer. Program "BODE" (see listing in Appendix F) is a GW BASICA program that will put a Bode diagram on the screen for any transfer function that fits the following general form:

$$\text{Transfer function} = \frac{a_0 + a_1 s + a_2 s^2 + a_3 s^3}{b_0 + b_1 s + b_2 s^2 + b_3 s^3}$$

Any of the coefficients a_0, a_1, a_2, a_3, b_0, b_1, b_2, or b_3 can have a value of 0 in a particular transfer function.

The program generates frequency response data over a range of frequencies from 1.0×10^{-6} to 5.6×10^5 rad/s. The following frequency bases are used for each power of 10: 1.0, 1.8, 3.2, and 5.6. These seem like odd numbers, but they happen to be even increments on a logarithmic scale. The Bode diagram uses a logarithmic scale for frequency, and 3.2 is just about the midpoint between 1.0 and 10.0. Also, 1.8 is just about the midpoint between 1.0 and 3.2; and 5.6 is about the midpoint between 3.2 and 10.0. This makes it easy to plot Bode diagrams on linear graph paper.

The program runs on PC compatibles with color monitors. A zoom command allows the user to expand a portion of the gain, angle, or frequency range to fill the entire graph. An unzoom command returns the original scale. The program also has an option to make a printed copy of the Bode data table.

Example 5.17

Use program "BODE" to generate a table of frequency response data for the self-regulating liquid tank described in Example 5.16. Program "BODE" is available on disk (see Preface), and the program listing is in Appendix F. The transfer function is as follows:

$$\frac{H}{Q_{in}} = \frac{2000}{1 + 1590s}$$

Solution

The input and a portion of the data table produced by the program are shown below.

TRANSFER FUNCTION COEFFICIENTS

$$TF = \frac{A(0) + A(1)S + A(2)S^2 + A(3)S^3}{B(0) + B(1)S + B(2)S^2 + B(3)S^3}$$

A(0) = 2000, A(1) = 0, A(2) = 0, A(3) = 0

B(0) = 1, B(1) = 1590, B(2) = 0, B(3) = 0

TRANSFER FUNCTION COEFFICIENTS

A(0) . . A(3):	2.00E+03	0.00E+00	0.00E+00	0.00E+00
B(0) . . B(3):	1.00E+00	1.59E+03	0.00E+00	0.00E+00

BODE DATA TABLE

Frequency, W (radian/second)	Gain (decibel)	Phase (degrees)
1.0E-06	66.0	-0.1
1.8E-06	66.0	-0.2
3.2E-06	66.0	-0.3
5.6E-06	66.0	-0.5
1.0E-05	66.0	-0.9
1.8E-05	66.0	-1.6
3.2E-05	66.0	-2.9
5.6E-05	66.0	-5.1
1.0E-04	65.9	-9.0
1.8E-04	65.7	-15.8
3.2E-04	65.0	-26.7
5.6E-04	63.5	-41.8
1.0E-03	60.5	-57.8
1.8E-03	56.5	-70.5
3.2E-03	51.8	-78.8
5.6E-03	46.9	-83.6
1.0E-02	42.0	-86.4
1.8E-02	37.0	-88.0
3.2E-02	32.0	-88.9
5.6E-02	27.0	-89.4
1.0E-01	22.0	-89.6
1.8E-01	17.0	-89.8
3.2E-01	12.0	-89.9
5.6E-01	7.0	-89.9

GLOSSARY

Bode diagram: A pair of graphs that share a common log frequency scale on the x-axis. The y-axis of one graph is the decibel gain of a component; the other is its phase angle. See Frequency response. (5.7)

Derivative: A mathematical expression of the rate of change of a variable. (5.2)

Differential equation: An equation that has one or more derivative terms. (5.2)

Frequency domain: A domain in which the equation that describes a component is a function of a frequency parameter (s). The equation defines the amplitude and phase angle of the output of the component as a function of the frequency, gain, and phase of a sinusoidal input signal. (5.4)

Frequency response: The set of values of the input-to-output gain and phase angle of a component when a sinusoidal input signal is varied over a range of frequencies. See Bode diagram. (5.7)

Functional Laplace transform: The Laplace transform of a time function such as $\sin \omega t$. (5.4)

Integral: A mathematical expression for the accumulation of an amount (or quantity) of a variable. (5.2)

Integro-differential equation: An equation that has at least one derivative and one integral term. (5.2)

Inverse Laplace transform: A mathematical transformation that converts the solution of a differential equation in the frequency domain into a solution in the time domain. (5.5)

Laplace transform: A mathematical transformation that converts equations that involve derivatives and integrals in the time domain into equations that involve only algebraic terms in the frequency domain. (5.4)

Linear differential equation: A differential equation that has only first degree terms (power = 1) in the dependent variable and its derivatives. (5.2)

Operational Laplace transform: The Laplace transform of a time domain operation (e.g., addition, differentiation, integration, etc.). (5.4)

Order of a differential equation: The order of the highest derivative that appears in the equation. (5.2)

Partial fraction expansion: The result of breaking a ratio of polynomials into a sum of terms so we can determine its inverse Laplace transform. (5.5)

Time constant (τ): A parameter that characterizes a group of components that are modeled by a linear, first-order differential equation with constant coefficients. The unit of the time constant is seconds, which explains the name "time" constant. (5.2)

Time domain: A domain in which the equation that describes a component is a function of time. The equation defines the size of the output as a function of time and the input signal. (5.4)

Transfer function: The frequency-domain ratio of the output of a component over its input, with all initial conditions set to zero. (5.6)

EXERCISES

5.1 The self-regulating liquid tank shown in Figure 5.1 has the following parameter values.

$$R_L = 2.16 \times 10^6 \text{ Pa·s/m}^3$$
$$C_L = 2.68 \times 10^{-4} \text{ m}^3/\text{Pa}$$
$$\rho = 1000 \text{ kg/m}^3 \text{ (water)}$$
$$g = 9.81 \text{ m/s}^2$$

Equation (5.6) is the differential equation that describes the relationship between the input flow rate (q_{in}) and the output level (h). Determine the numerical value of the coefficients of dh/dt, h, and q_{in}, and write Equation (5.6) using the numerical values.

5.2 Repeat Exercise 5.1 for the nonregulating liquid tank shown in Figure 5.2 and described by Equation (5.7). The tank has a diameter of 1.5 m and a height of 3 m. Write Equation (5.7) using a numerical value for A.

5.3 Repeat Exercise 5.1 for the electrical component shown in Figure 5.3 and described by Equation (5.10). Write Equation (5.10) with a numerical value for τ for the following parameter values.

$$R_L = 22 \text{ k}\Omega$$
$$C_L = 4.7 \text{ }\mu\text{F}$$

5.4 Repeat Exercise 5.1 for the liquid-filled thermometer shown in Figure 5.4 and described by Equation (5.11). Write Equation (5.11) with a numerical value for τ for the following parameter values.

Liquid in the bulb: mercury

Length of the bulb = 4.52 cm (outside)

Diameter of the bulb = 1.94 cm (outside)

Bulb material: glass

Thickness of the glass: 1.4 mm

Inside film coefficient: $h_i = 4000 \text{ W/m}^2\text{·K}$

Outside film: natural convection in still water

$T_w = 85°C$

$T_d = 20°C$

Neglect the stem of the thermometer and assume the bulb is a cylinder completely immersed in the liquid whose temperature is to be measured.

5.5 Repeat Exercise 5.1 for the control valve shown in Figure 5.5 and described by Equation (5.17). The diaphragm area is 0.0182 m². The mass of the moving parts (valve stem and diaphragm plate) is 0.561 kg. The valve stem position is 0 cm with an applied force of 375 N, and 2.5 cm with an applied force of 1885 N. The mechanical resistance produces a force of 12 N when the valve stem velocity is 1.25 m/s. Write Equation (5.17) with numerical values for A, m, R_m, and C_m.

5.6 Determine the frequency-domain function, $F(s)$, for each of the following time-domain functions, $f(t)$.

a. $f(t) = 7.8$ **b.** $f(t) = 3.2 \cos 1000t$

c. $f(t) = 120 \sin 25t$ **d.** $f(t) = 18t$

e. $f(t) = 16e^{-8t}$

f. $f(t) = 9e^{-3t} \sin 100t$

g. $f(t) = 8.2te^{-2.5t}$

h. $f(t) = 5e^{-7t} \cos 50t$

i. $f(t) = 45e^{-5(t-6)}$

j. $f(t) = 2 \sin(t - 6)$

k. $f(t) = 4.8e^{-5t} \cos(400t - 36°)$

l. $f(t) = 8\dfrac{d^2x}{dt^2} + 5\dfrac{dx}{dt}$, where $\dfrac{dx(0)}{dt} = 8$, $x(0) = -4$

m. $f(t) = 12 \displaystyle\int x \, dt + 17x$

5.7 Determine the time-domain function, $f(t)$, for each of the following frequency-domain functions, $F(s)$.

a. $F(s) = \dfrac{6.7}{s^2}$

b. $F(s) = \dfrac{25\omega}{s^2 + \omega^2}$

c. $F(s) = \dfrac{45}{s + 72}$

d. $F(s) = 345$

e. $F(s) = \dfrac{650}{(s + 8)^2}$

f. $F(s) = \dfrac{250\omega}{(s + 4)^2 + \omega^2}$

g. $F(s) = \dfrac{82}{s(5s + 1)}$

h. $F(s) = \dfrac{16(s + 5)}{(s + 5)^2 + \omega^2}$

i. $F(s) = \dfrac{28s}{s^2 + \omega^2}$

j. $F(s) = \dfrac{64/48°}{s + 8 - j16} + \dfrac{64/-48°}{s + 8 + j16}$

5.8 Complete the partial fraction expansion and find the inverse Laplace transformation of each of the following functions.

a. $\dfrac{4(s + 5)(s + 7)}{s(s + 3)(s + 6)}$

b. $\dfrac{2(s + 5)}{(s + 1)^2}$

c. $\dfrac{s + 2}{s^2 + 2s + 4}$

5.9 The dead-time process shown in Figure 4.9 is described by the following equation:

$$f_o(t) = f_i(t - t_d)$$

where f_o = output signal, kilogram/second

f_i = input signal, kilogram/second

t_d = dead-time lag, second

Determine the transfer function, $F_o(s)/F_i(s)$, from the equation above if the dead time, t_d, is 245 s.

5.10 Determine the transfer function, $I(s)/\Theta(s)$, for the temperature transmitter described by the following differential equation.

$$8.6\frac{di}{dt} + i = 5.7\theta$$

where i = output current signal, milliamperes

θ = input temperature signal, °Celsius

5.11 Determine the transfer function, $X(s)/I(s)$ for a process-control valve/electropneumatic converter described by the following differential equation.

$$0.0001 \frac{d^2x}{dt^2} + 0.02 \frac{dx}{dt} + x = 0.3i$$

where x = valve stem position, inch

i = current input signal to the converter, milliampere

5.12 Determine the transfer function, $\Theta(s)/X(s)$, for a tubular heat exchanger similar to the one shown in Figure 2.3 and described by the following differential equation.

$$25 \frac{d^2\theta}{dt^2} + 26 \frac{d\theta}{dt} + \theta = 125x$$

where x = valve position, inch

θ = temperature of the fluid leaving the heat exchanger, °Celsius

5.13 A manufacturing plant uses a liquid surge tank to feed a positive-displacement pump. The pump supplies a constant flow rate of liquid to a continuous heat exchanger. Determine the transfer function, $H(s)/Q(s)$, if the surge tank is described by the following equation:

$$h(t) = 0.5 \int q(t)\, dt$$

where $h(t)$ = level of liquid in the surge tank, meter

$q(t)$ = difference between the input flow rate and the output flow rate, cubic meter/second

t = time, second

5.14 The spring–mass–damping system shown in Figure 4.8 is described by the following differential equation.

$$m \frac{d^2x}{dt^2} + R \frac{dx}{dt} + Kx = f$$

where m = mass, kilogram

R = dashpot resistance, newton/(meter/second)

K = spring constant, newton/meter

x = position of the mass, meter

f = external force applied to the mass, newton

Determine the transfer function, $X(s)/F(s)$, if

$$m = 3.2 \text{ kg}$$
$$R = 2.0 \text{ N/(m/s)}$$
$$K = 800 \text{ N/m}$$

5.15 A PID controller is described by the following equation:

$$i = Ke + kT_d \frac{de}{dt} + \frac{K}{T_i} \int e\, dt$$

where i = current output of the controller, milliampere

e = error input signal, volt

K = gain setting of the controller

T_d = derivative action setting of the controller

T_i = integral action setting of the controller

Determine the transfer function, $I(s)/E(s)$, if K is 3.6, T_d is 0.008, and T_i is 2.2.

5.16 An armature-controlled dc motor is sometimes used in speed and position control systems (see Figures 2.6 and 2.15). The dc motor operation is described by the following equations:

$$e = Ri + L\frac{di}{dt} + K_e\omega$$

$$i = \frac{q}{K_t}$$

$$q = J\frac{d\omega}{dt} + b\omega$$

where e = armature voltage, volt

i = armature current, ampere

ω = motor speed, radian/second

q = motor torque, newton meter

J = moment of inertia of the load, kilogram meter^2

b = damping resistance of the load, newton·meter/(radian/second)

R = armature resistance, ohm

L = armature inductance, henry

K_e = back emf constant of the motor, volt/(radian/second)

K_t = torque constant of the motor, newton meter/ampere

A small permanent-magnet dc motor has the following parameter values:

$$J = 8 \times 10^{-4}\ \text{kg·m}^2$$
$$b = 3 \times 10^{-4}\ \text{N·m/(rad/s)}$$
$$R = 1.2\ \Omega$$
$$L = 0.020\ \text{H}$$
$$K_e = 5 \times 10^{-2}\ \text{V/(rad/s)}$$
$$K_t = 0.043\ \text{N·m/A}$$

Substitute the parameters above into the preceding equations to obtain the exact differential equations of the dc motor. Determine the transfer function, $\Omega(s)/E(s)$, by transforming all three equations into frequency-domain algebraic equations. Use algebraic operations to obtain the ratio of Ω/E, which is the desired transfer function.

5.17 Use the program "BODE" to generate frequency response data from the transfer functions obtained in Exercises 5.10, 5.11, 5.12, 5.14, 5.15, and 5.16. Construct Bode diagrams from the data.

PART TWO

Measurement

CHAPTER 6

Measuring Instrument Characteristics

OBJECTIVES

A measuring instrument has many important characteristics. The two characteristics that have the greatest effect on the performance of a control system are accuracy and speed of response (or frequency response). These two characteristics are not completely independent. If the measured variable is changing, the accuracy of the measurement depends on how quickly the measuring instrument can follow these changes. The speed of response of a measuring instrument is related to its frequency response. In Chapter 17 you will see that the frequency response of each component in the control loop is an important factor in the design of the controller.

The purpose of this chapter is to introduce you to the characteristics of measuring instruments and the terminology associated with measurement. After completing this chapter, you will be able to

1. Calculate the mean and standard deviation of a set of measurements
2. Determine measured accuracy and repeatability from a calibration report
3. Determine the combined hysteresis and dead band from a calibration curve
4. Explain the difference between repeatability and reproducibility
5. Explain the difference between static characteristics and dynamic characteristics
6. Draw straight lines on a calibration curve for independent linearity, terminal-based linearity, and zero-based linearity; then determine the maximum error for each line
7. Determine the 95% response time, the time constant, and the 10 to 90% rise time from the step response curve of a critical or overdamped measuring instrument
8. Determine the 10 to 90% rise time, peak percent overshoot, and 2% settling time from the step response of an underdamped measuring instrument
9. Determine the dynamic lag and the dynamic error from a graph of the ramp response of a measuring instrument
10. Use the computer program from Chapter 5 to obtain frequency response data and plot Bode diagrams of single- and dual-capacity sensors
11. Use the terms in the glossary at the end of this chapter in discussions about measuring instruments

6.1 INTRODUCTION

The purpose of a measuring instrument is to obtain the true value of the measured variable. The ideal measuring instrument would do this exactly, but in practice, this ideal is never achieved. There is always some uncertainty in the measurement of a variable. Indeed, there is even some uncertainty in the standards we use to calibrate a measuring instrument. For this reason, we begin the chapter with a brief review of statistics, the subject that deals with uncertainty.

The next three sections divide the characteristics of measuring instruments into three categories: operating characteristics, static characteristics, and dynamic characteristics. *Operating characteristics* include measurement details, operational details, and environmental effects. *Static characteristics* deal with accuracy when the value of the measured variable is constant or changing very slowly. In contrast, *dynamic characteristics* deal with the measurement of a variable whose value is changing rather quickly.

The selection of a measuring instrument for a particular application can be a complex and difficult procedure. In selecting a temperature transmitter, for example, a designer must consider the following possible primary elements: thermocouple, RTD, thermistor, IC sensor, liquid-filled element, vapor-filled element, gas-filled element, or radiation pyrometer. A set of selection criteria are presented in the form of questions that can be asked in the process of selecting a measuring instrument.

The final section is a glossary of terms presented in the chapter. The principal source document for the terminology used in this chapter is the Instrument Society of America standard S51.1, "Process Instrumentation Terminology."

6.2 STATISTICS

There is an uncertainty when we measure the value of a variable. This uncertainty occurs when repeated measurements under identical conditions give different results. For example, let's assume that five measurements of the temperature of a fluid result in the following measured values: 217, 214, 215, 215, and 216°C. Statistics cannot tell us what the true temperature is, but it can help us understand the uncertainty we are confronted with.

The individual measurements of a variable are called *observations*, and the entire collection of observations is called a *sample*. The simplest statistical measure of the sample is the *arithmetic average* or *mean*. The sample mean is an estimate of the expected value of the next observation. The mean is computed by summing the observations and dividing by the number of observations. The mean of a sample of n observations is given by the following equation:

$$\text{Sample mean} = \bar{x} = \frac{x_1 + x_2 + x_3 + \cdots + x_n}{n} \tag{6.1}$$

The mean gives us an estimate of the expected value of an observation, but it gives no idea of the dispersion or variability of the observations. For a measure of variability, we begin by computing the deviation between each observation and the mean.

$$\text{Deviation of observation } x_i = d_i = x_i - \bar{x}$$

The standard deviation, S_x, is a measure of variability, which is defined by the following equation.

$$S_x = \sqrt{\frac{d_1^2 + d_2^2 + d_3^2 + \cdots + d_n^2}{n-1}} \qquad (6.2)$$

The standard deviation gives us an idea of the variability of the observations in the sample. If the errors in measurement are truly random and we take a large number of observations, 68% of all observations will be within 1 standard deviation of the mean. Over 95% of all samples will be within 2 standard deviations of the mean, and almost all samples will be within 3 standard deviations of the mean.

Example 6.1

Compute the mean and standard deviation of the following temperature measurements: $217°C$, $214°C$, $215°C$, $215°C$, and $216°C$.

Solution

$$\bar{x} = \frac{217 + 214 + 215 + 215 + 216}{5} = 215.4°C$$

$$d_1^2 = (217 - 215.4)^2 = (1.6)^2 = 2.56$$
$$d_2^2 = (214 - 215.4)^2 = (-1.4)^2 = 1.96$$
$$d_3^2 = (215 - 215.4)^2 = (-0.4)^2 = 0.16$$
$$d_4^2 = (215 - 215.4)^2 = (-0.4)^2 = 0.16$$
$$d_5^2 = (216 - 215.4)^2 = (0.6)^2 = 0.36$$
$$S_x = \sqrt{\frac{2.56 + 1.96 + 0.16 + 0.16 + 0.36}{5-1}}$$
$$= 1.14°C$$

6.3 OPERATING CHARACTERISTICS

Operating characteristics include details about the measurement by, operation of, and environmental effects on the measuring instrument.

Measurement

A measuring instrument can measure any value of a variable within its *range* of measurement. The range is defined by the *lower range limit* and the *upper range limit*. As the names imply, the range consists of all values between the lower range limit and the upper range limit. The *span* is the difference between the upper range limit and the lower range limit.

$$\text{Span} = \text{upper range limit} - \text{lower range limit}$$

Resolution, dead band, and sensitivity are different characteristics that relate in different ways to an increment of measurement. When the measured variable is continuously varied over the range, some measuring instruments change their output in discrete steps rather than in a continuous manner. The *resolution* of this type of measuring instrument is a single step of the output. Resolution is usually expressed as a percent of the output span of the instrument. Sometimes the size of the steps varies through the range of the instrument. In this case, the largest step is the *maximum resolution*. The *average resolution*, expressed as a percent of output span, is 100 divided by the total number of steps over the range of the instrument.

$$\text{Average resolution } (\%) = \frac{100}{N} \tag{6.3}$$

where N represents the total number of steps.

The *dead band* of a measuring instrument is the smallest change in the measured variable that will result in a measurable change in the output. Obviously, a measuring instrument cannot measure changes in the measured variable that are smaller than its dead band. *Threshold* is another name for *dead band*.

The *sensitivity* of a measuring instrument is the ratio of the change in output divided by the change in the input that caused the change in output. Sensitivity and gain are both defined as a change in output divided by the corresponding change in input. However, *sensitivity* refers to static values, whereas *gain* usually refers to the amplitude of sinusoidal signals.

Operation

The *reliability* of a measuring instrument is the probability that it will do its job for a specified period of time under a specified set of conditions. The conditions include limits on the operating environment, the amount of overrange, and the amount of drift of the output.

Overrange is any excess in the value of the measured variable above the upper range limit or below the lower range limit. When an instrument is subject to an overrange, it does not immediately return to operation within specifications when the overload is removed. A period of time called the *recovery time* is required to overcome the saturation effect of the overload. The *overrange limit* is the maximum overrange that can be applied to a measuring instrument without causing damage or permanent change in the performance of the device. Thus one reliability condition is that the measured variable does not exceed the overrange limit.

Drift is an undesirable change over a specified period of time. *Zero drift* is a change in the output of the measuring instrument while the measured variable is held constant at its lower limit. *Sensitivity drift* is a change in the sensitivity of the instrument over the specified period. Zero drift raises or lowers the entire calibration curve of the instrument. Sensitivity drift changes the slope of the calibration curve. The reliability conditions specify an allowable amount of zero drift and sensitivity drift.

Environmental Effects

The environment of a measuring instrument includes ambient temperature, ambient pressure, fluid temperature, fluid pressure, electromagnetic fields, acceleration, vibration, and mounting position. The *operating conditions* define the environment to which a measuring instrument is subjected. The *operative limits* are the range of operating conditions that will not cause permanent impairment of an instrument.

Temperature effects may be stated in terms of the zero shift and the sensitivity shift. The *thermal zero shift* is the change in the zero output of a measuring instrument for a specified change in ambient temperature. The *thermal sensitivity shift* is the change in sensitivity of a measuring instrument for a specified change in ambient temperature.

6.4 STATIC CHARACTERISTICS

Static characteristics describe the accuracy of a measuring instrument at room conditions with the measured variable either constant or changing very slowly. *Accuracy* is the degree of conformity of the output of a measuring instrument to the ideal value of the measured variable as determined by some type of standard. Accuracy is measured by testing the measuring instrument with a specified procedure under specified conditions. The test is repeated a number of times, and the accuracy is given as the maximum positive and negative error (deviation from the ideal value). The *error* is defined as the difference between the measured value and the ideal value:

$$\text{Error} = \text{measured value} - \text{ideal value}$$

Accuracy is expressed in terms of the error in one of the following ways:

1. In terms of the measured variable (e.g., $+1°C/-2°C$)
2. As a percent of span (e.g., $\pm 0.5\%$ of span)
3. As a percent of actual output (e.g., $\pm 1\%$ of output)

The *repeatability* of a measuring instrument is a measure of the dispersion of the measurements (the standard deviation is another measure of dispersion). Accuracy and repeatability are not the same. Figure 6.1 uses the pattern of bullet holes in a target to illustrate the difference between repeatability and accuracy. Notice that a rifle which is repeatable but not accurate produces a tight pattern—but that the pattern is not centered on the bull's-eye. The distance from the center of the bull's-eye to the center of the pattern is called the *bias* or *systemic error*. A shooter who

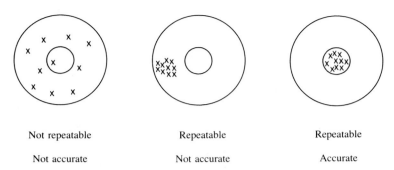

Not repeatable	Repeatable	Repeatable
Not accurate	Not accurate	Accurate

Figure 6.1   Rifle target patterns illustrate the difference between repeatability and accuracy.

is aware of the bias may adjust accordingly and produce an accurate and repeatable pattern on the next try. An experienced operator will make a similar adjustment to compensate for bias in a controller. Thus an automatic controller that is repeatable but not accurate may still be very useful.

Repeatability and reproducibility deal in slightly different ways with the degree of closeness among repeated measurements of the same value of the measured variable. *Repeatability* is the maximum difference between several consecutive outputs for the same input when approached from the same direction in full range traversals. *Reproducibility* is the maximum difference between a number of outputs for the same input, taken over an extended period of time, approaching from both directions. Reproducibility includes hysteresis, dead band, drift, and repeatability. The measurement of reproducibility must specify the time period used in the measurement. Reproducibility is obviously more difficult to determine because of the extended time period that is required.

The procedure of determining the accuracy of a measuring instrument is called *calibration*. The data from the calibration of a measuring instrument may be presented in tabular form in a *calibration report* or in graphical form in a *calibration curve*. The input is usually expressed as a percent of the input span. The output and the error are usually expressed as a percent of the ideal output span. Table 6.1 is an example of a calibration report.

The measured accuracy and reproducibility of the measuring instrument are taken directly from the calibration report. *Measured accuracy* is the maximum negative and positive errors in any of the readings. In Table 6.1 the maximum negative error is -0.35%, which occurs in cycle 1 at 30% on the upscale part of the cycle (i.e., the top half of the table). The maximum positive error is 0.29, which occurs in cycle 3 at 80% on the downscale side (bottom half of the table). The measured accuracy is -0.35 to $+0.29\%$.

Repeatability can be determined from a calibration report by finding the line of data that has the greatest dispersion between the readings for the three cycles. In Table 6.1, for example, the maximum difference between readings on a given line

Table 6.1 Calibration Report (Percent)

Input or Ideal Output	Cycle 1	Cycle 2	Cycle 3	Average	Error
0	−0.06	−0.05	−0.04	−0.05	−0.05
10	9.80	9.82	9.84	9.82	−0.18
20	19.69	19.70	19.72	19.70	−0.30
30	29.65	29.67	29.69	29.67	−0.33
40	39.70	39.71	39.72	39.71	−0.29
50	49.85	49.86	49.87	49.86	−0.14
60	60.02	60.03	60.06	60.04	0.04
70	70.14	70.16	70.17	70.16	0.16
80	80.21	80.23	80.23	80.22	0.22
90	90.19	90.22	90.23	90.21	0.21
100	100.08	100.09	100.11	100.09	0.09
90	90.24	90.25	90.27	90.25	0.25
80	80.26	80.28	80.29	80.28	0.28
70	70.22	70.24	70.27	70.24	0.24
60	60.12	60.13	60.14	60.13	0.13
50	49.96	49.97	49.98	49.97	−0.03
40	39.78	39.80	39.82	39.80	−0.20
30	29.73	29.75	29.77	29.75	−0.25
20	19.76	19.78	19.79	19.78	−0.22
10	9.84	9.85	9.86	9.85	−0.15
0	−0.05	−0.04	−0.04	−0.04	−0.04

occurs at an input of 70% in the downscale traversal. The output for cycle 1 is 70.22%, and for cycle 3 it is 70.27%. The difference, 0.05%, is the repeatability of the measuring instrument.

Figures 6.2 and 6.3 are two forms of the calibration curve. In Figure 6.2 the error is plotted versus the input to the measuring instrument. This gives a detailed view of the error over the input range of the instrument. In Figure 6.3, the output is plotted versus the input. This gives a good overall view of the performance of the instrument.

The calibration curves in Figures 6.2 and 6.3 both form a loop as the upscale and downscale curves follow different paths. This loop formed by an upscale–downscale cycle is called *hysteresis*. Hysteresis is defined as the dependence of the value of a variable on the history of past values and the current direction of traversal. Hysteresis is determined by subtracting the dead band from the maximum difference between corresponding upscale and downscale readings for a full cycle traversal (see Figure 6.3).

The ideal measuring instrument would produce a perfectly straight calibration curve. On a graph of percent input versus percent output, the ideal straight line would pass through the points (0,0) and (100,100). The ideal never actually occurs, and the closeness of the calibration data to a straight line is called the *linearity* of the measuring instrument. When linearity is stated, the method of determining the straight

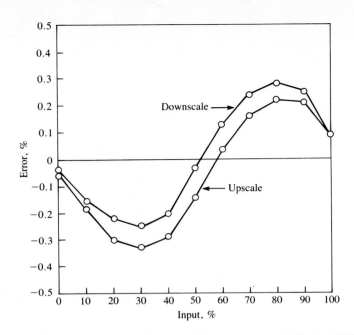

Figure 6.2   Calibration curve of the error from the calibration report in Table 6.1. Plotting the error instead of the output gives a more detailed view of the nature of the error over the range of the instrument.

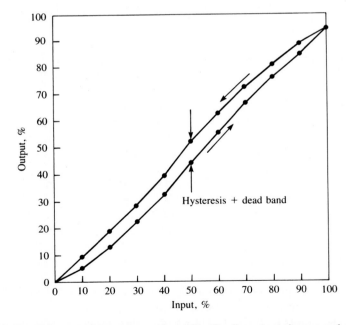

Figure 6.3   Calibration curve of output versus input. Plotting the output gives a good overview of the performance of the measuring instrument. The data in this graph were altered to emphasize the hysteresis to make a better illustration. Hysteresis is usually much less than that shown here. Table 6.1 and Figure 6.1 are more realistic examples of hysteresis in a measuring instrument.

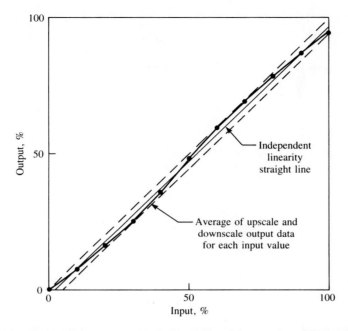

Figure 6.4   The calibration curve shown here is the average of the upscale and downscale outputs for each input value. The independent linearity straight line can be determined on the calibration curve. First determine the location of the two parallel lines that are the closest together while enclosing the entire calibration curve. These are the two dashed lines in the graph. The independent linearity straight line is the line midway between the two dashed lines.

line must also be specified. The most common types of linearity are *independent linearity*, *terminal-based linearity*, *zero-based linearity*, and *least-squares linearity*.

When calibration data are compared to a straight line, attention focuses on the deviation or error between each output reading and the output determined by the line for the same input value. All methods of picking the straight line attempt to minimize the amount of the error. Independent linearity is illustrated in Figure 6.4. The straight line is placed such that it minimizes the absolute value of the maximum error. A pair of parallel lines help to locate the independent linearity line on the calibration curve. The two lines must enclose the entire calibration curve, and they must be as close together as possible without including part of the calibration curve. The independent linearity line will be midway between the two parallel lines. The line does not necessarily pass through the origin, and it does not necessarily pass through either endpoint on the calibration curve.

The zero-based and terminal-based linearity lines are shown in Figure 6.5. The zero-based line passes through the minimum point of the calibration curve and has a slope such that it minimizes the maximum error. The terminal-based line is the easiest to construct. It simply connects the two endpoints on the calibration curve.

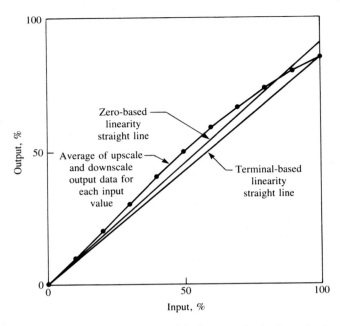

Figure 6.5 The zero-based and terminal-based straight lines can also be determined on the calibration curve. The terminal-based straight line simply connects the two endpoints of the calibration curve. The zero-based straight line passes through the zero point on the calibration curve and minimizes the absolute value of the maximum error.

The least-squares fit is the most sophisticated method, involving considerable computation, which is usually done on a computer. The least-squares line minimizes the sum of the squares of the deviations. The result is usually quite close to the independent linearity line.

Example 6.2

A tachometer-generator is a device used to measure the speed of rotation of gasoline engines, electric motors, speed control systems, and so on. The "tach" produces a voltage proportional to its rotational speed. Consider a tachometer-generator that has an ideal rating of 5.0 V per 1000 rpm, a range of 0 to 5000 rpm, and an accuracy of ±0.5%. If the output of the tach is 21 V, what is the ideal value of the speed? What are the minimum and maximum possible values of the speed?

Solution

The range of the tachometer is 0 to 25 V, corresponding to a speed range of 0 to 5000 rpm. The ideal value of the speed is equal to 200 times the output voltage.

$$\text{Ideal speed} = 21 \times 200 = 4200 \text{ rpm}$$

The accuracy is $\pm 0.5\%$ of full scale or $\pm 0.005 \times 5000 = \pm 25$ rpm. Thus the speed could be anywhere between $4200 - 25$ and $4200 + 25$ rpm.

$$\text{Ideal speed} = 4200 \text{ rpm}$$
$$4175 \leq \text{speed} \leq 4225 \text{ rpm}$$

6.5 DYNAMIC CHARACTERISTICS

Dynamic characteristics describe the performance of a measuring instrument when the measured variable is changing rapidly. Most sensors do not give an immediate, complete response to a sudden change in the measured variable. A measuring instrument requires a certain amount of time before the complete response is indicated by the output. The amount of time required depends on the resistance, capacitance, mass or inertance, and dead time of the instrument. Dynamic characteristics are stated in terms of the step response, ramp response, and frequency response of the measuring instrument.

Step Response

The response of a measuring instrument to a step change in the measured variable is often used to define its dynamic characteristics. A step change occurs when the measured variable is suddenly changed from one steady state value to a second steady state value. For example, a step change in temperature can be achieved by quickly moving the temperature probe from ice water to boiling water. This constitutes a step change from 0°C to 100°C. The step response curve is a plot of the output of the measuring instrument from the time the step change occurred until the time the output reaches its new steady state value. The response curve is normalized by expressing the initial steady state value as 0%, the final steady state value as 100%, and the output of the measuring instrument as a percent change. Terminology defined in terms of the percent of output change can easily be applied to step changes of any given size.

The step response of an instrument can be classified as overdamped, critically damped, or underdamped (see Figure 1.12). The step response of overdamped or critically damped instruments is stated in terms of response time and rise time. The step response of underdamped instruments is stated in terms of rise time, peak percent overshoot, and settling time.

A typical overdamped or critically damped step response curve is shown in Figure 6.6. *Response time* is the time required for the output to reach a designated percentage of the total change. The percentage is stated as a prefix. The 95% response time shown in Figure 6.6 is the time required for the output to reach 95% of the total change. A 98% response time (not shown) is the time required for the output to reach 98% of the total change. The 63.2% response time is given the special name *time constant* and symbol τ. This is the same time constant that we encountered several

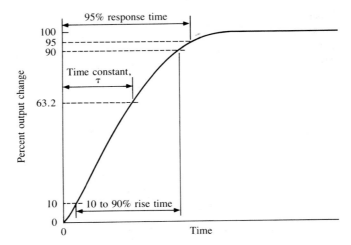

Figure 6.6 Typical step response curve for an overdamped or critically damped measuring instrument. The response is stated in terms of the response time and the rise time.

times in Chapter 5 [see Equations (5.6), (5.10), and (5.11)]. Some examples of time constants of temperature-measuring instruments are given below.

Typical Time Constants

Bare thermocouple (in air)	35 s
Thermocouple in glass well (in air)	66 s
Thermocouple in porcelain well (in air)	100 s
Thermocouple in iron well (in air)	120 s
Bare thermocouple (in still liquid)	10.0 s
Thermometer bulb (water flowing at 2 ft/min)	6.0 s
Thermometer bulb (water flowing at 60 ft/min)	2.4 s

Rise time is the time required for the output to go from a small percent change to a large percent change. The two percentages are stated as a prefix. Thus a 10 to 90% rise time is the time required for the output to go from 10% to 90% of the total change. A 5 to 95% rise time is the time required for the output to go from 5% to 95% of the total change. If the percentages are omitted, they are assumed to be 10% and 90%.

A typical underdamped step response curve is shown in Figure 6.7. The rise time is defined the same as it was for the overdamped or critically damped response. The peak percent overshoot (PPO) is defined as follows:

$$\text{PPO} = \frac{(\text{peak level}) - (\text{final ss level})}{(\text{final ss level}) - (\text{initial ss level})} * 100$$

For the normalized curve, the final steady state level is 100, the initial steady state level is zero, and the peak percent overshoot reduces to the following:

$$\text{PPO} = (\text{peak level}) - (\text{final ss level})$$

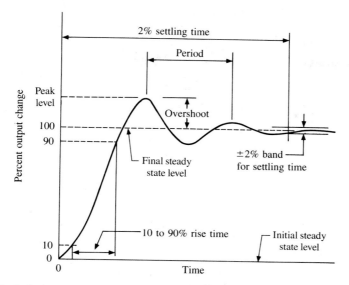

Figure 6.7   Typical step response curve for an underdamped measuring instrument. The response is stated in terms of the rise time, peak percent overshoot, and settling time.

The *settling time* is the time it takes for the response to remain within a small band above and below the 100% change line. The 2% settling time is shown in Figure 6.7.

Both first- and second-order differential equations are used to model the behavior of a measuring instrument. If the behavior of a measuring instrument is dominated by a single capacitance and a single resistance, it is modeled by a first-order differential equation. The transfer function and step response of this model were developed in Chapter 5 (see Figure 5.8 and Equations (5.6) and (5.25)). Using C as the input of the instrument, C_m as its output, G as its steady-state gain, and τ as its time constant, we have the following transfer function and step response for the first-order model of a measuring instrument:

Transfer function:

$$\frac{C_m(s)}{C(s)} = \frac{G}{1 + \tau s} \tag{6.4}$$

Step input:

$$C(s) = K/s \tag{6.5}$$

Step response:

$$c_m(t) = GK(1 - e^{-t/\tau}) \tag{6.6}$$

Note that the step input is shown in frequency domain and the step response is shown in time domain. Equation (6.6) produces a graph that is similar to Figure 6.6.

If the behavior of a measuring instrument is dominated by two capacitances and two resistances, it is modeled by a second-order differential equation. Using C, C_m, and G as before, we have the following transfer function for the measuring instrument:

$$\frac{C_m(s)}{C(s)} = \frac{G}{1 + (2\alpha/\omega_0^2)s + (1/\omega_0^2)s^2} \tag{6.7}$$

where ω_0 = resonant frequency, radian/second

α = damping coefficient, 1/second

We now proceed to determine the step response of the second order system defined by Equation (6.7). The first step is to multiply Equation (6.7) by the step response given in Equation (6.5). For simplicity, we also multiplied the numerator and denominator by ω_0^2 to get the following frequency domain expression for C_m:

$$C_m(s) = \frac{GK\omega_0^2}{s(\omega_0^2 + 2\alpha s + s^2)} \tag{6.8}$$

The next step is to complete the partial fraction expansion of Equation (6.8). The inverse Laplace transform can be obtained from the partial fraction expansion by using the Laplace transform pairs in Table 5.1. There are three cases to consider.

Case 1: Overdamped Response, $\alpha > \omega_0$

When α is greater than ω_0, the quadratic in the denominator of Equation (6.8) has two real, unequal roots, and the step response is overdamped. Let $-a$ and $-b$ represent the two roots. Equation (6.8) can then be written and expanded as follows:

$$C_m(s) = \frac{GK}{s(s+a)(s+b)} = GK\left(\frac{K_1}{s} + \frac{K_2}{s+a} + \frac{K_3}{s+b}\right)$$

Applying transform pairs 1 and 3 in Table 5.1 results in the following time domain equation for the output of the measuring instrument:

$$c_m(t) = GK(K_1 + K_2 e^{-at} + K_3 e^{-bt}) \tag{6.9}$$

Equation (6.9) defines an overdamped response similar to Figure 6.6.

Case 2: Critically Damped Response, $\alpha = \omega_0$

When α is equal to ω_0, the quadratic in the denominator of Equation (6.8) has two real, equal roots, and the step response is critically damped. Let a represent this root. Equation (6.8) can then be written and expanded as follows:

$$C_m(s) = \frac{GK}{s(s+a)^2} = GK\left[\frac{K_1}{s} + \frac{K_2}{(s+a)^2} + \frac{K_3}{s+a}\right]$$

Applying transform pairs 1, 3, and 4 in Table 5.1 results in the following time domain equation for the output of the measuring instrument:

$$c_m(t) = GK(K_1 + K_2 t e^{-at} + K_3 e^{-at}) \tag{6.10}$$

Equation (6.10) defines a critically damped response which is also similar to Figure 6.6.

Case 3: Underdamped Response, $\alpha < \omega_0$

When α is less than ω_0, the quadratic in the denominator of Equation (6.8) has two complex, conjugate roots, and the step response is underdamped. Let $(-\alpha + j\omega_d)$ and $(-\alpha - j\omega_d)$ represent these roots. Equation (6.8) can then be written and expanded as follows:

$$C_m(s) = \frac{GK}{s(s + \alpha - j\omega_d)(s + \alpha + j\omega_d)}$$

$$= GK\left[\frac{K_1}{s} + \frac{K_2\underline{/\theta}}{(s + \alpha + j\omega_d)} + \frac{K_2\underline{/-\theta}}{(s + \alpha - j\omega_d)}\right]$$

Applying transform pairs 1 and 10 in Table 5.1 results in the following time domain equation for the output of the measuring instrument:

$$c_m(t) = GK[K_1 + 2K_2e^{-\alpha t}\cos(\omega_d t + \theta)] \tag{6.11}$$

Equation (6.11) defines an underdamped response that is similar to Figure 6.7. The following is a summary of the important parameters of an underdamped second-order system.

Resonant frequency $= \omega_0$, radian/second

Damping coefficient $= \alpha$, 1/second

Damping ratio $= \zeta = \alpha/\omega_0$

Natural frequency $= \omega_d = \sqrt{\omega_0^2 - \alpha^2}$, radian/second

$\qquad\qquad\qquad = \omega_0\sqrt{1 - \zeta^2}$, radian/second

The natural frequency is also called the damped natural frequency.

Example 6.3

The following data were obtained from a thermometer with a protective well, which was plunged into a moving liquid at time $t = 0$ s. The thermometer and well were maintained at 50°C before the test. The liquid temperature was 150°C. Plot the step response curve and determine the rise time, the time constant, and the 95% response time.

Time (s)	Temperature (°C)
0	50
20	64
40	78
60	92
80	104
100	113
120	120
140	126
160	130
180	133
200	135
300	143
400	147
500	150

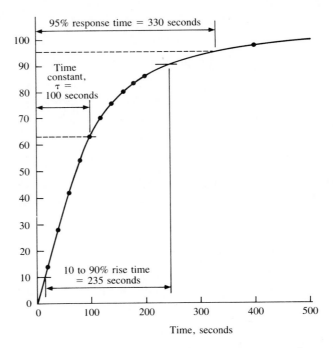

Figure 6.8 Step response curve for Example 6.3. The time constant, response time, and rise time are read directly from the graph.

Solution

The normalized response curve is plotted in Figure 6.8. The desired values can be read directly from the graph.

$$\text{Time constant, } \tau = 100 \text{ s}$$
$$95\% \text{ response time} = 330 \text{ s}$$
$$10 \text{ to } 90\% \text{ rise time} = 235 \text{ s}$$

Example 6.4

Determine the time constant of a mercury thermometer similar to Figure 5.4 and defined by Equation (5.11). The bulb has a diameter of 0.5 cm and a length of 1.5 cm. The outside film coefficient is estimated to be 30 W/m²·K. The inside film coefficient and the resistance of the thermometer wall are both negligible compared to the outside film coefficient. The thickness of the wall may also be neglected.

Solution

From Equation (5.11), the time constant is

$$\tau = R_T C_T$$

1. Determine the thermal resistance of the bulb.

$$\text{Bulb surface} = A = \frac{\pi d^2}{4} + \pi d L$$

$$A = \frac{\pi (0.005)^2}{4} + \pi (0.005)(0.015)$$

$$= 2.55\text{E}-04 \text{ m}^2$$

$$R_T = \frac{1}{Ah} = \frac{1}{(2.55\text{E}-04)(30)}$$

$$= 131 \text{ K/W}$$

2. Determine the thermal capacitance of the bulb.

$$\text{Bulb volume} = \frac{\pi d^2 L}{4}$$

$$= \frac{\pi (0.005)^2 (0.015)}{4}$$

$$= 2.95\text{E}-07 \text{ m}^3$$

From Appendix A, the density and specific heat of mercury are

$$\rho = 13{,}600 \text{ k/m}^3$$
$$S_h = 140 \text{ J/kg·K}$$

The thermal capacitance of the bulb is

$$C_T = \rho V S_h$$
$$= (1.36\text{E}+04)(2.94\text{E}-07)(1400)$$
$$= 0.56 \text{ J/kg·K}$$

3. Determine the time constant of the bulb.

$$\text{Time constant} = \tau = R_T C_T$$
$$= (131)(0.56) = 73.2 \text{ s}$$

Ramp Response

The ramp response of a measuring instrument, although not widely used, does give additional insight into the dynamic characteristics of a measuring instrument. Figure 6.9 shows a graph of the results of a ramp response test. The input temperature was increased at a steady rate until it was allowed to "settle in" at a higher value. The measured temperature lagged behind the input temperature and caught up some time after the input had settled in at the new temperature. There are two ways to view

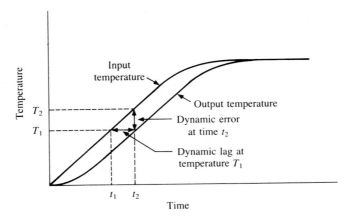

Figure 6.9   Typical ramp response curve showing the dynamic error and the dynamic lag.

the ramp response curve. One way leads us to the concept of dynamic error, the other to the concept of dynamic lag.

If we draw a horizontal line at a temperature of T_1 on the dynamic response graph, the line intersects the input curve at time t_1, and the output curve at time t_2. The difference between these two times, $t_2 - t_1$, is the *dynamic lag* at temperature T_1. Dynamic lag is the amount of time that elapses between the time the input reaches a certain temperature and the time the output reaches that same temperature.

If we draw a vertical line at a time of t_2 on the ramp response graph, the line intersects the output curve at temperature T_1, and the input curve at temperature T_2. The difference between these two temperatures is the *dynamic error* at time t_2. Dynamic error is the difference between the input temperature and the output temperature at a given time.

Example 6.5

A temperature sensor is used to measure the temperature of an oil bath. The input temperature is increasing at a constant rate of 1.2°C/min, and the dynamic lag is 2.3 min.

    a. What is the dynamic error in degrees Celsius?

    b. If the range of the temperature sensor is 75 to 125°C, what is the dynamic error in percent?

Solution

    a. Let T_1 represent the input temperature at time t_1. At time $t_2 = t_1 + 2.3$ min, the input temperature will be $T_2 = T_1 + (1.2)(2.3) = T_1 + 2.76$°C. Thus, at time t_2, the input temperature will be $T_1 + 2.76°C$ and the output temperature will be T_1. The dynamic error is the difference between the input temperature and the output temperature at time t_2.

$$\text{Dynamic error} = T_1 + 2.76 - T_1 = 2.76°C$$

b. The range of the sensor is $125 - 75 = 50°C$. The dynamic error in percent is $100 \times 2.76/50 = 5.52\%$.

$$\text{Dynamic error} = 5.52\%$$

Frequency Response

When it comes to control system analysis and design, frequency response and the transfer function are the most useful methods for defining the dynamic response of a measuring instrument. In Chapter 5 we introduced you to frequency response and presented a computer program that produces frequency response data from the transfer function. In this section we apply to measuring instruments the methods presented in Chapter 5.

The liquid-filled thermometer in Figure 5.4 has a transfer function that is typical of many measuring instruments. A component with this transfer function is called a first-order lag because the highest derivative in the time-domain equation is first order. The transfer function for the liquid-filled thermometer is given by Equation (5.28), which is reproduced here for convenience.

$$\frac{T_m}{T_a} = \frac{1}{1 + \tau s} \tag{5.28}$$

In Example 6.4, we calculated a time constant of 73.4 s for a mercury-filled thermometer. The frequency response of this component is plotted in Figure 6.10. Notice how the gain is near 0 decibels (dB) at frequencies below 0.002 rad/s and how it drops at a rate of 20 dB per decade for frequencies above 0.1 rad/s. The low-frequency gain curve is a straight line at 0 dB. The high-frequency gain curve is also a straight line with a slope of -20 dB per decade increase in frequency. Extend the two lines and you will see that they intersect at the point on the 0-dB line marked with a diamond and labeled ω_b. This point is called the *break point*, and the radian frequency of the break point is equal to the reciprocal of the time constant, τ.

$$\text{Break-point frequency} = \omega_b = \frac{1}{\tau} \tag{6.12}$$

All first-order lag components have frequency response curves just like the one in Figure 6.10. The only difference is the break-point frequency, ω_b, which is given by Equation (6.12).

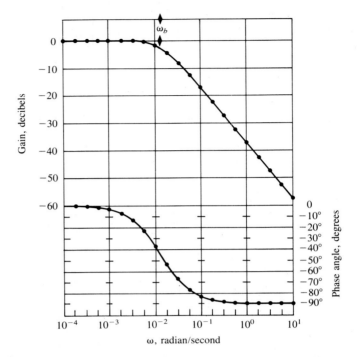

Figure 6.10 Bode diagram of a mercury-filled thermometer with a time constant of 73.4 s (Example 6.4).

Some measuring instruments have a more complex transfer function than the first-order lag we just discussed. Figure 6.11 shows a liquid-filled thermometer enclosed in a protective sheath (or well). The sheath presents a second thermal resistance and the fluid in the well is a second thermal capacitance. The two resistances and two capacitances form what is known as a two-capacity, interacting system. The differential equation for this system has both a second-order and a first-order derivative term, so this system is also called a *second-order lag*. The transfer function is given by Equation (6.13).

$$\frac{T_m}{T_a} = \frac{1}{1 + (\tau_1 + \tau_2 + \tau_2 R_1 / R_2)s + (\tau_1 \tau_2)s^2} \tag{6.13}$$

where $\tau_1 = R_1 C_1$ = time constant of the well

$\tau_2 = R_2 C_2$ = time constant of the thermometer

R_1 = thermal resistance between the measured fluid and the inside of the protective sheath, kelvin/watt

R_2 = thermal resistance between the inside of the protective sheath and the liquid inside the bulb, kelvin/watt

C_1 = thermal capacitance of the sheath, joule/kelvin

C_2 = thermal capacitance of the bulb, kelvin/watt

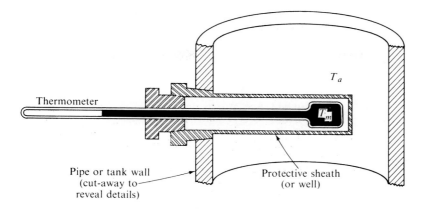

a) A cut-away view of a thermometer and
protective sheath

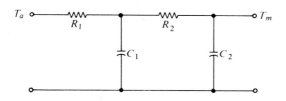

b) The equivalent electrical circuit for the
thermometer and protective sheath.

Figure 6.11 Temperature sensor in a protective sheath and the electrical equivalent.

The *resonant frequency* of a second-order lag is more complex, and more inter-esting, than that of a first-order lag, especially when the damping ratio, ζ, is less than 1. Figure 6.12 shows the Bode diagram of a second-order lag for values of ζ from 0.05 to 20. Notice that at low frequencies, the gain is near 0 dB, just like the first-order lag. At high frequencies, the second-order lag gain drops 40 dB per decade increase in frequency, twice the rate of the first-order lag. However, it is near the break point, ω_b, where the most interesting things happen. As the damping ratio drops below 1, the gain at the break point increases until it reaches 20 dB with a damping ratio of 0.05. A gain of 20 dB is a tenfold increase. In other words, the output is 10 times as large as the input at the resonant frequency when the damping ratio is 0.05. This phenomenon has many useful applications in electronic systems.

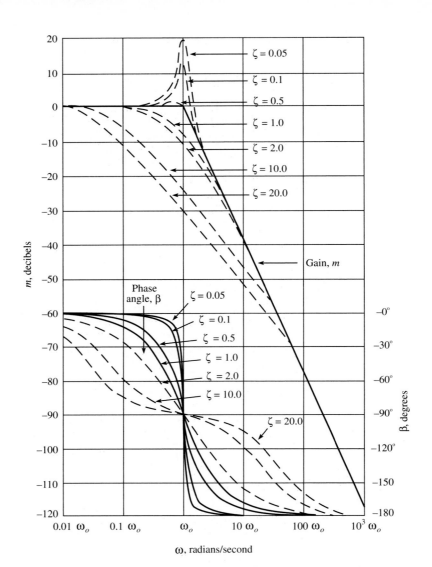

Figure 6.12   Bode diagram of second-order components.

6.6 SELECTION CRITERIA

Selecting the right measuring instrument for a job is complicated by the fact that usually there is no single correct choice. The designer must choose from among a number of good choices, each with its own set of advantages and disadvantages. We have already mentioned some of the choices for temperature measurement: thermocouple, RTD, thermistor, IC sensor, liquid-filled element, vapor-filled element, gas-filled element, or radiation pyrometer. Flow measurement presents the following set of choices for the primary element: magnetic, orifice, venturi, vortex, pitot tube, turbine, thermal, ultrasonic, and variable area. Similar choices exist for other measurements, such as force, velocity, level, pressure, position, and displacement. There is no set procedure for selecting a measuring instrument, and judgment is an important part of the process. Some questions the designer might consider in the process of selecting a measuring instrument are listed below.

A. Questions about the measured variable
  1. What is the measured variable?
  2. What accuracy is required?
     a. At 100% of the input range?
     b. At 75% of the input range?
     c. At 50% of the input range?
     d. At 25% of the input range?
  3. What fluid or solid is being measured?
  4. What are the upper and lower range limits?
  5. What overrange might occur?
  6. What type of dynamic changes might occur in the measured variable?
     a. What is the maximum rate of change?
     b. What is the maximum frequency at which changes occur?
B. Questions about the measuring instrument
  1. What is the primary sensor?
  2. What effect will the primary sensor have on the measured variable?
  3. What is the purpose of the output: indicate, transmit, record, totalize, other?
  4. Is the output signal analog or digital?
  5. What is the output signal: 3–15 psi, 4–20 mA, 0–5 V, 8-bit TTL digital, other?
  6. What type of signal conditioning is provided?
     a. Amplification or attenuation?
     b. Signal conversion: resistance to current, millivolt to current, analog to digital, digital to analog, etc.?
     c. Filters: low pass, high pass, band pass?
     d. Linearization of the signal?
     e. Square-root function?
     f. Other special functions?
     g. Conversion to engineering units?
  7. What is the operating principle of the transmitter?
  8. What environmental conditions will the instrument encounter?
  9. What are the thermal zero drift and thermal sensitivity drift ratings of the instrument?
  10. What are the long-term zero drift and sensitivity drift ratings of the instrument?
  11. What are the dead band, hysteresis, and linearity ratings of the instrument?

C. Questions about economics
1. What is the initial cost of the instrument?
2. What are the installation costs?
3. What are the maintenance costs?
4. What is the expected life of the instrument?
5. What is the anticipated replacement cost?

GLOSSARY

Accuracy: The degree of conformity of the output of a measuring instrument to the ideal value of the measured variable. (6.4)

Accuracy, measured: The maximum positive and negative errors in a calibration report. (6.4)

Bias: The difference obtained by subtracting the ideal value from the average of repeated measurements of a measured variable. (6.4)

Calibration: The procedure of determining the accuracy of a measuring instrument. (6.4)

Calibration curve: A graphical presentation of calibration data. (6.4)

Calibration report: A tabular presentation of calibration data. (6.4)

Damping coefficient (designated by α):   A number that determines the type of damping of a second-order system. The response is underdamped if $\alpha < \omega_0$, critically damped if $\alpha = \omega_0$, and overdamped if $\alpha > \omega_0$. (6.5).

Damping ratio (designated by ζ):   The ratio of damping coefficient (α) over resonant frequency (ω_0). The response is underdamped if $\zeta < 1$, critically damped if $\zeta = 1$, and overdamped if $\zeta > 1$. (6.5)

Dead band: The smallest change in the measured variable that will result in a measurable change in the output. (6.3)

Drift: An undesirable change over a specified period of time. (6.3)

Drift, sensitivity: A change over time in the sensitivity of a measuring instrument. (6.3)

Drift, zero: A change over time in the output of a measuring instrument when the measured variable is at the lower range limit. (6.3)

Dynamic error: The difference between the input temperature and the output temperature at a given time. (6.5)

Dynamic lag: The amount of time that elapses between the time the input reaches a certain temperature and the time the output reaches that same temperature. (6.5)

Error: The difference determined by subtracting the ideal value from the measured value. (6.4)

Hysteresis: The dependence of the value of a variable on the history of past values and the current direction of traversal. (6.4)

Linearity: The closeness of the calibration curve to a straight line. (6.4)

Linearity, independent: Comparison of the calibration data with a straight line that minimizes the maximum value of the deviation between the data and the line. (6.4)

Linearity, least-squares: Comparison of calibration data with the straight line that minimizes the sum of the squares of the deviations between the data and the straight line. (6.4)

Linearity, terminal-based: Comparison of the calibration data with a straight line that connects the two ends of the calibration curve. (6.4)

Linearity, zero-based: Comparison of the calibration data with a straight line that passes through the zero point of the calibration curve and minimizes the maximum value of the deviation between the data and the line. (6.4)

Natural frequency (designated by ω_d): The frequency of oscillation of an under-damped second-order system. (6.5)

$$\omega_d = \sqrt{\omega_0^2 - \alpha^2} = \omega_0\sqrt{1 - \zeta^2}$$

Operating conditions: The environment in which a measuring instrument operates. (6.3)

Operative limits: The end values of the range of operating conditions that will not cause permanent impairment of a measuring instrument. (6.3)

Overrange: Any excess in the value of the measured variable above the upper range limit or below the lower range limit. (6.3)

Overrange limit: The maximum overrange that can be applied to a measuring instrument without causing damage or permanent change in performance. (6.3)

Overshoot: The maximum height of the step response curve measured above the 100% change line. (6.5)

Range: The values of the measured variable that can be measured by the measuring instrument. The range includes all values between the *lower range limit* and the *upper range limit*. (6.3)

Reliability: The probability that a component will do its job for a specified time period under a specified set of conditions. (6.3)

Repeatability: The maximum deviation from the average of repeated measurements of the same static variable. (6.4)

Reproducibility: The maximum difference between a number of outputs for the same input, taken over an extended period of time, approaching from both directions. Reproducibility includes hysteresis, dead band, drift, and repeatability. (6.4)

Resolution: A single step of output in a measuring instrument whose output changes in discrete steps. (6.3)

Resonant frequency (designated by ω_0): The break-point frequency of an under-damped second-order component. (6.5)

Response time: The time required for the output to reach a designated percentage of the total change, after a step change in input. (6.5)

Rise time: The time required for the output to go from a small percent change to a large percent change, after a step change in input. Unless otherwise specified, the change is from 10% to 90%. (6.5)

Sensitivity: The ratio of the change in output divided by the change in input that caused the change in output. (6.3)

Settling time: The time it takes for the response of an underdamped component to remain within a small band above and below the 100% change line. (6.5)

Span: The size of the range, equal to the upper range limit minus the lower range limit. (6.3)

Systemic error: Another name for bias. (6.4)

Thermal sensitivity shift: A change in the sensitivity of a measuring instrument caused by a specified change in the ambient temperature. (6.3)

Thermal zero shift: A change in the zero output of a measuring instrument caused by a specified change in the ambient temperature. (6.3)

Time constant (designated by τ): The time required for the output of a component to reach 63.2% of the total change after a step change in input. (6.5)

EXERCISES

6.1 A pressure sensor is tested for repeatability by increasing the input pressure from 0 to 15 psi 10 times, recording the output reading each time the input reaches 15 psi. The output readings are as follows:

15.45, 15.53, 15.61, 15.42, 15.55
15.47, 15.51, 15.59, 15.46, 15.60

a. Calculate the mean and standard deviation of the 10 readings.

b. Determine the measured accuracy, bias, and repeatability of the readings.

c. Use 3 standard deviations to estimate the range of readings you would expect in a very large number of test runs.

6.2 Explain the difference between repeatability and reproducibility.

6.3 A pressure gage with a range from 3 to 15 psi was tested for dead band. When the input pressure was increased from 9.00 to 9.05 psi, the output remained stationary at 9.10 psi. When the input was increased from 9.00 psi to 9.06 psi, the output changed from 9.10 psi to 9.11 psi. Express the dead band in psi, as a percent of the lower input value, and as a percent of the span.

6.4 The following data are the average upscale and downscale values from a calibration report. Both the input and the output are expressed in terms of percent of span. Plot the calibration curve and determine the combined hysteresis and dead band.

Upscale		Downscale	
Input	Output	Input	Output
0	2.5	100	90.2
10	7.2	90	84.8
20	13.4	80	78.0
30	20.6	70	70.0
40	29.1	60	61.3
50	38.3	50	52.2
60	47.8	40	43.1
70	57.1	30	33.9
80	67.0	20	24.0
90	77.5	10	14.0
100	90.2	0	2.5

6.5 Explain the difference between static characteristics and dynamic characteristics.

6.6 The following data are the averages of the upscale and downscale readings from a calibration report. Draw the calibration curve and then draw the straight lines for independent linearity, terminal-based linearity, and zero-based linearity. Determine the maximum error for each straight line.

Input	Average Output	Input	Average Output
0	2.8	60	61.1
10	9.6	70	70.0
20	17.4	80	76.8
30	26.5	90	83.2
40	38.7	100	88.8
50	50.5		

6.7 A potentiometer is used to measure the position of a shaft. The input range of shaft positions is from $-160°$ to $+160°$. The corresponding output range of the potentiometer is from -20 to $+20$ V. The accuracy is $\pm 1\%$. If the output is $+8$ V, what is the ideal position? What are the minimum and maximum possible positions?

6.8 Determine the sensitivity of the position sensor in Exercise 6.7 in volts per degree.

6.9 A potentiometer has 800 turns of wire and a full-scale output of 20 V. Determine the average resolution in volts and as a percentage of the range.

6.10 The following results were obtained from the calibration of a force transducer.

Input Force (N)	Output Voltage (V)
0	0.06
2	0.63
4	1.20
6	1.77
8	2.35
10	2.94
12	3.55
14	4.17
16	4.80
18	5.43
20	6.06

Draw the calibration curve and then draw the straight lines for independent linearity, terminal-based linearity, and zero-based linearity. Determine the maximum error for each straight line.

6.11 Determine the sensitivity of the force transducer of Exercise 6.10 in volts per newton at 20% and 80% of full scale. (*Hint:* Divide the change in output voltage between 2 and 6 N by the change in force of $6 - 2 = 4$ N to determine the sensitivity at 20%.)

6.12 The following data were obtained from a temperature-measuring means that was plunged from a liquid bath maintained at 50°C into a second bath maintained at 100°C. Plot the response curve and determine the 95% response time, the time constant, and the 10 to 90% rise time.

Time (s)	Temperature (°C)	Time (s)	Temperature (°C)
0	50.0	45	91.7
5	57.5	50	93.0
10	65.0	60	95.1
15	71.5	70	96.8
20	76.7	80	98.0
25	81.0	90	98.9
30	84.7	100	99.4
35	87.5	110	99.8
40	90.0	120	100.0

6.13 The following data were obtained from a step response test of an underdamped component. Plot the response curve and determine the 10 to 90% rise time, the overshoot, and the 2% settling time.

Time (s)	Output (%)	Time (s)	Output (%)
0	0.0	30.0	92.0
5	36.5	32.6	96.0
10	74.5	34.0	100.0
13.5	100.0	36.0	103.0
15	110.0	39.5	105.0
18.4	120.0	42.5	103.0
21.8	110.0	44.4	100.0
24.0	100.0	50.0	97.8
25.0	96.0	55.0	100.0
28.4	91.0	60.0	101.0

6.14 A speed sensor is used to measure the speed of an electric motor. The actual speed is increasing at a constant rate of 10 rpm per millisecond. The dynamic lag is 4 ms. What is the dynamic error in rpm? If the range of the speed sensor is 0 to 5000 rpm, what is the dynamic error, in percent?

6.15 Determine the time constant of an alcohol-filled thermometer. The bulb has a diameter of 0.25 cm and a length of 0.85 cm. The film coefficient is estimated to be 12 W/m²·K. Ignore the resistance and thickness of the bulb wall.

6.16 Use the program "BODE" from Chapter 5 to generate frequency response data for the thermometer in Exercise 6.15. Use the data to construct a Bode diagram.

6.17 A liquid-filled thermal element is enclosed in a protective well. The following thermal resistances and capacitances were computed for the element and well.

$$R_1 = 125 \text{ K/W}$$
$$C_1 = 0.45 \text{ J/kg·K}$$
$$R_2 = 65 \text{ K/W}$$
$$C_2 = 0.72 \text{ J/kg·K}$$

The transfer function is given by Equation (6.5). Determine the coefficients of s^2 and s, and write the transfer function. Use the program "BODE" to generate frequency response data and plot the Bode diagram.

6.18 A measuring instrument has a resonant frequency (ω_0) of 5 radians/second, a damping coefficient (α) of 3 1/s, and a steady state gain (G) of 0.2 mA/degree Celsius. Determine the following:

a. The transfer function, C_m/C (see Equation (6.7)).

b. The frequency domain response, C_m, to a step change in input, C, from 0°C to 100°C (i.e., multiply the transfer function by $C = 100/s$). Put the result in the form of Equation (6.8).

c. Determine the roots of the quadratic term in the denominator of the equation you obtained in step b. Verify that the response is underdamped.

d. Complete the partial fraction expansion of C_m.

e. Use transform pairs 1 and 10 in Table 5.1 to get the time domain expression, c_m, for the step response.

f. Use a spreadsheet or a program similar to the following to generate a table of $c_m(t)$ versus t values from the time domain expression you obtained in part e.

```
10   FOR T = 0 TO 3 STEP .05
20   CM(T) = 10 + 12*EXP(-4*T)*COS(6*T + 110.03/57.3)
30   LPRINT USING "          ##.##     ##.##"; T, CM(T)
40   NEXT T
50   END
```

g. From the table you produced in part (f), determine the 10 to 90% rise time, the peak percent overshoot, and the 2% settling time. You can do this by plotting a graph and reading the values from the graph, or you can get the values by interpolation directly from the data table.

CHAPTER 7

Signal Conditioning

OBJECTIVES

A signal conditioner prepares a signal for use by another component. The input to a signal conditioner is usually the output from a sensor (or primary element). The operations performed by a signal conditioner include isolation, impedance conversion, noise reduction, amplification, linearization, and conversion.

The purpose of this chapter is to give you the ability to analyze, specify, or design signal conditioning systems. After completing this chapter, you will be able to

1. Sketch and explain the ideal and practical models of the operational amplifier

2. Explain common-mode voltage, common-mode gain, and common-mode rejection ratio

3. Specify the common-mode rejection ratio required to satisfy a given limitation on the common-mode component of the output voltage

4. Design an inverting or noninverting amplifier that will produce a given output voltage range (with a lower range limit of 0), given the input voltage range (also with a lower limit of 0) and the value of R_{in}

5. Design a summing amplifier that will produce a given output voltage range (with a nonzero lower range limit), given the input voltage range and the value of R_{in}

6. Design a differential amplifier or an instrumentation amplifier with a specified gain

7. Determine the acquisition time of a sample-and-hold circuit, given the value of R_{in} and C

8. Discuss the purposes of isolation and impedance transformation; name and briefly describe a circuit used for impedance transformation; name and briefly describe two methods used for galvanic isolation

9. Design a low-pass filter or a notch filter that will produce a specified attenuation factor at a given frequency

10. Construct a table or a graph that defines the input/output relationship of a linearizer, given the nonlinear input/output relationship of the measuring instrument

11. Construct a piecewise-linear approximation of a nonlinear input/output function

12. Determine the temperature versus output voltage calibration curve for a resistance temperature detector (RTD)/Wheatstone bridge circuit, given the temperature range, the resistance versus temperature characteristic of the RTD, and $R_3/(R_3 + R_4)$ for the bridge

13. Given the number of bits in an analog-to-digital converter, determine the quantization error in percent of full scale

7.1 INTRODUCTION

A measuring instrument uses some characteristic of the material and construction of a primary element to convert the value of a measured variable into an electrical or mechanical signal. The signal produced by the primary element may be a small voltage, a change in resistance, a force, a pressure, or some other phenomenon that is related to or depends on the value of the measured variable. The output of the primary element usually requires some additional signal conditioning to make it usable by another component.

The operational amplifier is the heart of many modern signal conditioning modules. For this reason, our study of signal conditioning begins with the operational amplifier and its application in various signal conditioning circuits. You should complete the section on op-amp circuits before moving on to the section on signal conditioning.

One of the objectives of a measuring instrument is to minimize the effect of the measurement on the measured variable. In some situations, this may mean electrical isolation of the measuring instrument from the process with an isolation amplifier. In other situations, it may mean providing a high input impedance with a voltage follower. Thus we begin the section on signal conditioning with isolation and impedance conversion.

The industrial process is a harsh environment for making precise measurements. There are many potential sources of electrical interference, and noise is always a concern when electrical signals are transmitted. Noise reduction with various types of filters is another job for signal conditioning.

Linearization is the next topic. The ideal measuring instrument output is a linear signal that goes from 0 to 100% as the measured variable goes from the lower range limit to the upper range limit. Sometimes the signal is close to the ideal. At other times, the signal is very nonlinear. An orifice flow element, for example, produces a pressure drop that is proportional to the square of the flow rate. Linearization of this type of signal is a very desirable goal. Thermocouples also produce a nonlinear signal, although not nearly as nonlinear as the flow orifice. Linearization of thermocouple signals is also desirable.

Changing the level of a signal is the simplest form of signal conditioning, but certainly not the least important. Amplification, or sometimes attenuation, matches the output of the measuring instrument to the other components in the control system.

A number of primary elements convert changes in the measured variable into small changes in the resistance of the element. Strain gages and resistance temperature detectors (RTDs) are two examples. A bridge circuit is usually used to measure small resistance changes.

Conversion is another aspect of signal conditioning. The conversion may be from one analog signal to another; it may be from an analog signal to a digital signal; or it may be from a digital signal to an analog signal. The analog-to-analog conversions include voltage-to-current conversion and current-to-voltage conversion. The analog-to-digital converter (ADC) does the analog-to-digital conversion, and the digital-to-analog converter (DAC) goes the other way with the signal.

Once a signal is converted to digital, a whole new realm of digital conditioning can be applied. Our study concludes with a look at digital conditioning, including linearization, filtering, calibration tables, and conversion to engineering units.

7.2   THE OPERATIONAL AMPLIFIER

An *operational amplifier* is a high-gain dc amplifier with two inputs and one output. The output is equal to the difference between the voltages on the two inputs multiplied by the gain of the amplifier. The gain of different op amps ranges from 10^4 to 10^7, with 10^5 being a typical value. An op amp requires both a positive and a negative voltage supply, labeled $+V_{cc}$ and $-V_{cc}$, respectively. Two more inputs are provided for offset null adjustments, bringing the total number of terminals to seven. Op amps come in several different IC packages, including the popular eight-lead MINIDIP package.

The operating characteristics of an operational amplifier circuit depend almost entirely on components external to the amplifier. The gain, input impedance, output impedance, and frequency response depend on the external resistors and capacitors in the circuit, not on the gain, input impedance, or output impedance of the amplifier. This means that the behavior of an op-amp circuit can be made to fit a particular application by proper selection and placement of a few resistors and capacitors. Also, the stability of these components can be selected to meet the requirements of the application. With this versatility and ease of design, it is not hard to understand why the op amp has become such a popular component in control systems.

Op-Amp Equivalent Circuit

Figure 7.1 shows the circuit symbol and an equivalent circuit of an operational amplifier. The two inputs are called the *inverting input* and the *noninverting input*. We will use the symbol v_1 for the voltage at the inverting input, v_2 for the voltage at the noninverting input, v_{out} for the voltage at the output terminal, V_{cc} for the positive supply voltage, and $-V_{cc}$ for the negative supply voltage. The currents of interest are the current entering the inverting input, i_1, the current entering the noninverting input, i_2, and the current leaving the output terminal, i_{out}. Notice that $i_1 = -i_2$, and we only needed to define one of the two currents. However, it is useful in analyzing op-amp circuits to be able to refer to both i_1 and i_2.

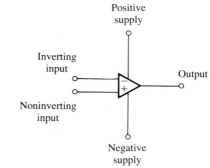

a) Operational amplifier circuit symbol

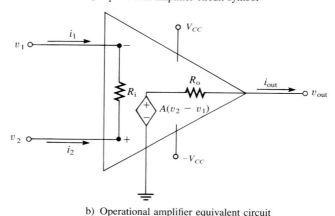

b) Operational amplifier equivalent circuit

Figure 7.1 Circuit symbol and equivalent circuit for an operational amplifier. Typical values for a 741 op amp are $R_i = 2$ MΩ, $R_o = 75$ Ω, and $A = 10^5$. In many applications, the op amp can be considered to be ideal with R_i equivalent to an open circuit, R_o equal to zero, and A approaching infinity.

The gain of the op amp, A, determines the output voltage as follows:

$$v_{out} = A(v_2 - v_1) \tag{7.1}$$

The output voltage of an op amp is limited by the supply voltages. When the output reaches its positive or negative limit, we say that the op amp has saturated. For most op amps, the saturation voltage level is about 80% of the supply voltage.

$$-V_{sat} < v_{out} < +V_{sat} \tag{7.2}$$
$$-V_{sat} = -0.8V_{cc} \text{ (approximately)}$$
$$+V_{sat} = +0.8V_{cc} \text{ (approximately)}$$

When the op amp is not saturated, its output voltage is between the two saturation voltages, and we say that it is operating in its linear region. The high gain of the op amp and the relatively low supply voltage (usually ± 20 V) combine to limit the difference between v_1 and v_2 to very small values. Let us illustrate with an example.

Example 7.1

Determine the value of $v_2 - v_1$ that will saturate an op amp if the gain, A, is 10^5, and the supply voltages are -20 V and $+20$ V.

Solution

The amplifier will saturate at 80% of the supply voltages, about -16 V and $+16$ V. At $+16$ V, the difference is

$$v_2 - v_1 = \frac{16}{10^5} = 0.16 \text{ mV}$$

At -16 V, the difference is

$$v_2 - v_1 = \frac{-16}{10^5} = -0.16 \text{ mV}$$

Ideal Op-Amp Model

Figure 7.2 shows a graph of the open-loop response of an op amp. The sloping straight line in the center of the graph is called the *linear operating region*. The horizontal lines on either side make up the *saturation region*. Some op-amp applications

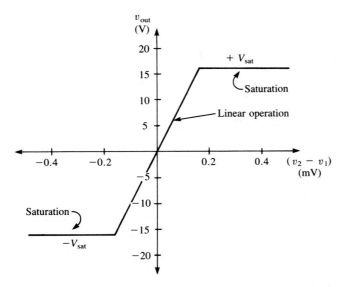

Figure 7.2   Open-loop response of an op amp with a gain of 10^5 and supply voltages of ± 20 V. The straight line in the middle of the graph is called the linear operating region. The straight lines at the top and bottom of the graph are called the saturation region.

utilize the saturation region, but most op-amp circuits are designed to operate in the linear region.

Five simplifying assumptions are used to define the model of an ideal op amp operating in the linear region. These assumptions greatly simplify the analysis of linear op-amp circuits, often with negligible effect on the accuracy of the results. The five assumptions are as follows:

Assumption	Result
1. Infinite gain, $A = \infty$	$v_1 = v_2$
2. Infinite input resistance, $R_i = \infty$	$i_1 = i_2 = 0$ A
3. Zero output resistance	$R_o = 0\ \Omega$
4. Infinite slew rate	No frequency limit
5. No offsets	Ignore offset error

Example 7.1 showed us that the difference between v_1 and v_2 is only a fraction of a millivolt. Assumption 1 states that we can neglect this difference in most applications. The practical result of assumption 1 is stated in Equation 7.3:

$$v_1 = v_2 \tag{7.3}$$

The internal resistance between the two input terminals, R_i, is large, ranging from 50 kΩ to over 100 MΩ. A typical value of R_i is 2 MΩ. The small value of $v_2 - v_1$ and the large value of R_i combine to limit i_2 to extremely small values. For the conditions in Example 7.1 with $R_i = 2$ MΩ, the value of i_2 at positive saturation is

$$i_2 = \frac{v_2 - v_1}{R_i}$$

$$= \frac{1.6 \times 10^{-4}}{2 \times 10^6} = 8 \times 10^{-11}\ \text{A}$$

The value of current i_2 is so small that we say it is *virtually* zero.

Assumption 2 states that we can neglect currents i_1 and i_2 in most applications. The practical result of assumption 2 is stated in Equation (7.4):

$$i_1 = i_2 = 0\ \text{A} \tag{7.4}$$

Assumption 3 is stated in Equation (7.5):

$$R_o = 0\ \Omega \tag{7.5}$$

The *slew rate* is the maximum rate at which the output of an op amp can change. Assumption 4 states that there is no such limit and the output can change instantaneously from one value to another. Assumption 5 states that there are no offset errors in the op-amp circuit.

The five assumptions define what is called the *ideal* operational amplifier. We will use this model frequently in analyzing op-amp circuits.

Common-Mode Rejection Ratio

In an ideal op amp, the output depends only on the difference between v_2 and v_1, not on the level of the two voltages. Consider the following two sets of conditions:

$$\text{Condition 1: } v_1 = -50 \ \mu V$$
$$v_2 = +50 \ \mu V$$
$$A = 10^5$$

$$\text{Condition 2: } v_1 = 99.95 \text{ mV}$$
$$v_2 = 100.05 \text{ mV}$$
$$A = 10^5$$

In both conditions, $v_2 - v_1 = 100 \ \mu V$ and $A(v_2 - v_1) = 10.0$ V. Thus the output of an ideal op amp is 10.0 V for both conditions 1 and 2.

In real life, operational amplifiers are not ideal, and the output will be different for the two conditions. The reason for the difference is due to slight differences in the way the op amp handles the two inputs. The inputs are amplified by slightly different gains, with the result that the common level of the two signals is amplified and added to the output. We call this common level the *common-mode voltage*, v_c. The common-mode voltage is simply the average of v_1 and v_2:

$$v_c = \frac{v_1 + v_2}{2} \tag{7.6}$$

In a practical op amp, the common-mode voltage is multiplied by the common-mode gain, A_c, and then added to the output of the op amp. The output consists of the following two components:

$$v_{out} = A_c v_c + A(v_2 - v_1) \tag{7.7}$$

Let's continue our example of the two conditions by assuming a common mode gain, A_c, of 10. The outputs including the common-mode component are:

$$\text{Condition 1: } v_c = \frac{50 + -50}{2} = 0 \ \mu V$$

$$v_{out} = A_c v_c + A(v_2 - v_1)$$
$$= 10(0) + 10^5[50 - (-50)] \times 10^{-6} = 10.0 \text{ V}$$

$$\text{Condition 2: } v_c = \frac{99.5 + 100.5}{2} = 100 \text{ mV}$$

$$v_{out} = A_c v_c + A(v_2 - v_1)$$
$$= 10(10^{-1}) + 10^5[100.05 - 99.05] \times 10^{-3} = 11.0 \text{ V}$$

Thus a common mode of only 0.1 V results in a 10% increase in the output voltage. If the common mode is raised to 1 V, the op amp will saturate. Clearly, common-mode voltage can be a problem in applications where it is present. The ability of an op amp to minimize the influence of the common-mode voltage is measured by the ratio of the differential gain of the op amp, A, divided by the common-mode gain,

A_c. This ratio is called the *common-mode rejection ratio (CMRR)*:

$$\text{CMRR} = \frac{A}{A_c} \tag{7.8}$$

The *common mode rejection* (CMR) is the logarithm of CMRR expressed in decibel units:

$$\text{CMR} = 20 \log_{10}\left(\frac{A}{A_c}\right) \tag{7.9}$$

The CMRR and CMR of our example are

$$\text{CMRR} = \frac{10^5}{10} = 10^4$$

$$\text{CMR} = 20 \log_{10}(10^4) = 80 \text{ dB}$$

Example 7.2

In a certain application of an op amp, a differential voltage $(v_2 - v_1)$ of 120 μV has a common-mode voltage that varies from 0 to 2 V. The amplifier differential gain, A, is 10^5. The common-mode error must not be greater than 1.5% of the output when the common-mode voltage is 0. Specify the common-mode rejection ratio for this application.

Solution

First, determine the output with no common-mode voltage; second, determine the maximum allowable output voltage with a common mode of 2 V; third, determine the common-mode gain; and fourth, determine the value of CMRR.

$$v_{\text{out}}\Big|_{v_c=0} = A(v_2 - v_1) = (1E+05)(120E-06) = 12 \text{ V}$$

$$v_{\text{out}}\Big|_{v_c=2} = 12\left(\frac{101.5}{100}\right) = 12.18 \text{ V}$$

$$A_c = \frac{12.18 - 12}{v_c} = \frac{0.18}{2} = 0.09$$

$$\text{CMRR} = \frac{1E+05}{0.09} = 1.11 \times 10^6$$

$$\text{CMR} = 121 \text{ dB}$$

7.3 OP-AMP CIRCUITS

This section covers a number of op-amp circuits that are used in control systems. The op-amp model used in the circuits assumes that $i_1 = i_2 = 0$, and $R_o = 0$. For the first several circuits, the op-amp model does not assume that $v_1 = v_2$. This is done to give a better understanding of the effect of the amplifier gain on the circuit equation.

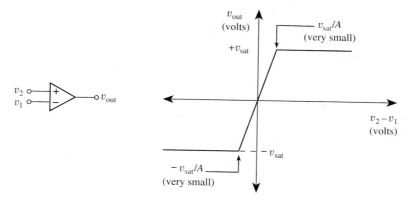

Figure 7.3   An op-amp comparator signals which of two input voltages is greater. Since open-loop gain A is very large, both $-V_{sat}/A$ and V_{sat}/A will be very small. As long as $|V_2 - V_1|$ exceeds a fraction of a millivolt, $V_{out} = V_{sat}$ when $V_2 > V_1$, and $V_{out} = -V_{sat}$ when $V_2 < V_1$.

Comparator

A *comparator* is a circuit that accepts two input voltages and indicates which voltage is greater. The comparator shown in Figure 7.3 is an op amp in the open-loop configuration, with no feedback or additional components. It is the simplest operational-amplifier circuit.

When the two input voltages are exactly equal, the theoretical output of the comparator is 0 V. However, as soon as the difference between the input voltages exceeds a fraction of a millivolt, the output of the comparator saturates. When saturation occurs and v_2 is greater than v_1, the comparator output is $+V_{sat}$. When saturation occurs and v_2 is less than v_1, the comparator output is $-V_{sat}$.

Multiple comparators may be used with a voltage divider circuit to provide switching at multiple voltage levels. Figure 7.4 shows a high-low level indicator that illustrates how this is done. The voltage divider formed by resistors R_1, R_2, and R_3 forms the two voltages v_a and v_b such that $v_a > v_b$. In this application, the op-amp supply voltages are set such that $-V_{sat} = 0$ V and $+V_{sat} = 5$ V to form logic 0 and logic 1 signals. The operation of the circuit is such that the high-level indicator is

Figure 7.4   Two comparators are used in this high-low level indicator. In this application, the comparator output switches between 0 V (logic 0) and 5 V (logic 1). Notice that R_1, R_2, and R_3 form a voltage divider such that $V_a > V_b$. When V_{in} is greater than V_a, the HI indicator is on and the LO indicator is off. When V_{in} is less than V_b, the HI indicator is off and the LO indicator is on. When V_{in} is between V_a and V_b, both indicators are off.

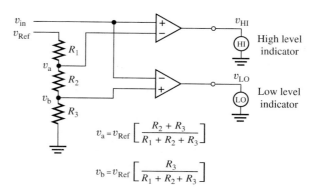

ON only when v_{in} is greater than v_a, and the low-level indicator is ON only when v_{in} is less than v_b.

Voltage Follower

A *voltage follower* is another simple application of an op amp. The output voltage is equal to the input voltage. At this point, the obvious question is: If the voltage is not changed, why use it? The answer is that although the voltage is the same, the impedance is not.

We can look at impedance changes from two perspectives. Consider placing or not placing a voltage follower between a primary element and a signal conditioning amplifier. If the voltage follower is not used, the primary sensor "sees" the input impedance of the amplifier. This might be 50 kΩ. With the voltage follower in place, the primary sensor sees the input impedance of the op amp, which could be 100 MΩ. So the effect of the voltage follower is to increase the impedance as seen by the primary sensor. This higher load impedance is a decided advantage because it reduces the current output of the primary sensor, thus reducing self-heating errors and non-linearities caused by high current levels.

Now let's examine the second perspective. The amplifier also sees an impedance when it looks toward the primary element. If the voltage follower is not used, the amplifier sees the Thévenin equivalent of the primary element. This could be 1, 10, or 100 kΩ. With the voltage follower in place, the amplifier sees the output impedance

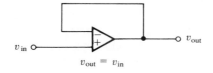

$$v_{out} = v_{in}$$

a) Voltage follower

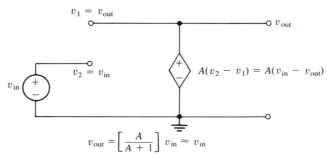

$$v_{out} = \left[\frac{A}{A+1}\right] v_{in} \approx v_{in}$$

b) Equivalent circuit

Figure 7.5 An op-amp voltage follower has unity voltage gain and a very high input impedance (typically 100 MΩ). A voltage follower may be inserted between the primary element and the remainder of the measuring instrument, thereby reducing the current produced by the primary element. The smaller primary element current reduces errors by self-heating and loading effects.

of the op amp, which could easily be less than 100 Ω. Just as a high load impedance is good for the primary element, a low source impedance is good for the amplifier. The advantage of the voltage follower is that it transforms impedances in both directions.

A circuit diagram and an equivalent circuit of a voltage follower are shown in Figure 7.5. Notice that the output terminal is connected to the inverting input, making $v_1 = v_{out}$. Also, the input voltage is connected to the noninverting input, making $v_2 = v_{in}$. Rather than using the assumption that $v_1 = v_2$, we will use Equation (7.1) for the output of the op amp.

$$v_{out} = A(v_2 - v_1) = A(v_{in} - v_{out})$$
$$v_{out}(A + 1) = Av_{in}$$
$$v_{out} = \left(\frac{A}{A + 1}\right)v_{in} \tag{7.10}$$

From Equation (7.10), we see that the output is slightly smaller than the input. Let's see just how small. If $A = 10^5$, then $A/(A + 1) = 10^5/(10^5 + 1) = 0.99999$. If $A = 10^5$, then $v_{out} = 0.99999v_{in}$.

Inverting Amplifier

The *inverting amplifier* changes the sign and the level of the input signal. It can either increase, decrease, or not change the size of the signal, based on the values of two resistors, one between the input signal and the inverting input, the other between the inverting input and the output terminal. The gain of the inverting amplifier is the ratio of the second resistor over the first. The circuit diagram and equivalent circuit are shown in Figure 7.6.

Figure 7.6 An inverting amplifier has a gain that can be less than 1, equal to 1, or greater than 1. The equivalent circuit assumes that R_i is an open circuit and R_o is zero ohms. The inverting input, v_1, is a virtual zero (actually $-v_{out}/A$).

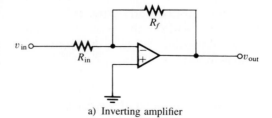

a) Inverting amplifier

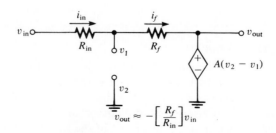

b) Equivalent circuit

The analysis of the equivalent circuit of the inverting amplifier begins with the observation that the current through resistor R_{in} is equal to the current through resistor R_f, and voltage v_2 is equal to zero. Then we apply Ohm's law to replace the currents by voltage drops divided by resistance values.

$$i_{in} = i_f \quad \text{and} \quad v_2 = 0$$

$$\frac{v_{in} - v_1}{R_{in}} = \frac{v_1 - v_{out}}{R_f}$$

By Equation (7.1), $v_{out} = A(v_2 - v_1) = -Av_1$. Thus $v_1 = -v_{out}/A$ and

$$\frac{v_{in} - (-v_{out}/A)}{R_{in}} = \frac{-v_{out}/A - v_{out}}{R_f}$$

$$\frac{v_{in}}{R_{in}} = -\frac{v_{out}}{R_f}\left(\frac{R_f}{AR_{in}} + \frac{1}{A} + 1\right) \tag{7.11}$$

The first two terms enclosed in parentheses are so small that they are virtually zero, and the entire term in parentheses can be assumed to be equal to 1. For example, if $A = 10^5$ and $R_f = 100R_{in}$, then

$$\left(\frac{R_f}{AR_{in}} + \frac{1}{A} + 1\right) = \left(\frac{100}{10^5} + \frac{1}{10^5} + 1\right) = 1.001$$

If we assume that the term in parentheses is equal to 1, we obtain the following equation for the output voltage of an inverting amplifier:

$$v_{out} = -\frac{R_f}{R_{in}}v_{in} \tag{7.12}$$

Example 7.3

A primary element produces an output voltage that ranges from 0 to 100 mV as the measured variable goes from the lower range limit to the upper range limit. Design an inverting amplifier that will take the output of the primary element and produce an output range of 0 to −5 V.

Solution

The required gain is $5/0.1 = 50$. If we choose the value of R_{in} to be 1 kΩ, the required value of R_f is

$$R_f = 50(\text{gain}) = 50(1000) = 50 \text{ k}\Omega$$

Noninverting Amplifier

The *noninverting amplifier* circuit can increase the size of the signal, but it cannot decrease the size. In the extreme case, it can leave the size of the signal unchanged, but that reduces the circuit to the simple voltage-follower circuit.

Figure 7.7 shows the circuit diagram and the equivalent circuit for a noninverting amplifier. Notice that the positions of the input voltage and the ground connection

Figure 7.7   A noninverting amplifier
has a gain that cannot be less than 1.

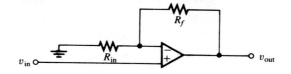

a) Noninverting amplifier

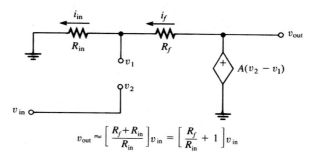

$$v_{out} \approx \left[\frac{R_f + R_{in}}{R_{in}}\right] v_{in} = \left[\frac{R_f}{R_{in}} + 1\right] v_{in}$$

b) Equivalent circuit

are interchanged from their positions in the inverting amplifier. The analysis of the
equivalent circuit is similar to the analysis of the inverting amplifier.

$$v_2 = v_{in}$$

$$v_{out} = A(v_2 - v_1) = A(v_{in} - v_1)$$

$$v_1 = v_{in} - \frac{v_{out}}{A}$$

$$i_{in} = i_f$$

$$\frac{v_1}{R_{in}} = \frac{v_{out} - v_1}{R_f}$$

$$\frac{v_{in} - v_{out}/A}{R_{in}} = \frac{v_{out} - v_{in} + v_{out}/A}{R_f}$$

$$\frac{v_{in}}{R_{in}} + \frac{v_{in}}{R_f} = \frac{v_{out}/A}{R_{in}} + \frac{v_{out}}{R_f} + \frac{v_{out}/A}{R_f}$$

$$v_{out}\left(\frac{R_f}{R_{in}A} + 1 + \frac{1}{A}\right) = \left(\frac{R_f + R_{in}}{R_{in}}\right) v_{in} \qquad (7.13)$$

Once again, for large values of A and values of R_f that are no more than 100 times
as large as R_{in}, the terms enclosed in parentheses on the left-hand side of the equation
are virtually equal to 1. With that simplifying assumption, we get the following
equation for the noninverting amplifier circuit:

$$v_{out} = \left(\frac{R_f + R_{in}}{R_{in}}\right) v_{in} = \left(\frac{R_f}{R_{in}} + 1\right) v_{in} \qquad (7.14)$$

Example 7.4

Repeat the amplifier design from Example 7.3 using the noninverting amplifier.

Solution

The required gain is 50. We will use a 1-kΩ resistor for R_{in}. The gain is given by the first term enclosed in parentheses in Equation (7.14).

$$50 = \frac{R_f + 1000}{1000} = 50$$

$$R_f = 49 \text{ k}\Omega$$

Summing Amplifier

A *summing amplifier* adds two or more input signals forming an output that is the inverse of the sum. Each input may be multiplied by a weighting factor that is formed by the ratio of two resistors. We will use the ideal op-amp model to develop the equation that defines the output of the summing amplifier circuit.

Figure 7.8 shows the circuit diagrams for two-input and three-input summing amplifiers. We will develop the equation for the output of the three-input summing

Figure 7.8 A summing amplifier forms the inverted and weighted sum of its inputs. The weighting factor is the feedback resistance, R_f, divided by the input resistor, R_a, R_b, or R_c.

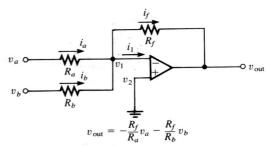

$$v_{out} = -\frac{R_f}{R_a}v_a - \frac{R_f}{R_b}v_b$$

a) Two-input summing amplifier

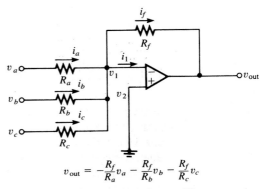

$$v_{out} = -\frac{R_f}{R_a}v_a - \frac{R_f}{R_b}v_b - \frac{R_f}{R_c}v_c$$

b) Three-input summing amplifier

amplifier. The development begins by applying Kirchhoff's current law at the inverting input terminal.

$$i_a + i_b + i_c - i_f - i_1 = 0$$

From Equation (7.4), $i_1 = 0$ and

$$i_f = i_a + i_b + i_c$$

From Equation (7.3), $v_1 = v_2 = 0$ (v_2 is grounded). Now apply Ohm's law at each resistor:

$$\frac{-v_{out}}{R_f} = \frac{v_a}{R_a} + \frac{v_b}{R_b} + \frac{v_c}{R_c}$$

$$v_{out} = -\left(\frac{R_f}{R_a}\right)v_a - \left(\frac{R_f}{R_b}\right)v_b - \left(\frac{R_f}{R_c}\right)v_c \qquad (7.15)$$

The resistor ratio R_f/R_a is the weighting factor for input voltage v_a, ratio R_f/R_b is the weighting factor for input voltage v_b, and so on.

Example 7.5

Design a circuit that produces an output voltage that is the average of three input voltages. An inversion of the output signal is permissible.

Solution

The design can be implemented with a three-input summing amplifer with weighting factors of $\frac{1}{3}$ on each input. We will select a 3.33-kΩ resistor for R_f and 10-kΩ resistors for R_a, R_b, and R_c.

Integrator

An *integrator* circuit produces an output that is proportional to the integral of the input voltage. An inverting amplifier can be converted to an integrator by replacing resistor R_f by a capacitor. Figure 7.9 shows an integrator circuit and a graphical interpretation of the integral.

The graph in Figure 7.9 is a plot of the input voltage, v_{in}, plotted versus time. The area under the graph between time t_1 and time t_2 is shaded. This shaded area is equal to the integral of the input voltage from time t_1 to time t_2.

$$\text{Shaded area} = \int_{t_1}^{t_2} v_{in}\, dt$$

If the input voltage represented the speed of travel of an object, the integral from t_1 to t_2 would represent the distance traveled during the time interval from t_1 to t_2. If the input voltage represented the flow rate of liquid into a tank, the integral would represent the amount of liquid that flowed into the tank from time t_1 to time t_2. In

Figure 7.9   An integrator produces an output that is proportional to the integral of the input voltage. On a graph of input voltage versus time, the integral between time t_1 and time t_2 is equal to the area under the graph between time t_1 and time t_2. The speedometer of a car displays the speed of the car, while the odometer displays the integral of the speed (i.e., the distance traveled).

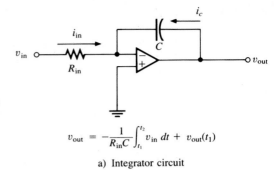

$$v_{out} = -\frac{1}{R_{in}C}\int_{t_1}^{t_2} v_{in}\,dt + v_{out}(t_1)$$

a) Integrator circuit

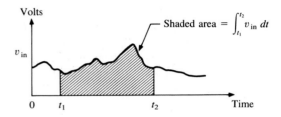

b) Graphical interpretation of an integral

general, the integral of a rate of change of quantity is the amount of that quantity that changes between the time limits of the integration.

Refer to the integrator circuit in Figure 7.9. Applying Kirchhoff's current law at the inverting input terminal, and applying Equation (7.4) (i.e., $i_1 = 0$), we get the following equation:

$$i_{in} = -i_c$$

The current through a capacitor is equal to the capacitance value times the rate of change of the voltage across the capacitor terminals. The voltage across the capacitor is $v_{out} - v_1 = v_{out}$ since $v_1 = v_2 = 0$. Thus

$$i_c = C\frac{dv_{out}}{dt}$$

$$\frac{v_{in}}{R_{in}} = -C\frac{dv_{out}}{dt}$$

$$dv_{out} = -\left(\frac{1}{R_{in}C}\right)v_{in}\,dt$$

Integrating the equation above gives the equation for the output of the integrator as a function of the input, v_{in}.

$$v_{out}(t_2) = \frac{-1}{R_{in}C}\int_{t_1}^{t_2} v_{in}(t)\,dt + v_{out}(t_1) \tag{7.16}$$

Example 7.6

The input to an integrator is a constant 100 mV. The input resistance is 10 kΩ and the capacitance is 1 μF.

    a. Find the expression for the output voltage at time t_2 as a function of t_1, t_2, and $v_{out}(t_1)$.

    b. If $t_1 = 5$ s and $v_{out}(t_1) = +10$ V, find the time t_2 when the op amp reaches saturation at -16 V (i.e., when $v_{out}(t_2) = -16$ V).

Solution

    a. $v_{out}(t_2) = -\dfrac{1}{R_{in}C} \displaystyle\int_{t_1}^{t_2} v_{in}(t_1)\, dt + v_{out}(t_1)$

        $v_{out}(t_2) = -\dfrac{1}{10^4 \cdot 10^{-6}} \displaystyle\int_{t_1}^{t_2} 0.1\, dt + v_{out}(t_1)$

        $v_{out}(t_2) = -100(0.1t)\Big|_{t_1}^{t_2} + v_{out}(t_1)$

        $v_{out}(t_2) = -10(t_2 - t_1) + v_{out}(t_1)$

    b. $-16 = -10(t_2 - 5) + 10$

        $t_2 = (16 + 10)/10 + 5 = 7.6$ s

Differentiator

A *differentiator* circuit produces an output that is proportional to the rate of change (derivative) of the input voltage. An integrator can be converted to a differentiator by replacing resistor R_f by a capacitor. Figure 7.10 shows a differentiator and a graphical interpretation of the derivative.

    Unfortunately, the most rapidly changing part of the input signal is often unwanted noise spikes or other forms of high frequency noise. In this case, the differentiator magnifies the noise even more than the useful signal, making the output very jittery. The solution to the noise problem is to add a resistor in series with the capacitor to limit the amplification of high frequency signals. Sometimes this series resistance is provided by the source impedance.

Differential Amplifier

A *differential amplifier* is a circuit that amplifies the difference between two input voltages, neither of which is equal to zero. The common-mode rejection ratio is a major concern in a differential amplifier. The circuit diagram of a differential amplifier circuit is shown in Figure 7.11.

    We begin the development of the differential amplifier equation by recalling that $v_1 = v_2$ and by applying the voltage-divider rule to resistors R_b and R_g.

$$v_1 = v_2 = v_b\left(\frac{R_g}{R_b + R_g}\right)$$

Figure 7.10 A differentiator produces an output voltage that is inverted and proportional to the rate of change of the input voltage. On a graph of input voltage versus time, the derivative is equal to the slope of the graph.

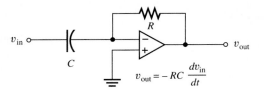

$$v_{out} = -RC \frac{dv_{in}}{dt}$$

a) Differentiator circuit

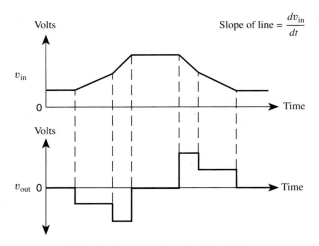

Slope of line $= \dfrac{dv_{in}}{dt}$

b) Graphical interpretation of a derivative

For resistors R_a and R_f, we have the following equation:

$$\frac{v_a - v_1}{R_a} = \frac{v_1 - v_{out}}{R_f}$$

$$v_{out} = v_1 \left(\frac{R_f}{R_a} + 1 \right) - \left(\frac{R_f}{R_a} \right) v_a$$

$$v_{out} = \left(\frac{R_g}{R_a} \right) \left(\frac{R_f + R_a}{R_b + R_g} \right) v_b - \left(\frac{R_f}{R_a} \right) v_a \tag{7.17}$$

If $R_a = R_b$, and $R_f = R_g$, Equation (7.17) reduces to

$$v_{out} = \frac{R_f}{R_a} (v_b - v_a) \tag{7.18}$$

Instrumentation Amplifier

An *instrumentation amplifier* (IA) is essentially a differential amplifier with high input impedance, high common-mode rejection, balanced differential inputs, and gain determined by a user-selected external resistor. In order to provide a guaranteed level of performance, all components except the external gain resistor are inside the IA package. The high input impedance of the IA minimizes the current draw from the input

Figure 7.11 A differential amplifier is used to measure the difference between two voltages. The circuit equation simplifies considerably when $R_a = R_b$ and $R_f = R_g$.

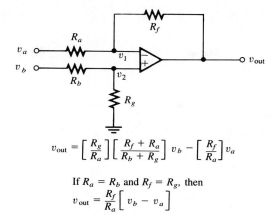

$$v_{\text{out}} = \left[\frac{R_g}{R_a} \right] \left[\frac{R_f + R_a}{R_b + R_g} \right] v_b - \left[\frac{R_f}{R_a} \right] v_a$$

If $R_a = R_b$ and $R_f = R_g$, then

$$v_{\text{out}} = \frac{R_f}{R_a} \left[v_b - v_a \right]$$

circuit, thus reducing self-heating and loading effects on the input circuit. Figure 7.12 shows a three-op-amp instrumentation amplifier. The input stage consists of two op amps in a modified voltage follower configuration. The second stage is the differential amplifier shown in Figure 7.11 with balanced resistors (i.e., $R_b = R_a$ and $R_g = R_f$).

Assuming ideal op amps in Figure 7.12, the same current (i) flows through resistors R'_1, R_e, and R_1. By Ohm's law,

$$i = \frac{v_b - v_a}{R'_1 + R_e + R_1} = \frac{v_b - v_a}{R_1 + R_e + R_1} = \frac{v_b - v_a}{2R_1 + R_e}$$

and

$$i = \frac{v_{ib} - v_{ia}}{R_e}$$

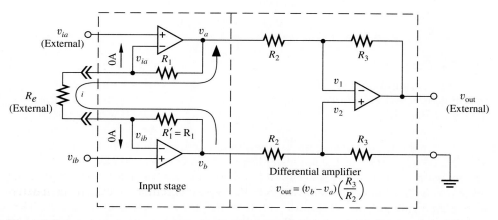

Figure 7.12 A three-op-amp instrumentation amplifier consists of an input stage and a differential amplifier. The user can adjust the size of the external resistor to obtain a wide range of gains.

Equating the right side of the two equations for i, we have

$$\frac{v_b - v_a}{2R_1 + R_e} = \frac{v_{ib} - v_{ia}}{R_e}$$

$$v_b - v_a = \left(\frac{2R_1 + R_e}{R_e}\right)(v_{ib} - v_{ia})$$

Finally, replacing $v_b - v_a$ in Equation (7.18) by the right side of the above equation gives us the following equation for the instrumentation amplifier:

$$v_{out} = \left(\frac{2R_1 + R_e}{R_e}\right)\left(\frac{R_3}{R_2}\right)(v_{ib} - v_{ia}) \tag{7.19}$$

The user can adjust the size of external resistor R_e to obtain a wide range of gains without increasing the common-mode error signal.

Example 7.7

An instrumentation amplifier similar to Figure 7.12 has the following precision resistor values:

$$R_1 = 1000\ \Omega,\ R_2 = 1000\ \Omega,\ \text{and}\ R_3 = 1000\ \Omega$$

Determine the value of the external resistor, R_e, that will result in a gain of 1000.

Solution

$$\text{IA gain} = \left(\frac{2R_1 + R_e}{R_e}\right)\left(\frac{R_3}{R_2}\right)$$

$$1000 = \left(\frac{2(1000) + R_e}{R_e}\right)\left(\frac{1000}{1000}\right)$$

$$1000R_e = 2000 + R_e$$

$$999R_e = 2000$$

$$R_e = 2000/999 = 2.002\ \Omega$$

Function Generator

Some signal conditioning applications require a component with an output that is a nonlinear function of its input. Components of this type are called *function generators*. The inverting amplifier can be converted into a function generator by replacing one of its two resistors with a component that has a nonlinear volt–ampere relationship.

A logarithmic amplifier is one example of a function generator. A transistor with a grounded base serves as the nonlinear element. The transistor replaces the feedback resistor with the collector connected to the summing junction and the emitter connected to the output terminal of the op amp. The logarithmic amplifier utilizes the logarithmic volt–ampere relationship of the transistor to generate an output voltage that is proportional to the natural logarithm of the input voltage.

Linearization of the signal from a differential pressure flow transmitter is another example of an application of a function generator. The flow transmitter output voltage (v_F) is equal to a constant (K_F) multiplied by flow rate (Q) squared, as shown in Equation (7.20):

$$v_F = K_F Q^2 \tag{7.20}$$

In other words, the flow transmitter output voltage is proportional to the flow rate squared. We would prefer a voltage that is proportional to the flow rate, and for that we need a function generator whose output is equal to the square root of its input. With an input voltage equal to $K_L Q^2$, a square root function generator will produce an output voltage equal to $\sqrt{K_L}(Q)$, which is the desired linear signal. Figure 7.13a shows a square-root function generator used to linearize the output signal from a flow transmitter.

Figure 7.13 A nonlinear function generator implements a square-root function to linearize the output of a flow transmitter.

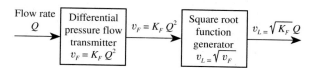

a) A square root function generator is used to linearize the output voltage of a differential pressure flow transmitter.

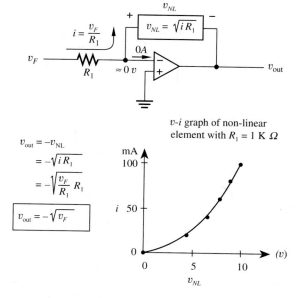

b) An inverting amplifier is used to make a square root function generator.

An op-amp circuit for the square-root function generator is shown in Figure 7.13b. The circuit is an inverting amplifier with a nonlinear element in the feedback path. The nonlinear element has the following volt–ampere relationship:

$$v_{NL} = \sqrt{iR_1} \tag{7.21}$$

Recall that the summing junction of the op amp is a virtual 0 V. Thus the voltage across R_1 is v_F, and current i is easily determined by Ohm's law:

$$i = \frac{v_F}{R_1} \tag{7.22}$$

The output voltage v_{out} is also easily determined:

$$v_{out} = -v_{NL} \tag{7.23}$$

Combining Equations (7.21), (7.22), and (7.23) results in the following equation for v_{out} in terms of v_F:

$$v_{out} = -\sqrt{v_F} \tag{7.24}$$

Example 7.8

A square root function generator similar to Figure 7.13b uses a 1 kΩ resistor for R_1. Draw a volt–ampere graph of the nonlinear element for values of current i from 0 to 100 mA.

Solution

$$v_{NL} = \sqrt{iR_1} = \sqrt{1000i}$$

i (mA)	0	20	40	60	80	100
v_{NL} (volts)	0	4.47	6.32	7.75	8.94	10.00

The volt–ampere graph is shown in Figure 7.13b.

7.4 SIGNAL CONDITIONING

The type of signal conditioning required for a particular measurement depends on the electrical characteristics of the primary element and the component that will receive the signal. Typical conditioning includes galvanic isolation, common-mode isolation, impedance transformation, amplification, noise reduction, signal conversion, linearization, compensation, and calibration.

In the trend toward distributed control, the signal conditioning circuitry has moved close to the sensor on the process floor. Signal conditioning is done in the individual measuring transmitters, and it is done in signal conditioning systems using plug-in modules for various functions. The plug-in modules are rack mounted in card cages that can accept 4, 8, or 16 plug-in modules. The following is a brief description of some typical plug-in modules.*

A *millivolt/thermocouple* module accepts low-level dc voltage signals with a span as low as 2 mV or as high as 55 mV. The zero can be adjusted from −5 to 25 mV. The module converts the input to a high-level output such as a 4- to 20-mA current signal. The circuit provides isolation from dc and ac common-mode voltages of several hundred volts, and a common-mode rejection ratio of 140 dB at 60 Hz. The module provides noise rejection, linearization, cold junction compensation, and break detection for thermocouple inputs. The module can be factory calibrated for a specified zero and span.

A *resistance temperature detector* (*RTD*) module accepts input from 100-Ω platinum or 10-Ω copper RTD sensors. The input signal is converted to a high-level input such as a 4- to 20-mA current signal. The module provides common-mode isolation from ac and dc voltages of several hundred volts, and a CMR of 120 dB at 60 Hz. It also provides excitation for the RTD sensor, lead wire compensation, noise reduction, linearization, and break detection. A typical 98% response time is 0.25 s.

A *frequency input* module accepts pulse signals from digital tachometers, turbine flow meters, and other sensors that produce a series of pulses. The frequency of the input pulses is converted to a high-level voltage or current signal. The frequency range is selectable from as low as 25 Hz to as high as 12 kHz. The 98% response time varies from 10 s for the low-frequency range to 0.2 s for the high-frequency range.

The remainder of this section covers details of various signal conditioning operations.

Isolation and Impedance Conversion

To quote Lord Kelvin: "The act of measuring something destroys that which is measured." Consider the problem of measuring the temperature of a small container of water with a thermometer. Let's assume that the water temperature is 40°C and the temperature of the thermometer is 22°C before the measurement takes place. The measurement consists of placing the thermometer in the water and reading the thermometer after a temperature equilibrium has been reached between the water and the thermometer. The equilibrium temperature depends on the ratio of the heat capacities of the thermometer and the water. If the heat capacities are equal, the equilibrium temperature will be 31°C, the average. Whatever the equilibrium temperature is, it will be less than 40°C and greater than 22°C. Thus the act of measuring the temperature of the water has changed the temperature.

One purpose of isolation and impedance conversion is to avoid, or at least minimize, the destruction of "that which is measured." An equally important purpose is to protect the measuring instrument from "that which is measured." For example, a high common-mode voltage can easily destroy an op-amp circuit unless the circuit is isolated from the high voltage.

---

* *Series 1800 Analog Signal Conditioning System* (Wixom, Mich.: Acromag).

Impedance transformation is one method of protecting the measured variable and the measuring instrument from each other. The voltage follower, introduced in Section 7.3, does this job quite well (refer to Section 7.3 for further details). When galvanic isolation is required, special amplifiers called *isolation amplifiers* are used. There are two methods that can be used to obtain galvanic isolation: transformer coupling and optical coupling. A typical isolation amplifier with transformer coupling has a capacitance of about 10 pF between the input and the output circuits, and a CMRR of about 120 dB. Optical coupling uses a light beam to transmit the signal from the source circuit to the receiver circuit. This makes it possible to remove all electrical connections between the two circuits.

Noise Reduction

Control systems exist in an environment filled with high levels of electromagnetic energy just waiting to produce noise in electric signal lines. The best noise reduction system is one that prevents the noise from ever getting to the signal. Noise prevention consists of careful grounding of signal lines and shielding of cables to keep the signals as free of noise as possible. Despite the best noise prevention effort, some noise will appear in the control signals. Special circuits called *filters* are designed to reduce the level of the noise in the signals. Actually, filters can be used to reduce everything in a specific range (or band) of frequencies. The terms *low-pass filter*, *band-pass filter*, *high-pass filter*, and *notch filter* name various filters according to the band of frequencies they allow to pass through unaffected.

Before going on, we should pause a moment and consider the frequency components of a signal. Everyone is familiar with the fact that we can tune a radio to receive different stations. Each station is assigned a carrier frequency on which it superimposes its voice and music signals. A radio antenna receives the carrier signals from all stations in its vicinity, but the radio receiver is sensitive to only one carrier at a time, depending on the position of the tuner. You select a different station by moving the tuner to a different position.

In the discussion of filters, we view control signals as a collection of signal components with different frequencies. The term *component*, as used in the following discussion, refers to a frequency component of a control signal. A filter chooses to accept certain frequencies and reject other frequencies. The Bode diagram tells us which frequency components the filter reduces and how much each component is reduced. In this section we construct Bode diagrams of a low-pass filter and a notch filter.

A low-pass filter does not affect the components of a signal that are below a certain frequency called the break-point frequency, f_b. All components of the signal that are above the break point are reduced in amplitude, and the higher the frequency, the greater the reduction. Low-pass filters are very common in signal conditioning because most of the useful information in the signal is in the low-frequency components. Since noise tends to occur at the higher frequencies, a low-pass filter can often be designed to reduce the noise without affecting the information content of the signal.

A high-pass filter is just the opposite of the low-pass filter. It does not affect the components of the signal that are above the break-point frequency. All components of the signal that are below the break point are reduced in amplitude, and the lower the frequency, the greater the reduction. High-pass filters do not make much sense in control, for the same reasons that low-pass filters are useful.

A band-pass filter has two frequency values, called the *half-power frequencies*, that are separated by a frequency range called the *bandwidth* of the filter. The band-pass filter has little or no effect on components of the signal that are between the two half-power frequencies. All components of the signal that are outside the half-power frequencies are reduced in amplitude, and the farther they are from the half-power frequency, the greater the reduction. The radio tuner is an adjustable band-pass filter that accepts one station and rejects all other stations.

A notch filter is just the opposite of the band-pass filter. The notch filter also has two half-power frequencies separated by the bandwidth. However, the notch filter reduces all components of the signal between the two half-power frequencies and does not affect the components on either side. The maximum reduction of the signal occurs at the midpoint between the two half-power frequencies. Much of the noise in a process signal occurs at particular frequencies, such as the 60-Hz noise generated by electric power lines. Notch filters are good for removing a particular frequency component, such as the 60-Hz noise component of a signal. Low-pass filters and notch filters are the most common noise reduction circuits.

Three low-pass filter circuits are shown in Figure 7.14. The first circuit is a passive *RC* low-pass filter. This is the same circuit we analyzed in Chapter 5 (see Figure 5.3). The time-domain equation and transfer function were also developed in Chapter 5. The two equations are given below with the letters e and E replaced by the letters v and V.

$$\tau \frac{dv_{out}}{dt} + v_{out} = v_{in} \tag{5.10}$$

$$\frac{V_{out}}{V_{in}} = \frac{1}{1 + \tau s} \tag{5.27}$$

where $\tau = RC$.

One problem with the passive circuit in Figure 7.14a is that the equation of the circuit changes when a load resistor is connected to the output terminal. The break-point frequency is thus dependent on the load resistor—a very undesirable situation because it means we cannot design the filter until we know the exact size of the load resistance. Any time the load resistance changes, the filter must also be changed.

The active filter in Figure 7.14b uses a voltage follower to isolate the filter circuit from the load resistance. The active filter break-point frequency is not affected by the load resistance value. If the input impedance of the source, v_{in}, is a problem, a voltage follower can be used to isolate the input as well. Equations (5.10) and (5.27) also apply to the active circuit in Figure 7.14b.

A two-stage active low-pass filter is shown in Figure 7.14c. This circuit has twice the reduction power of the single-stage circuit. We will use the frequency domain to develop the transfer function of the two-stage filter from the transfer functions of each stage.

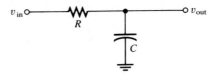

a) Passive low-pass filter

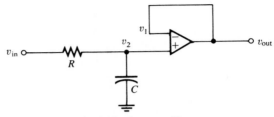

b) Active low-pass filter

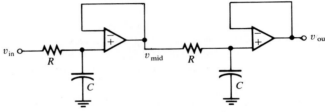

c) Two-stage active low-pass filter

Figure 7.14 The passive RC circuit (a) is the simplest low-pass filter. Adding the voltage follower (b) isolates the filter from the load resistor. Adding a second isolated stage results in a two-stage active low-pass filter (c) with twice the attenuation of the single-stage filter.

$$\frac{V_{out}}{V_{in}} = \left(\frac{V_{mid}}{V_{in}}\right)\left(\frac{V_{out}}{V_{mid}}\right)$$

$$\frac{V_{out}}{V_{in}} = \left(\frac{1}{1 + \tau s}\right)\left(\frac{1}{1 + \tau s}\right)$$

$$\frac{V_{out}}{V_{in}} = \frac{1}{(1 + \tau s)(1 + \tau s)} \tag{7.25}$$

$$\frac{V_{out}}{V_{in}} = \frac{1}{1 + 2\tau s + \tau^2 s^2} \tag{7.26}$$

Example 7.9

Use the program "BODE" from Appendix F to generate frequency response data for a two-stage active low-pass filter with a time constant, $\tau = 0.01$. Plot the Bode diagram of the filter and determine the attenuation at a frequency of 60 Hz (i.e., $\omega = 377$ rad/s).

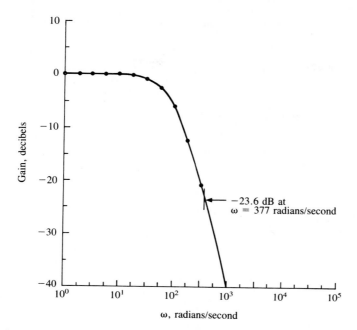

Figure 7.15   Bode diagram for the two-stage active filter circuit with a time constant $\tau = 0.01$ s (Example 7.9). The attenuation at 60 Hz (377 rad/s) is 23.6 dB.

Solution

The Bode diagram is shown in Figure 7.15. The gain at 377 rad/s is -23.6 dB, so the attenuation is 23.6 dB. The gain can also be computed from the transfer function by substituting $j377$ for ω.

$$\text{Gain (at } \omega = 377) = \frac{1}{[1 + j377(0.01)][1 + j377(0.01)]}$$

$$\text{Gain} = \frac{1}{(3.90 \underline{/75^\circ})(3.90 \underline{/75^\circ})} = 0.0657 \underline{/-150^\circ}$$

$$\text{Decibel gain} = 20 \log_{10}(0.0657) = -23.6$$

The design of a low-pass filter consists of determining the filter time constant, τ [i.e., the coefficient of s in the denominator of Equation (5.27) or Equation (7.25)]. The following rules are intended for attenuation factors of 10 or more. For a single-stage low-pass filter, the time constant (τ) is equal to the attenuation factor (A) divided by the radian frequency (ω_s) of the component you wish to attenuate. For a two-stage low-pass filter, the time constant is equal to the square root of the attenuation factor divided by ω_s. These two rules are summarized in the following equations:

$$\text{Single-stage filter: } \tau = \frac{A}{\omega_s} \tag{7.27}$$

$$\text{Two-stage filter: } \tau = \frac{\sqrt{A}}{\omega_s} \tag{7.28}$$

where $\tau = $ filter time constant, second

$$A = \frac{\text{amplitude before filtering}}{\text{amplitude after filtering}}$$

$\omega_s = $ frequency to be filtered, radian/second

A notch filter produces a V-shaped curve on the Bode amplitude diagram. The notch of the V is located at the frequency of the component to be attenuated. Equation (7.29) gives the transfer function of a notch filter.

$$\frac{v_{\text{out}}}{v_{\text{in}}} = \frac{(1 + \tau_s s)(1 + \tau_s s)}{(1 + k\tau_s s)\left(1 + \dfrac{\tau_s}{k} s\right)} = \frac{1 + 2\tau_s s + \tau_s^2 s^2}{1 + \tau_s(k + 1/k)s + \tau_s^2 s^2} \tag{7.29}$$

where $\tau_s = \dfrac{1}{\omega_s}$

$\omega_s = $ frequency to be filtered, radian/second

$k = $ number determined by the attenuation factor

A value of $k = 20$ produces an attenuation factor, A, of slightly more than 10. Values of k for other attenuation factors can be determined experimentally, using the program "BODE" to verify the attenuation factor obtained.

Example 7.10

Design a notch filter that will attenuate a 60-Hz signal component by a factor of 10. Use the program "BODE" to generate frequency response data and plot the Bode diagram amplitude curve.

Solution

The first step is to determine the coefficients in Equation (7.29) and multiply the numerator and denominator binomials to prepare the equation for use in the program "BODE".

$$\omega_s = 2\pi(60) = 377 \text{ rad/s}$$

$$\tau_s = \frac{1}{377} = 0.002653 \text{ s}$$

$$k\tau_s = 20(0.002653) = 0.05305 \text{ s}$$

$$\frac{\tau_s}{20} = \frac{0.002653}{20} = 0.0001326 \text{ s}$$

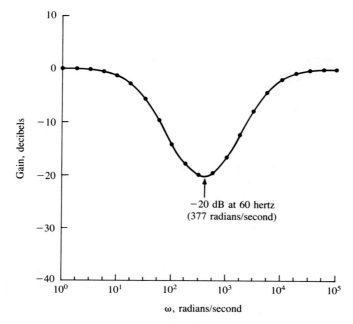

Figure 7.16 Bode diagram of a notch filter designed to produce at least 20 dB of attenuation at 60 Hz (377 rad/s). See Example 7.10 for details.

$$\frac{V_{out}}{V_{in}} = \frac{(1 + 0.002653s)(1 + 0.002653s)}{(1 + 0.05305s)(1 + 0.0001326s)}$$

$$\frac{V_{out}}{V_{in}} = \frac{1 + 5.31 \times 10^{-3}s + 7.04 \times 10^{-6}s^2}{1 + 5.32 \times 10^{-2}s + 7.04 \times 10^{-6}s^2}$$

The Bode diagram is shown in Figure 7.16.

Linearization

The ideal measuring instrument produces a linear calibration curve in which the output goes from 0% to 100% of the output range as the input goes from 0% to 100% of the input range. Some primary elements, such as a platinum resistance detector, produce a very linear signal, and the right amount of amplification will produce a nearly ideal calibration curve. Other primary elements produce nonlinear outputs, and the signal must be linearized to produce a nearly ideal calibration curve. We will examine the *linearization* of two nonlinear signals. The first nonlinear signal involves the measurement of liquid flow rate with an orifice and differential pressure (Δp) transmitter. The second involves the measurement of the volume of a round liquid tank that is resting on its side.

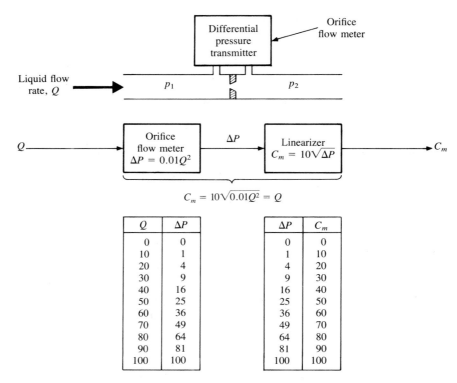

Figure 7.17   An orifice flow meter produces a Δp signal that is proportional to the square of the flow rate, Q. By forming the square root of its input signal, the linearizer produces an output that is proportional to the flow rate, Q.

Figure 7.17 illustrates the measurement and linearization of liquid flow rate when the flow is turbulent. The flow meter consists of the orifice and differential pressure transmitter, which produces an output, Δp, that is proportional to the square of the flow rate, Q. If we express both signals in percent of their full-scale range, the equation for the output of the flow meter is given by Equation (7.30).

$$\Delta p = p_1 - p_2 = 0.01Q^2 \tag{7.30}$$

The table on the left in Figure 7.17 shows corresponding values of Δp and Q over the range of the measurement. Notice the obvious nonlinearity of the output, Δp. At a flow rate of 20%, Δp is only 4%; at a flow rate of 50%, Δp is only 25%. One way to handle a nonlinear signal is to plot the calibration curve on nonlinear graph paper. A square-root scale will linearize the signal from the Δp transmitter. However, the preferred method of handling the signal is to linearize the signal with a signal conditioning component that we will call a linearizer.

The output of the Δp transmitter, Δp, is the input to the linearizer. The output of the linearizer c_m is also expressed as a percent of the full-scale range. The design goal of the linearizer is to produce a percent output that is exactly equal to the percent flow rate signal, Q, as given in the equation

$$c_m = Q \qquad\qquad (7.31)$$

The mathematical description of the linearizer is the inverse of the function that describes the flow meter. We can obtain the inverse function by solving Equation (7.30) for Q and then substituting c_m for Q in the resulting inverse function.

$$0.01Q^2 = \Delta p$$
$$Q^2 = 100\,\Delta p$$
$$Q = \sqrt{100\,\Delta p}$$
$$c_m = 10\sqrt{\Delta p} \qquad\qquad (7.32)$$

The table on the right in Figure 7.17 shows corresponding values of Δp and c_m over the range of measurement. Tracing a few signals through the two tables will verify Equation (7.31). For example, a flow rate of 20% produces a Δp of 4%, which in turn produces a c_m of 20%. The input/output curves of the flow meter and the linearizer are plotted in Figure 7.18. A function and its inverse have an interesting and useful graphical feature. The two curves produce a pattern that is symmetrical about the line $y = x$. We can use this symmetry to plot the graph of the inverse of any nonlinear calibration curve. When we do this, we are producing the curve that defines the linearizer for that nonlinear signal. Another way of defining the linearizer is to produce a table of values similar to the table on the right in Figure 7.17. This table was constructed by switching the two columns from the table on the left and changing the heading on the flow rate column from Q to c_m.

> The linearizer can be defined mathematically by the inverse of the function that defines the primary element. It can be defined with a table by switching the columns of the primary element table and renaming the left column. It can be defined graphically by constructing the curve that completes the symmetry about the line $y = x$.

Once the linearizer is defined, by one of the three methods above, we are faced with the problem of implementing the definition in an electric circuit. An operational amplifier makes an excellent function generator. However, it requires an electrical component that has a nonlinear volt–ampere characteristic that matches either the primary element function or its inverse. Refer to Section 7.3 for details on the op-amp circuits for generating functions. Another approach to implementing the linearizer is to construct an approximate inverse function by a set of straight lines, each forming a portion of the nonlinear curve. This type of function is called a *piecewise-linear function*. Figure 7.18 uses a five-step piecewise-linear function to approximate the function $c_m = 10\sqrt{\Delta p}$. Each of the five lines covers 20% of the c_m scale. For simplicity, only one of the break points is not on the ideal curve. The point at $c_m = 20\%$ was moved from $\Delta p = 4$ to $\Delta p = 3.2$. All other endpoints are on the ideal curve. Table 7.1 of corresponding values at the midpoints illustrates the accuracy of the approximation. The errors above $c_m = 35\%$ can be reduced to about 0.5% by moving the

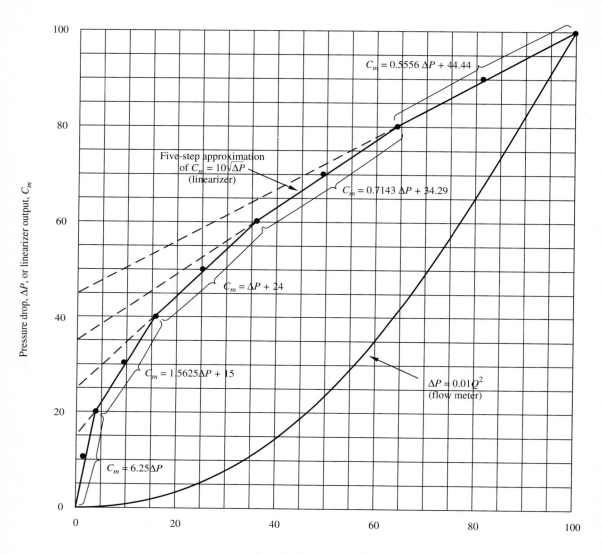

Figure 7.18 A piecewise-linear approximation is one way to implement a nonlinear function. Here the inverse of the square function, $\Delta p = 0.01Q^2$, is formed by a five-segment piecewise-linear approximation. With careful selection of the lines, the error in the top 65% of the curve can be limited to about 0.5% of full-scale range.

Table 7.1 Comparison of the Mid-points of the Line Segments with the Corresponding Ideal Values[a]

Δp	Ideal c_m	Approximate c_m	Percent Error
1	10	6.25	3.75
9	30	29.06	0.94
25	50	49.00	1.00
49	70	69.29	0.71
81	90	89.44	0.56

[a] All values are in percent of full-scale range.

endpoints until the maximum error is minimized. This will be left as an exercise for the student.

Example 7.11

Construct a five-step, piecewise-linear approximation of the function defined by the input/output table below. Locate the endpoints on the curve at the following input values: 0, 20, 40, 60, 80, and 100. Compute the slope and intercept of each line segment, write the equation for each segment, and construct an error table similar to Table 7.1. Move the endpoints to reduce the largest errors and write the equations and the error table for the second set of line segments.

Input	Output	Input	Output	Input	Output
0	3.0	35	33.2	70	72.8
5	7.0	40	38.7	75	77.1
10	11.0	45	44.6	80	81.2
15	15.2	50	50.4	85	85.2
20	19.5	55	56.4	90	89.3
25	23.8	60	62.1	95	93.2
30	28.3	65	67.8	100	97.0

Solution

We will proceed as follows:

1. Use $m = (y_2 - y_1)/(x_2 - x_1)$ to compute the slope of each line segment.
2. Use $b = y_1 - mx_1$ to compute the intercept of each segment.
3. Use $y = mx + b$ to write the equation of each line.

The results are summarized in the following two tables.

Endpoints and Equations of the Line Segments

Line No.	(x_1, y_1)	(x_2, y_2)	Equation
1a	0, 3.0	20, 19.5	$y = 0.825x + 3.0$
2a	20, 19.5	40, 38.7	$y = 0.960x + 0.30$
3a	40, 38.7	60, 62.1	$y = 1.170x - 8.10$
4a	60, 62.1	80, 81.2	$y = 0.955x + 4.80$
5a	80, 81.2	100, 97.0	$y = 0.790x + 18.00$

Midpoint Error for Each Line Segment	Input	Ideal Output	Approximate Output	Error
	10	11.0	11.25	−0.25
	30	28.3	29.10	−0.80
	50	50.4	50.40	0.00
	70	72.8	71.65	1.15
	90	89.3	89.10	0.20

The two large errors are in line segments 2a and 4a. One way to reduce the error is to make these two segments smaller. We chose to move the second endpoint from 20 to 30 and the fourth endpoint from 60 to 70. The results are as follows:

Endpoints and Equations of the Line Segments	Line No.	(x_1, y_1)	(x_2, y_2)	Equation
	1b	0, 3.0	30, 28.3	$y = 0.843x + 3.0$
	2b	30, 28.3	40, 38.7	$y = 1.040x - 2.90$
	3b	40, 38.7	70, 72.8	$y = 1.137x - 6.77$
	4b	70, 72.8	80, 81.2	$y = 0.840x + 14.00$
	5b	80, 81.2	100, 97.0	$y = 0.790x + 18.00$

Midpoint Error for Each Line Segment	Input	Ideal Output	Approximate Output	Error
	15	15.2	15.65	−0.45
	35	33.2	33.50	−0.30
	55	56.4	55.77	0.63
	75	77.1	77.05	0.10
	90	89.3	89.10	0.20

Further reduction is possible with a careful selection of the endpoints. Sometimes moving the endpoints off the line helps to reduce the error. A large, accurate graph is very helpful in determining the endpoints.

In the flow example, the equation of the inverse function was easy to determine. In the second example, the inverse function is more difficult to determine, so we resort to graphical and tabular methods to define the linearizer. Figure 7.19a illustrates a liquid storage tank resting on its side with the axis in a horizontal position. The input to the sensor is the level, h, of the liquid in the tank. The output of the sensor is a nonlinear volume signal designated by V_{NL} (the NL subscript indicates that we consider the output of the sensor to be a nonlinear measurement of the volume in the tank). The graph of h versus V_{NL} forms an S-shaped curve as shown below the sensor in Figure 7.19c. The output from the sensor, V_{NL}, is the input to the linearizer. The output of the linearizer, V_{LIN}, is such that a graph of h versus V_{LIN} is the straight line at the bottom in Figure 7.19e. The required graph of the linearizer is the reverse S-shaped curve shown under the linearizer in Figure 7.19d.

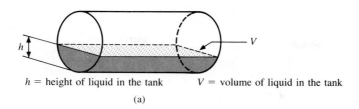

h = height of liquid in the tank V = volume of liquid in the tank

(a)

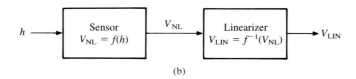

(b)

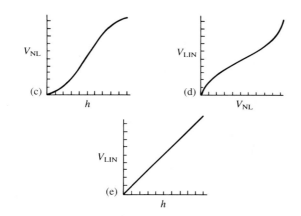

Figure 7.19   A nonlinear signal results when liquid level is used to measure the volume of liquid in a round tank resting on its side. The graph of the liquid level, h, versus the volume in the tank forms an S-shaped curve. The linearizer must form a reverse S-shaped curve to linearize the signal.

The equation that defines the volume in the tank as a function of the height, h, is quite complex. So complex, in fact, that we will use an equation that defines the volume of the tank from empty to half full. Then we will use the obvious symmetry of the tank to complete the curve for the top half. In the following equations, h is the height of the tank, r is the radius of the tank, and V_{NL} is the output of the level sensor. The following equations define the volume of liquid in the bottom half of the tank.

$$k = \frac{h}{r}$$

$$\alpha = \tan^{-1}\left(\frac{\sqrt{k(2-k)}}{k-1}\right) \tag{7.33}$$

$$V_{NL} = \frac{50(\alpha - \sin \alpha)}{\pi} \tag{7.34}$$

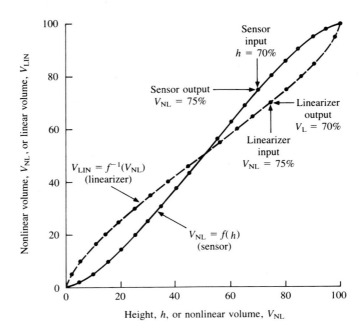

Figure 7.20 Input versus output curves for the sensor and the linearizer for measurement of the volume of liquid in a round tank resting on its side.

More accurate graphs of V_{NL} versus h and V_{LIN} versus V_{NL} are shown in Figure 7.20. Table 7.2 was used to construct the two curves in Figure 7.20.

Amplification and Analog-to-Analog Conversion

Changing the level of an analog signal is accomplished by either an inverting amplifier, a noninverting amplifier, or a differential amplifier. Sometimes it is necessary to convert the signal from a voltage to a current or from a current to a voltage. Figure 7.21 shows simple op-amp converter circuits.

Table 7.2 Values of Height, h, Versus Volume, V_{NL}, for the Sensor in Figure 7.19[a]

h	V_{NL}	h	V_{NL}	h	V_{NL}	h	V_{NL}
0	0.0	25	19.6	55	56.4	80	85.8
5	1.9	30	25.2	60	62.7	85	90.6
10	5.2	35	31.2	65	68.8	90	94.8
15	9.4	40	37.3	70	74.8	95	98.1
20	14.2	45	43.6	75	80.4	100	100.0
		50	50.0				

[a] Values are in percent of full-scale range.

Figure 7.21 Voltage-to-current and current-to-voltage converters using the ideal op amp model.

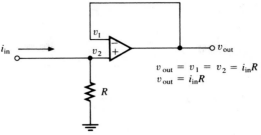

$$v_{out} = v_1 = v_2 = i_{in}R$$
$$v_{out} = i_{in}R$$

a) Current-to-voltage converter

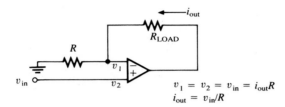

$$v_1 = v_2 = v_{in} = i_{out}R$$
$$i_{out} = v_{in}/R$$

b) Voltage-to-current converter

The current-to-voltage converter in Figure 7.21a is a voltage follower with a resistor connected from the noninverting input to ground. The input current, i_{in}, passes to ground through the resistor, making voltage v_2 equal to $i_{in}R$. The output terminal is connected to the input, making $v_{out} = v_1$. Assuming that the op amp is ideal makes $v_1 = v_2$, and by proper substitution, we get the following equation for the current-to-voltage converter:

$$v_{out} = i_{in}R \qquad \qquad (7.35)$$

Example 7.12

Design a current-to-voltage converter that converts a 20-mA input signal into a 5-V output signal.

Solution

Equation (7.35) applies to this problem.

$$v_{out} = i_{in}R$$

$$R = \frac{v_{out}}{i_{in}} = \frac{5}{0.02} = 250 \ \Omega$$

The converter is the circuit in Figure 7.21a with $R = 250 \ \Omega$.

The voltage-to-current converter in Figure 7.21b is a noninverting amplifier with the load resistor placed in the feedback path where r_f is normally placed. The output current, i_{out}, passes to ground through load resistor R_{LOAD} and scaling resistor R, making voltage v_1 equal to $i_{out}R$. The input voltage is connected to the noninverting terminal, making $v_2 = v_{in}$. For an ideal op amp, $v_1 = v_2$, and we get the following equation for the voltage-to-current converter.

$$i_{out} = \frac{v_{in}}{R} \tag{7.36}$$

Example 7.13

Design a voltage-to-current converter that converts a 120-V dc input signal into a 20-mA current signal.

Solution

Equation (7.36) applies to this problem.

$$i_{out} = \frac{v_{in}}{R}$$

$$R = \frac{v_{in}}{i_{out}} = \frac{120}{0.02} = 6 \text{ k}\Omega$$

The converter is the circuit shown in Figure 7.21b with $R = 6$ kΩ.

Bridge Circuits

A number of primary elements convert changes in the measured variable into small changes in the resistance of the element. Strain gage force transducers, strain gage pressure transducers, and resistance temperature detectors are three examples. A bridge circuit is the traditional method of measuring small changes in the resistance of an element. The operation of a bridge falls into two categories, balanced and unbalanced operation. In the balanced operation, the resistance of the sensor is determined from the values of three other resistors whose values are known with precision. In the unbalanced operation, the change in the sensor resistance from a base value produces a small difference between two voltages. A differential amplifier is used to amplify the difference between the two voltages.

A balanced *Wheatstone bridge* is shown in Figure 7.22. The unknown resistance of the sensor is labeled R_s. Resistors R_2 and R_4 have fixed values for a given measurement. Resistor R_3 can be continuously adjusted over a calibrated range of values. The value of R_3 is adjusted until the meter in the center reaches a null position, indicating that $v_a = v_b$ and $i_{ab} = 0$. When this occurs, the bridge is said to be *balanced*. When the bridge is balanced, current i_1 passes through resistors R_s and R_2. Also, current i_3 passes through resistors R_3 and R_4. Since $v_a = v_b$, it follows that the voltages across R_s and R_3 are equal, as well as the voltages across R_2 and R_4. Thus

$$i_1 R_s = i_3 R_3$$
$$i_1 R_2 = i_3 R_4$$

Figure 7.22   Wheatstone bridge circuit. When the circuit is in the balanced condition, $R_s = R_2 R_3 / R_4$.

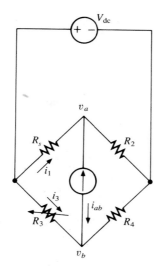

Dividing the first equation by the second yields

$$\frac{R_s}{R_2} = \frac{R_3}{R_4} \tag{7.37}$$

$$R_s = \frac{R_2}{R_4} R_3 \tag{7.38}$$

If the value of R_3 and the ratio R_2/R_4 are known with precision, the value of R_s can be determined accurately.

An unbalanced Wheatstone bridge and instrumentation amplifier circuit are shown in Figure 7.23. The values of resistors R_2, R_3, and R_4 are fixed such that the bridge is balanced when R_s is at a predetermined base value. We will call this base resistance value R_{bal} for balanced resistance value. Equation (7.35) expresses R_{bal} in terms of resistors R_2, R_3, and R_4.

$$R_{bal} = \frac{R_2 R_3}{R_4} \tag{7.39}$$

The purpose of an unbalanced bridge circuit is to produce an output voltage that is proportional to the difference between R_s and R_{bal}. To simplify and normalize the unbalanced bridge equation, we introduce two additional parameters: ε and α. The first parameter, ε, is the fractional difference between R_s and R_{bal} as defined by Equation (7.40):

$$\varepsilon = \left(\frac{R_s - R_{bal}}{R_{bal}} \right) \tag{7.40}$$

Notice that 100ε is the percent difference between R_s and R_{bal}.

The second parameter, α, is the voltage divider ratio for the voltage across resistor R_3 as defined by Equation (7.41).

$$\alpha = \frac{R_3}{R_3 + R_4} \tag{7.41}$$

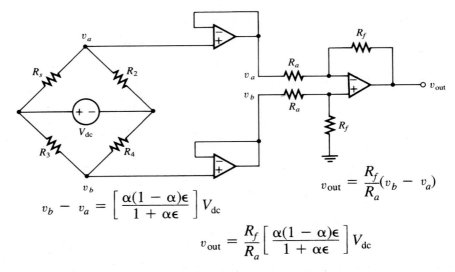

$$v_b - v_a = \left[\frac{\alpha(1 - \alpha)\epsilon}{1 + \alpha\epsilon} \right] V_{dc}$$

$$v_{out} = \frac{R_f}{R_a}(v_b - v_a)$$

$$v_{out} = \frac{R_f}{R_a} \left[\frac{\alpha(1 - \alpha)\epsilon}{1 + \alpha\epsilon} \right] V_{dc}$$

Figure 7.23 Unbalanced Wheatstone bridge and instrumentation amplifier circuit. The purpose of the circuit is to produce an output voltage that is proportional to the difference between R_s and R_{bal}.

The bridge equation is developed by applying the voltage divider rule to write equations for v_a, v_b, and $v_b - v_a$, reducing the equations, and using the parameters defined in Equations (7.39) to (7.41).

$$v_a = \left(\frac{R_2}{R_s + R_2} \right) V_{dc}$$

$$v_b = \left(\frac{R_4}{R_3 + R_4} \right) V_{dc}$$

$$\frac{v_b - v_a}{V_{dc}} = \left(\frac{R_4}{R_3 + R_4} - \frac{R_2}{R_s + R_2} \right) \tag{7.42}$$

We now proceed to convert the right-hand side of Equation (7.42) to a function of α and ϵ. The first step is to combine the two terms into a single fraction:

$$\frac{v_b - v_a}{V_{dc}} = \frac{R_s R_4 + R_2 R_4 - R_2 R_3 - R_2 R_4}{(R_3 + R_4)(R_s + R_2)}$$

In the numerator, use $-R_2 R_4$ to cancel $R_2 R_4$, replace $R_2 R_3$ by $R_{bal} R_4$, and factor out R_4. In the denominator, replace R_2 by $R_{bal} R_4/R_3$. Equation (7.39) was used for both replacements.

$$\frac{v_b - v_a}{V_{dc}} = \frac{R_4(R_s - R_{bal})}{(R_3 + R_4)(R_s + R_{bal} R_4/R_3)}$$

Now multiply the numerator and denominator by $R_3(R_3 + R_4)/R_{bal}$ and rearrange the terms as follows:

$$\frac{v_b - v_a}{V_{out}} = \left[\frac{R_3}{R_3 + R_4}\right]\left[\frac{R_4}{R_3 + R_4}\right]\left[\frac{(R_s - R_{bal})/R_{bal}}{\left(\frac{R_sR_3 + R_{bal}R_4}{(R_3 + R_4)R_{bal}}\right)}\right]$$

By Equation (7.41):

$$\frac{R_3}{R_3 + R_4} = \alpha$$

and

$$\frac{R_4}{R_3 + R_4} = (1 - \alpha)$$

By Equation (7.40):

$$(R_s - R_{bal})/R_{bal} = \varepsilon$$

Adding and subtracting $R_{bal}R_3$ in the last term in the denominator changes nothing, but does facilitate its reduction.

$$\frac{R_sR_3 + R_{bal}R_4}{(R_3 + R_4)R_{bal}} = \frac{R_sR_3 + R_{bal}R_4 + R_{bal}R_3 - R_{bal}R_3}{(R_3 + R_4)R_{bal}}$$

$$= \frac{(R_3 + R_4)R_{bal} + R_3(R_s - R_{bal})}{(R_3 + R_4)R_{bal}}$$

$$= 1 + \alpha\varepsilon$$

Finally,

$$\frac{v_b - v_a}{V_{dc}} = \alpha(1 - \alpha)\left(\frac{\varepsilon}{1 + \alpha\varepsilon}\right) \tag{7.43}$$

For very small values of ε, Equation (7.39) reduces to

$$\frac{v_b - v_a}{V_{dc}} = \alpha(1 - \alpha)\varepsilon \tag{7.44}$$

Table 7.3 gives the percent difference between the result using Equation (7.43) and the result using Equation (7.44). For values of ε less than or equal to 0.01, the difference is less than 1%.

Table 7.3 Values of the Percent Difference Between Results From Equation (7.43) and Results From Equation (7.44)

ε	α					
	0.0	0.2	0.4	0.6	0.8	1.0
0.0001	0.000	0.002	0.004	0.006	0.008	0.010
0.0010	0.000	0.020	0.040	0.060	0.080	0.100
0.0100	0.000	0.200	0.398	0.596	0.794	0.990
0.1000	0.000	1.961	3.846	5.660	7.407	9.091

7.5 DATA SAMPLING AND CONVERSION

Digital control systems are used in a wide variety of industries to increase productivity and efficiency. Industries that use digital control systems include food processing, oil refining, chemical processing, steel production, automobile manufacturing, and many others. Digital control systems are also used in many of the products we buy. In automobiles, for example, a microprocessor-based system controls fuel flow and spark timing to achieve optimum engine performance.

Most of the physical parameters in a process are analog signals. Digital systems must be able to convert these analog signals into digital form for processing by the computer. This is the job of the *data acquisition system* shown in Figure 7.24. In a sense, the data acquisition system is the "eyes and ears" of the digital computer. The computer must also be able to convert its digital control actions into analog form for input to the analog actuators. This is the job of the *data distribution system*, also

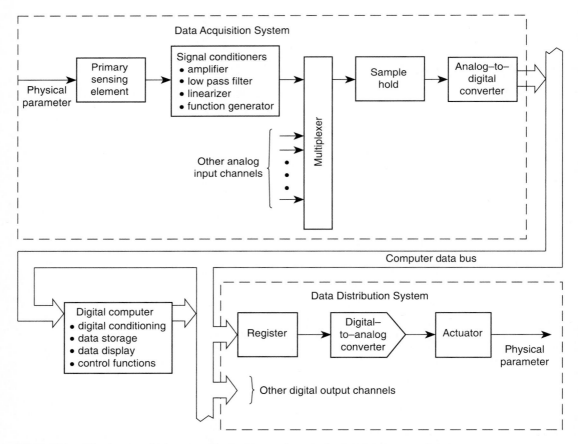

Figure 7.24 The data acquisition system is the "eyes and ears" of a digital control system; the data distribution system is its "hands and arms."

shown in Figure 7.24. In a sense, the data distribution system is the "hands and arms" of the digital computer. Data acquisition and distribution involve two very important concepts: *data sampling* and *data conversion*.

Data Sampling

Data sampling is a process in which a switch connects momentarily to an analog signal in a sequence of pulses separated by evenly spaced increments of time called the *sampling interval*. You can also visualize data sampling as the result obtained by multiplying the analog signal by a series of pulses with an amplitude of 1, a very narrow pulse width, and a period equal to the sampling interval. Figure 7.25a shows an

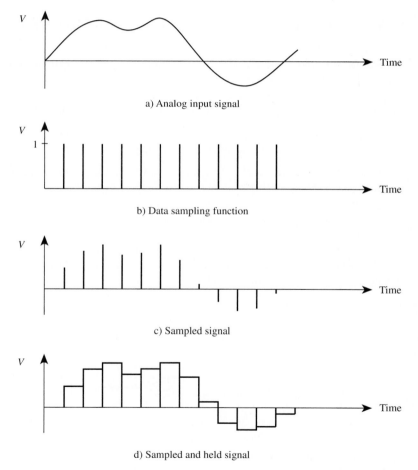

Figure 7.25   A sampled signal (c) may be viewed as the product of the analog input signal (a) multiplied by the data sampling function (b). If the data samples are held until the next sample (d), the resulting signal is a stairstep approximation of the analog signal.

analog signal; Figure 7.25b shows a series of sampling pulses; and Figure 7.25c shows the sampled signal. If the sampling switch is replaced by a *sample-and-hold* circuit, each sampled value is held until the beginning of the next sampling interval, as shown in Figure 7.25d. The sampled and held signal gives a stairstep approximation of the original signal.

The sample-and-hold circuit is used to hold a sample of an analog signal as it is converted to a digital signal by an analog-to-digital converter. Refer to Figure 7.26 for the sample-and-hold circuit diagram. The sample-and-hold device has a signal input, a control input, and an output. It operates in two modes: the sampling or tracking mode when the switch is closed, and the holding mode when the switch is open. The ideal sample and hold takes a sample in zero time and holds the signal value indefinitely with no degradation of the signal. This ideal circuit is called a zero-order sample and hold.

A practical sample-and-hold circuit varies from the ideal in a number of parameters. The following are some parameters of a practical sample-and-hold circuit.

1. The *acquisition time* is the time from the instant the sample command is given until the output is within a specified band of the input. This is determined by the size of the input resistor and the holding capacitor. These two elements form a first-order lag with a time constant equal to the product of the resistance times the capacitance value.

$$\text{Time constant} = \tau = R_{in}C$$

An acquisition time equal to five times the time constant is enough for the output to reach 99.3% of the total change to the new input value. This assumes that the input is constant during the sampling period.

2. The *aperture time* is the time it takes for the switch to open. It is the time between the hold command and the time the switch is completely open.

3. The *decay rate* is the rate of change of the output in the holding mode. It is caused by leakage through the capacitor and the small current that enters the op amp through the inverting input.

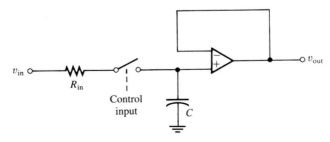

Figure 7.26 A sample-and-hold circuit has a signal input, a control input, and an output. It operates in two modes: the sampling or tracking mode when the switch is closed, and the holding mode when the switch is open.

Data sampling raises two obvious questions:

1. How often must we sample a signal to get all the information in the original signal?
2. What happens if we do not sample often enough?

The Nyquist criterion answers the first question, and the explanation of the Nyquist criterion will reveal the answer to the second question. The *Nyquist criterion* may be stated as follows:

All the information in the original signal can be recovered if it is sampled at least twice during each cycle of the highest frequency component.

If f_h is the highest frequency that occurs in the original signal, then the minimum *sampling rate* is given by the following equation:

$$\text{Minimum sampling rate} = f_s(\text{min}) = 2f_h \qquad \textbf{(7.45)}$$

We can gain some insight into the Nyquist criterion by considering the frequency spectra of the original signal and the sampled signal. Figure 7.27a shows a typical *frequency spectrum* of an analog signal. A frequency spectrum is simply a plot of the maximum voltage of each possible component of a signal versus the frequency of that component. The main point in Figure 7.27a is that there are no components of the original signal with a frequency greater than f_h.

The data sampling function is represented in Figure 7.25b as a series of pulses spaced f_s seconds apart. This sampling function can be represented by an infinite series of sinusoidal components called a Fourier series. This series has components with frequencies of $f_s, 2f_s, 3f_s, \ldots, nf_s, \ldots$ etc. If we visualize the sampling process as the multiplication of the original signal by the sampling function, then, in effect, we have amplitude modulation of the original signal with carrier frequencies of $f_s, 2f_s, 3f_s, \ldots$ etc. The original signal is shifted and folded around each carrier as shown in Figure 7.27b. The original signal can be recovered from the sampled signal shown in Figure 7.27b with a low-pass filter that removes all frequency components greater than f_h.

Figure 7.27c shows what happens when the sampling rate is less than f_h. Part of the folded signal around f_s overlaps part of the original signal. The overlapped portion of the folded signal becomes part of the original signal. It cannot be removed by a low-pass filter. This overlap of the folded signal is called *aliasing*. The effect of aliasing is that one or more frequency components are added to the original signal. Figure 7.28 illustrates how an *alias frequency* is produced by an insufficient sampling rate. The obvious solution to the aliasing problem is either to increase the sampling rate or filter the original signal to remove all components with a frequency greater than $f_s/2$.

Data Conversion

Data converters are the interface between the analog and digital domains. The *analog-to-digital converter* (*ADC* or *A/D*) converts analog signals into a digital code.

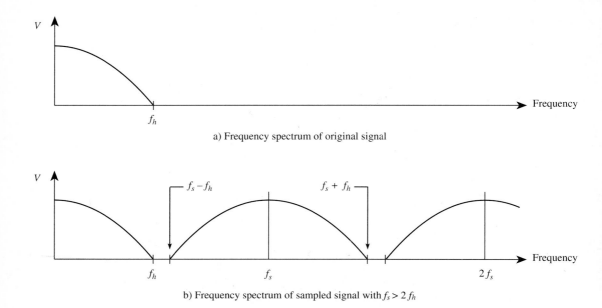

a) Frequency spectrum of original signal

b) Frequency spectrum of sampled signal with $f_s > 2f_h$

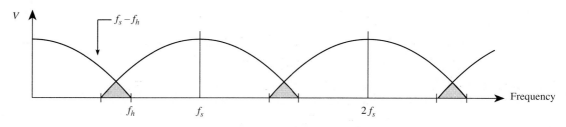

c) Frequency spectrum of sampled signal with $f_s < 2f_h$

Figure 7.27   In a sampled signal, the frequency spectrum of the original signal is both subtracted from and added to the sample frequency, f_s, and all multiples of f_s. If $f_s > 2f_h$, the original signal can be recovered by filtering out all frequencies above f_s. When $f_s < 2f_h$, the difference frequency spectrum overlaps the original signal and filtering will not recover the original signal.

The *digital-to-analog converter* (*DAC* or *D/A*) converts digital codes into analog voltage levels.

The input to an *A/D converter* is an analog voltage that can have any value from 0 to its full-scale range (*FS*). The output, however, can have only a finite number of output codes, each defining a state of the ADC. Figure 7.29 shows the transfer function of an ideal 3-bit A/D converter with 8 output states. The output states are assigned 3-bit binary codes from 000 to 111. We will use Figure 7.29 to help explain several important points concerning A/D converters.

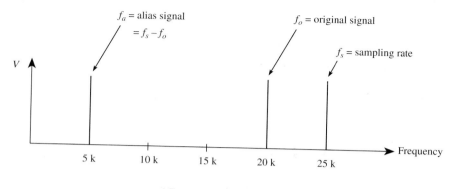

a) Frequency spectrum

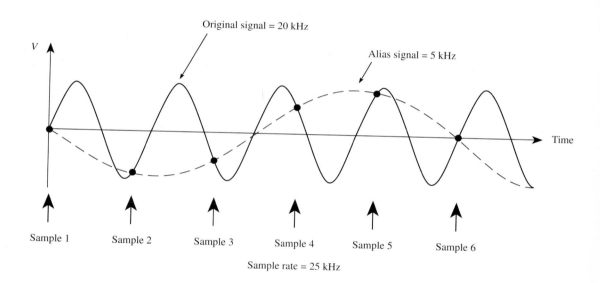

b) Voltage vs time diagram

Figure 7.28   Sampling a 20 kHz original signal at a sampling rate of 25 kHz results in a 5 kHz alias component in the sampled signal. In the frequency spectrum, the alias appears as the difference frequency obtained by subtracting the original frequency from the sampling rate.

The resolution of an A/D converter is defined as the number of output states and is often expressed as the number of bits in the output code. If n is the number of bits, then 2^n is the number of output states. Thus the 3-bit ADC in Figure 7.29 has $2^3 = 8$ states. An 8-bit ADC has $2^8 = 256$ states, and a 10-bit ADC has $2^{10} = 1024$ states. Between each pair of steps is a transition that makes the change from one output state to another. There is always one less transition than the number of states. The 3-bit ADC has $8 - 1 = 7$ transitions. An n-bit ADC has $2^n - 1$ transitions.

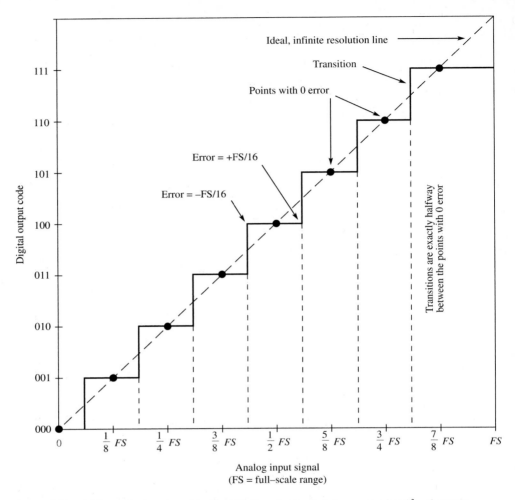

Figure 7.29   The transfer function of an ideal 3-bit analog-to-digital converter has $2^3 = 8$ output states and $2^3 - 1 = 7$ transitions. Each step represents a range of analog input values of $FS/8$. The quantization error is 0 in the center of each step and ranges from $-FS/16$ at the front edge of the step to $+FS/16$ at the rear of the step.

At any state, there is a small range of input voltages that forms a step that extends from one transition to the next. This range of values is called the quantization size or "width" of the code. The ideal width of each interior step is designated as V_{LSB}, where LSB signifies Least Significant Bit. Resolution is sometimes expressed in terms of V_{LSB}. The value of V_{LSB} is equal to the full-scale range of the input divided by the number of output states:

$$\text{Resolution} = V_{LSB} = FS/2^n \tag{7.46}$$

Notice that the center of each step intersects the ideal line that extends from the lower left corner to the upper right corner of the graph. These intersections are marked with dots, and they are the only points on the graph that have zero error. On each step, there is an error called the *quantization error* that ranges from $-V_{LSB}/2$ on the left end of the step to $+V_{LSB}/2$ on the right end. In Figure 7.29 the quantization error is $\pm FS/16$.

The input to a *D/A converter* is a finite number of digital input codes. The output is an equal number of discrete output voltage levels. The resolution of a DAC is sometimes specified as the number of bits in the input code (the same as for an ADC). Figure 7.30 shows the ideal transfer function of a 3-bit D/A converter.

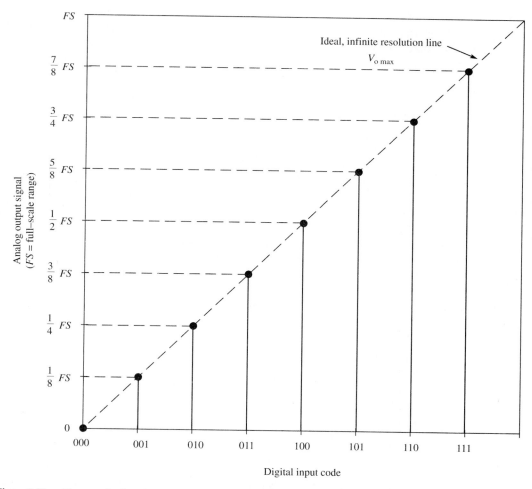

Figure 7.30 The transfer function of an ideal 3-bit digital-to-analog converter has only 8 possible input states and therefore, only 8 possible output voltage levels. The 8 dots on the graph define the input/output relationship for the 8 states.

Unlike the A/D converter, the analog signal from a DAC does not range over an infinite number of values between 0 and *FS*. Rather, there is a discrete analog output value for each digital input code. As the input code is traversed from 000 to 111, the analog output steps from point to point on the ideal, infinite resolution line. Each step of the analog output is a change in voltage of V_{LSB}, which is also used to express the resolution of a DAC [see Equation (7.46)].

The top step in Figure 7.30 represents the maximum output voltage of the DAC. We will designate this voltage as V_{omax}. The step size (and resolution) of a DAC can also be expressed in terms of V_{omax} as defined in Equation (7.47):

$$\text{Resolution} = V_{\text{LSB}} = V_{\text{omax}}/2^{n-1} \tag{7.47}$$

As Equations (7.46) and (7.47) indicate, resolution is often looked upon as the "step size" when the input binary word goes from one binary state to another. It is important to note that the maximum output, V_{omax}, will never reach the full-scale value, *FS*. However, as the resolution of the DAC increases, V_{omax} does approach the value of *FS*. In fact, *FS* is the limiting value of V_{omax} as the resolution approaches infinity.

Digital-to-Analog Converters

The function of a digital-to-analog converter is to produce an output voltage that is a weighted sum of the non-zero bits in the digital input code. The least significant bit (bit 0) is given a weighting of 1. The weighting of each successive bit is two times the weighting of the previous bit. Thus a 4-bit DAC will have weighting factors of 1, 2, 4, and 8; an 8-bit DAC will have weighting factors of 1, 2, 4, 8, 16, 32, 64, and 128. The bit with the highest weighting factor is called the most significant bit (MSB). The weighting of the MSB of an *n*-bit DAC is 2^{n-1}.

The sum of all the weighting factors is the weighting given to the highest input code (the code with all bits equal to 1). An interesting fact about these weighting factors is that they form a series in which the sum of any number of terms is always 1 less than two times the largest term. For example:

$$1 + 2 + 4 + 8 = 15 = (2 \cdot 8) - 1$$
$$1 + 2 + 4 + 8 + 16 + 32 + 64 + 128 = 255 = (2 \cdot 128) - 1$$

The binary-weighted digital-to-analog converter uses a summing amplifier to form the weighted sum of all the non-zero input bits. The input resistances of the summing amplifier provide the required weighting factors. Figure 7.31 shows a 4-bit *binary-weighted DAC*. The feedback resistance (R_f), the MSB input resistance (R), and the input voltage level used for a binary 1 determine the output voltage level of the MSB. If we let V_1 represent the input voltage of binary 1, then Equation (7.48) defines the voltage level of the MSB. In Figure 7.31, the value of 5 V is used for V_1. Equation (7.49) gives the full scale voltage, and Equation (7.50) gives the LSB voltage.

$$V_{\text{MSB}} = -R_f V_1/R \tag{7.48}$$
$$V_{\text{FS}} = 2 \cdot V_{\text{MSB}} \tag{7.49}$$
$$V_{\text{LSB}} = V_{\text{MSB}}/2^{n-1} \tag{7.50}$$

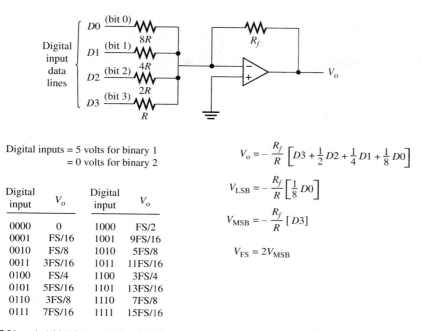

Digital inputs = 5 volts for binary 1
        = 0 volts for binary 2

Digital input	V_o	Digital input	V_o
0000	0	1000	FS/2
0001	FS/16	1001	9FS/16
0010	FS/8	1010	5FS/8
0011	3FS/16	1011	11FS/16
0100	FS/4	1100	3FS/4
0101	5FS/16	1101	13FS/16
0110	3FS/8	1110	7FS/8
0111	7FS/16	1111	15FS/16

$$V_o = -\frac{R_f}{R}\left[D3 + \frac{1}{2}D2 + \frac{1}{4}D1 + \frac{1}{8}D0\right]$$

$$V_{LSB} = -\frac{R_f}{R}\left[\frac{1}{8}D0\right]$$

$$V_{MSB} = -\frac{R_f}{R}[D3]$$

$$V_{FS} = 2V_{MSB}$$

Figure 7.31   A 4-bit binary-weighted D/A converter. Weighting is provided by input resistors R, $2R$, $4R$, and $8R$. Each digital input ($D0 - D3$) has a value of 0 or 5 V, depending on the corresponding bit in the input code. The output is the weighted sum of only those inputs that have a value of 5 V.

Although simple to understand, there are many disadvantages to the binary-weighted DAC. Each input resistor has a different resistance based upon the ratio of R-$2R$-$4R$-$8R$, and so on. It is difficult to build IC resistors that can be accurately matched at ratios greater than 20:1. This limits binary-weighted DACs to five bits or less. Precision resistors or 10-turn potentiometers could be used, but that would be expensive. In addition, each binary bit position is represented by a different Thévenin equivalent resistance, and the devices that provide the inputs to the DAC will see different loads.

The *R-2R ladder-type DAC* eliminates both of the disadvantages of the binary-weighted DAC. It is a resistive network consisting of only two different values of resistors, and the resultant analog output is the weighted sum of the applied digital input word. Figure 7.32 shows an *R-2R* DAC, and Figure 7.33 shows a development of the output equation of the *R-2R* DAC. This development uses the superposition principle and Thévenin's theorem to develop four equivalent circuits, one for each of the four bits in the digital input word.

Analog-to-Digital Converters

The function of an analog-to-digital converter is to produce a digital code word that accurately represents the level of the analog input voltage. In this section, we begin

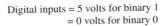

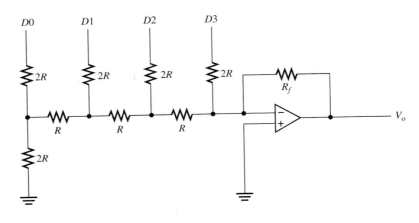

$$V_o = -\frac{R_f}{2R}\left[D3 + \frac{1}{2}D2 + \frac{1}{4}D1 + \frac{1}{8}D0\right]$$

$$V_{LSB} = -\frac{R_f}{2R}\left[\frac{1}{8}D0\right]$$

$$V_{MSB} = -\frac{R_f}{2R}[D3]$$

$$V_{FS} = 2V_{MSB}$$

Figure 7.32 A 4-bit *R-2R* D/A converter uses only 2 resistor sizes to provide the four weighting factors. The digital inputs are 0 V for a binary 0 or 5 V for a binary 1. The output is the weighted sum of only those inputs that have a value of 5 V.

with examples of five analog-to-digital conversion techniques and end with a discussion of the major considerations in the selection of an analog-to-digital converter. Here are the five A/D conversion techniques.

1. The *binary counter technique* uses a binary counter connected to a D/A converter and a comparator to generate the digital word that represents an analog input. A series of clock pulses advances the counter through its binary states, producing a stairstep voltage output from the D/A converter. Both the analog input and the DAC output are input to a comparator. At the instant that the DAC output passes the analog input, the output of the comparator switches to a high level. The logic circuit latches the count in the counter. Since the latched count is the one that made the DAC output equal to the analog input, it is an accurate measure of the analog input.

   Although simple, the counter-type ADC is relatively slow. The conversion time depends on the level of the analog signal and the number of bits in the digital word. In the worst case, an 8-bit counter type ADC would require 256

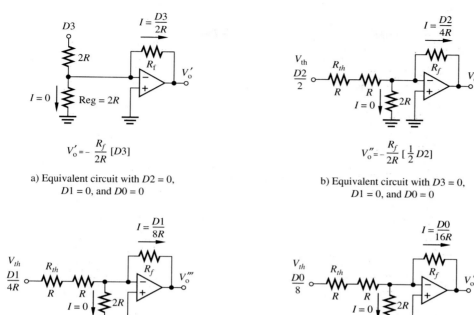

$$V_0' = -\frac{R_f}{2R}[D3]$$

a) Equivalent circuit with $D2 = 0$,
$D1 = 0$, and $D0 = 0$

$$V_0'' = -\frac{R_f}{2R}[\tfrac{1}{2}D2]$$

b) Equivalent circuit with $D3 = 0$,
$D1 = 0$, and $D0 = 0$

$$V_0''' = -\frac{R_f}{2R}[\tfrac{1}{4}D1]$$

c) Equivalent circuit with $D3 = 0$,
$D2 = 0$, and $D0 = 0$

$$V_0'''' = -\frac{R_f}{2R}[\tfrac{1}{8}D0]$$

d) Equivalent circuit with $D3 = 0$,
$D2 = 0$, and $D1 = 0$

$$V_0 = V_0' + V_0'' + V_0''' + V_0'''' = -\frac{R_f}{2R}\left[D3 = \tfrac{1}{2}D2 + \tfrac{1}{4}D1 + \tfrac{1}{8}D0\right]$$

Figure 7.33   Superposition and Thévenin's theorem are used to construct the four equivalent circuits used to obtained the output equation of the R-$2R$ D/A converter.

clock pulses to complete the conversion. For a 12-bit ADC, the worst case would require 4096 clock pulses.

A counter-type analog-to-digital converter is illustrated in Figure 7.34. This is a relatively easy circuit to construct using the components specified in the schematic. The timing diagram is quite useful to help explain the conversion operation. The 555 clock waveform is shown at the top of the timing diagram. The output of the clock is fed into the binary counter so that each clock pulse will cause the counter to increase its count by one bit. The output of the counter will sequence from 0000 to 1111 in the same sequence as the binary inputs in Figure 7.31. When the counter reaches 1111, the next clock pulse restores the 0000 count and the sequence starts over. The counter output is shown in binary form on the second section of the timing diagram.

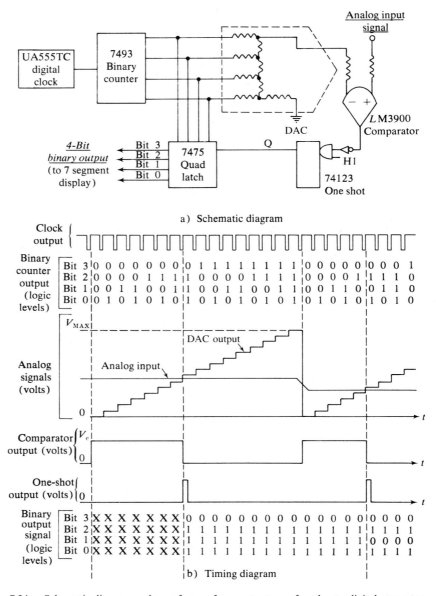

Figure 7.34 Schematic diagram and waveforms of a counter type of analog-to-digital converter.

The binary output of the counter is fed into a binary ladder DAC producing the stairstep output shown in the third section of the timing diagram. The analog input signal (the signal that is to be converted to digital) is superimposed on the stairstep waveform to show the relative size of the two inputs to the comparator. The output of the comparator is positive when the analog input signal is greater than the analog signal from the DAC. When the relative magnitude of the two analog signals reverses, the output of the comparator reverses also.

Notice that the comparator output switches from $+V_c$ to 0 at the point that the analog input signal crosses the stairstep output from the DAC. The one-shot circuit converts the $+V_c$ to 0 change in the comparator output into a pulse that is used to trigger the quad latch. Each time the quad latch is triggered, the binary signal from the counter is transferred to the output of the latch and held there until the next triggering pulse. The result is that the output of the quad latch reflects the count for which the analog input signal was approximately equal to the stairstep voltage. The binary output signal from the analog-to-digital converter is shown on the bottom line of the timing diagram.

2. The *successive approximation technique* sequentially compares a series of binary weighted values with the analog input to generate a digital word that represents the analog input voltage. Successive approximation is a relatively fast technique, completing the conversion in just n steps, where n is the resolution in bits. This technique is best compared to the process of weighing a person on a balance scale. The person adjusting the scale adds weights until the scale is balanced, and then announces the weight.

    The successive approximation technique selects the most significant bit that is less than the analog value being converted, and then adds all successive lesser bits for which the total of accepted bits does not exceed the analog value. When all bits have been considered, the analog value is approximated by the sum of the bits accepted. This is one of the more commonly used techniques when medium to high speed conversion is required. Figure 7.35 shows a 4-bit successive approximation A/D converter.

3. The *single slope technique* uses a fixed-rate, variable-time ramp voltage to generate a digital word that represents the analog input voltage. The ramp voltage is generated by an integrator with a reference voltage input. A comparator compares the ramp voltage to the analog input voltage and signals the instant when the ramp voltage equals the analog input voltage. The time required for the ramp voltage to equal the input voltage is measured by a clock counter. When the two values are nearly equal (depending on the resolution), the counter is disabled and the digital word in the counter's output register represents the analog value. The counter is read and then reset to prepare for the next measurement. This technique is one of many integrating analog-to-digital converters. Figure 7.36 shows a single-slope ADC.

4. The *dual slope technique* uses a variable-slope, fixed-time ramp followed by a fixed-slope, variable-time ramp to generate the digital word that represents

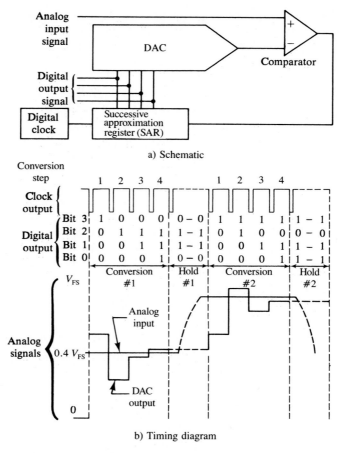

a) Schematic

b) Timing diagram

Figure 7.35 A 4-bit successive approximation A/D converter. A conversion takes place in four steps, each initiated by a negative clock pulse. The four steps for conversion 1 are as follows:

1. The SAR outputs digital word 1000, producing a DAC output of $V_{FS}/2$ V. This is higher than the analog input, so bit 3 is reset to 0 and latched for the duration of the conversion.
2. The SAR outputs word 0100, producing a DAC output of $V_{FS}/4$ V. This is lower than the analog input, so bit 2 is set to 1 and latched for the duration.
3. The SAR outputs word 0110, producing a DAC output of $3V_{FS}/8$ V. This is lower than the analog input, so bit 1 is set to 1 and latched for the duration.
4. The SAR outputs digital word 0111, producing a DAC output of $7V_{FS}/16$ V. This is higher than the analog input, so bit 0 is reset to 0 and latched for the duration

The final digital output is 0110.

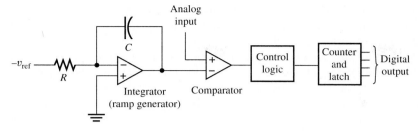

a) Schematic diagram

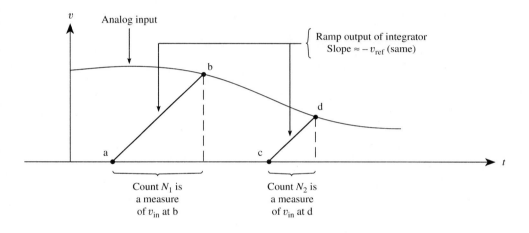

b) Waveform of two conversions
- conversion 1 begins at a and ends at b
- conversion 2 begins at c and ends at d

Figure 7.36   A single-slope A/D converter. Conversion begins (at point a or c) with the counter reset and the integrator output at 0 V. The ramp has the same slope every time, so the count in the counter when the ramp reaches v_{in} is proportional to v_{in}. Thus count N_1 is proportional to v_{in} at point b and count N_2 is proportional to v_{in} at point d.

the analog input voltage. The variable-slope ramp is generated by an integrator with the analog input voltage as input. The fixed-slope ramp is generated by the same integrator with a reference voltage as input. A timer counter is used to time the first ramp and measure the second ramp.

The dual-slope ADC has a high degree of accuracy with relatively little effect from time or temperature drifts. This is due to the inherent self-calibrating characteristic of the dual-slope method. It also has excellent noise rejection.

However, the dual-slope converter is slow, so it is used primarily where accuracy is important and speed is not a primary consideration. Figure 7.37 shows a dual-slope A/D converter.

5. The *flash* or *parallel technique* compares the analog input voltage to a set of reference voltages to generate the digital word that represents the analog input. The analog voltage to be converted is input to the array of comparators. All of the comparators whose voltage reference is less than the analog value being converted will have a high output; those whose reference is greater will

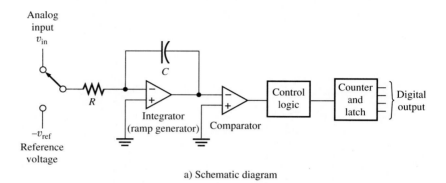

a) Schematic diagram

b) Waveform of two conversions with $v_{in1} > v_{in2}$

Figure 7.37 A dual-slope A/D converter. The first slope (ab or de) is proportional to the input voltage and continues for the same time interval every time (while the counter steps through a full count). Thus the voltage levels at b and e are proportional to v_{in1} and v_{in2} respectively. The second slope is the same every time, so the count required to reach 0 V at c and f is proportional to v_{in1} and v_{in2} respectively. The count at c is a measure of v_{in1} and the count at f is a measure of v_{in2}.

have a low output. The outputs of the comparators are the input to a priority encoder that generates the desired digital code word. The main advantage of the flash converter is its very short conversion time: it is very fast.

The flash converter requires a reference voltage and a comparator for every state of the digital output except zero. Thus a 3-bit flash converter will have seven reference voltages and seven comparators. A 4-bit flash converter will need 15 reference voltages and comparators. In general, an n-bit flash converter will require $2^n - 1$ reference voltages and comparators. The large number of comparators required for a reasonable resolution makes the flash technique fairly expensive. Figure 7.38 shows a 3-bit flash converter.

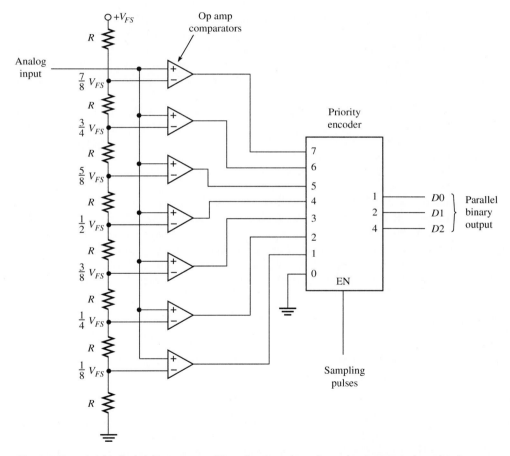

Figure 7.38   A 3-bit flash A/D converter. The reference voltage for each comparator is set by the resistive voltage-divider network. The output of each comparator is connected to an input of the priority encoder. The encoder is sampled by a pulse on the Enable input, and a 3-bit binary code representing the value of the analog input appears on the encoder's outputs. The binary code is determined by the highest-order input having a **HIGH** level.

Selection of an Analog-to-Digital Converter

In addition to the usual factors of cost, size, and availability, the control system designer has the following three factors to consider in the selection of an ADC.

1. *Resolution* is specified as the number of bits in the digital code of the output states. It determines the number of output states, the size of the LSB, and the quantization error ($\pm 1/2$LSB). The quantization error ranges from 0.19% for 8-bit resolution to 0.00076% for 16-bit resolution. The designer must specify the minimum resolution that reduces the quantization error to an acceptable level.

2. *Accuracy* includes other factors besides resolution, such as gain error, offset error, linearity, and missing codes. *Gain error* is a change in the slope of the infinite resolution line from the ideal shown in Figure 7.29. The line still passes through the 0 volt, 000 digital code point. *Offset error* is a displacement left or right of the infinite resolution line with no change in its slope. *Linearity error* is a deviation of the infinite resolution line from a straight line. *Missing codes* is the absence of one or more expected codes in the output as the input is traversed over its full range.

    When considering the accuracy of an ADC, the designer's major concern is missing codes. The gain and offset errors can be hardware adjusted or software compensated. Linearity can also be compensated, but not as easily as gain and offset. Missing codes, however, cannot be restored.

3. *Conversion speed* is determined by how fast the analog signal changes, and it dictates the type of ADC selected. If the analog signal varies at a very slow speed, there is little need for a fast converter that requires fast, expensive components. If the analog signal varies at a moderate speed, the converter will have to operate faster, requiring faster, more expensive components and conversion techniques. If the analog signal varies at a high speed, both conversion techniques and component speed are of paramount importance. In general, the conversion speed requirement will dictate the type of converter selected.

Example 7.14

An analog-to-digital converter has a resolution of 4 bits and a quantization error of $\pm\frac{1}{2}$ bit. The input voltage is 5 V at the center of the top step. Determine the quantization error in volts and percent of the full-scale range.

Solution

The binary number 0000 represents 0 V and the number 1111 represents 5 V. There will be $2^4 - 1$ or 15 steps between 0000 and 1111. A 1-bit change in the binary number is equal to a $5/15 = 0.333$-V change in the voltage it represents. The quantization error is $\pm\frac{1}{2}$ bit, which is $\pm\frac{1}{2}(0.333) = \pm 0.166$ V. The percent accuracy is given in the following equation:

$$\text{Percent accuracy} = \pm\left(\frac{0.166}{5}\right)(100) = \pm 3.33\%$$

Digital Conditioning

Signal conditioning is not always completed when the analog signal is converted to digital. Filtering, linearization, calibration, and conversion to engineering units can all be accomplished by a microprocessor working with the digitized samples. Digital signal processing makes it possible to extract additional information from the signal. Vision systems, for example, involve high-speed processing of an array of data samples.

Digital filtering consists of computing some type of weighted average of the current sample and previous samples. Digital linearization and calibration is accomplished by storing factory calibration data in some type of read-only memory for use by the microprocessor. Additional information, such as ambient temperature or static pressure, can be included in the calibration data. The result is a precise correction for all the nonlinearities of the sensor throughout the range of temperatures and pressures for which the measuring instrument is rated. Digital calibration produces accuracies that cannot be achieved by analog calibration, with the added benefit of using the actual engineering units for the calibrated signal.

GLOSSARY

Acquisition time: The time from the instant the sample command is given until the output is within a specified band of the input. (7.5)

ADC: Abbreviation of analog-to-digital converter. (7.5)

A/D converter: Abbreviation of analog-to-digital converter. (7.5)

Alias frequency: One or more frequency components that are added to the original signal when the signal is sampled at an insufficient sampling rate. (7.5)

Analog-to-digital converter: A device that produces a digital code word that accurately represents the level of an analog input voltage. (7.5)

Aperture time: The time it takes a sample-hold switch to open, measured from the time the hold command is given until the switch is completely open. (7.5)

Band-pass filter: A circuit that does not change components within a specified band of frequencies and attenuates components outside of the specified band. (7.4)

Binary counter ADC: An analog-to-digital converter that uses a binary counter and a D/A converter to generate a digital word that represents an analog input voltage. (7.5)

Binary-weighted DAC: A digital-to-analog converter that uses a summing amplifier to form the weighted sum of all the non-zero bits in the digital input word. (7.5)

CMM: Abbreviation of common-mode rejection. (7.2)

CMMR: Abbreviation of common-mode rejection ratio. (7.2)

Common-mode rejection: The logarithm of the common-mode rejection ratio expressed in decibel units. (7.2)

Common-mode rejection ratio: The ratio of the differential gain (A) of an operational amplifier divided by its common mode gain (A_c). (7.2)

Comparator: A circuit that accepts two input voltages and signals which one is greater. (7.3)

DAC: Abbreviation of digital-to-analog converter. (7.5)

D/A converter: Abbreviation of digital-to-analog converter. (7.5)

Data acquisition system: A system that conditions a number of analog signals and converts them into digital form for processing by a computer. (7.5)

Data distribution system: A system that converts the digital representation of a number of signals into analog form for output to analog actuators. (7.5)

Data sampling: A process in which a switch connects momentarily to an analog signal in a sequence of pulses separated by evenly spaced increments of time called the sampling interval. (7.5)

Decay rate: The rate of change in the output of a sample-hold circuit when it is in the hold mode. (7.5)

Differential amplifier: A circuit that amplifies the difference between two input voltages. (7.3)

Differentiator: A circuit that produces an output voltage that is equal to the rate of change of the input voltage. (7.3)

Digital-to-analog converter: A device that produces an output voltage that is an accurate analog representation of the input digital code word. (7.5)

Dual-slope ADC: An analog-to-digital converter that uses a variable-slope, fixed-time ramp followed by a fixed-slope, variable-time ramp to generate a digital code word that represents an analog input voltage. (7.5)

Flash or parallel ADC: An analog-to-digital converter that uses a series of reference voltage/comparator pairs to generate a digital code word that represents an analog input voltage. (7.5)

Frequency spectrum: A plot of the maximum voltage of each possible component of a signal versus the frequency of that component. (7.5)

Function generator: An inverting amplifier that produces a nonlinear relationship between its input and output voltages. (7.3)

High-pass filter: A circuit that does not change components above its break-point frequency and attenuates components below its break-point. (7.4)

Instrumentation amplifier: A differential amplifier with voltage followers on each input. (7.3)

Integrator: A circuit in which the change in the output voltage over some interval of time is equal to the integral of the input voltage over that same time interval. (7.3)

Inverting amplifier: A circuit that produces an output that is equal to $-G$ times the input where G can be any value greater than 0. (7.3)

Isolation amplifier: An amplifier that uses transformer or optical coupling to electrically separate the input and output circuits. (7.4)

Linearization: The process of conditioning a nonlinear calibration signal to produce a linear calibration signal. (7.4)

Linear operating region: The middle operating region of an operational amplifier where the output is equal to the gain times the difference between its two input voltages $(v_2 - v_1)$. (7.2)

Low-pass filter: A circuit that does not change components below its break-point frequency and attenuates components above its break-point. (7.4)

Noninverting amplifier: A circuit that produces an output voltage that is equal to $+G$ times its input voltage, where G can be any value greater than 1. (7.3)

Notch filter: A circuit that rejects components within a specified band of frequencies and does not change components outside of that band. (7.4)

Nyquist criterion: A sampling criterion that states that all the information in the original signal can be recovered if it is sampled at least twice during each cycle of the highest frequency component. (7.5)

Operational amplifier: A very high-gain amplifier that has two input lines and one output line. The output voltage is equal to the product of the gain times the difference between the two input voltages. (7.2)

R-2R ladder DAC: A digital-to-analog converter that uses two resistor sizes in a resistive ladder network to form the weighted sum of all the non-zero bits in the digital input word. (7.5)

Sample-and-hold: A circuit that uses a sampling switch to sample an analog voltage and a capacitor to hold the sample until the next closure of the sampling switch. (7.5)

Sampling interval: The time between the momentary closures of a sampling switch. Numerically equal to the reciprocal of the sampling rate. (7.5)

Sampling rate: The rate of momentary closures of a sampling switch. Numerically equal to the reciprocal of the sampling interval. (7.5)

Saturation region: Two regions where the output of an operational amplifier has a constant value ($-V_{\text{sat}}$ on the left or $+V_{\text{sat}}$ on the right). (7.2)

Single-slope ADC: An analog-to-digital converter that uses a fixed-rate, variable-time ramp voltage to generate a digital word that represents the analog input voltage. (7.5)

Slew rate: The maximum rate at which the output voltage of an operational amplifier can change. (7.2)

Successive approximation ADC: An analog-to-digital converter that sequentially compares a series of binary-weighted values with the analog input to generate a digital code word that represents the analog input voltage. (7.5)

Summing amplifier: A circuit with several inputs that produces an output voltage that is the inverted, weighted sum of the input voltage. (7.3)

Voltage follower: A circuit with a very high input impedance, a very low output impedance, and an output voltage that is equal to the input voltage. (7.3)

Wheatstone bridge: A circuit used to measure the value of an unknown resistor in which the unknown resistor and three other resistors are connected in a diamond configuration. (7.4)

EXERCISES

7.1 Determine the values of $v_2 - v_1$ that will saturate an op amp if the gain, A, is 2×10^6, and the supply voltages are $+18$ V and -16 V.

7.2 Determine the values of $v_2 - v_1$ that will saturate an op amp if the common-mode voltage is 8 V, the common-mode gain is 0.1, the differential mode gain is 200,000, and the supply voltages are $+20$ V and -20 V.

7.3 An op amp has a differential voltage of 80 μV and a common-mode voltage that varies from 0 to 5 V. The amplifier gain is 200,000. Specify the CMRR required to limit the common-mode error to a maximum of 2% of the output when the common-mode voltage is 0 V.

7.4 An op amp has an open-loop gain, A, of 50,000.

a. Use Equation (7.12) to compute the value of R_{in} that will result in $v_{\text{out}} = -15$ V when $v_{\text{in}} = 6$ V and $R_f = 10$ kΩ.

b. Use Equation (7.11) to compute v_{out} for the values of v_{in}, R_{in}, and R_f from part (a).

c. Determine the percent difference between the values of v_{out} in parts (a) and (b).

7.5 A primary element produces an output voltage that ranges from 0 to 22 mV. Design an inverting amplifier that produces an output voltage with a range of 0 to -4 V. Use a 1-kΩ resistor for R_{in}.

7.6 Repeat Exercise 7.5, but change the output voltage range to 0 to $+4$ V.

7.7 A primary element produces an output voltage that ranges from 0 to 22 mV. Design a summing amplifier that will produce an output voltage with a range of -1 to -5 V. Use a 1-kΩ resistor for R_{in}.

7.8 Given a 10-μF capacitor, compute the value of R_{in} for an integrator with the following defining equation:

$$v_{out} = -\int_{t_1}^{t_2} v_{in}\, dt + v_{out}(t_1)$$

7.9 Design an instrumentation amplifier with a gain of 100. Use Equation (7.18) and a 100-kΩ resistor for R_f.

7.10 Determine the time constant and the 99.3% acquisition time of a sample-and-hold circuit if $C = 100\ \mu$F and $R = 1$ kΩ.

7.11 Design (a) a single-stage low-pass filter and (b) a two-stage low-pass filter that will attenuate a 159.2-Hz signal by an attenuation factor of 100. Use the program "BODE" to generate frequency response data and plot a Bode magnitude diagram for each filter. Compare the two Bode diagrams and comment.

7.12 Design a notch filter that will attenuate a 159.2-Hz signal by an attenuation factor of 20. Use program "BODE" and plot a Bode magnitude diagram for the notch filter. Compare the notch filter Bode diagram with the two low-pass filter diagrams from Exercise 7.11.

7.13 Change the endpoints of the line segments in Figure 7.18 to minimize the maximum error. Construct a table similar to Table 7.1 showing the error at the midpoints and the endpoints of the line segments.

7.14 Construct a five-step piecewise-linear approximation of the function defined by the input/output table below. Construct a graph similar to Figure 7.18 and a table similar to Table 7.1.

Measuring Instrument Input/Output Table[a]

Input	Output	Input	Output	Input	Output
0	2.9	35	31.9	70	69.8
5	6.8	40	36.7	75	75.2
10	10.8	45	41.8	80	80.3
15	14.8	50	47.2	85	85.1
20	18.9	55	52.8	90	89.5
25	23.1	60	58.5	95	93.5
30	27.4	65	64.2	100	97.1

[a] All values are percent of full-scale range.

7.15 A platinum RTD has the following resistance versus temperature characteristic:

Temperature, T (°C)	Resistance, R_s (Ω)
0	100.0
25	109.8
50	119.8
75	129.6
100	139.3

An unbalanced Wheatstone bridge is to be used to convert the RTD resistance into a voltage signal. The balance resistance value $R_{bal} = 119.8$ Ω, so the bridge output will be 0 V at 50°C.

Your assignment is to design the unbalanced bridge circuit, an instrumentation amplifier, and a summing amplifier that will give a full-scale output from 1 to 5 V as the temperature goes from 0 to 100°C. As part of your design assignment, you are to complete the following table of results from various steps of the design process.

T (°C)	R_s (Ω)	ε	$v_b - v_a$	v_1	v_{out}
0	100.0				
25	109.8				
50	119.8				
75	129.6				
100	139.3				

Your design must take into account the self-heating effect of the current through the resistance element. In the steady state, the heat dissipated by the current through the resistance element causes the probe temperature (T_s) to be slightly above the temperature of the surrounding fluid (T_f). The resistance element actually measures T_s, resulting a self-heating effect error equal to $T_s - T_f$. The self-heating effect is expressed in terms of the *dissipation constant* of the probe, in mW/°C. The heat dissipated by the probe is equal to the product of the dissipation constant times the heating effect error ($T_s - T_f$). The heat generated by the probe current is equal to v_s^2/R_s (where v_s is the voltage across the probe and R_s is the resistance of the probe). In the steady-state condition, the heat dissipated is equal to the heat generated, thus:

$$v_s^2/R_s = \text{(dissipation constant)(self-heating effect error)}/1000$$

The division by 1000 is necessary to convert the units of the right side from milliwatts to watts to match the units of the left side.

The design specifications provide the following information relative to the self-heating effect and the selection of the resistance values of the bridge circuit.

(1) Self-heating effect of the probe: 8 mW/°C

(2) Maximum self-heating effect
error for balanced bridge at 50°C: 0.1°C

(3) Bridge voltage, V_{dc}:   1 V

(4) Balance resistance value, R_{bal}:   119.8 Ω

(5) Resistance ratio, R_3/R_{bal}:   1

Complete the following design steps.

(1) Determine the resistances of R_2, R_3, and R_4, and the value of α for the bridge.

    a. Use $v_s^2/R_{bal} = (8\ \text{mW/°C})(0.1°C)/1000$ to compute the voltage across the resistance element, v_s, when the bridge is balanced.

    b. Use v_s from the step above, $V_{dc} = 1$ V, and $R_{bal} = 119.9$ Ω to compute the resistance of R_2. As a safety factor, multiply your result by 1.1 and round up to the nearest whole ohm. Use the rounded value for R_2.

    c. Use design specification 5 to determine the resistance of R_3.

    d. Use Equation (7.39) to determine the resistance of R_4.

    e. Use Equation (7.41) to determine α.

(2) Determine the values of ε.

    a. Use Equation (7.40) to determine the value of ε for the following temperature values: 0, 25, 50, 75, 100°C.

    b. Complete the ε column in the results table.

(3) Determine the values of $v_b - v_a$.

    a. Use Equation (7.43) to determine the value of $v_b - v_a$ for the following temperature values: 0, 25, 50, 75, 100°C.

    b. Complete the $v_b - v_a$ column in the results table.

(4) Design the instrumentation amplifier.

    a. Figure 7.12 is a diagram of the instrumentation amplifier. The first step in the design is to switch the two input voltages to give an inverted input. This is necessary to cancel the inversion of the summing amplifier that follows the instrumentation amplifier. With the inputs reversed and with the output labeled v_1 instead of v_{out}, the output of the instrumentation amplifier is given by the following equation.

$$v_1 = -(R_f/R_a)(v_b - v_a)$$

    b. Compute the gain, R_f/R_a, that will result in a span of 4 volts for v_1 as the temperature goes from 0 to 100°C. This is the gain of the instrumentation amplifier. Use a 1 kΩ resistor for R_a and determine the value of R_f to give the required gain.

c. Complete the v_1 column in the results table. The values are defined by the equation for v_1 in step 4a. Use the gain you computed in step 4b.

(5) Design the summing amplifier.

a. Figure 7.8a is a diagram of the two-input summing amplifier. Replace the upper input in Figure 7.8a by v_1, the output of the instrumentation amplifier. Replace the lower input by v_c, an offset voltage to be provided. Rename resistors R_f and R_a in Figure 7.8a as R_1, and rename resistor R_b as R_c. With these name changes, the output of the summing amplifier is given by the following equation:

$$V_{out} = -v_1 - (R_1/R_c)v_c$$

b. Use a -1 V supply for v_c, a 1 kΩ resistor for R_c, and compute the resistance of R_1 that will result in an output voltage of 1 V when $T = 0°C$.

c. Complete the v_{out} column in the results table.

(6) Determine the terminal-based linearity of the output voltage, v_{out}.

7.16 Determine the minimum sampling rate to get all the information from a signal that has a maximum frequency of 1200 Hz.

7.17 Determine the alias frequency that is produced when a 16 kHz signal is sampled at a rate of 20 kHz.

7.18 The logic levels of the input to the 4-bit binary-weighted DAC in Figure 7.31 are 5 V for a logic 1 and 0 V for a logic 0. The resistance values are $R_f = 2$ kΩ and $R = 1$ kΩ.

a. Determine V_{LSB}, V_{MSB}, and V_{FS}.

b. Determine v_0 for each of the following binary inputs:

0101   1011   1111   0111

7.19 Use the successive approximation technique outlined in Figure 7.35 to convert each of the following analog voltages to 4-bit digital words. Use $V_{FS} = 20$ V.

a. 4.2 V   b. 19.6 V

c. 9.25 V   d. 14.71 V

7.20 Determine the quantization error in volts and percent of the full-scale range for each of the following ADCs. The input voltage is 10 V at the center of the top step.

a. Resolution = 4 bits; error = $\pm\frac{1}{2}$ bit

b. Resolution = 8 bits; error = ± 1 bit

c. Resolution = 10 bits; error = $\pm\frac{1}{2}$ bit

d. Resolution = 12 bits; error = $\pm\frac{1}{2}$ bit

CHAPTER 8

Microprocessors and Communication

OBJECTIVES

Microprocessors and communication systems have made computer-integrated manufacturing (CIM) possible. Microprocessors provide intelligence everywhere in the control system. Communication networks link field units into "islands of automation" and provide the nervous system of the automated factory.

The purpose of this chapter is to provide you with an understanding of the microprocessors, buses, interfaces, and communication networks used in control systems. After completing this chapter, you will be able to

1. Name and describe six microprocessors used in control

2. Sketch diagrams of a microcomputer and a microprocessor

3. Discuss each of the following aspects of a microcomputer:

   a. The number of lines in the data bus

   b. The number of lines in the address bus

   c. The logical and physical memory spaces

   d. Virtual memory and virtual machine operation

4. Name and describe four board-level buses

5. Describe the RS-232 and RS-449 serial interfaces

6. Describe the IEEE-488 parallel data bus

7. Explain baseband, carrierband, and broadband networks

8. Discuss MAP networks and explain MAP

9. Name and describe seven functions of a protocol

10. Name and describe two access protocols

11. Describe the HDLC protocol

12. Use the terms listed at the end of this chapter in discussions about microcomputers and communication

8.1 INTRODUCTION

Communication between the various sensors and control loops in a plant has become increasingly important as industry moves toward the automated factory. Intelligence has moved closer to the machines and processes, giving us access to information in a way not possible in the past. "With all the intelligence now available in the control devices and systems in plants, the urge to communicate cannot be denied, and the benefits that result are too great to be ignored."*

The microprocessor has played a major role, if not the major role, in moving industry into the communication age. Microprocessors are used in every part of a control system, bringing intelligence to field devices such as the smart transmitter. Microprocessors are also a vital part of the communication process, as the source device, the destination device, and the controlling device. A general knowledge of the microprocessors involved in control will be very useful in understanding communication systems. For this reason, the study of communication begins with an overview of the microprocessors most often found in control.

The microprocessor is just one of several components in a microcomputer system. Additional components consist of RAM and ROM memory, digital I/O ports, analog I/O ports, and peripheral device interfaces. Industry has developed several standardized interface buses to help the control engineer design and develop microcomputer systems for special applications. With these standardized buses, the engineer can use a modular approach to design and develop the system. Section 8.3 reviews four popular board-level buses.

With the rise of the personal computer, serial interfaces and parallel interfaces have become household words (at least in some households). Every computer printer requires an interface to communicate with a personal computer. Industry has developed standards for serial and parallel interfaces for data transfers between devices such as computers, printers, and sensors. An organization also needs the ability to share information among many users. A network is a communication channel that provides two-way communication between many users. If the users are located within an area of a mile or so, this type of communication channel is called a local area network (LAN). In Sections 8.4 and 8.5 we examine the RS-232C and RS-449 standards for serial interfaces, the IEEE-488 standard for the parallel data bus, and the MAP network for communication among all devices and units in an automated factory.

Communication requires adherence to a set of rules that will assure the success of a transfer of information from a source to a destination. That some rules are necessary is obvious—if, for example, the source sends a message when the destination is not "listening," the message will be lost. The set of rules that govern the transfer of information is called a *protocol*. A protocol is sometimes referred to as *handshaking*, or "the please and thank you routine." Industry has made, and is continuing to make,

---

* George J. Blickley, *Control Engineering*, October 1987.

a major effort to develop an effective protocol for factory automation. In Section 8.6 we review some existing protocols.

8.2 MICROPROCESSORS

A typical microcomputer consists of a processing and control unit, memory units, and an input/output interface (see Figure 8.1). The units in the system are served by three groups of signal lines called *buses*. They are called *external buses* to differentiate them from three corresponding buses inside the microprocessor. The microprocessor is a single chip that handles the processing and control functions of the computer. Other chips provide the memory and input/output functions of the system. Microprocessors are classified as 8-bit, 16-bit, or 32-bit according to the number of lines in the external *data bus*. This determines the number of bits the processor can transfer in one machine cycle. A group of 8 data bits is called a *byte*. In discussions about microprocessors, a group of 16 bits is called a *word*, and a group of 32 bits is called a *double word* (in discussions about minicomputers and mainframe computers, a word might refer to 24, 32, 48, 64, or 128 bits).

The number of lines (bits) in the external *address bus* is another important consideration in a microcomputer system. The number of external address bits determines the number of memory bytes the computer can select (sometimes called the *physical address space* of the microprocessor). A microprocessor with a 16-bit address bus has a physical address space of $2^{16} = 65,536$ bytes (64 kilobytes). With an address bus of 20 bits, the physical address space is $2^{20} = 1,048,576$ bytes (1 megabyte or 1 MB). With an address bus of 32 bits, the physical address space is 4,294,967,296 bytes (4 gigabytes or 4 GB). Some microprocessors do not bring out all the internal address lines. An MC68000, for example, has a 32-bit internal address bus, but the

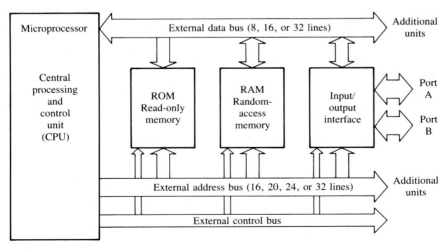

Figure 8.1 Typical microcomputer.

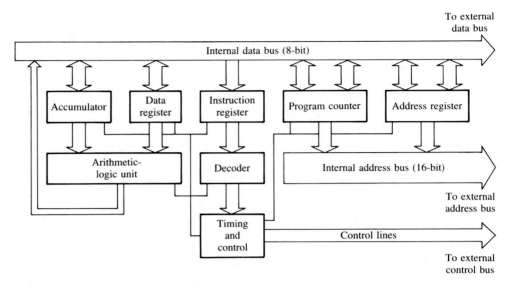

Figure 8.2   Diagram representing most of the elements of a microprocessor. The stack pointer (not shown) would look just like the address register.

external address bus has only 24 bits. The term *logical address space* refers to the internal addressing capability of the microprocessor. Thus the MC68000 has a logical address space of 4 gigabytes and a physical address space of 16 megabytes.

Figure 8.2 is representative of the microprocessors used in control systems. The buses are called internal buses to differentiate them from the external buses shown in Figure 8.1 (some microprocessors do not bring out all of the internal data and address lines).

A microcomputer executes a program by fetching the instructions from memory one after another, executing each instruction before fetching the next. Instructions have two parts, an operation code and an operand. The *op code* specifies the operation to be performed, and the *operand* specifies one of the numbers the operation will be performed on (the other number is already in the accumulator).

Let's trace one instruction through the microprocessor in Figure 8.2, explaining the function of each part along the way. The fetch operation begins when the program counter places the address of the next instruction on the address bus. The microprocessor fetches the instruction from memory and then increments the contents of the program counter to be ready for the next instruction.

The processor places the op code of the instruction in the instruction register and the operand in the address register. The decoder deciphers the op code and signals the control unit to carry out the operation. When an operation is to be performed on two numbers, one number will already be in the accumulator, placed there by a previous instruction. The other number is identified by the operand, which is now in the address register. The processor puts the contents of the address register on the address bus and gets the operand from memory. In our representative diagram, the second operand is placed in the data register.

The arithmetic-logic unit performs the operation on the contents of the accumulator and the data register and places the result on the data bus, where it can be placed in the accumulator, the data register, or one of the other registers. This completes the instruction and the processor moves on to fetch and execute the next instruction.

Eight-Bit Microprocessors

The 8-bit microprocessor is well suited for many control applications. Even though it has been superseded by more powerful 16-bit and 32-bit processors, the 8-bit microprocessor is by no means obsolete. We would not use an 18-wheel truck to move a small load that could be carried by a pickup truck. Neither would a control engineer use a 32-bit MC68020 or 80386 in a control application that could be handled by an 8-bit 8085 or Z80. In both situations, the use of excessive equipment would be inefficient and more expensive than necessary. In this section we examine four popular 8-bit microprocessors: the Intel 8080 and 8085, the Motorola 6800, and the Zilog/Mostek Z80. These four have established a standard 8-bit microprocessor that is illustrated in the external model in Figure 8.3.

The first things a user needs to know about a microprocessor are the number of data bits it can handle and the physical address space. These are defined by the external model in Figure 8.3. The next thing of interest to the user is the *programming model*, also shown in Figure 8.3. Together, the external model and the programming model define the tools the microprocessor provides to do a job. As shown by the external model, all four microprocessors handle data in groups of 8 bits and have a physical address space of 64 kilobytes (65,536 bytes).

The programming model and the instruction repertoire define the major differences in the four processors. In this discussion, we focus only on the programming models. All processors have a program counter, a stack pointer, a flag or condition code register, at least one additional address register, an accumulator, and at least one additional data register. However, the 8080, 8085, and Z80 have four data registers compared with one (accumulator B) for the 6800. The Z80 also has two 16-bit index registers, an interrupt register, and a refresh register. The refresh register is used with dynamic RAM, a type of memory that must be accessed every few milliseconds in order to recharge the memory cell. This is in contrast to static RAM, which retains its content as long as the power remains.

The M68000 Family Microprocessors

The Motorola M68000 family is a series of 8-, 16-, and 32-bit microprocessors that all share the same 32-bit internal architecture and user programming model (see Figure 8.4). The user programming model defines the internal registers that are available to the user in all members of the M68000 family. The model includes eight 32-bit data registers, eight 32-bit address registers, a 32-bit program counter, and an 8-bit condition code register. The data registers are all general-purpose registers that

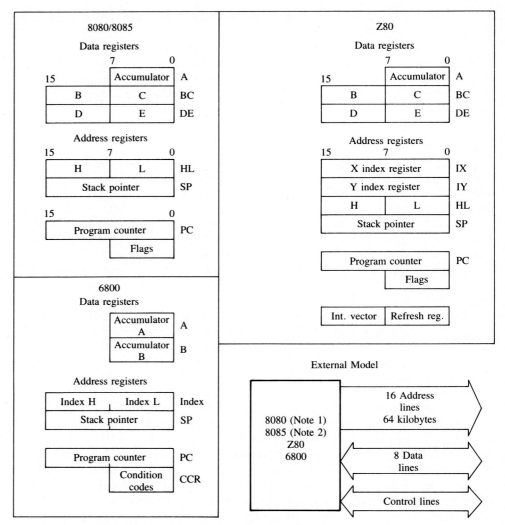

Notes: 1. The 8080 multiplexes 8 control signals on the data lines at the beginning of each instruction cycle. Control signals must be latched externally.

2. The 8085 multiplexes the low 8 address bits on the 8 data lines at the beginning of each instruction cycle. Low 8 address bits must be latched externally.

Figure 8.3 Programming models and external model of four popular 8-bit microprocessors: the Intel 8080 and 8085, the Zilog/Mostek Z80, and the Motorola 6800.

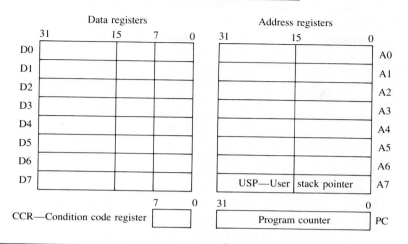

Figure 8.4 is shown with the following content:

M68000 Family User Programming Model

Data registers

Address registers

	31	15	7	0
D0				
D1				
D2				
D3				
D4				
D5				
D6				
D7				

31	15	0	
			A0
			A1
			A2
			A3
			A4
			A5
			A6
USP—User	stack pointer		A7

CCR—Condition code register [7 0]

Program counter PC [31 0]

Supervisor Programming Model Supplement

68000/008:   ISP, SR
68010/012:   ISP, SR, VBR, SFC, DFC
68020:   ISP, SR, VBR, SFC, DFC, MSP, CACR, CAAR
68030:   ISP, SR, VBR, SFC, DFC, MSP, CACR, CAAR, MMU

Description of Registers

ISP A7′   32-bit interrupt stack pointer
MSP A7″   32-bit master stack pointer
VBR   32-bit vector base register
SR   16-bit status register (includes CCR from user model)
SFC   3-bit source function code
DFC   3-bit destination function code
CACR   32-bit catch control register
CAAR   32-bit catch address register
MMU   Memory management unit—6 registers (2 64-bit, 3 32-bit, 1 16-bit)

Figure 8.4   All members of the Motorola 68000 family share a common user programming model. A supervisor programming model supplement increases the capability of newer models while maintaining upward compatibility.

can be accessed as an 8-bit byte, a 16-bit word, or a 32-bit word, as indicated by the vertical lines in the user programming model. Data registers can be used as the source or the destination of all arithmetic, logical, and data movement operations. The address registers can be accessed as either a 16-bit address or a 32-bit address, but all address calculations are carried out to the full 32-bits. Only one of the address registers is dedicated—A7 is the user stack pointer for subroutine calls and other stack operations. The M68000 family features a large linear address space, a simple but powerful instruction set, and a rich complement of flexible addressing modes.

The use of the 32-bit data and address registers sets the stage for upward compatibility of all members of the family from the 16-bit MC68000 and 8-bit MC68008 to the 32-bit MC68030. As new members were added to the family, all new features were incorporated in a supplementary model called the supervisor programming model supplement (SPMS) (see Figure 8.4). In the first two members of the family (MC68000 and MC68008), the SPMS includes only the 32-bit interrupt stack pointer and the 16-bit status register. Figure 8.4 lists the registers included in the SPMS for each member of the family. Notice that each model includes all the features of its immediate predecessor—assuring the upward mobility of all members of the family.

The external configurations of the M68000 family members are illustrated in Figure 8.5. The major external differences are in the number of external data lines and the physical address space determined by the number of external address lines. The number of data lines classifies the member as an 8-bit processor (MC68008), a 16-bit processor (MC6800, MC6810, and MC6812), or a 32-bit processor (MC68020 and MC68030). The number of address lines determines the *physical memory space* of the processor. This varies from 1 megabyte in one version of the MC68008 to 4 gigabytes in the MC68020 and MC68030. Notice also the increase in processor operating speeds from the original MC68000 to the MC68030.

One significant improvement in the M68000 family is the ability of the newer members to act as a *virtual machine*. A virtual machine uses hardware and software to emulate a processor with 4 gigabytes of physical memory and a full complement of peripherals, even though the machine actually has much less physical memory and few of the peripherals. In a sense, the software and hardware emulation produces an image of the full complement machine. This image is called *virtual memory*. A mass storage device such as a large-capacity disk drive stores the virtual image, and only a portion of the image is in the *physical memory* at any given time. The MC68000 and MC68008 have no virtual machine capabilities; the remaining four members all have virtual machine capability. The progression to MC68030 brought additional hardware capabilities, including two on-chip instruction caches and a memory management unit that translates logical addresses into physical addresses to implement the emulation of the virtual memory.

A short example will explain how a virtual machine works. In this example, the physical memory consists of five pages with 4 kilobytes in each page (in the 68000, page size can vary from 256 bytes to 32 kilobytes). The logical (internal) memory is the same size as the physical memory. In this example, the user program requires 10 pages of memory, so a virtual image of 10 pages is stored on a disk drive. Let's identify the five pages of physical memory as P0...P4 and the 10 pages of virtual memory as V0...V9.

When the program starts, the first five pages of virtual memory are loaded into the physical memory, and everything is fine as long as the processor is reading and writing from the first five pages of *virtual memory*. A write to virtual page V3 becomes a write to physical page P3; a read from virtual page V4 is actually a read from physical page P4. Then it happens—the processor attempts to write to virtual page V5. A fault signal indicates that the desired page is not in the physical memory. The memory management unit takes over and temporarily suspends the write operation. The MMU decides which physical page to use (let's arbitrarily say that it chooses

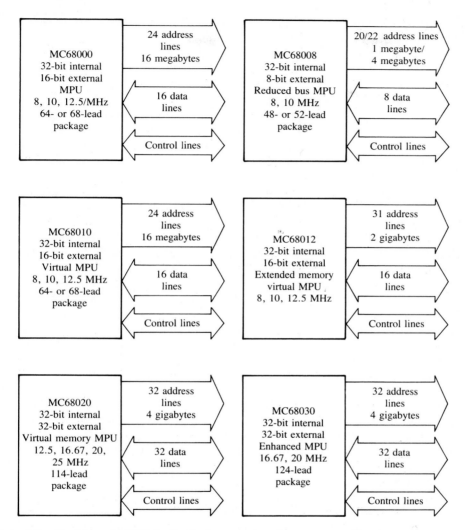

Figure 8.5 The Motorola 68000 family of microprocessors shares a common 32-bit internal architecture and instruction set. Different members of the family bring out different numbers of address lines and data lines, as indicated in the diagrams.

page P0). The MMU stores the contents of physical page P0 in file V0 on the disk. This is an important step, because the contents of P0 may have been changed by previous operations, so it is not the same as it was when loaded from the disk. With the contents of P0 saved, the MMU next loads file V5 from the disk into physical page P0. The MMU translates the virtual address from page V5 into a corresponding physical address on page P0. Finally, the MMU signals the processor to continue the instruction. The processor uses the translated address to write to physical page P0.

A virtual machine is very useful for the development and testing of new software. With a virtual machine emulation, a software engineer can develop and test the software for a new system without the expense and effort required to build a complete system.

The 8086/8088/80X86 Family of Microprocessors

The Intel 8086/8088/80X86 family is a series of 8-, 16-, and 32-bit microprocessors that share a common internal architecture and user programming model. The members of the family include the 8086, 8088, 80186, 80188, 80286, and 80386. All members of the family except the 80386 have a 16-bit internal bus structure. The 80386 has a 32-bit internal bus structure. The user programming model is shown in Figure 8.6. The crosshatched portion is unique to the 80386; the remainder of the model is common to all members of the family. The model includes an instruction pointer (program counter in the other models), a flags register, and a stack pointer. These are all 16-bit registers that are extended to 32 bits in the 80386. The four data registers are general-purpose registers identified by the letters A, B, C, and D. They can be used as 16-bit registers with the names AX, BX, CX, and DX, or they can each be divided into two 8-bit registers with the names AH/AL, BH/BL, CH/CL, and DH/DL. The 80386 can use the full 32 bits with the names EAX, EBX, ECX, and EDX.

The four address registers are 16-bit general-purpose registers identified by the names SI, DI, BP, and SP. The 80386 can use the full 32 bits with the names ESI, EDI, EBP, and ESP. Any named portion of a data or address register can be specified as the operand in almost all instructions.

The memory is organized into variable-length segments that are addressed using a two-component address. One component points to the beginning of the segment, the other defines the offset from the beginning of the segment to the location of the byte defined by the two components. The four 16-bit segment registers (CS, SS, DS, and ES) define four different memory segments for code, data, stack, and extra data. In the 8086/8088/80186/80188 processors, a 20-bit physical address is formed by shifting the 16-bit segment base left 4 bits and adding the 16-bit offset to it. In the 80286, the two components define a 24-bit physical address. The 80386 uses a 16-bit segment selector and a 32-bit offset to define a 32-bit physical address.

The external models of the family are shown in Figure 8.7. A lack of pins on the chip forced a multiplexing of data and control signals on some of the address lines. The net result is an 8-bit external data bus for the 8088 and 80188; a 16-bit external data bus for the 8086, 80186, and 80286; and a 32-bit external data bus for the 80386. The 8086, 8088, 80186, and 80188 all have a 20-bit external address bus and a 1-megabyte physical address space. The 80286 has a 24-bit external address bus and a 16-megabyte physical address space. The 80386 has a 32-bit address bus and a 4-gigabyte address space.

Both the 80286 and the 80386 provide on-chip support of virtual memory with a memory management unit that includes address translation, multitasking hardware, and a protection mechanism. The 80386 has a logical address space of 2^{46} bytes (64 terabytes) for virtual memory. See the section on the M68000 for further explanation of virtual memory and virtual machine operation.

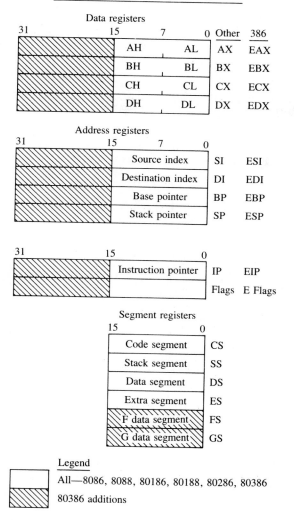

Figure 8.6 Programming model for Intel's 16/32-bit family of microprocessors. All members of the family share a common 16-bit architecture, which is extended to a 32-bit architecture for the 80386 (crosshatched portion).

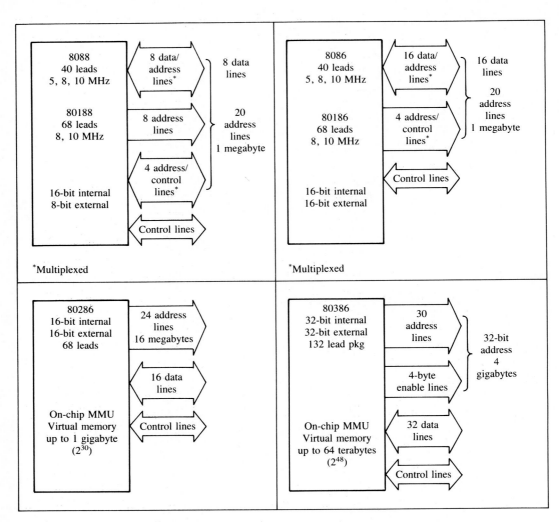

Figure 8.7   External models for Intel's 16/32-bit family of microprocessors. All members of the family are upwardly compatible.

8.3 BOARD-LEVEL BUSES

Communication occurs at many different levels in control systems. One level of communication is the transfer of information between the components of a micro-computer system (see Figure 8.1). The external data, address, and control buses form a communication path between the microprocessor, the RAM unit, the ROM unit, and the input/output unit. Digital data are passed from one unit to another at mega-

Table 8.1 Microprocessor System
Buses

Bus	Standard	Pins	Data	Address
IBM PC		62	8	20
Multibus I	IEEE-796	86	16	20
Multibus II	IEEE-P896.2	96	32	32
Q-bus		72	16	16
STD Bus	IEEE-P961	56	8	16
S-100	IEEE-696	100	8	16
VMEbus	IEEE-P1014	96	32	32

hertz speeds over parallel channels on the bus. The distance the signals travel is quite short, usually confined within a 10- or 20-inch card rack.

When microprocessors first arrived, engineers designed a new bus structure for each new microcomputer system. This bus design consumed a considerable portion of the engineer's time, and soon became a case of "reinventing the wheel" for each new system design. It was not long before enterprising designers began using existing buses for new microcomputer systems. The following is a brief chronology of the development of standard buses for microcomputer systems.

In 1975, MITS developed a 100-pin bus for their Altair 8800 microcomputer, which used an Intel 8080 microprocessor. Soon a number of companies adopted the Altair bus, which became known as the S-100 Bus. In 1976, Intel introduced a single-board microcomputer called the SBC 80/10. In 1977, Intel formalized the specification of SBC 80/10 and gave it a new name: "multibus." In 1978, Pro-log and Mostek introduced the STD Bus. In 1981, IEEE standardized the S-100 Bus as IEEE-696, and Mostek, Motorola, and Signetics introduced the VMEbus. In 1983, Intel introduced Multibus II, and the original Multibus became known as Multibus I. By 1985, the IBM PC had become popular as a bus structure for control applications. Bus structures will, no doubt, continue to improve as more powerful microprocessors become available. Table 8.1 summarizes the buses most often used in control systems. Descriptions of Multibus I, STD Bus, VMEbus, and Multibus II follow.

Multibus I

Multibus is the result of the successful effort by Intel Corporation to put an entire microcomputer system on a single board. The size of the Multibus card reflects that intention (see Figure 8.8). The card is a rather large 6 in. deep by 12 in. wide. The Institute of Electrical and Electronic Engineers (IEEE) has standardized Multibus under the specification IEEE-796.

True to its name, the Multibus specification defines four different bus structures. The largest bus is the *multibus system bus*, which uses an 86-pin edge connector on the backplane edge of the board. The second bus is the *local bus extension* or *LBX bus*, which uses a 60-pin edge connector, also on the backplane edge of the card. The third bus is the *SBX expansion bus*, which uses a 36- or 44-pin connector socket mounted on the Multibus board. The fourth bus is the *multichannel I/O bus*, which

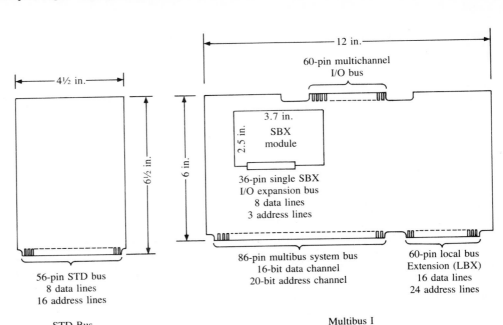

Figure 8.8 The small size of the STD Bus card is good for customizing I/O, as the designer can select the functions needed on a card-by-card basis. The large size of the Multibus I card is better suited for applications that require more computer power.

uses a 60-pin edge connector on the front edge of the card, opposite the backplane edge.

The multibus system bus was originally designed to provide an 8-bit data channel and a 16-bit address space. It has since been expanded to provide a 16-bit data channel and a 20-bit address space. Additional lines on the bus provide control commands and handshake signals.

The purpose of the local bus extension (LBX) is to provide a direct high-speed channel between the processor board and external memory units. The bus has 16 data lines which can transfer data at a rate of 19 megabytes per second. It also has 24 address lines, which means the processor can access up to 16 megabytes of external memory at almost the same speed as the on-board memory. The separate LBX bus means that the processor can access the off-board memory without disturbing the system bus.

The SBX expansion bus is intended for expansion of the I/O capabilities in smaller increments than adding another 6- by 12-in. Multibus card. The SBX connector comes in two sizes to accommodate two board sizes, the single SBX board and the double SBX board. The single SBX board is 2.5 in. deep by 3.7 in. wide and connects to the Multibus board through a 36-pin connector. The double SBX board is 2.5 in. deep by 7.5 in. wide and connects to the Multibus board through a 44-pin connector. Connection to the external I/O device is made through an edge connector on the SBX board.

The multichannel I/O bus provides a path for direct memory access (DMA), a method used to transfer large blocks of data from an external source directly into the on-board memory without involving the processor.

STD Bus

From a control engineer's perspective, the letters STD stand for "Simple To Design," and that describes the philosophy behind the development of the STD Bus. The STD Bus was intended to provide a modular card approach to the design of 8-bit microprocessor systems. The bus has 16 address lines and 8 data lines. However, the address space has been increased to 20 bits by multiplexing the 4 high-order address bits on the data bus. The STD Bus has also been extended to accommodate 16-bit processors.

The size of the card, 4.5 in. wide by 6.5 in. deep, is in keeping with the modular concept of design (see Figure 8.8). Industry has responded to the modular concept by providing a wide variety of STD modules to the system designer. The following is a sample of the cards offered by one supplier of STD board modules.

An 8088 CPU card

A 64K or 128K static RAM card with battery backup

A bank select ROM card with 576K of on-card memory

A 96-channel digital I/O card

A mechanical relay card with eight independent relays

A solid-state relay card with eight independent relays

A card with eight optically isolated inputs

A 16-channel analog input card with multiplexed ADC

A four-channel DAC card (digital-to-analog converter)

One indication of the success of the STD Bus is the number of processors that are available on an STD card. They include the 8080, Z80, NSC800, 8088, 80188, 8085, 6800, 6809, 6502, and 68008.

The cards that make up a system are mounted in a card cage with a slot and backplane connector for each card. Card cages come in a variety of sizes based on the number of slots available. One vendor has 11 different sizes, ranging from four slots to 24 slots. The corresponding pins on the backplane connectors are wired together so that the electrical connections are the same for every slot. Any card can go in any slot. Figure 8.9 illustrates a six-module STD Bus system.

The applications of microprocessors in control systems are usually I/O bound. That means that the number and variety of input and output devices is the limiting factor on the design, development, and operation of the system. An 8-bit microprocessor is more than adequate for many of these applications. The STD Bus is a good choice for many of the I/O-intensive real-time applications in control systems. The

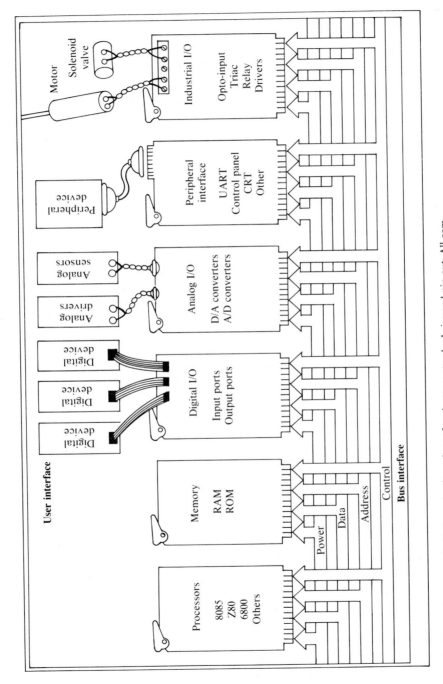

Figure 8.9 Systems are built from circuit cards conforming to a standard size and pin-out. All communicate on a common 56-line bus—the STD Bus. (Reprinted from *Control Engineering* magazine with permission. Copyright Cahners Publishing Co., 1983.)

designer can produce an economical design by choosing the functions needed on a module-by-module basis.

VMEbus

VMEbus is a three-bus structure consisting of the VMEbus system bus, the VMXbus memory expansion bus, and the VMSbus serial bus. The VMEbus specification defines two card sizes that meet the DIN Eurocard standard (Figure 8.10). The single card measures 100 mm wide by 160 mm deep (3.94 in. by 6.3 in.) and has a DIN 96-pin connector that mates with the P1 socket in the top half of the backplane. The double card measures 233.3 mm wide by 160 mm deep (9.19 in. by 6.3 in.) and has two DIN 96-pin sockets that mate with the P1 and P2 sockets on the backplane. Figure 8.11 illustrates a typical VME card cage for rack mounting a combination of single and double VME cards.

The P1 upper backplane connector plus 32 lines from the P2 lower backplane connector comprise the VMEbus. The P1 connector provides 16 data lines, 24 address lines, and all the control lines required for the operation of a 16-bit microprocessor such as the MC68000 or MC68010. The P2 connector provides the additional data, address, and control lines required for the operation of a 32-bit microprocessor such as the MC68020 or 80386. The remaining 64 pins on the P2 connector may be used

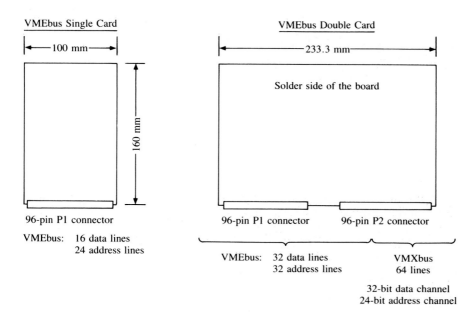

Figure 8.10   The VMEbus uses the DIN Eurocard standard. The single VME card uses one connector, P1, which mates with the P1 socket in the backplane. The double card has two connectors, P1 and P2, which mate with the P1 and P2 sockets in the backplane.

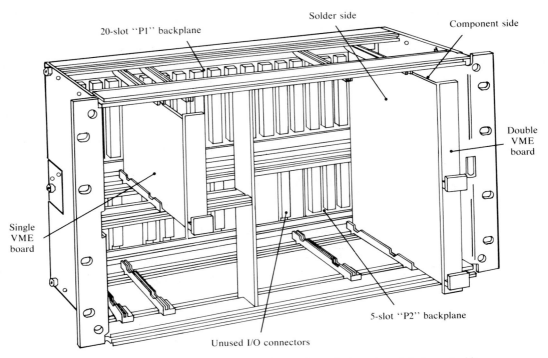

Figure 8.11 Typical VME card cage for 19-in. rack mounting allows for insertion of single or double height "Eurocards." Backplane connectors are pin and socket type for better mechanical and electrical performance than commonly used edge connectors. (Reprinted from *Control Engineering* magazine with permission. Copyright Cahners Publishing Co., 1984.)

for I/O lines or they may be used to provide a VMXbus. The VMEbus has a master/slave asynchronous communication structure with seven levels of priority interrupt and four levels of arbitration. The VMEbus can handle multiprocessing with any mix of 8-, 16-, and 32-bit processors. The bus automatically senses whether 8, 16, or 32 data lines are required and adjusts accordingly.

The VMXbus uses the remaining 64 lines on the P2 connector to provide a private, high-speed communication path between a CPU card and its associated memory cards. The VMXbus provides 32 data lines and 24 address lines that give it a 16-megabyte address space. The VMXbus transfers data at speeds in excess of 48 megabytes per second.

The VMSbus is a high-speed serial bus that provides rapid communication of short messages between system components. The VMSbus lines are located on the P1 connector, making it available to both single and double cards.

Applications for the VMSbus include process control, factory automation, digital communications, image processing, computer-aided design (CAD), computer-aided manufacturing (CAM), computer-aided engineering (CAE), data acquisition, and robotics. The following are some examples of VMSbus modules.

MC68000/68010 CPU board with 1-megabyte on-board memory

2-megabyte dynamic RAM memory board

128-kilobyte static RAM memory board

Multiprotocol serial communication processor board

Winchester-floppy drive controller board

Ethernet 2.0-compatible controller board (LAN)

Graphics controller board

Digital input and output boards

Analog input and output boards

A VMEbus multiprocessor system is illustrated in Figure 8.12.

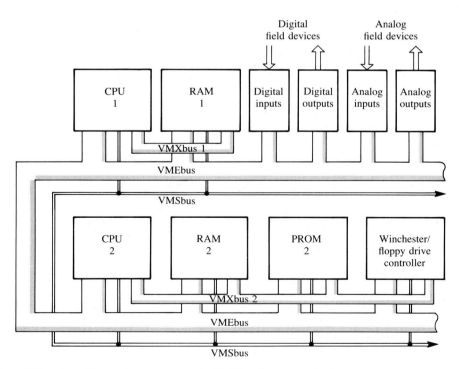

Figure 8.12   A multiprocessor system uses the VMEbus for communication between the processors, memory units, and I/O units; the VMXbus for high-speed transfers between the processors and their associated memory units; and the VMSbus for serial transmission of short messages between all units.

Multibus II

Intel introduced Multibus II in 1983 as an open-system extension of Multibus to support 32-bit microprocessors. The new system consists of five separate buses. The *parallel system bus* (PSB) replaces the multibus system bus. The new *local bus extension* (LBX II) is an upgraded version of the original LBXbus. The *SBX expansion bus* and the *multichannel I/O bus* are the same as they were in the original Multibus system. The *serial system bus* (SSB) is a new addition for Multibus II.

The mechanical configuration was changed to the Eurocard mechanical standard with its more reliable pin and socket connectors. The specification defines two board sizes, as illustrated in Figure 8.13. The single board is 100 mm wide by 220 mm deep (3.94 in. by 8.66 in.) with a single 96-pin P1 backplane connector. The double card is 233.3 mm wide by 220 mm deep (9.19 in. by 8.66 in.) with two 96-pin backplane connectors, P1 and P2. The card cage is similar to the VMSbus card cage except for an increased depth to accommodate the deeper Multibus II cards.

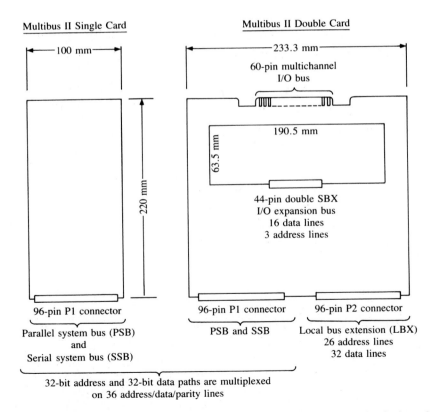

Figure 8.13 Multibus II uses the Eurocard standard and defines two card sizes. The single card uses the P1 backplane connector. The double card uses the P1 and P2 backplane connectors.

The parallel system bus (PSB) is a synchronous bus that is located on the P1 connector. It can support multiprocessing with 8-, 16-, and 32-bit microprocessors. The data channel can be 8, 16, or 32 bits, and four different address spaces are defined. The memory address space uses 32 bits, the I/O address space is 16 bits, the message address space can be 16 or 32 bits, and the interconnect address space is 16 bits. The address and data signals are multiplexed on 36 address/data/parity lines.

The LBX II bus is located on the P2 connector. It has 32 data lines and 26 address lines for a total address space of 64 megabytes. The LBX II bus can handle up to four memory cards with the same speed as the on-board memory.

The SSB Bus is located on the P1 connector, where it uses two wires in the message address space. Its purpose is very similar to the VMSbus—rapid communication of short messages between system components.

A wide variety of processor boards, memory boards, I/O boards, communication boards, and data acquisition boards is available for Multibus systems.

8.4   COMMUNICATION INTERFACES

Communication is the orderly transmission of information from a sender to a receiver. A conversation between two individuals is a form of communication; so is reading a book, listening to a radio, or watching a program on television. Another form of communication takes place between electronic components such as computers, controllers, measuring instruments, display terminals, and printers. In this form of communication, information in the form of digital (or analog) signals is transmitted over a path called a *communication channel*.

If the communication channel consists of a single path, it is called a *serial channel*. If the channel consists of many paths, it is called a *parallel channel* or a bus. Parallel transfer is faster than serial transfer, but it is more expensive because more lines are required. Parallel data lines also have more crosstalk problems. Within a computer, data are transferred in parallel to maximize speed (the address bus and the data bus of a microcomputer system are examples of parallel communication channels). Beyond the computer, parallel transfer usually involves distances of a few feet and rarely more than 50 or 60 ft. The remainder of this chapter is concerned with the serial transfer of digital data (with the exception of the IEEE-488 data bus).

In either serial or parallel, the data may be transmitted synchronously or asynchronously. *Synchronous transmission* means that a timing signal is used to make the sender and receiver act together. *Asynchronous transmission* uses control signals between the sender and receiver to make sure the receiver accepts the data before the sender terminates the transmission. This procedure is sometimes referred to as "handshaking" or the "please-and-thank-you" routine.

A *communication channel* is a path for electrical transmission of information from a sending station to a receiving station. Communication channels are divided into three categories: simplex, half-duplex, or full-duplex. A *simplex channel* can communicate in only one direction. One station is always the sender; the other station is always the receiver. A *half-duplex channel* can communicate in both directions, but only one direction at a time. This is the normal mode of communication between two individuals. A *full-duplex channel* can communicate in both directions at once. Although this mode does not work well between individuals, it is very effective between electronic components.

Information in digital form consists of a collection of binary digits that represent messages, numbers, or commands. A typical message might look like this: 0110 1001 1100 1011 1000 1000 1010 0011 0001 . . . . The sending device must first convert the binary information into a group of signals; then it sends the information to the receiver, one signal at a time. The simplest conversion the sending device can use is a binary voltage signal for each binary digit (one voltage level for a binary 1 and another voltage level for a binary 0). However, other conversion methods may be used to increase the speed at which the information is transmitted. For example, a signal with four voltage levels can be used for each pair of binary digits as follows:

Voltage Level	Binary Digits
1	00
2	01
3	10
4	11

In a similar manner, a signal with eight voltage levels can be used for each three binary digits. In general, the number of voltage levels (L) required to represent N binary digits is given by

$$L = 2^N \tag{8.1}$$

A voltage signal with 16 levels can be used to represent four binary digits. A serial path that uses a signal with 16 levels is similar to a 4-bit parallel bus. Both can transmit four binary digits at a time. The difference is the parallel bus requires four paths to do the job, while the 16-level signal does it on one path.

Two important characteristics of a communication channel are the maximum number of bits it can transmit in 1 second (BPS_{max}); and the maximum number of signals it can transmit in 1 second (baud).

BPS_{max} is the maximum number of binary digits a communication channel can transmit in 1 second.

Baud rate is the maximum number of signals a communication channel can transmit in 1 second.

If N is the number of binary digits represented by each signal, then

$$BPS_{max} = N(\text{baud rate}) \tag{8.2}$$

A communication channel has a frequency response that is similar to a band-pass filter. For example, a voice-grade telephone line will pass frequency components between 300 and 3300 Hz. It has a bandwidth of $3300 - 300 = 3000$ Hz. Frequency components below 300 Hz or above 3300 Hz will not pass through the channel. This band-pass characteristic of a communication channel has two significant consequences.

The first consequence has to do with the relationship between the bandwidth and the values of BPS_{max} and baud rate for the channel. In 1948, Claude Shannon showed that the maximum capacity of a channel in bits/second is proportional to the bandwidth of the channel. In other words, the bandwidth of the channel sets an upper limit on the bit transfer rate (BPS_{max}) of a sending device.

The second consequence is that the digital signal must be frequency translated so that it falls between the band limits of the communication channel (see Figure 8.14). It has been shown that a digital signal is composed of many frequency components. The components in the signal vary according to the pattern of 1's and 0's in the signal. For example, at 100 BPS, the pattern 110011001100... would produce a square wave with frequency components of 25, 75, 125, 175, 225, 275 Hz, and so on. The power in the components diminishes rapidly as the frequency increases. Figure 8.14 shows the power envelope of a digital signal that has a bit transfer rate of 100 BPS. Most of the power is in components with a frequency of less than 200 Hz. A rule of thumb is that most of the power in a digital signal is in components with frequencies less than twice the bit transfer rate ($2 \times BPS$).

Notice the positions of the digital signal envelope and the lower band limit of the channel in Figure 8.14. The communication channel will not pass the digital

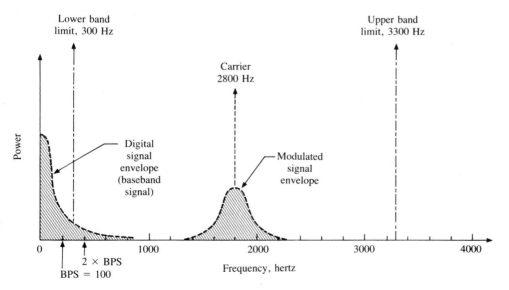

Figure 8.14   The digital signal on the left has a bit transfer rate (BPS) of 100 bits/second. Most of the power in the signal is in components with frequencies below 200 Hz ($2 \times BPS$). The signal is translated by a modulator so that the power in the modulated signal is in components with frequencies between the band limits of the communication channel.

signal because most of the power is below the lower band limit of the channel. A process called *modulation* can be used to translate the digital signal. Amplitude modulation is the simplest type of modulation. An amplitude modulator forms the product of two signals: the digital signal, which is called the *baseband signal*, and a sinusoidal signal, called the *carrier*. Amplitude modulation of a baseband signal produces two signals: one equal to the carrier plus the baseband signal, the other equal to the carrier minus the baseband signal. The two signals are called the *upper sideband* and the *lower sideband*. The envelopes of the two sideband signals are shown on either side of the carrier in Figure 8.14. Notice that the envelopes are well within the band-pass limits of the signal. The communication channel will pass the modulated signal.

> Modulation of a digital signal translates a baseband signal to a frequency range that can be transmitted on a communication channel.

RS-232C Serial Interface

In 1969, the Electronic Industries Association published RS-232C, a specification that describes the signals in a 25-pin connector used for serial transmission of digital data. RS-232 has been almost universally accepted as the "standard" serial port for digital equipment. The full title of the specification is "Interface Between Data Terminal Equipment and Data Communication Equipment Employing Serial Binary Data Interchange." The term *data terminal equipment* refers to the sending or receiving equipment that is using the RS-232C serial port. This includes computers, terminals, printers, plotters, and so on. The term *data communication equipment* refers to a device called a *modem* that modulates the signal at the sending end of the communication channel and demodulates the signal at the receiving end. The term *modem* is a contraction of "modulator/demodulator." Figure 8.15 illustrates an RS-232C communication circuit.

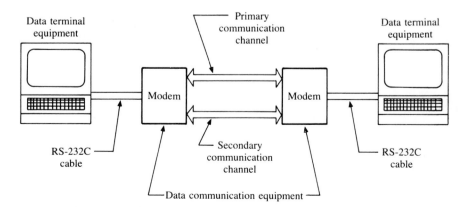

Figure 8.15   RS-232C communication circuit.

The RS-232C specification defines two communication channels, the primary channel and the secondary channel. Each channel has two carrier signals, one for each direction. The secondary channel is slower than the primary channel and may not be used in a particular application. The RS-232C specification describes the electrical characteristics of the signals in detail, but leaves many choices to the designer of the modem. Consequently, RS-232C does not guarantee compatibility between all terminals and all modems. RS-232C does not apply to communication speeds greater than 20,000 bits/second, and the maximum cable length is 50 ft. The signals are restricted to two voltage regions: a positive region from 3 to 15 V dc, and a negative region from -3 to -15 V dc. Signals must pass through the transition region from -3 to 3 V in less than 1 ms. Table 8.2 lists the signal names, pin assignments, and directions.

Table 8.2   RS-232C Signals and Pin Assignments

Pin	Signal Name	Direction
1	Protective Ground (Frame)	
2	Transmitted Data	Terminal ⟶ modem
3	Received Data	Terminal ⟵ modem
4	Request to Send	Terminal ⟶ modem
5	Clear to Send	Terminal ⟵ modem
6	Data Set Ready	Terminal ⟵ modem
7	Signal Ground (Common Return)	
8	Carrier Detect	Terminal ⟵ modem
9		
10		
11		
12	Secondary Carrier Detect	Terminal ⟵ modem
13	Secondary Clear to Send	Terminal ⟵ modem
14	Secondary Transmitted Data	Terminal ⟶ modem
15	Transmit Clock (DCE Source)	Terminal ⟵ modem
16	Secondary Received Data	Terminal ⟵ modem
17	Receive Clock	Terminal ⟵ modem
18		
19	Secondary Request to Send	Terminal ⟶ modem
20	Data Terminal Ready	Terminal ⟶ modem
21	Signal Quality Detector	Terminal ⟵ modem
22	Ring Indicator	Terminal ⟵ modem
23	Data Rate Selector	Terminal ⟶ modem
24	Transmit Clock (DTE Source)	Terminal ⟶ modem
25		

Source: EIA Standard RS-232C (Washington, D.C.: Electronic Industries Association, 1969).

The following is a brief description of how the signals are used to transmit a message. When the sending terminal is ready to transmit a message, it turns on *Request to Send,* and the sending modem responds by turning on its outgoing primary carrier. The receiving modem detects the carrier, turns on its carrier, and turns on *Carrier Detect* to alert the receiving terminal that a message is forthcoming. The sending modem, 750 ms after sensing the receiver's carrier, turns on *Clear to Send,* enabling transmission of data. When the sending terminal senses that Clear to Send is on, it begins to send signals on its *Transmitted Data* line. The sending modem modulates the outgoing carrier with each signal. The receiving modem demodulates each signal and puts the demodulated signal on *Received Data.* The receiving terminal receives an exact replica of the transmitted message.

RS-449 Serial Interface

Between 1975 and 1977, the Electronic Industries Association published the following three specifications for an improved serial interface.

RS-422 defines balanced receivers and drivers.

RS-423 defines unbalanced receivers and drivers.

RS-449 defines the signals that make up the interface.

The RS-449 interface consists of two connectors: a nine-pin connector for the secondary channel signals, and a 37-pin connector for all other signals. The RS-449 specification includes all the RS-232C signals plus a few additional signals. The new signals provide features such as local loopback and remote loopback, which help pinpoint the location of a malfunction.

The addition of balanced drivers and receivers is the major improvement in the interface circuits. A balanced driver uses a separate return line for each signal. This is in contrast with the unbalanced circuits used in RS-232C, in which all signals share a common return line. Electrical improvements in the circuits allow higher operating speeds and longer cable lengths than RS-232C. Table 8.3 lists the maximum cable lengths and data rates for balanced and unbalanced lines.

IEEE-488 Parallel Data Bus

The IEEE-488 data bus was originally developed in 1970 by Hewlett-Packard for use with its own minicomputers and instruments. The interface was well received

Table 8.3 RS-449 Maximum Cable Lengths and Data Rates

Maximum Cable Length (ft)	Data Rate (bits/second)	
	Unbalanced Drivers and Receivers	Balanced Drivers and Receivers
4000	Below 900	Below 90,000
400	10,000	1,000,000
40	100,000	10,000,000

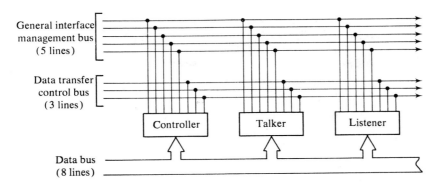

The bus controller uses the management and transfer control lines to handle bus traffic. The open-ended bus connects up to 15 devices, but total cable distance cannot exceed 66 ft. Device addresses are set by five-bit jumpers or switches on the back panel of each instrument.

Figure 8.16 IEEE-488 parallel data bus. (From J. Washburn, "Communications Interface Primer—Part II," *Instruments and Control Systems*, April 1978, pp. 62–63. Copyright Chilton Co., 1978.)

and has been built into a wide range of instruments, peripherals, and calculators. In 1975, the Institute of Electrical and Electronic Engineers made it a standard with the designation IEEE-488. The International Electro-technical Commission also adopted the IEEE-488 standard with the designation IEC Standard 625-1. Industry often refers to the standard as the General Purpose Interface Bus (GPIB), and the names GPIB, HP-1B, IEEE-488, and IEC Standard 625-1 are used synonymously. Figure 8.16 describes the IEEE-488 interface.

The IEEE-488 standard defines three device types: talkers, listeners, and controllers. *Talkers* are devices that send data, commands, and status information to listeners. *Listeners* are identified by an address so that a talker can send data to a specific listener. Talkers include sensing instruments of various types. Listeners include recording instruments such as printers and plotters. *Controllers* manage the communication between the talkers and the listeners.

IEEE-583 CAMAC Interface

The CAMAC (or IEEE-583) interface was originally designed for nuclear instrumentation laboratories, where instruments were constantly swapped among systems. IEEE-583 is a completely specified interface system. The standard defines the mechanical configuration, the electrical connectors, the data transmission paths, and the protocol. The mechanical configuration consists of a "crate" containing slots for 25 modules. The two slots on the right are occupied by a controller that manages all CAMAC communications. Five types of controllers are available: one for direct access to a computer, one for parallel transmission, two for different types of serial transmission, and one for stand-alone operation. Figure 8.17 illustrates these different possibilities with a CAMAC interface.

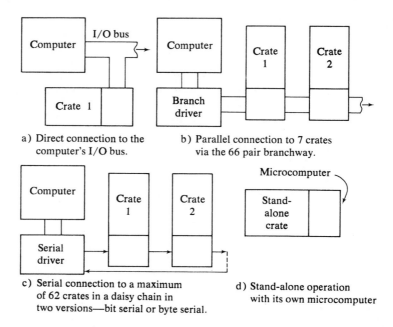

Figure 8.17 CAMAC's different configurations. (From J. Washburn, "Communications Interface Primer—Part II," *Instruments and Control Systems*, April 1978, p. 61. Copyright Chilton Co., 1978.)

8.5 LOCAL AREA NETWORKS

A network is a communication channel that connects a large number of user stations to one or more central stations. The central stations provide access to large host computers, central databases, graphic printers, and other devices that are too expensive to be dedicated to a single station. If the stations are separated by considerable distance, such as branch offices in different cities, the network is called a *wide area network*. If all stations are located within a radius of about a mile, the network is called a *local area network* (LAN). Local area networks are an important means of communication in corporations, universities, government agencies, and many other organizations. Most of the effort in factory automation is directed at establishing a communication path between various units and devices in the plant. Indeed, the local area network is the central nervous system of the automated factory.

Local area networks usually use coaxial cable to provide the communication channel or channels. Some local area networks provide a single communication channel, others provide many channels. A single-channel LAN may use an unmodulated *baseband* signal or a modulated *carrierband* signal. A multichannel LAN uses a *broadband* system that has many channels, each with its own carrier frequency. In a large local area network, a broadband network forms the *backbone* that interconnects many single channel subnets via interconnects called *bridges* or *gateways*.

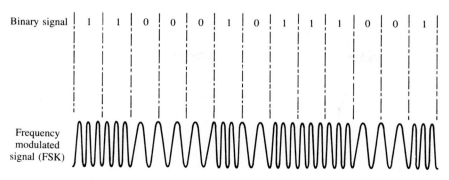

Figure 8.18 Example of a frequency-modulated binary signal using frequency shift keying (FSK).

IEEE-802.4 Single-Channel Systems

A single-channel network, whether baseband or carrierband, uses the entire band-width of the coaxial cable. Many single-channel networks are carrierband systems that use a version of frequency modulation called *frequency shift keying* (FSK). Two frequencies are selected, one for transmitting a binary 0, the other for transmitting a binary 1. If F0 is the frequency used for a 0, and F1 is the frequency used for a 1, the modulated signal appears to alternate between the two frequencies with a con-stant amplitude. A binary signal and the FSK modulated signal would appear as shown in Figure 8.18.

Carrierband subnets are used to interconnect process controllers, programmable logic controllers, and other intelligent components in a factory network. The subnets are the branches of the factory network, each connecting to the broadband backbone network through an interconnecting bridge or gateway. All nodes in a subnet trans-mit and receive at the same frequency. The IEEE-802.4 networking standard specifies two carrierband signals: phase-continuous FSK and phase-coherent FSK.

An 802.4 *phase-coherent system* uses a data rate of either 5 or 10 megabits/second. The binary signal is directly encoded using a frequency equal to the data rate for a binary 1, and a frequency equal to twice the data rate for a binary 0. The data rates and signal frequencies are listed below.

Data Rate (megabits/second)	Frequency (MHz)	
	Binary 1	Binary 0
5	5	10
10	10	20

The modulated signal will always consist of a full cycle of one of the two signals. A binary 1 is represented by one full cycle of the lower frequency; a binary 0 is repre-sented by two full cycles of the higher frequency.

An 802.4 *phase-continuous system* has a data rate of 1 Mbit/second and uses a Manchester-encoded signal to modulate a 5-MHz carrier. The carrier is modulated to 6.25 MHz for a logic 1 and to 3.75 MHz for a logic 0. In the Manchester encoder, the signal bits are allocated equal time intervals, each equal to the reciprocal of the data transfer rate in bits/second. For the 802.4 phase-continuous system, the time interval for each bit is 1 μs. The value of each bit (0 or 1) is converted into a signal transition at the center of the interval. The signal is changed from 0 to 1 to represent a binary 1; and from 1 to 0 to represent a binary 0. At the end of each interval, the signal goes to the level required to encode the next bit. A Manchester-encoded message contains both the data and the synchronizing clock signal (the clock is defined by the signal transition).

IEEE-802.4 Broadband Networks

A broadband network provides many frequency channels that can accommodate several local area networks plus voice and even video communications, all operating on the same cable. Each LAN uses two frequency channels. All LAN nodes transmit on one frequency and receive on another frequency. A device called a *headend repeater* translates the signal from the transmitting frequency to the receiving frequency. Figure 8.19 depicts a broadband network serving two LANs. In the diagram, node A1 is sending a message to node A2 on LAN-A, and node B3 is sending a message to node B2 on LAN-B. Node A1 modulates its signal on frequency F1. The LAN-A headend demodulates the F1 signal and remodulates the signal on frequency F2. The F2 signal goes to all nodes, but only node A2 accepts the message (the message is addressed to node A2). Node B3 modulates its signal on frequency F3. The LAN-B headend demodulates the F3 signal and remodulates the signal on frequency F4. All nodes receive the F4 signal, but only B2 accepts the message.

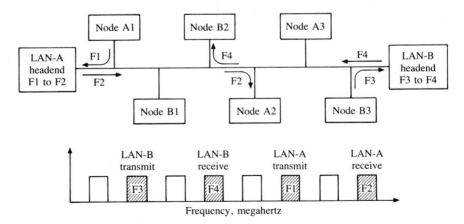

Figure 8.19 A broadband network can support more than one LAN. In the diagram, node A1 is sending a message to node A2 on LAN-A, while node B3 is sending a message to node B2 on LAN-B. Each LAN is serviced by a headend that translates the transmitted signal (F1 or F3) into a receivable signal (F2 or F4).

The IEEE-802.4 standard defines a multilevel duobinary AM/phase-shift keying broadband network. The frequency channels are predefined, and the signal is encoded to control modulation. The 802.4 token-bus protocol specifies the channel usage. The transmit and receive frequency channels can be changed or programmed. The standard designates three data rates: 1, 5, and 10 megabits/second.

The Open Systems Interface

In 1979, the International Standards Organization (ISO) began developing a communication network model that became known as the *open systems interface* (OSI). A number of LANs are based on the OSI model, including Ethernet and MAP (*Manufacturing Automation Protocol*). The OSI model consists of the following seven functional levels.

Open Systems Interface

Level	Name	Function
7	Application	Upper-level services
6	Presentation	Upper-level services
5	Session	Upper-level services
4	Transport	Upper-level services
3	Network	Routing between cables
2	Data link	Routing between nodes
1	Physical	Routing between nodes

Each layer is a module responsible for providing networking services to the level above it. The functions fall into one of three distinct divisions. The first two levels of the OSI model (physical and data link) deal with transmitting messages between nodes on the same cable. Level 3 (network) covers the transmission of messages from one cable to another cable. The upper four levels (transport, session, presentation, and application) handle upper-level services such as data formatting and access security.

The physical level of the OSI model is concerned with the data rate and the type of cable, connectors, and signaling methods used to transmit and receive the message. The data link level defines the formation of data frames, priorities, and procedures (protocol) to assure an orderly transfer of messages.

The network level specifies the routing of messages across a bridge or gateway from one cable to another cable. A *bridge* is used to connect two cables that have the same protocol. A *gateway* is used to connect two cables that have different protocols.

The transport level provides reliable end-to-end data transfer. The session level translates and synchronizes names and addresses. The presentation and application levels are concerned with factory management.

MAP Networks

The Manufacturing Automation Protocol (MAP) is an emerging international standard for a multilevel communication network for factory automation. The purpose of a MAP network is to link together all the controllers, computers, workstations,

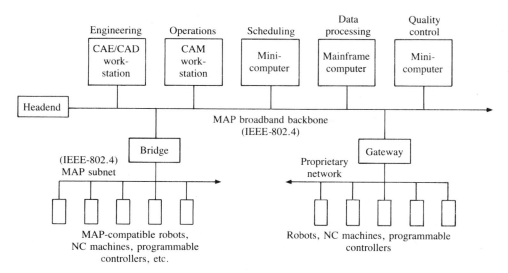

Figure 8.20   A MAP factory network uses bridges for MAP subnets and gateways for proprietary networks.

production machines, and offices in an entire factory. The impetus for the development of MAP came in the late 1970s, when General Motors observed that half of their automation budget was used for custom interfaces for incompatible devices. The vast number of communication systems used by different vendors of automation equipment had created a modern-day Tower of Babel. In 1980, General Motors and Boeing established a task force to develop a public-domain communication standard for the factory environment. The specification was called the Manufacturing Automation Protocol, or simply MAP.

The MAP architecture is based on the seven-level OSI model. For the physical level, MAP specifies a 10-megabit/second IEEE-802.4 broadband network for the backbone, and 5-megabit/second IEEE-802.4 carrierband subnets. MAP specifies a token-passing protocol for the network. The backbone network connects various central resources such as mainframe computers, CAD/CAM workstations, data bases, application computers, bridges to factory subnets, gateways to proprietary networks, and wide area network (WAN) gateways to other sites. Most of the messages on the backbone network are high-volume database information movements. The carrierband subnets connect the robots, NC machines, programmable logic controllers, and process controllers on the plant floor. Some of the subnets are proprietary LANs that connect to the backbone via a gateway. Figure 8.20 shows a typical MAP network.

Ethernet

In 1980, Digital Equipment Corporation, Intel, and Xerox jointly published the specification for the Ethernet LAN. Ethernet follows the seven-level OSI model. The physical layer specifies a 10-megabit/second Manchester-encoded baseband signal. The data link layer operates under an access protocol called *carrier sense multiple*

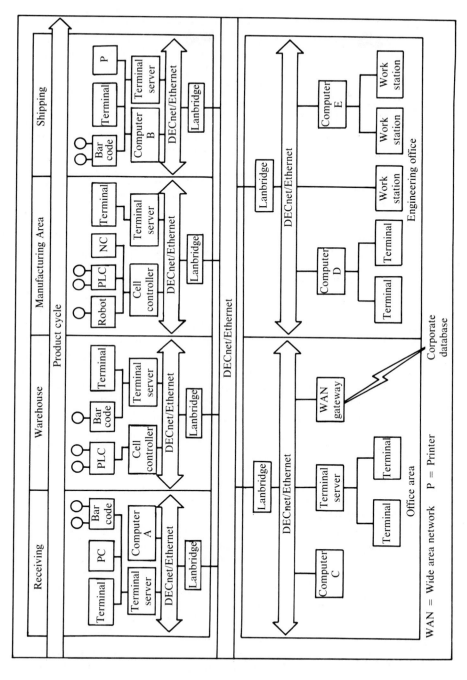

Figure 8.21 A DECnet/Ethernet baseband cable backbone extends a manufacturing network to all areas of a plant—even to a remote corporate data base. (Reprinted from *Control Engineering* magazine with permission. Copyright Cahners Publishing Co., 1987.)

331

access with collision detection (CSMA/CD). It also handles error detection and addressing of the source and destination. Ethernet was integrated with DECnet, a proprietary network used with DEC computers. Figure 8.21 shows an application of a DECnet/Ethernet network.

8.6 COMMUNICATION PROTOCOLS

A *protocol* is a set of rules that govern the operation of the devices that share a communication channel. The rules define the following functions:

1. *Framing.* Messages are transmitted in units called frames. A typical frame consists of a header section, a data section, and a trailer section. The header section defines the beginning of the frame and other information, such as the source address, the destination address, the length of the data section, and the purpose of the frame. The data section contains the message to be transmitted. The trailer section contains an error-checking number and an end-of-frame mark.

2. *Error control.* The transmitter uses a predefined algorithm to compute a frame check sequence (a number) from the characters in the message. The cyclic redundancy check (CRC) is one such algorithm. In CRC, the entire message block is treated as a large binary number. This large number is divided by a predefined 16-bit number, and the remainder of this division (also 16 bits) is the frame check sequence that is included in the trailer. The receiver repeats the division process on the received message and checks its remainder with the frame check sequence. If the two numbers are not identical, an error has occurred, and a retransmission is requested. The probability that an error will be undetected by CRC is extremely low.

3. *Sequence control.* The messages are numbered in the order in which they occur. This sequence number is used to prevent the loss of messages, eliminate duplicate messages, and identify messages for retransmission. Sequencing is a type of error control.

4. *Access control.* Access control defines how each node on the network obtains access to the network to transmit data. There are two methods of access control: polling and contention. In the *polling* method, the nodes on the network take turns being invited to transmit a message. In the *contention* system, each node competes for access to the network.

5. *Transparency.* The flag that indicates the end of a frame is a particular bit pattern, such as 01111110. It is possible that some messages will include the same bit pattern as the flag. Somehow the receiver must be able to tell the difference between a 01111110 in the message and the 01111110 flag that ends the message. The technique for accomplishing this objective is called *transparent text transmission*.

6. *Flow control.* Flow control involves starting and stopping the transmission of messages on a communication channel. It includes initiating transmissions after a channel has been idle, returning the channel to idle when there are no messages to transmit, and stopping transmission when the receiver cannot handle more information.

7. *Configuration control.* The configuration of a network consists of the addresses of the active nodes in the system. Occasionally, a node will be removed from a network or a new node will be added to the network. When this happens, the network must be reconfigured to incorporate the changes.

The CSMA/CD Access Protocol

The carrier sense multiple access with collision detection (CSMA/CD) protocol allows a station to transmit whenever it is ready and it does not detect the presence of another station's carrier on the communication channel. If two stations start to transmit at the same time, a collision has occurred. When a collision occurs, a collision detection signal is sent to all stations. The two transmitting stations stop transmitting, wait for a random period of time, and then retry the transmission. The station delay is determined by an algorithm that uses a random number generator, so it is very unlikely that two stations will have the same delay time.

The CSMA/CD protocol works well as long as there are not a lot of collisions. As the volume of message traffic increases, so does the probability of collisions. As the channel approaches 50% of capacity, stations begin to experience long delays in obtaining access. Another disadvantage of a contention protocol is that it is not possible to calculate a worst-case time for a node to transmit its message over the network. For these reasons, a CSMA/CD system may not be the best choice for real-time environments such as process control.

The Token-Passing Access Protocol

The *token-passing* protocol is a polling-type access protocol. A special bit pattern called the token is passed from station to station in ascending order of the station address. The station with the highest address passes the token to the station with the lowest address, so the token continues to circulate through all stations in an unending loop.

The access rule is very simple. Only the station with the token can access the communication channel to transmit a message, and only one frame (or packet) can be transmitted before passing the token to the next station. If a station has several frames to transmit, it sends one frame each time it receives the token. If a station receives the token and has no message to transmit, it immediately passes the token to the next station. The size of a frame is defined, so it is possible to calculate the worst-case access time (i.e., every station is transmitting a full-sized frame). For this reason, token-passing systems are considered more practical for real-time control applications.

The HDLC Protocol

The High Level Data Link Control (HDLC) is a standard communication link protocol established by the International Standards Organization (ISO) in 1977. HDLC very closely resembles the Synchronous Data Link Control (SDLC) introduced by IBM in 1974. Both HDLC and SDLC are bit-oriented, full-duplex communication protocols. HDLC has several different operation modes. The following is a description of the host computer/multiterminal operating mode. The host computer is called the primary station. The terminals are called the secondary stations.

Opening flag	Secondary station address	Control	Information	Frame check sequence	Closing flag
01111110	8 bits	8 bits	Variable length	16 bits	01111110

Figure 8.22   HDLC information frame format.

A primary station controls the network and issues commands to the secondary stations according to a preassigned order of priority. The secondary stations respond by sending appropriate responses. The secondary stations can initiate a transmission only when given permission to do so by the primary station.

The HDLC uses three types of frames: an information frame used to transfer data, a supervisory frame used for control signals, and a nonsequenced frame used to initialize and control secondary stations. Figure 8.22 shows the HDLC frame format.

The 8-bit flag (01111110) marks the beginning and the end of a frame. The flag pattern must never occur within the frame. If it does, the receiver will interpret the pattern as an end-of-frame flag and ignore the remainder of the message. A technique called *zero stuffing* is used to achieve transparent text transmission. When the transmitter encounters a sequence of five 1's in the frame data, it inserts a zero as the next bit in the message. Each time the receiver encounters five 1's followed by a zero, it removes the zero, knowing that it was stuffed by the transmitter. When the receiver encounters a zero followed by six 1's and a zero, it recognizes the end-of-frame flag, knowing that the transmitter would have stuffed a zero if it was not the flag.

The address byte identifies a secondary station. The control byte determines the type of frame (information, supervisory, or nonsequenced) and additional control information. The frame check sequence is the CRC error-checking number.

GLOSSARY

Address bus: A group of signal lines that carry the address of one byte in the memory space of a computer. (8.2)

Asynchronous: A method of communication that does not use a timing signal to make the sending and receiving stations work together. (8.4)

Backbone: A broadband network that connects various central resources, such as mainframe computers, CAD/CAM workstations, databases, application computers, bridges to factory subnets, gateways to proprietary networks, and wide area network (WAN) gateways to other sites. (8.5)

Baseband: A network that uses the entire bandwidth of the cable for a single unmodulated signal. (8.5)

Baud rate: The maximum number of signals a communication channel can transmit in 1 second. A signal may represent one or more binary bits in the message. (8.4)

Bit transfer rate (BPS): The number of bits a communication channel transmits in 1 second. (8.4)

Bridge: A device used to carry messages between two distinct networks that use the same protocol. (8.5)

Broadband: A network that provides many channels that can accommodate several local area networks plus other signals, such as voice and video. (8.5)

Bus: A group of lines that carry data and signals between units of a computer system. (8.2)

Carrierband: A network that uses the entire bandwidth of the cable for a single carrier-modulated signal. (8.5)

Communication channel: A path for electrical transmission of information from a sending station to a receiving station. (8.4)

Contention: An access method in which the nodes transmit messages on a first-come, first-served basis. (8.6)

CSMA/CD: An access protocol that allows a station to transmit whenever it is ready, and it does not detect the presence of another carrier on the line. Collision detection is used to prevent two stations from transmitting at the same time. (8.6)

Data bus: A group of lines that carry data between units of a computer system. (8.2)

Full-duplex: Simultaneous communication in both directions. (8.4)

Gateway: A device used to carry messages between two distinct networks that use different protocols. (8.5)

Half-duplex: Communication in both directions, but only one direction at a time. (8.4)

Headend repeater: A device, located at the end of a network cable, that converts signals from one frequency to another frequency. (8.5)

Logical memory space: The amount of memory that can be addressed on the internal address lines of a microprocessor. (8.2)

Manufacturing Automation Protocol (MAP): A protocol for a factory communication network that uses a broadband backbone network serving a number of carrierband subnets. (8.5)

Modulation: The alteration of a sinusoidal carrier by a data signal for the purpose of transmitting the data signal. (8.4)

Open systems interface (OSI): A seven-layer model for communications protocols developed by the International Standards Organization. (8.5)

Parallel channel: A channel that provides multiple paths to transmit all bits of a binary message unit at the same time. (8.4)

Physical memory: The memory that is actually available to a microprocessor. (8.2)

Physical memory space: The amount of memory that can be addressed on the external address lines of a microprocessor. (8.2)

Polling: An access method in which the nodes take turns being invited to transmit a message. (8.6)

Programming model: A model of a central processing unit that shows the internal registers available to the programmer. (8.2)

Protocol: A set of rules that govern the operation of the devices that share a communication channel. (8.1)

Serial channel: A channel that transmits binary messages one bit at a time. (8.4)

Simplex: Communication in only one direction. (8.4)

Synchronous: A method of communication that uses a timing signal to make the sending and receiving stations work together. (8.4)

Token passing: An access protocol that uses polling with a circulating bit pattern called a token. (8.6)

Virtual machine: A computer that uses hardware and software to emulate a processor with a physical memory equal to the logical memory and a full complement of peripheral devices. (8.2)

Virtual memory: An image, created on a large-capacity disk, that represents all the logical memory available to a microprocessor. Peripherals may also be included in the image. (8.2)

EXERCISES

8.1 Name and describe the main parts of a microcomputer system.

8.2 A certain microcomputer can transfer a byte of data to a peripheral device in 150 μs. How long will it take to transmit a 650-character message if each character uses one byte?

8.3 On what basis is a microprocessor classified as 8-bit, 16-bit, or 32-bit?

8.4 A special-purpose microprocessor has a 28-bit internal address bus and an 18-bit external address bus. Determine the logical address space and the physical address space of the microprocessor.

8.5 Describe the level of communication provided by board-level buses.

8.6 Summarize the differences between Multibus I and Multibus II.

8.7 Summarize the differences between Multibus I and the STD Bus.

8.8 The size of the STD Bus card makes it suitable for what type of microcomputer applications?

8.9 A 16-level signal is used to transmit messages on a 300-baud line. Determine the data transfer rate in bits/second.

8.10 An amplitude modulator has two inputs and one output. The inputs are the baseband signal, B, and the carrier, C. The output of the modulator, M, is equal to the product of the two inputs: $M = BC$. Use a trigonometric identity to show that if B and C are as given below, then M is the sum of three components with frequencies of 17,000, 18,000, and 19,000 rad/s.

$$B = 1 + 2 \cos 1000t$$

$$C = \cos 18,000t$$

8.11 Outline the sequence of signals used to transmit a message on an RS-232C serial communication channel.

8.12 Explain the difference between baseband, carrierband, and broadband networks.

8.13 Sketch the output of a Manchester encoder for the following input message if the data transfer rate is 1 megabit/second:

$$\text{message} = 0 \quad 1 \quad 1 \quad 0 \quad 1 \quad 1 \quad 0 \quad 1$$

8.14 In the OSI model, a data block starts at level 7 in the transmitter and moves down through the levels until it reaches transmitter level 1, the physical level. The message is transmitted to the receiver at level 1. Then it makes its way up to level 7 in the receiver. On the transmitter side, each level receives a data unit from the level above it. Levels 7 through 2 add a header to the data unit they received to form a new, larger data unit for the level below. Level 2 also adds a trailer to form the data unit for level 1. The result is a pyramid that shows the overhead produced by the OSI model. Sketch the pyramid using the following units for the headers and trailers.

Level 7 adds header: AH (application header)

Level 6 adds header: PH (presentation header)

Level 5 adds header: SH (session header)

Level 4 adds header: TH (transport header)

Level 3 adds header: NH (network header)

Level 2 adds header: F (flag), A (address), and C (control) and trailer: FCS (frame correction sequence) and F (flag)

Level 1 transmits the frame (unit) it receives from level 2.

8.15 Name and explain two methods of controlling access to a communication network.

CHAPTER 9

Position, Motion, and Force Sensors

OBJECTIVES

In the late eighteenth century, a Scottish engineer named James Watt greatly improved the steam engine with several inventions. One of Watt's inventions was a governor that regulated the speed of the steam engine. Since that early beginning, the control of position or speed has become a major branch of control technology sometimes referred to as servo control. Numerical control of machine tools requires precise positioning of a workpiece and exact control of the speed and feed of the cutting tool. Robotic arms move in specified paths that require sensing and control of position and speed. Some sequential controllers must sense the presence of a part and may even identify the part and determine its exact location. The power-steering unit in a car uses the position of the steering wheel as a setpoint and positions the wheels of the car accordingly.

The purpose of this chapter is to give you an entry-level ability to discuss, select, and specify position, motion, and force sensors. The examples selected represent a reasonable cross section of the variety of sensors used to measure these quantities. After completing this chapter, you will be able to

1. Make a list of considerations in the selection of sensors to measure position, displacement, speed, acceleration, and force

2. Describe the major features of the following position sensors: potentiometers, LVDTs, synchros, resolvers, optical encoders, proximity sensors, and photoelectric sensors

3. Describe the major features of dc tachometers, ac tachometers, and optical tachometers

4. Explain how a digital velocity signal can be obtained from the pulse output of an optical encoder

5. Describe an accelerometer

6. Describe a strain gage force sensor and a pneumatic, null-balance force sensor

7. Use an example spec sheet to select a proximity sensor

9.1 INTRODUCTION

A *sensor* is a device that converts a *measurand* (parameter to be measured) into a signal in a different form. The measurand is the input to the sensor and the signal produced by the sensor is the output. Sensors are also called *transducers* or *primary elements*. The output of the sensor may be a force, displacement, voltage, electrical resistance, or some other physical quantity. Usually, a signal conditioner is required to convert the sensor output into an electrical (or pneumatic) signal suitable for use by a controller or display device. The sensor and its signal conditioner comprise the two parts of a measuring transmitter (Figure 9.1). Typical input/output graphs of the sensor, signal conditioner, and measuring transmitter are included in Figure 9.1.

The types of sensors are too numerous to include all of them in a couple of chapters. Examples of several types of sensors are included for each of the more common controlled variables. Our purpose is to illustrate the basic principles of sensors with a series of specific examples. In a control system, the function of the transducer is to provide a signal that is a measure of the controlled variable. For this reason, the examples are classified by the variable that is measured rather than by the manner in which the output is developed.

9.2 POSITION AND DISPLACEMENT MEASUREMENT

Sensing Methods

Position and displacement measurement is divided into two types: linear and angular. *Linear* position or displacement is measured in units of length. *Angular* position or displacement is measured in radians or degrees. The primary standard of length is the international meter, which is defined in terms of the wavelength of the red-orange line of krypton 86. All other units of length are defined in terms of the primary standard. The meter is the basic unit of length in the International System of Units (SI), and the radian is the unit of angle. A radian is equal to $180/\pi$ degrees (approximately $57.3°$). Linear and angular displacement are illustrated in Figure 9.2.

In continuous processes, displacement sensors are used to measure the thickness of a sheet, the diameter of a rod, the separation of rollers, or some other dimension of the product. In discrete-parts manufacturing, position sensors are used to measure the presence of a part, to identify a part, to determine the position of a part, or to measure the size of a part. The acronym PIP refers to *presence, identification, and position* measurement of parts in a manufacturing operation.

Sensors used to measure the size of a product or part usually have a sensing shaft that is in mechanical contact with the object to be measured. The mechanical connection between the sensing shaft and the object is of paramount importance—the sensor actually measures the position of the sensing shaft. A number of methods are used to translate the position of the sensing shaft into a measurable quantity. For example, movement of the shaft may cause a change in capacitance, self-inductance, mutual inductance, or electrical resistance. The shaft might also be coupled to an encoder, a device that converts the movement into a digital signal.

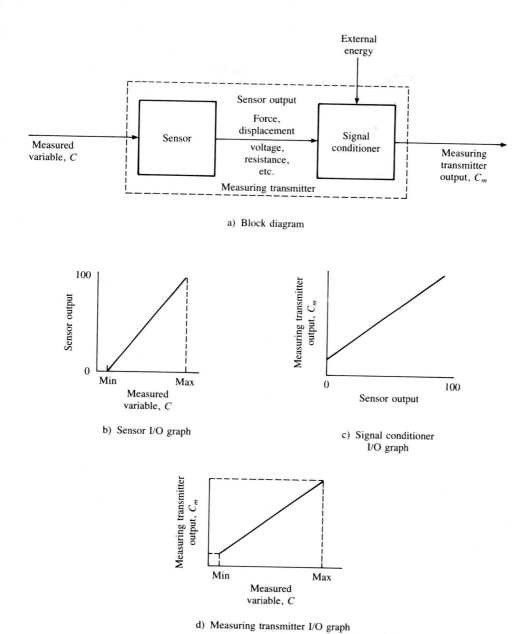

a) Block diagram

b) Sensor I/O graph

c) Signal conditioner
I/O graph

d) Measuring transmitter I/O graph

Figure 9.1　A measuring transmitter consists of two parts: a sensor (or primary element) and a signal conditioner.

Figure 9.2   Linear displacement can be converted into angular displacement of a wheel (c). The angular displacement in radians ($\Delta\theta$) is equal to the linear displacement (ΔS) divided by the radius of the wheel $\Delta\theta = \Delta S/r$.

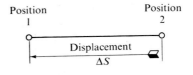

a) Linear displacement

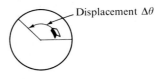

b) Angular displacement

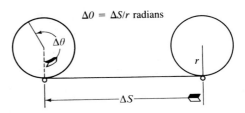

c) Conversion of linear displacement to angular displacement

The simplest PIP sensor, a lever-actuated switch, also makes contact with the object to be measured. However, many PIP sensors do not touch the object to be measured. Some noncontacting sensors use changes in self-inductance, reluctance, or capacitance to detect the presence of an object. Other noncontacting sensors use the interruption or blocking of a light beam to sense the presence of an object, to identify the object, or to measure the position of the object.

Potentiometers

A *potentiometer* consists of a resistance element with a sliding contact that can be moved from one end to the other. Potentiometers are used to measure both linear and angular displacement, as illustrated in Figure 9.3. The resistance element produces a uniform drop in the applied voltage, E_s, along its length. As a result, the voltage of the sliding contact is directly proportional to its distance from the reference end.

Figure 9.3 Two types of potentiometric displacement sensors: (a) linear; (b) angular. In both types, E_{out} is a measure of the position of the sliding contact.

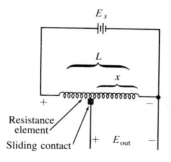

a) A linear displacement potentiometer

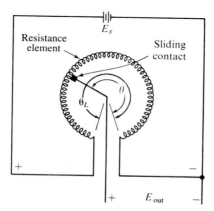

b) An angular displacement potentiometer

Linear potentiometer: $E_{out} = \left(\dfrac{x}{L}\right) E_s$

Angular potentiometer: $E_{out} = \left(\dfrac{\theta}{\theta_L}\right) E_s$

When the resistive element is wirewound, the resolution of the potentiometer is determined by the voltage step between adjacent loops in the element. If there are N turns in the element, the voltage step between successive turns is $E_T = E_S/N$, where E_S is the full-scale voltage. Expressed as a percentage of the full-scale output, the percentage resolution is given by the following relationship.

$$\text{Resolution (\%)} = \frac{100 E_T}{E_S} = \frac{100(E_S/N)}{E_S}$$

or

$$\text{Resolution (\%)} = \frac{100}{N} \qquad\qquad \textbf{(9.1)}$$

Potentiometers are subject to an error whenever a current passes through the lead wire connected to the sliding contact. This error is called a *loading error* because it is caused by the load resistor connected between the sliding contact and the reference point. A potentiometer with a load resistor is illustrated in Figure 9.4. If R_p is the resistance of the potentiometer and a is the proportionate position of the sliding contact, then aR_p is the resistance of the portion of the potentiometer between the sliding contact and the reference point. The load resistor, R_L, is connected in parallel with resistance aR_p. The equivalent resistance of this parallel combination is $(R_L)(aR_p)/(R_L + aR_p)$.

The resistance of the remaining portion of the potentiometer is equal to $(1 - a)R_p$, and the equivalent total resistance is the sum of the last two values.

$$R_{EQ} = (1 - a)R_p + \frac{aR_L R_p}{R_L + aR_p}$$

The output voltage, E_{out}, may be obtained by voltage division as follows:

$$E_{out} = \left(\frac{aR_L R_p/(R_L + aR_p)}{(1 - a)R_p + aR_L R_p/(R_L + aR_p)} \right) E_s$$

$$E_{out} = \left(\frac{a}{1 + ar - a^2 r} \right) E_S$$

where

$$r = \frac{R_p}{R_L}$$

The loading error is the difference between the loaded output voltage (E_{out}) and the unloaded output voltage (aE_S).

$$\text{Loading error} = aE_S - E_{out}$$

$$= aE_S - \left(\frac{a}{1 + ar - a^2 r} \right) E_S$$

$$= \left(\frac{a^2 r(1 - a)}{1 + ar - a^2 r} \right) E_S \quad \text{volts}$$

Figure 9.4   A loading error is produced in a potentiometer when a load resistor is connected between the sliding contact and the reference terminal.

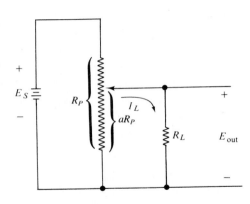

The loading is usually expressed as a percentage of the full-scale range, E_S.

$$\text{Loading error (\%)} = 100\left[\frac{a^2r(1-a)}{1+ar(1-a)}\right] \qquad (9.2)$$

Example 9.1

The potentiometer in Figure 9.4 has a resistance of 10,000 Ω and a total of 1000 turns. Determine the resolution of the potentiometer and the loading error caused by a 10,000-Ω load resistor when $a = 0.5$.

Solution

The resolution is given by Equation (9.1).

$$\text{Resolution} = 100/N = 100/1000 = 0.1\%$$

The loading error is given by Equation (9.2).

$$\begin{aligned}
\text{Loading error} &= 100\left[\frac{a^2r(1-a)}{1+ar(1-a)}\right] \\
&= 100\left[\frac{(0.5)^2(1)(0.5)}{1+(0.5)(1)(0.5)}\right] \\
&= 10\%
\end{aligned}$$

Linear Variable Differential Transformers

The *linear variable differential transformer (LVDT)* is a rugged electromagnetic transducer used to measure linear displacement. A diagram of an LVDT is shown in Figure 9.5. It consists of a single primary winding located between two secondary windings

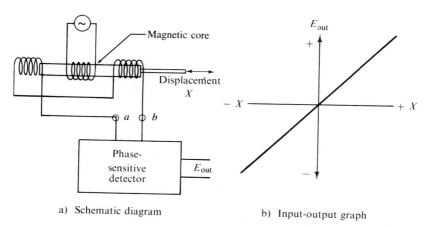

a) Schematic diagram   b) Input-output graph

Figure 9.5   The linear variable differential transformer (LVDT) and phase-sensitive detector produce a dc voltage proportional to the displacement of the movable magnetic core.

on a hollow cylindrical form. A movable magnetic core provides a variable coupling between the windings.

The sensing rod is attached to the magnetic core and moves the core in response to the displacement that is to be measured. An ac voltage is applied to the primary winding, and the transformer coupling results in an ac voltage across each secondary winding. When the magnetic core is in the center position, the two secondary voltages cancel each other at terminals $a-b$ (i.e., at the input to the phase-sensitive detector). When the core is moved to one side, the voltage in the secondary coil with more coupling becomes larger, and the other secondary voltage becomes smaller. An ac voltage appears at terminals $a-b$. We will call this ac voltage v_{a-b}. The magnitude of v_{a-b} is proportional to the displacement of the core from the null position. When the core is moved to the same distance on the other side of the null position, the ac voltage at terminals $a-b$ is equal to $-v_{a-b}$. In other words, the amplitude of the ac voltage is proportional to the amount of displacement and the phase depends on the direction of the displacement. A positive displacement produces a $0°$ phase angle. A negative displacement produces a $180°$ phase angle.

The phase-sensitive detector converts the ac secondary voltage into a dc voltage, E_{out}. The magnitude of the dc voltage is proportional to the amplitude of the ac voltage. The sign of the dc voltage is positive if the ac phase angle is $0°$, and negative if the ac phase angle is $180°$. The result is the overall input/output graph illustrated in Figure 9.5b. In some ac systems, the ac secondary voltage is used as the error signal. The phase-sensitive detector is not required in these systems.

Synchro Systems

A *synchro* is a rotary transducer that converts angular displacement into an ac voltage, or an ac voltage into an angular displacement. Three different types of synchros are used in angular displacement transducers: control transmitter, control transformer, and control differential.

Synchros are used in groups of two or three to provide a means of measuring angular displacement. For example, a control transmitter and control transformer form a two-element system that measures the angular displacement between two rotating shafts. The displacement measurement is then used as an error signal to synchronize the two shafts. The term *electronic gears* is sometimes used to describe this type of system, because the two shafts are synchronized as if they were connected by a gear drive. The addition of a control differential forms a three-element system that provides adjustment of the angular relationship of the two shafts during operation.

A two-element synchro system is shown in Figure 9.6. The control transmitter is designated CX and the control transformer is designated CT. Both the transmitter and the transformer have an H-shaped rotor with a single winding. Connections to the rotor winding are made through slip rings on the shaft. The stators each have three coils spaced $120°$ apart and connected in a Y configuration.

An ac voltage is applied to the rotor winding of the control transmitter. This voltage induces an ac voltage in the three stator windings, which are uniquely determined by the angular position of the rotor. The voltages induced in the transmitter are applied to the transformer stator windings, which, in turn, induce a voltage in the

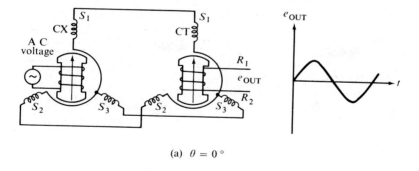

(a) $\theta = 0°$

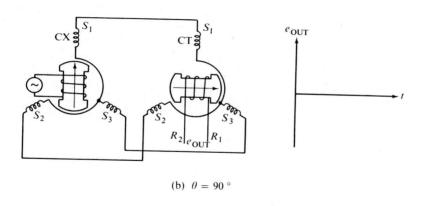

(b) $\theta = 90°$

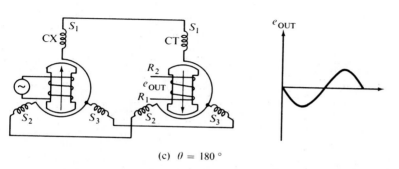

(c) $\theta = 180°$

Figure 9.6   A two-element synchro system measures the phase difference between two rotating shafts.

transformer rotor winding. The transformer rotor voltage is uniquely determined by the relative position of the two rotors, as shown in Figure 9.6. A graph of the transformer rotor voltage (e_{out}) is included for each of the three relative rotor positions shown. The amplitude of e_{out} is a maximum when the angular displacement of the two rotors is 0° or ±180°, and zero when the angular displacement is ±90°. Notice the change in sign of the ac voltage between the 0 and 180° positions.

A graph of the amplitude of the output voltage (e_{out}) versus the angular displacement (θ) is shown in Figure 9.7. The output voltage is described by the following mathematical relationship:

$$e_{out} = (E_m \cos \theta) \sin \omega t \tag{9.3}$$

where E_m = maximum amplitude

θ = angular displacement

ω = radian frequency of the ac voltage applied to the transmitter rotor

The sign and magnitude of the amplitude term ($E_m \cos \theta$) are uniquely determined by the angular displacement θ, as shown in Figure 9.7. The graph is reasonably linear for values of θ from 20 to 160°. The operating point is located at the center of this linear region (i.e., $\theta = 90°$). The magnitude of e_{out} is proportional to the amount of angular displacement. The sign or phase of e_{out} is determined by the direction of the angular displacement.

A control differential may be added to the system in Figure 9.8 to provide remote adjustment of the operating point. The control differential has a three-pole rotor, with three windings connected in a Y configuration. Connections to the other end of the rotor windings are made through three slip rings on the shaft. The stator also has three windings connected in a Y. The control differential is connected between the transmitter and the transformer, as shown in Figure 9.7. The output voltage is now given by the following relationship:

$$e_{out} = [E_m \cos (\theta + \theta_d)] \sin \omega t \tag{9.4}$$

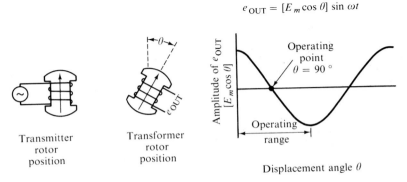

$e_{OUT} = [E_m \cos \theta] \sin \omega t$

Transmitter rotor position

Transformer rotor position

Amplitude of e_{OUT} [$E_m \cos \theta$]

Operating point $\theta = 90°$

Operating range

Displacement angle θ

Figure 9.7   The amplitude of e_{out} varies as the cosine of the angular displacement between the two shafts.

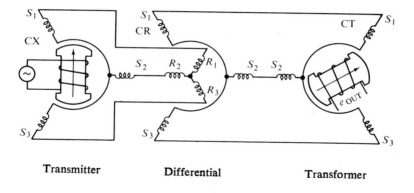

Transmitter   Differential   Transformer

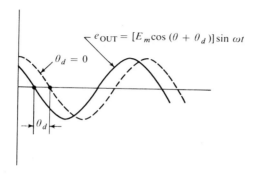

Figure 9.8   A three-element synchro system provides remote adjustment of the operating point.

The angle θ_d is the relative displacement of the differential rotor. When θ_d is zero, the three-element system is not different from the two-element system (the dashed output curve). However, when θ_d is not zero, the entire curve is displaced by an angular amount equal to θ_d. In other words, the control differential provides a means of adjusting the operating point simply by moving the differential rotor. This is especially useful when the transmitter and transformer rotors are connected to rotating shafts.

An application of a three-element synchro system is illustrated in Figure 9.9. The process consists of two rotating rolls that must be synchronized with the capability to adjust their angular relationship during operation. The control is accomplished by regulating the speed of the slave roll to maintain the desired angular relationship. The three-element synchro system acts as the measuring transmitter, setpoint, and error detector. The setpoint is the angular position of the differential rotor. It represents the desired angular relationship between the two driven rolls. The transmitter and transformer combination compares the angular relationship of the two rolls with the desired relationship, and produces an ac output signal proportional to the difference (i.e., the error signal). The converter-amplifier acts as a high-gain proportional

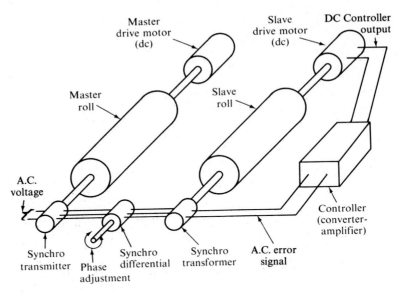

Figure 9.9 A three-element synchro system is used to synchronize two shafts with adjustment of the angular displacement between the two shafts while they are rotating.

controller. The gain is sufficiently high that the proportional offset is maintained within acceptable limits.

Example 9.2

The synchro system in Figure 9.9 operates at a frequency of 400 Hz. The maximum amplitude of the transformer rotor voltage is 22.5 V. Determine the ac error signal produced by each of the following pairs of angular displacement.

    a. $\theta = 90°$, $\theta_d = 0°$
    b. $\theta = 60°$, $\theta_d = 0°$
    c. $\theta = 135°$, $\theta_d = -15°$
    d. $\theta = 100°$, $\theta_d = -45°$

Solution

Equation (9.4) gives the relationship between θ, θ_d, and e_{out}.

$$e_{out} = E_m \cos(\theta + \theta_d) \sin \omega t$$

At 400 Hz, $\omega = 2\pi(400) = 2570$ rad/s. The maximum amplitude, $E_m = 22.5$ V.

    a. $\cos(\theta + \theta_d) = \cos(90 + 0) = \cos 90 = 0$

       $e_{out} = 0$ V

    b. $\cos(\theta + \theta_d) = \cos(60 + 0) = \cos 60 = 0.5$

       $e_{out} = (22.5)(0.5) \sin 2570t$
           $= 11.25 \sin 2570t$ V

c. $\cos(\theta + \theta_d) = \cos(135 - 15) = \cos(120) = -0.5$

$e_{out} = (22.5)(-0.5)\sin 2750t$

$\qquad = -11.25 \sin 2570t$ V

or

$e_{out} = 11.25 \sin(2570t + 180)$ V

d. $\cos(\theta + \theta_d) = \cos(100 - 45) = \cos(55) = 0.574$

$e_{out} = (22.5)(0.574)\sin 2570t = 12.9 \sin 2570t$

Resolvers

A *resolver* is a rotary transformer that produces an output signal that is a function of the rotor position. Figure 9.10 shows the position of the coils in a resolver. The two rotor coils are placed 90° apart. The two stator coils are also placed 90° apart. Either pair of coils can be used as the primary with the other pair forming the secondary. The following equations define the secondary voltages in terms of the primary voltages when the rotor coils are used as the primary.

$$E_1 = K(E_3 \cos\theta - E_4 \sin\theta) \qquad (9.5)$$
$$E_2 = K(E_4 \cos\theta + E_3 \sin\theta) \qquad (9.6)$$

When a resolver is used as a sensor, one of the rotor windings is shorted as shown in Figure 9.11. If E_4 is the shorted coil, Equations (9.5) and (9.6) simplify to the following.

$$E_1 = KE_3 \cos\theta \qquad (9.7)$$
$$E_2 = KE_3 \sin\theta \qquad (9.8)$$

Equations (9.7) and (9.8) define the output of the resolver shown in Figure 9.11. This output is reasonably linear for values of θ over a range of $\pm35°$. The excitation voltage, E_3, is a sinusoidal voltage that can be represented as follows:

$$E_3 = A \sin\omega t$$

Figure 9.10 A resolver is a transformer-type sensor that produces a signal that is a trigonometric function of shaft position, θ. The two rotor coils are placed 90° apart. The two stator coils are also 90° apart.

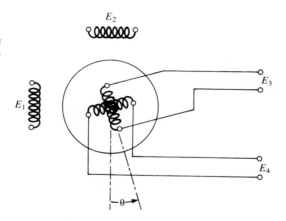

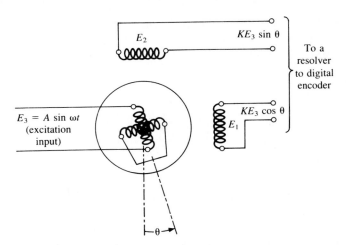

Figure 9.11 A resolver used as a sensor has one of its rotor coils shorted. An ac voltage is applied to the other rotor coil. The two output voltages, E_1 and E_2, are trigonometric functions of the displacement angle θ.

Output voltage, E_1, is essentially a sinusoidal voltage whose amplitude varies according to the cosine of the angular position of the rotor. Output voltage, E_2, is also a sinusoidal voltage, but its amplitude varies according to the sine of the angular position of the rotor.

The resolver position sensor requires a signal conditioning circuit that can convert the two voltages, E_1 and E_2, into a usable signal representing the position of the rotor, θ. If a digital signal is required, the signal conditioner must also convert the signal from an analog form to a digital form. We will refer to a signal conditioner that performs both functions as a *resolver-to-digital converter*.

One problem with a resolver is the necessity of brushes and slip rings to bring the excitation voltage to coil E_3 on the rotor. The brushes are subject to wear and must be protected from the dirty environment encountered in industry. The brushless resolver has been developed to solve this problem. A brushless resolver uses a transformer to couple the excitation voltage to the rotor coil, eliminating the need for brushes and slip rings.

Optical Encoders

An *encoder* is a device that provides a digital output in response to a linear or angular displacement. The resolver and digital converter discussed in the preceding section is an angular encoder. In this section we describe another type of encoder, the optical encoder.

An optical encoder has four main parts: a light source, a code disk, a light detector, and a signal conditioner. This section deals with the first three parts. Position encoders can be classified into two types: incremental encoders and absolute encoders.

An *incremental encoder* produces equally spaced pulses from one or more concentric tracks on the code disk. The pulses are produced when a beam of light passes through accurately placed holes in the code disk. Each track has its own light beam; thus an encoder with three tracks will have three light sources and three light sensors. Figure 9.12 illustrates an optical encoder with three tracks. Each track has a series of equally spaced holes in an otherwise opaque disk. The inside track has only one hole,

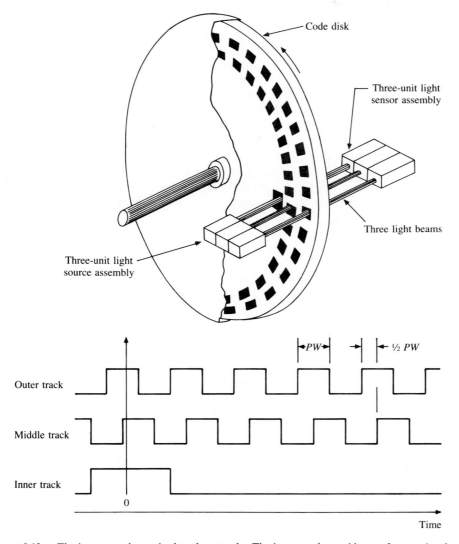

Figure 9.12 The incremental encoder has three tracks. The inner track provides a reference signal to locate the home position. The middle track provides information about the direction of rotation. In one direction, the middle track lags the outside track; in the other direction, the middle track leads the outside track.

which is used to locate the "home" position on the code disk. The other two tracks have a series of equally spaced holes that go completely around the code disk. The holes in the middle track are offset from the holes in the outside track by one-half the width of a hole. The purpose of the offset is to provide directional information. The diagram of the track pulses in Figure 9.12 was made with the disk rotating in the counterclockwise direction. Notice that the pulses from the outer track lead the pulses from the inner track by one-half the pulse width. If the direction is reversed to a clockwise direction, the pulses from the middle track will lead the pulses from the outer track by the same one-half of the pulse width.

The primary functions of the signal conditioner for an incremental encoder are to determine the direction of rotation and count pulses to determine the angular displacement of the code disk. The pulse count is a digital signal, so an analog-to-digital converter is not required for an encoder.

An angular, incremental encoder can be used to measure a linear distance by coupling the encoder shaft to a tracking wheel as shown in Figure 9.13. The wheel rolls along the surface to be measured, and the signal conditioner counts the pulses. The total displacement that can be measured in this manner is limited only by the capacity of the counter in the signal conditioner. The incremental encoder simply rotates as many times as the application requires. The measured displacement is obtained from the total pulse count as given by the following equation:

$$x = \frac{\pi d N_T}{N_R} \tag{9.9}$$

where x = measured displacement, meter
   d = diameter of the tracking wheel, meter
   N_T = total pulse count
   N_R = number of pulses in one revolution

An *absolute encoder* produces a binary number that uniquely identifies each position on the code disk. Absolute encoders may have from 6 to 20 tracks. Each track produces one bit of the binary number according to the code that is established by the hole pattern in the code disk. Figure 9.14 shows an absolute encoder with seven tracks that form the natural binary representation of 128 unique positions on the code disk. The number of unique positions on the code disk is related to the number of bits in the binary number (which is equal to the number of tracks on the code disk). This also establishes the resolution of the encoder according to the following equations:

$$\text{Number of positions} = 2^N \tag{9.10}$$
$$\text{Resolution} = 1 \text{ part in } 2^N \tag{9.11}$$

where N = number of tracks = number of bits in the number

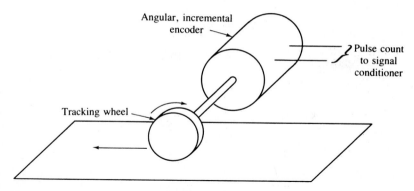

Figure 9.13 An incremental encoder coupled to a tracking wheel is used to measure linear displacement.

There are a number of binary codes that could be used in an encoder. The three most popular codes are the natural binary code, the Gray code, and the BCD code. Figure 9.15 shows the pattern for these three codes for the numbers from 0 to 10. The Gray code is a popular code for counters because only one bit changes each time the count increases by one. Refer to Chapter 3 for further discussion of binary codes.

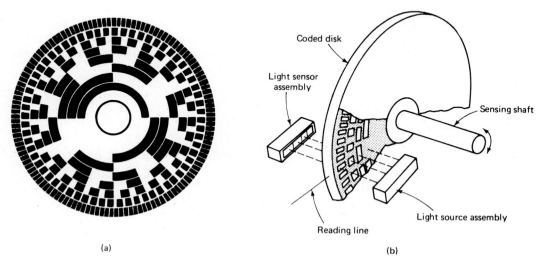

(a)   (b)

Figure 9.14 Absolute optical encoder: (a) typical code disk; (b) encoder elements. [From H. Norton, *Sensor and Analyzer Handbook* (Englewood Cliffs, N.J.: Prentice-Hall, Inc., 1982), Fig. 1-44, p. 107.]

Arabic number	(Natural) Binary Digital number 8 4 2 1	(Natural) Binary Code pattern $2^3\ 2^2\ 2^1\ 2^0$	Gray (Binary) Digital number $G_3\ G_2\ G_1\ G_0$	Gray (Binary) Code pattern	BCD Digital number Tens 8 4 2 1	BCD Digital number Units 8 4 2 1	BCD Code pattern Tens 2^0	BCD Code pattern Units $2^3\ 2^2\ 2^1\ 2^0$
0	0 0 0 0		0 0 0 0		0 0 0 0	0 0 0 0		
1	0 0 0 1		0 0 0 1			0 0 0 1		
2	0 0 1 0		0 0 1 1			0 0 1 0		
3	0 0 1 1		0 0 1 0			0 0 1 1		
4	0 1 0 0		0 1 1 0			0 1 0 0		
5	0 1 0 1		0 1 1 1			0 1 0 1		
6	0 1 1 0		0 1 0 1			0 1 1 0		
7	0 1 1 1		0 1 0 0			0 1 1 1		
8	1 0 0 0		1 1 0 0			1 0 0 0		
9	1 0 0 1		1 1 0 1		0 0 0 0	1 0 0 1		
10	1 0 1 0		1 1 1 1		0 0 0 1	0 0 0 0		

Figure 9.15 Digital code structure for absolute encoders. [From H. Norton, *Sensor and Analyzer Handbook* (Englewood Cliffs, N.J.: Prentice-Hall, Inc., 1982), Fig. 1-45, p. 108.]

Example 9.3

An incremental encoder is used with a tracking wheel to measure linear displacement as shown in Figure 9.13. The tracking wheel diameter is 5.91 cm, and the code disk has 180 holes in the outside track and 180 holes in the middle track. Determine the linear displacement per pulse and the displacement measured by each of the following total pulse counts.

    a. $N_T = 700$
    b. $N_T = 2220$

Solution

The linear displacement per pulse can be determined by dividing Equation (9.9) by N_T.

$$\text{Displacement per pulse} = \frac{\pi d}{N_R}$$

$$= \frac{\pi(0.0591)}{180}$$

$$= 0.00103 \text{ m}$$

$$= 0.103 \text{ cm}$$

$$\text{Total displacement} = 0.00103 N_T \qquad \text{meter}$$

a. $N_T = 700$

   $x = (0.103)(700) = 0.722 \text{ m} = 72.2 \text{ cm}$

b. $N_T = 2220$

   $x = (0.103)(2220) = 2.29 \text{ m} = 229 \text{ cm}$

Example 9.4

An absolute encoder is to be used for measurements that require a resolution of at least 1 minute of arc. Determine the number of bits required to meet the specified resolution.

Solution

First determine the number of minutes in a full circle.

$$N = (60)(360) = 21,600 \text{ min/cycle}$$

Next find the smallest power of 2 that is larger than 21,600.

$$2^{14} = 16,384$$
$$2^{15} = 32,768$$

The encoder must have 15 bits to have a resolution of at least 1 minute of arc.

Proximity Sensors

Proximity sensors are switches that sense the presence of an object without actually touching the object. In this section we deal with inductive and capacitive proximity sensors. The next section covers photoelectric proximity sensors.

   Inductive proximity sensors are used to sense the presence of metal parts. They are also used in automated equipment to sense position and motion. Two or more proximity sensors can be used for simple part identification (e.g., differentiating between short, medium, and long parts).

   An inductive proximity sensor consists of a coil with a ferrite core, an oscillator/ detector circuit, and a solid-state switch. The oscillator creates a magnetic field in front of the sensor, centered on the axis of the coil. When a metal object enters the magnetic field, the amplitude of the oscillation diminishes due to a loss of energy to the target. The detector senses the change in amplitude and actuates a solid-state switch. When the metal object leaves the magnetic field, the oscillation returns to full amplitude, and the solid-state switch returns to its deactivated state.

   The *sensing range* of a proximity sensor is the distance from the sensing face within which a standard target will be detected. The *operate point* is the point at which the switch is activated as a standard target enters the magnetic field. The *release point* is the point at which the switch is deactivated as the target leaves the magnetic field. The release point is farther from the sensing face than the operate point. The difference between the operate point and the release point is called the

Table 9.1 Representative Specifications for Inductive Proximity Sensors

Sensing Range (mm)	Diameter (mm)	Length (mm)	Switching Speed (Hz)	Typical Part Number
1	8	40	5000	IP–1
2	12	40	1000	IP–2
5	18	40	400	IP–3
10	30	50	200	IP–4
20	47	60	40	IP–5

hysteresis of the sensor. A typical hysteresis value is 15% of the sensing range. Table 9.1 lists representative specifications for inductive proximity sensors.

Both the sensing range and the switching speed depend on the size and material of the target. The ratings are based on a square, mild steel plate 1 mm thick, with sides equal to the diameter of the sensing face or three times the sensing range, whichever is larger. The sensing range must be reduced for nonferrous, metal targets such as aluminum, brass, copper, and stainless steel. The reduction factors range from 0.25 to 1.00, depending on the material and the type of sensor.

Capacitive proximity sensors can detect the presence of metal objects or nonmetallic materials such as glass, wood, paper, rubber, plastic, water, and milk. Any material that has a dielectric constant of 1.2 or greater can be detected with a capacitive sensor. The larger the dielectric constant, the easier the material is to detect. Materials with a high dielectric constant can be detected at greater distances than those with a low dielectric constant. Materials with a high dielectric constant can even be detected through a container made of material with a lower dielectric constant. Table 9.2 gives the dielectric constants of a few selected materials.

The following are a few applications of capacitive proximity sensors.

1. Detecting the flow of liquid into a container
2. Sensing the level of liquid through a glass or plastic container
3. Detecting the level of liquids or powdered materials in a bin or container
4. Detecting the presence of a sheet of material on a belt conveyor
5. Detecting a tear in a sheet of paper traveling on a belt conveyor

Table 9.2 Dielectric Constants

Material	Dielectric Constant	Material	Dielectric Constant
Cereal	3–5	Salt	6
Ethanol	24	Sand	3–5
Flour	2.5–3	Sugar	3
Gasoline	2.2	Water	80
Nylon	4–5	Wood (dry)	2–6
Paper	1.6–2.6	Wood (wet)	10–30

Typical capacitive proximity sensors range in size from a diameter of 16 mm to a diameter of 95 mm. The sensing ranges for water are 5 mm for the smaller unit and 70 mm for the larger unit. The sensing distances for metals are the same as those for water. The sensing distance is reduced for other materials. For glass the reduction factor is 0.3; for wood it varies from 0.5 to 0.8, depending on the moisture content. Manufacturers of capacitive proximity sensors recommend physical tests to ensure detection of a particular material.

Photoelectric Sensors

Photoelectric sensors use a beam of light to detect the presence of objects that block or reflect the light beam. A light source provides the beam of light, and a photo-transistor detects the presence or absence of light from the source. Both incandescent lamps and infrared LEDs are used as the light source. Incandescent lamps provide a wide spectrum of visible light with relatively high power output, and the visible beam is easy to align. LEDs have a much longer life, generate less heat, and are less susceptible to shock and vibration damage. The LED has the added advantage that it can be modulated for increased noise immunity. Modulation means the LED is pulsed on and off at a very high frequency. The receiver is tuned to receive the modu-lated infrared light while rejecting unmodulated light. Modulation greatly reduces the influence of background lighting on the receiver. A modulated LED is called an *emitter* and its phototransistor is called a *receiver*.

Figure 9.16 illustrates five different ways of using a light beam to detect the pres-ence of an object. The five methods are direct scan, retroreflective scan, diffuse scan, convergent beam scan, and specular scan. In the *direct scan* method, the object to be detected passes between the light source on one side and the receiver on the other side. The object is detected when it breaks the light beam. Direct scan has the greatest range (up to 100 ft) and can handle the dirtiest environment. In the *retroreflective scan* method, the light source and the receiver are mounted in the same sensing unit. A special retroreflective target is used to reflect the light beam back to the receiver. The double distance the beam must travel and reflector losses give the retroreflective scan method a range of 10 to 30% of the range of the direct scan method.

The *diffuse scan* method is similar to the retroreflective method, but without the reflective target. The light beam strikes the surface of the object and is diffused in all directions. A portion of the diffused light reaches the receiver and actuates the switch-ing action. In the direct and retroreflective methods, the target breaks the beam. In the diffuse method, the target makes the beam. Photoelectric proximity sensors use the diffuse scan method. The *convergent beam* method is a variation of the diffuse scan method in which special lenses are used to focus the light beam on a point located a fixed distance from the light source. The convergent beam sensor will only detect objects that are near the point of focus. It is used to detect parts at a certain range while ignoring the background. Applications include broken wire detection and edge detection.

The *specular scan* method is used only when the object has a mirrorlike finish. The emitter and the receiver are mounted at equal angles from a line perpendicular

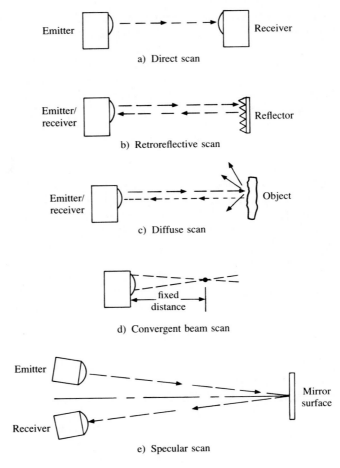

Figure 9.16   Five methods used with photoelectric sensors.

to the mirror surface of the object to be detected. When the object is present, it reflects the light beam back to the receiver (angle of incidence equals the angle of reflectance). The specular scan method is used to detect the difference between a dull and shiny surface. An example is detecting the presence or absence of a foil wrapping on a package.

9.3 VELOCITY MEASUREMENT

Sensing Methods

Velocity is the rate of change of displacement or distance. It is measured in units of length per unit time. Velocity is a vector quantity that has both magnitude (speed) and direction. A change in velocity may constitute a change in speed, a change in direction, or both.

Angular velocity is the rate of change of angular displacement. It is measured in terms of radians per unit time or revolutions per unit time. Angular velocity measurement is more common in control systems than linear velocity measurement. When linear velocity is measured, it is often converted into an angular velocity and measured with an angular velocity transducer. Three methods of measuring angular velocity are considered in this section: dc tachometers, ac tachometers, and optical tachometers.

DC Tachometers

A *tachometer* is an electric generator used to measure angular velocity. A *brush-type dc tachometer* is illustrated in Figure 9.17. The coil is mounted on a metal cylinder called the *armature*. The armature is free to rotate in the magnetic field produced by the two permanent-magnet field poles. The two ends of the coil are connected to opposite halves of a segmented connection ring called the *commutator*. There are two segments on the commutator for each coil on the armature (only one is shown in Figure 9.17). For example, an armature with 11 coils would have a commutator with 22 segments.

The two carbon brushes connect the lead wires to the commutator segments. The brushes and commutator act as a reversing switch that reverses the coil connection once for each 180° rotation of the armature. This switching action converts the ac voltage induced in the rotating coil into a dc voltage. In other words, the commutator and brush constitute an ac-to-dc converter.

The tachometer produces a dc voltage that is directly proportional to the angular velocity of the armature. This voltage is based on the following fact: A voltage is induced in a conductor when it moves through a transverse magnetic field. If the

Figure 9.17 Tachometer generator.

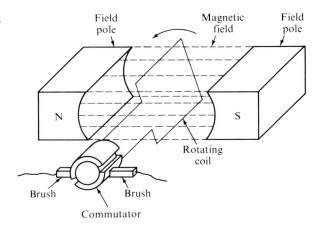

conductor, magnetic field, and velocity are mutually perpendicular, the induced voltage (E_L) is equal to the length of the conductor (L) times the flux density (B) times the velocity of the conductor (V).

$$E_L = LBV$$

In a tachometer, the velocity is perpendicular to the magnetic field only twice during each rotation. The velocity (V) in the above equation is replaced by $(V \sin \theta)$, which is the component perpendicular to the magnetic field. This means that a sinusoidal voltage is induced in each coil. However, when several coils are spaced evenly around the armature, the rectified voltage is very nearly equal to $E_L = LBV$.

The velocity of the conductors may be expressed in terms of the average radius (R), and the angular velocity (ω) in radians per second, or the angular velocity (S) in revolutions per minute.

$$V = R\omega = R2\pi S/60$$
$$E_L = 2\pi RBLS/60$$

Finally, the tachometer has N conductors of length L connected in series. The total voltage is the sum of the identical voltages induced in each conductor:

$$E = NE_L = \left(\frac{2\pi RBNL}{60}\right)S$$

The term enclosed in parentheses is called the EMF constant of the tachometer. It is designated by K_E and has units of volts per revolution per minute, or volts/rpm. Equations (9.12) and (9.13) define the electromotive force (EMF) constant and the output voltage for a dc tachometer.

$$E = K_E S = \frac{30 K_E \omega}{\pi} \tag{9.12}$$

$$K_E = \frac{2\pi RBNL}{60} \tag{9.13}$$

where E = tachometer output, volt

K_E = EMF constant, volt/rpm

S = angular velocity, revolution/minute

ω = angular velocity, radian/second

R = average radius, meter

B = flux density of the magnetic field, weber/square meter

N = effective number of conductors

L = length of each conductor, meter

A harsh industrial environment can be very hard on brush-type tachometers. Particulate contaminants can cause excessive wear in the brushes. Gaseous contaminants build up films on the commutator which cause inaccuracies. A sealed enclosure results in excessive heat buildup and thermal drift problems. A *brushless dc tachometer* solves these problems by reversing the positions of the permanent magnet and the coil. The armature is the permanent magnet and the coil is stationary. The brushes and commutator are not required because there are no electrical connections necessary to the armature. However, additional circuitry is required to sense the position of the armature and provide appropriate solid-state switching to produce a dc output. The solid-state switching circuit serves the same function as the brushes and commutator.

Example 9.5

A dc tachometer has the following specifications:

$$R = 0.03 \text{ m}$$
$$B = 0.2 \text{ Wb/m}^2$$
$$N = 220$$
$$L = 0.15 \text{ m}$$

Determine K_E and the output voltage at each of the following speeds:

$$S = 1000, 2500, \text{ and } 3250 \text{ rpm}$$

Solution

$$K_E = \frac{2\pi RBNL}{60}$$
$$= \frac{2\pi(0.03)(0.2)(220)(0.15)}{60}$$
$$= 0.0207 \text{ V/rpm}$$

For $S = 1000$ rpm,

$$E = (0.0207)(1000) = 20.7 \text{ V}$$

For $S = 2500$ rpm,

$$E = (0.0207)(2500) = 51.8 \text{ V}$$

For $S = 3250$ rpm,

$$E = (0.0207)(3250) = 67.3 \text{ V}$$

AC Tachometers

An ac tachometer is a three-phase electric generator with a three-phase rectifier on its output. The ac tachometer works well at high speeds, but the output becomes nonlinear at low speeds, due to the voltage drop across the rectifiers (about 0.7 V). For this reason, ac tachometers are usually limited to speed ranges of 100 to 1, compared with 1000 to 1 for dc tachometers. The ac tachometer has no brushes and has the same ability to withstand a contaminated environment as the brushless dc generator.

Optical Tachometers

An incremental encoder connected to a rotating shaft produces a sequence of pulses from which a digital velocity signal can be easily obtained. The major signal conditioning requirement is a timed counter. For example, assume that an incremental encoder has 1000 holes in the outside track, and the counter produces a new total every 10 ms. A shaft speed of 600 rpm (10 revolutions per second) will produce $10 \times 1000 = 10,000$ pulses per second. The counter will count $0.01 \times 10,000 = 100$ pulses during a 10-ms interval. Thus a count of 100 corresponds to an angular velocity of 600 rpm. Equations (9.14) and (9.15) define the relationship between the shaft speed and the timed count for an optical tachometer.

$$S = \frac{60C}{NT_c} \tag{9.14}$$

$$C = \frac{SNT_c}{60} \tag{9.15}$$

where S = shaft speed, revolution/minute
  N = number of pulses per shaft revolution
  C = total count during time interval T_c
  T_c = counter time interval, second

When a speed measurement is obtained from an absolute encoder, the track with the greatest number of holes (least significant digit) is used in the same manner as an incremental encoder. Optical encoders can handle very large dynamic ranges with extremely high accuracy and excellent long-term stability.

Example 9.6

An incremental encoder has 2000 pulses per shaft revolution.

  a. Determine the count produced by a shaft speed of 1200 rpm if the timer count interval is 5 ms.
  b. Determine the speed that produced a count of 224 for a timer count interval of 5 ms.

Solution

  a. Equation (9.15) applies:

$$C = \frac{(1200)(2000)(0.005)}{60} = 200$$

  b. Equation (9.14) applies:

$$S = \frac{(60)(224)}{(2000)(0.005)} = 1344 \text{ rpm}$$

9.4 ACCELERATION MEASUREMENT

Sensing Methods

Acceleration is the rate of change of velocity. The measurement of linear acceleration is based on Newton's law of motion: $f = Ma$ [i.e., the force (f) acting on a body of mass (M) is equal to the product of the mass times the acceleration (a)]. Acceleration is measured indirectly by measuring the force required to accelerate a known mass (M). The units of linear acceleration are meter/second^2.

Angular acceleration is the rate of change of angular velocity. The measurement of angular acceleration is usually obtained by differentiating the output of an angular velocity transducer. Angular acceleration is expressed in terms of radian/second^2 or revolution/second^2.

Accelerometers

A schematic diagram of a linear accelerometer is shown in Figure 9.18. The *accelerometer* is attached to the object whose acceleration is to be measured and undergoes the same acceleration as the measured object. The mass M is supported by cantilever springs attached to the accelerometer frame. Motion of the mass M is damped by a viscous oil surrounding the mass.

The accelerometer is a spring–mass–damping system similar to the control valve shown in Figure 5.5 and the second-order process shown in Figure 15.9a. A second-order system is characterized by its resonant frequency (f_0) and its damping ratio (ζ), as determined by the following equations:

$$f_0 = \frac{1}{2\pi} \sqrt{\frac{K}{M}} \qquad (9.16)$$

$$\zeta = \sqrt{\frac{b^2}{4KM}} \qquad (9.17)$$

Figure 9.18   Linear accelerometer.

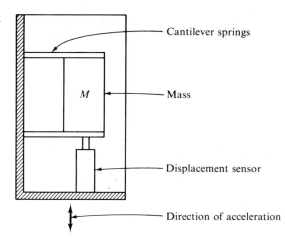

where f_0 = resonant frequency, hertz

          ζ = damping ratio

          K = spring constant, newton/meter ($K = 1/C_m$)

          M = mass, kilogram

          b = damping constant, newton·second/meter

Consider the situation in which the accelerometer frame in Figure 9.18 is accelerated upward at a constant rate. The mass M will deflect the cantilever springs down until the springs exert a force large enough to accelerate the mass at the same rate as the frame. When this occurs, the spring force (Kx) is equal to the accelerating force ($f = Ma$).

$$Kx = Ma$$

or

$$x = \frac{M}{K} a \qquad\qquad (9.18)$$

where x = displacement of the mass, meter

          M = mass, kilogram

          K = spring constant, newton/meter

          a = acceleration, meter/second^2

Equation (9.18) indicates that the displacement x is proportional to the acceleration and may be used as a measure of the acceleration.

However, most accelerometers are used to measure accelerations that change with time. The response of the accelerometer depends on the frequency with which the measured acceleration changes. If the frequency is well below the resonant frequency (f_0), Equation (9.18) is accurate and the displacement may be used as a measure of the acceleration. At frequencies well above the resonant frequency, the mass remains stationary. The displacement is equal to the displacement of the accelerometer frame. It is not a measure of the acceleration. At frequencies near the resonant frequency, the displacement of the mass is greatly exaggerated. Again, the displacement is not a measure of the acceleration. In conclusion,

> An accelerometer must have a resonant frequency considerably larger than the frequency of the acceleration it is measuring.

Thompson* has shown that the error in Equation (9.18) is less than 0.5% if the resonant frequency (f_0) is at least 2.5 times as large as the measured frequency when the damping ratio is 0.6.

---

* W. T. Thompson, *Mechanical Vibrations* (Englewood Cliffs, N.J.: Prentice-Hall, Inc., 1948).

Example 9.7

The accelerometer in Figure 9.18 has the following specifications.

$$M = 0.0156 \text{ kg}$$
$$K = 260 \text{ N/m}$$
$$b = 2.4 \text{ N·s/m}$$
$$x_{max} = \pm 0.3 \text{ cm}$$

Determine the following.

a. The maximum acceleration that can be measured
b. The resonant frequency, f_0
c. The damping ratio, ξ
d. The maximum frequency for which Equation (9.18) can be used with less than 0.5% error

Solution

a. The maximum acceleration may be determined by using Equation (9.18):

$$x = \frac{Ma}{K}$$

$$a_{max} = \frac{x_{max}K}{M}$$

$$= \frac{(0.003)(260)}{0.0156}$$

$$= 50 \text{ m/s}^2$$

b. The resonant frequency is given by Equation (9.16):

$$f_0 = \frac{1}{2\pi} \sqrt{\frac{260}{0.0156}}$$

$$= 20.6 \text{ Hz}$$

c. The damping ratio is given by Equation (9.17):

$$\xi = \sqrt{\frac{2.4^2}{(4)(260)(0.0156)}}$$

$$= 0.595$$

d. The maximum frequency for an error less than 0.5% is $f_0/2.5$:

$$f_{max} = \frac{20.6}{2.5} = 8.25 \text{ Hz}$$

9.5 FORCE MEASUREMENT

Sensing Methods

Force is a physical quantity that produces or tends to produce a change in the velocity or shape of an object. It has a magnitude and a direction which are both defined by Newton's second law of motion.

$$f = Ma \qquad (9.19)$$

where f = force applied to mass M, newton

M = mass, kilogram

a = acceleration of mass M, meter/second^2

The magnitude of force f is equal to the product of the magnitudes of M and a. The direction of force f is the same as the direction of acceleration a.

Equation (9.19) does not mean that there are no forces on a body if it is not accelerating, only that there is no net unbalanced force. Two equal and opposite forces applied to a body will balance, and no acceleration will result. All methods of force measurement use some means of producing a measurable balancing force. Two general methods are used to produce the balancing force: the null balance method and the displacement method.

A beam balance is an example of a null balance force sensor. The unknown mass is placed on one pan. Accurately calibrated masses of different sizes are placed on the other pan until the beam is balanced. The unknown mass is equal to the sum of the calibrated masses in the second pan.

A spring scale is an example of a displacement type of force sensor. The unknown mass is placed on the scale platform, which is supported by a calibrated spring. The spring is displaced until the additional spring force balances the force of gravity acting on the unknown mass. The displacement of the spring is used as the measure of the unknown force.

Two force sensors are covered in this section. The *strain gage load cell* is a *displacement* type of force sensor. The unknown force is applied to an elastic member. The displacement of the elastic member is converted to an electric signal proportional to the unknown force. The *pneumatic force transmitter* is a *null balance* type of force sensor. The unknown force is balanced by the force produced by air pressure acting on a diaphragm of known area. The air pressure is proportional to the unknown force and is used as the measured value signal.

Strain Gage Force Sensors

Strain is the displacement per unit length of an elastic member. For example, if a bar of length L is stretched to length $L + \Delta L$, the strain, ϵ, is equal to $\Delta L/L$. A *strain gage* is a means of converting a small strain into a corresponding change in electrical resistance. It is based on the fact that the resistance of a fine wire varies as the wire is stretched (strained).

There are two general types of strain gages: bonded and unbonded. *Bonded strain gages* are used to measure strain at a specific location on the surface of an elastic member. The bonded strain gage is cemented directly onto the elastic member at the point where the strain is to be measured. The strain of the elastic member is transferred directly to the strain gage, where it is converted into a corresponding change in resistance. *Unbonded strain gages* are used to measure small displacements. A mechanical linkage causes the measured displacement to stretch a strain wire. The change in resistance of the strain wire is a measure of the displacement. The displacement is usually caused by a force acting on an elastic member.

> The unbonded strain gage measures the total displacement of an elastic member. The bonded strain gage measures the strain at a specific point on the surface of an elastic member.

The *gage factor* of a strain gage is the ratio of the unit change in resistance to the strain.

$$G = \frac{\Delta R/R}{\Delta L/L} \tag{9.20}$$

where G = gage factor

ΔR = change in resistance, ohm

R = resistance of the strain gage, ohm

ΔL = change in length, meter

L = length of the strain gage, meter

The gage factor is usually between 2 and 4. The effective length (L) ranges from about 0.5 cm to about 4 cm. The resistance (R) ranges from 50 to 5000 Ω.

The stress on an elastic member is defined as the applied force divided by the unit area. If f is the applied force and A is the cross-sectional area, the stress is equal to f/A. In elastic materials, the ratio of the stress over the strain is a constant called the *modulus of elasticity* (E).

$$E = \frac{S}{\epsilon} \tag{9.21}$$

where E = modulus of elasticity, newton/square meter

S = stress, newton/square meter

ϵ = strain, meter/meter

An example of a strain gage force sensor is illustrated in Figure 9.19. The cantilever beam is the elastic member (a diving board is a familiar example of a cantilever beam). The unknown force is applied to the end of the beam. The strain produced by the unknown force is measured by a bonded strain gage cemented onto the top of the beam. The center of the strain gage is located a distance L units from the end of the beam.

The cantilever beam assumes a curved shape that approximates a semicircle. The top surface is elongated, while the bottom surface is compressed. Halfway between

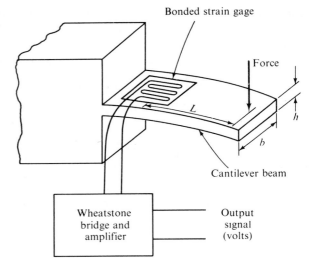

Figure 9.19 The strain gage load cell is an example of a displacement-type force sensor.

these two surfaces is a neutral surface, in which there is no displacement. The stress at any point on the top surface of the cantilever beam is given by the following equation:

$$S = \frac{6fL}{bh^2} \qquad (9.22)$$

where S = stress, newton/meter

   f = applied force, newton

   L = distance from the point to the end of the beam, meter

   b = width of the cantilever beam, meter

   h = height of the cantilever beam, meter

Equations (9.20), (9.21), and (9.22) can be combined to obtain the following expression for the unit change in resistance produced by the unknown force f:

$$\frac{\Delta R}{R} = \left[\frac{6GL}{bh^2 E} \right] f \qquad (9.23)$$

where ΔR = change in resistance of the strain gage, ohm

   R = unstrained resistance of the strain gage, ohm

   G = gage factor of the strain gage

   L = distance from the center of the strain gage to the end of the beam, meter

   b = width of the cantilever beam, meter

   h = height of the cantilever beam, meter

   E = modulus of elasticity of the cantilever beam, newton/square meter (see Table 9.3)

Table 9.3   Modulus of Elasticity of Common Metals

Material	E (N/m^2)
Aluminum	6.9×10^{10}
Beryllium	2.9×10^{11}
Copper	1.1×10^{11}
Gold	7.8×10^{10}
Steel	2.1×10^{11}
Lead	1.8×10^{10}
Molybdenum	3.5×10^{11}
Nickel	2.1×10^{11}
Silicon	7.1×10^{10}
Silver	7.7×10^{10}
Tungsten	4.1×10^{11}
Zinc	7.9×10^{10}

Pneumatic Force Transmitter

A pneumatic force transmitter is illustrated in Figure 9.20. The unknown force f is balanced by the force of the air pressure against the effective area of the diaphragm. The ball and nozzle is arranged such that the balance of the two forces is automatic. For example, suppose that the force f increases. The force rod moves upward, reducing the opening between the ball and nozzle. The pressure in the diaphragm chamber increases and restores the balanced condition. The air pressure (p) in the diaphragm chamber is determined by the following equation:

$$f = (p - 3)A \tag{9.24}$$

where f = unknown force, pound

   p = air pressure, pound/square inch

   A = effective area of the diaphragm, square inch

The $(p - 3)$ term simply indicates that 3 psi corresponds to a force of zero. The signal range is from 3 to 15 psi.

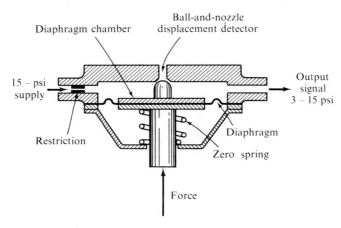

Figure 9.20   The pneumatic force transmitter is an example of a null balance force sensor.

Example 9.8

The strain gage force sensor in Figure 9.19 has the following specifications:

> *Cantilever Beam*
>
> Material: steel
>
> $E = 2 \times 10^{11}$ N/m^2
>
> Maximum allowable stress $= 3.5 \times 10^8$ N/m^2
>
> $b = 1$ cm
>
> $h = 0.2$ cm
>
> $L = 4$ cm
>
> *Strain Gage*
>
> Gage factor $= 2$
>
> Nominal resistance $= 120 \, \Omega$

Determine the maximum force that can be measured and the change in resistance produced by the maximum force.

Solution

The maximum force is obtained by substituting the maximum allowable stress into Equation (9.22) and solving for f_{max}.

$$f_{max} = \frac{S_{max}bh^2}{6L}$$

$$= \frac{(3.5E+08)(0.01)(0.002)^2}{(6)(0.04)}$$

$$= 58.3 \text{ N}$$

The change in resistance is obtained from Equation (9.23).

$$\Delta R = R\left[\frac{6GL}{bh^2E}\right]f$$

$$= 120\left[\frac{(6)(2)(0.04)}{(0.01)(0.002)^2(2E+11)}\right](58.3)$$

$$= 0.42 \, \Omega$$

Example 9.9

The pneumatic force transducer in Figure 9.20 has an effective area of 2.1 in^2. Determine the force range of the transmitter.

Solution

Equation (9.24) may be used.

$$f = (p-3)A = (15-3)(2.1) = 25.2 \text{ lb}$$

GLOSSARY

Absolute encoder: A device that produces a binary code that uniquely identifies the angular position of a sensing disk. (9.2)

Accelerometer: A device that measures the acceleration of an object to which it is attached. (9.4)

Bonded strain gage: A sensor that is bonded to an elastic member to measure the strain at that specific location on the surface of the elastic member. (9.5)

Convergent beam method: A variation of the diffuse scan method of photoelectric detection in which special lenses are used to focus the light beam on a fixed location. (9.2)

Diffuse scan method: A method of photoelectric detection in which the light source and the receiver are mounted in the same sensing unit. The light beam strikes the object and is diffused in all directions, but enough light reaches the receiver to actuate the switch. (9.2)

Direct scan method: A method of photoelectric detection in which the object to be detected passes between the emitter and the receiver. (9.2)

Encoder: A device that provides a digital output in response to a linear or angular displacement. (9.2)

Gage factor: The ratio of the unit change in resistance of a strain gage to the strain it is measuring. (9.5)

Incremental encoder: A device that, when rotated, produces equally spaced pulses from one or more concentric tracks on a code disk. (9.2)

Load cell: A sensor that measures force. (9.5)

Loading error: An error in a potentiometer caused by the current through the load resistor and equal to the difference between the loaded output voltage and the no-load output voltage. (9.2)

LVDT: Abbreviation for linear variable differential transformer, an electromagnetic transducer used to measure linear displacement. (9.2)

Operate point: The point at which a proximity sensor actuates when a standard target approaches the sensing face. (9.2)

Photoelectric sensor: A sensor that uses a beam of light to detect the presence of an object. The two parts of a photoelectric sensor are the emitter and the receiver. (9.2)

PIP: Abbreviation for presence, identification, and position measurement of parts in a manufacturing operation. (9.2)

Potentiometer: A resistance element with a sliding contact that can be moved from one end to the other. (9.2)

Proximity sensor: A device that senses the presence of an object without actually touching the object. (9.2)

Release point: The point at which a proximity sensor deactivates when a standard target moves away from the sensing face. (9.2)

Resolver: A rotary transformer that produces an output signal that is a function of the rotor position. (9.2)

Retroreflective scan method: A method of photoelectric detection in which the light source and the receiver are mounted in the same sensing unit. A special retroreflective target, mounted on the object to be detected, reflects the light beam from the source back to the receiver. (9.2)

Sensing range: The distance from the sensing face of a proximity sensor within which a standard target will be detected. (9.2)

Specular scan method: A method of photoelectric detection that is used only when the object has a mirror-like finish. The light source and the receiver are mounted such that the light beam reflects from the object into the receiver. (9.2)

Strain gage: A sensor that measures the displacement per unit length of an elastic member that is under stress. (9.5)

Synchro: A rotary transducer that converts angular displacement into an ac voltage or vice versa. The three types of synchro are the transmitter, the transformer, and the differential. (9.2)

Tachometer: An electric generator used to measure angular velocity. (9.3)

Unbonded strain gage: A sensor that is attached to an elastic member at two points to measure the total displacement between the two attachment points. (9.5)

EXERCISES

9.1 Determine the number of turns required to produce a potentiometer with a resolution of 0.2%.

9.2 The potentiometer in Figure 9.4 has a resistance of 100,000 Ω. Determine the loading error caused by the following values of R_L and a.

    a. $R_L = 1000\ \Omega$; $a = 0.25, 0.5, 0.75$

    b. $R_L = 10{,}000\ \Omega$; $a = 0.25, 0.5, 0.75$

    c. $R_L = 100{,}000\ \Omega$; $a = 0.25, 0.5, 0.75$

9.3 The synchro system in Figure 9.6 operates at a frequency of 60 Hz. The maximum amplitude of the transformer rotor voltage is 6.2 V. Determine the ac error signal produced by each of the following angular displacements.

    a. $\theta = 75°$ **b.** $\theta = 45°$

    c. $\theta = 150°$ **d.** $\theta = 110°$

9.4 For the synchro system in Problem 9.3, determine the angular displacement that will produce each of the following ac error signals.

    a. $3.1 \sin 377t$ **b.** $-4.8 \sin 377t$

    c. $5.5 \sin 377t$ **d.** $2.7 \sin (377t + 180°)$

9.5 The synchro system in Figure 9.9 operates at a frequency of 400 Hz. The maximum amplitude of the transformer rotor voltage is 22.5 V. Determine the ac error signal produced by each of the following pairs of angular displacements.

    a. $\theta = 60°$, $\theta_d = -60°$ **b.** $\theta = -30°$, $\theta_d = -20°$

    c. $\theta = 45°$, $\theta_d = 20°$ **d.** $\theta = -18°$, $\theta_d = -17°$

9.6 Equations (9.5) and (9.6) define the stator voltages (E_1 and E_2) of a resolver in terms of the rotor voltages (E_3 and E_4).

$$E_1 = K(E_3 \cos \theta - E_4 \sin \theta) \tag{9.5}$$
$$E_2 = K(E_4 \cos \theta + E_3 \sin \theta) \tag{9.6}$$

Assume that $K = 1$ and show that Equations (9.5) and (9.6) can be rearranged to define the rotor voltages in terms of the stator voltages as follows:

$$E_3 = E_1 \cos \theta + E_2 \sin \theta$$
$$E_4 = E_2 \cos \theta - E_1 \sin \theta$$

Hint: Use the following equivalent forms of $\cos \theta$ and $\sin \theta$ during the manipulation of the equations.

$$\cos \theta = \frac{x}{\sqrt{x^2 + y^2}} \qquad \sin \theta = \frac{y}{\sqrt{x^2 + y^2}}$$

9.7 Which of the following would you consider for angular measurement over a range of $\pm 160°$?

a. potentiometer   **b.** synchro

c. resolver   **d.** absolute encoder

9.8 Determine the number of positions in an absolute encoder with 16 tracks.

9.9 An incremental encoder is used with a 10-cm-diameter tracking wheel. The encoder has 2000 holes in the outside track and also in the middle track. Determine the linear displacement per pulse and the number of bits required in the counter to measure a distance of 6.3 m.

9.10 An absolute encoder is to be used for measurements that require a resolution of at least 0.1 minute of arc. Determine the number of bits required to meet the specified resolution.

9.11 Choose a proximity sensor from Table 9.1 to be used to sense the key on a rotating shaft. The output of the sensor will be used to count revolutions of the shaft and to measure the speed of the shaft. Consider the following in making your selection.

(1) Use a 0.8 safety factor on the sensing range.

(2) The maximum shaft speed is 250 rpm.

(3) Judgment dictates a minimum distance of 2 mm between the sensing head and the key.

9.12 The North Star Engineering Company specializes in photoelectric systems for inspection and control. The company just received a request to bid on a system to check the label placement on 2-liter plastic bottles. The bottles are 30 cm high and have a 3-cm-diameter white cap. The label wraps around the middle of the bottle and should extend from 6 cm above the bottom to 18 cm above the bottom. Tests show that the model PCS-1 convergent beam sensor can detect the presence of a bottle when a cap is at the point of focus (2 cm from the lens). The model PDS-2 direct scan sensor can see through the bottle and its contents but not through the label. Write a proposal with appropriate sketches to show how one PCS-1 and two PDS-2 sensors could be used to check the placement of labels.

9.13 A dc tachometer has the following specifications:

$$R = 0.025 \text{ m}$$
$$B = 0.22 \text{ Wb/m}^2$$
$$N = 120$$
$$L = 0.25 \text{ m}$$

Determine K_E and construct a calibration curve for a velocity range of 0 to 5000 rpm.

9.14 An incremental encoder has 5000 pulses per shaft revolution.

a. Determine the count produced by a shaft speed of 1800 rpm if the counter interval is 4 ms.

b. Determine the speed that produced a count of 1018 for a timer interval of 4 ms.

9.15 The accelerometer in Figure 9.18 has the following specifications:

$$M = 0.012 \text{ kg}$$
$$K = 320 \text{ N/m}$$
$$X_{max} = \pm 0.25 \text{ cm}$$

Determine the following:

a. The maximum acceleration that can be measured.

b. The resonant frequency, f_0.

c. The damping constant, b, required to produce a damping ratio of 0.6.

d. The maximum frequency for which Equation (9.18) can be used with less than 0.5% error.

9.16 The strain gage force transducer in Figure 9.19 has the following specifications:

Cantilever Beam

Material: steel

$E = 2 \times 10^{11} \text{ N/m}^2$

Maximum allowable stress $= 5.0 \times 10^8 \text{ N/m}^2$

$b = 1.25 \text{ cm}$

$h = 0.25 \text{ cm}$

$L = 6 \text{ cm}$

Strain Gage

Gage factor $= 2$

Nominal resistance $= 200 \ \Omega$

Determine the maximum force which can be measured and the change in resistance produced by the maximum force.

9.17 The pneumatic force transducer in Figure 9.20 is to have an input of 0 to 50 lb force and an output signal range of 3 to 15 psi. Determine the required effective area.

9.18 The resolver shown in Figure 9.11 is used to measure the angular displacement, θ. Notice that E_4 is shorted, E_3 is a sinusoidal input voltage, and voltages E_1 and E_2 depend on the angular position θ as given in Equations (9.7) and (9.8). Notice that E_1 and E_2 are sinusoidal voltages whose amplitudes vary as $\cos \theta$ and $\sin \theta$, respectively.

a. Assume that $E_3 = 10 \cos 2000\pi t$ volts and $K = 1$. By Equations (9.7) and (9.8), the amplitude of E_1 is $10 \cos \theta$ and the amplitude of E_2 is $10 \sin \theta$. Sketch graphs of the amplitudes of E_1 and E_2 versus θ for values of θ from $-90°$ to $+90°$.

b. Assume that voltage E_2 has been converted to a dc voltage equal to $10 \sin \theta$ volts. Design a five-step, piecewise-linear function that will input the voltage $10 \sin \theta$ volts and will output the voltage $\theta/9$ volts.

9.19 An optical encoder is used with a 10 cm diameter tracking wheel to measure linear displacement. The encoder generates 256 pulses per revolution (N_R).

a. Determine the total pulse count (N_T) produced by the measurement of a linear displacement of 2 m.

b. Determine the minimum number of bits required to store the count produced by the measurement of a distance of 20 m.

9.20 The strain gage force transducer in Figure 9.19 has the following specifications:

Cantilever Beam

Material: beryllium

$E = 2.9 \times 10^{11}$ N/m²

Maximum allowable stress $= 10.0 \times 10^8$ N/m²

$b = 2.1$ cm

$h = 0.4$ cm

$L = 12$ cm

Strain Gage 2

Gage factor $= 2$

Nominal resistance $= 300 \ \Omega$

Determine the maximum force that can be measured and the change in resistance produced by the maximum force.

CHAPTER 10

Process Variable Sensors

OBJECTIVES

In Chapter 9 we examined the sensors used to measure position, motion, and force. In this chapter we examine the sensors used to measure temperature, pressure, flow rate, and level. Some of these sensors are already familiar to you. For example, the home thermostat contains a temperature sensor that measures the room air temperature. Water pumps contain a pressure sensor that actuates a switch to turn the pump on or off. Water meters measure the amount of water you use, and every home has one or two level sensors.

The purpose of this chapter is to give you an entry-level ability to discuss, select, and specify temperature, flow rate, pressure, and level sensors. After completing this chapter, you will be able to

1. Make a list of major considerations in the selection of sensors to measure temperature, flow rate, pressure, and level

2. Describe the following types of temperature sensors: bimetallic, filled thermal system, resistance, thermistor, thermocouple, and radiation pyrometer

3. Compute the coefficients of a quadratic calibration equation for a resistance temperature detector, given the resistance values at the minimum, maximum, and midpoint temperatures

4. Describe the following types of flow rate sensors: differential pressure, turbine, vortex shedding, and magnetic

5. Describe the following types of pressure measurement: strain gage diaphragm, capacitance diaphragm, Bourdon tube, and bellows

6. Describe the following types of level sensors: displacement float, static pressure, and capacitance

10.1 TEMPERATURE MEASUREMENT

Sensing Methods

Temperature is a measure of the degree of thermal activity attained by the particles in a body of matter. When two adjacent bodies of matter are at different temperatures, heat is transmitted from the warmer body to the cooler body until the two bodies are at the same temperature (two bodies at the same temperature are said to be in thermal equilibrium). The standard temperature scales are illustrated in Figure 10.1. The Celsius and Kelvin scales are the common and absolute temperature scales of the SI system of units.

In the act of measuring the temperature of a body, heat is transmitted between the thermometer and the body until the two are in equilibrium. The thermometer actually measures the equilibrium temperature—not the initial temperature of the body. Thus the measuring process imposes a change in the original temperature of the body. Except when the thermometer and the measured body are at the same temperature before the measurement, 100% accuracy in temperature measurement is impossible to attain. Additional errors are introduced by heat loss from the portion of the thermometer that is not immersed in the measured body. Careful consideration is required to minimize these measurement errors.

Differential Expansion (Bimetallic) Thermostats

A bimetallic element consists of two strips of different metals bonded together to form a leaf, coil, or helix. The two metals must have different coefficients of thermal expansion so that a change in temperature will deform the original shape. A bimetallic thermometer is formed by attaching a scale and indicator to the bimetallic element such that the indicator displacement is proportional to the temperature. A *bimetallic thermostat* is formed by replacing the dial and indicator with a set of contacts. The

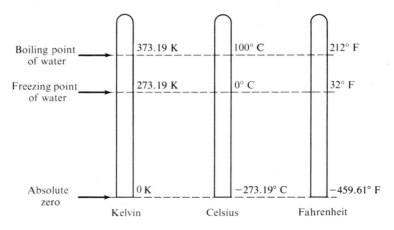

Figure 10.1   Standard temperature scales.

Figure 10.2 Bimetallic thermostat.

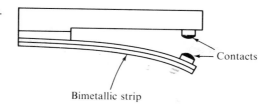

Contacts

Bimetallic strip

bimetallic thermostat is frequently used in on–off temperature control systems. Figure 10.2 is a schematic diagram of a bimetallic thermostat.

Filled Thermal Systems

A *filled thermal system* (FTS) uses a bulb filled with a liquid, gas, or vapor as the temperature sensor. A small-bore tube called a capillary connects the bulb to a spiral, helix, or bellows pressure element that converts the pressure into a usable signal, a temperature indication, or a temperature recording. The Scientific Apparatus Makers Association (SAMA) has classified filled thermal systems into four major categories according to the type of filling substance.

> *Class I. Liquid-filled thermal systems* use the thermal expansion of a liquid to measure temperature. The filling fluid is usually an organic liquid, such as xylene, which has a very large coefficient of thermal expansion. Class I systems have the range −87 to 371°C. Desirable features include small bulb sizes, narrow spans, and relatively low cost. Less desirable features include a short capillary and compensation difficulties.

> *Class II. Vapor-filled thermal systems* use the vapor pressure of a volatile liquid to measure temperature. The filling medium is in both the liquid and the gaseous form, and the interface must occur in the sensing bulb. A *class IIA* system can only measure temperatures above ambient—the capillary and pressure element contain the vapor while the bulb contains some vapor and all the liquid. A *class IIB* system can only measure temperatures below ambient—the capillary and pressure element contain liquid while the bulb contains some liquid and all the vapor. A *class IIC* system can measure temperatures above and below ambient, but the vapor and liquid must change places during a transition from one side of ambient to the other side. This transition causes a delay that makes class IIC unsuitable for measuring temperatures that pass through ambient. A *class IID* system uses a second nonvolatile fluid to overcome the transition problem of the class IIC system. The nonvolatile fluid fills the pressure element, the capillary, and part of the sensing bulb. The bulb contains all the volatile fluid (both the vapor and the liquid portions). The nonvolatile fluid acts as a seal and transmitter of pressure between the sensing bulb and the pressure element. Desirable features of class II systems include a long capillary, a short response time (4 to 5 s), and no need to compensate for the temperature of the capillary and pressure element. Less desirable features include a nonlinear output and no overrange capacity.

Class III. Gas-filled thermal systems use the fact that the pressure of a confined gas is proportional to its absolute temperature. Nitrogen is usually used as the fill, except for extremely low temperatures, when helium is preferred. Class III systems have the temperature range −268 to 760°C. Different filling pressures are used to obtain different temperature ranges. Desirable features of class III systems include no elevation effect due to the weight of the fluid, a large temperature range, and a large overrange (150 to 300%). Less desirable features include large bulb sizes, large required span, and low power in the pressure element.

Class V. Mercury-filled thermal systems use the thermal expansion of mercury to measure temperature. (*Note:* There is no SAMA class IV.) Mercury is separated from the other liquid-filled systems (class I) because of the unique characteristics of mercury for temperature measurement. Desirable features of class V include a very linear scale, easy compensation, rapid response, plenty of power in the pressure element, and excellent accuracy. The major objection to mercury is the possibility of mercury contamination on accidental breakage of the filled system.

A class III temperature transmitter is illustrated in Figure 10.3. The primary element consists of an inert gas sealed in a bulb that is connected by a capillary to a bellows pressure element. The bulb is immersed in the liquid to be measured until a thermal equilibrium is reached. The inert gas responds to the temperature change with a corresponding change in internal pressure. The primary-element bellows converts the gas pressure into an upward force on the right-hand side of the force beam. The air pressure in the feedback bellows produces a balancing upward force on the left-hand side of the force beam.

The feedback bellows pressure is regulated by the combination of the restriction in the supply line and the relative position of the nozzle and force beam. Air leaks from the nozzle at a rate that depends on the clearance between the force beam and the end of the nozzle. The restriction in the supply line is sized such that the feedback bellows pressure is 3 psi when the nozzle clearance is at maximum. The bellows pressure will rise to 15 psi when the nozzle is completely closed by the force beam. The left end of the force beam and the nozzle form what is called a flapper and nozzle displacement detector. This is a very sensitive detector which produces a 3- to 15-psi signal, depending on a very small displacement of the flapper.

The arrangement of the force beam is such that it automatically assumes the position that results in a balance-of-forces condition. For example, assume a rise in the measured temperature. The gas pressure increases the primary element bellows, thereby increasing the force on the right-hand side of the force beam. The increased force tends to rotate the force beam about the flexure, which acts as a fulcrum. This moves the left end of the force beam closer to the nozzle, increasing the feedback bellows pressure. The increase in the feedback bellows pressure increases the balancing force on the left-hand side of the force beam. The force beam is balanced by a feedback bellows pressure that is always proportional to the primary element bellows pressure. The feedback bellows pressure is an accurate measure of the primary element gas pressure, and it is used as the output signal of the pressure transducer. A typical input/output graph is illustrated in Figure 10.3.

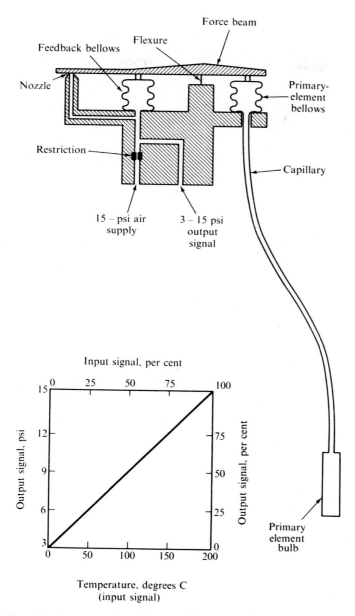

Figure 10.3   Class III temperature transmitter and its input/output graph.

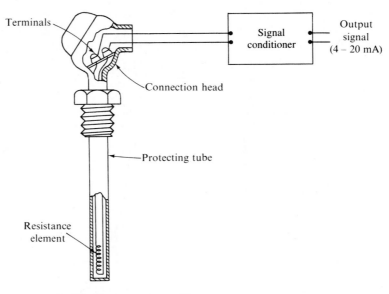

Figure 10.4   Typical RTD temperature transmitter.

Resistance Temperature Detectors

Resistance temperature sensors use a temperature-induced change in the resistance of a material to measure temperature. The electrical resistance of most metals increases as the temperature increases. Resistance temperature detectors (*RTDs*) use this property to measure temperature. A typical RTD temperature transmitter is illustrated in Figure 10.4. The sensing element is a wirewound resistor located in the end of a protecting tube. Platinum and nickel are the metals most often used to construct the sensing element. Platinum is noted for its accuracy and linearity; nickel is noted for its modest cost and relatively large change in resistance for a given change in temperature. The signal conditioner is actually a resistance-to-current converter. Table 10.1 lists temperature, resistance, and output signals for a typical RTD temperature transmitter. Desirable features of RTDs include a wide temperature range (-240 to $649°C$), high accuracy, excellent repeatability, and good linearity.

Table 10.1   Typical Values for a Platinum RTD Temperature Transmitter

Temperature (°C)	Resistance (Ω)	Output Signal (mA)
0	100.0	4
25	109.9	8
50	119.8	12
75	129.6	16
100	139.3	20

Example 10.1

The resistance of a platinum RTD is approximated by the following equation:

$$R = R_0(1 + a_1 T + a_2 T^2)$$

where
R = resistance at $T°C$, ohm
R_0 = resistance at $0°C$, ohm
T = temperature, $°C$
a_1, a_2 = constants

Determine the values of R_0, a_1, and a_2 for the platinum RTD described in Table 10.1. Use the resistance values at 0, 50, and 100°C to find R_0, a_1, and a_2. Check the accuracy of the equation at 25°C.

Solution

There are three unknowns: R_0, a_1, and a_2. Therefore, three equations are required to determine the three values. These equations are obtained by substituting the following three sets of temperature and resistance values from Table 10.1: $(0°, 100 \ \Omega)$, $(50°, 119.8 \ \Omega)$, and $(100°, 139.3 \ \Omega)$.

$$100.0 = R_0(1 + 0a_1 + 0^2 a_2) \tag{a}$$
$$119.8 = R_0(1 + 50a_1 + 50^2 a_2) \tag{b}$$
$$139.3 = R_0(1 + 100a_1 + 100^2 a_2) \tag{c}$$

Equation (a) is easily solved for R_0.

$$R_0 = 100 \ \Omega \tag{a'}$$

Substitute 100 for R_0 into Equations (b) and (c), and simplify.

$$50a_1 + 2500a_2 = 0.198 \tag{b'}$$
$$100a_1 + 10{,}000a_2 = 0.393 \tag{c'}$$

Equations (b') and (c') are easily solved for a_1 and a_2.

$$a_1 = 0.00399$$
$$a_2 = -6 \times 10^{-7}$$

The equation for R is

$$R = 100(1 + 0.00399T - 6 \times 10^{-7}T^2)$$

Check the equation at 25°C:

$$R = 100[1 + (0.00399)(25) - (6E-07)(25)^2]$$
$$= 109.9 \ \Omega$$

The value predicted by the equation is the same as the value in Table 10.1.

Thermistors

A *thermistor* is a semiconductor that has a large change in resistance with changes in temperature. Most thermistors have a negative temperature coefficient, and in some cases, the resistance decreases over 5% for each 1°C increase in temperature. Thermistors are made from mixtures of a number of metal oxides that are sintered at high temperature into a semiconductor element. Desirable features of thermistors include small size, fast response, narrow span, and high sensitivity. Less desirable features include a very nonlinear calibration graph, poor high-temperature stability, unsuitability for large spans, and high impedance.

Thermocouples

A *thermocouple* consists of two dissimilar wires that are connected only at each end. The two ends are referred to as the hot junction and the cold junction. When the two junctions are at different temperatures, a small voltage [called an electromotive force (EMF)] is generated in the circuit formed by the two wires. The EMF generated is approximately proportional to the temperature difference. Thus a thermocouple is a sensor that converts temperature into a voltage signal. Figure 10.5 illustrates a thermocouple temperature transmitter, and Table 10.2 lists typical temperature, EMF, and output signal values.

The material used in the two wires determines the type of thermocouple. The most common combinations are: type E (chromel–constantan), type J (iron–constantan), type K (chromel–alumel), type R-S (platinum–platinum rhodium), and type T (copper–constantan). The EMF values generated by an iron–constantan thermocouple with a cold junction at 0°C are included in Table 10.2.

The signal conditioner in Figure 10.5 is essentially a millivolt-to-current converter. It converts the millivolt signal from the thermocouple into an electric current

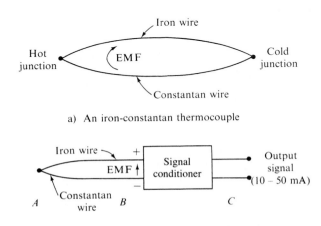

a) An iron-constantan thermocouple

b) A typical thermocouple and signal conditioner

Figure 10.5 Typical thermocouple temperature transmitter.

Table 10.2   Typical Values for an Iron–Constantan Thermocouple Temperature Transmitter

Temperature (°C)	EMF (mV)	Output Signal (mA)
0	0.00	10.0
50	2.61	19.7
100	5.28	29.6
150	8.01	39.7
200	10.77	50.0

signal suitable for transmission to a remote controller. The conditioner also compensates for ambient changes in the cold junction temperature. Thermocouple burnout protection is a safety feature required of most T/C signal conditioners. This feature provides maximum output from the transmitter whenever the thermocouple is open circuited. In a control system, the burnout protection will turn off the heat source if the thermocouple is open circuited for any reason.

Desirable features of thermocouples include their very small size, low cost, ease of installation, ruggedness, extremely wide range (near absolute zero to 2700°C), reasonable accuracy, and fast response time.

Radiation Pyrometers

A *radiation pyrometer* measures the temperature of an object by sensing the thermal radiation emanating from the object, without contacting the object. An optical system collects the visible and infrared energy coming from the object and focuses it on a detector. The detector converts the energy into an electrical signal, which is a complex function of the absolute temperature of the object.

The pyrometric sensing method is based on radiation laws that define the total heat radiated from an ideal *blackbody* and the relationship between the surface temperature and the wavelength spectrum of the radiated energy. These laws show that the total heat radiated from a blackbody varies as the fourth power of the absolute temperature of the surface. They also show that the wavelength of the radiated energy decreases as the surface temperature increases. This is evidenced by the changing glow of an object as its temperature increases.

Most objects emit less energy than a blackbody. The *emittance* of a body is the ratio of the radiant energy emitted from the body to the radiation emitted by a blackbody at the same temperature. The emittance of most objects is considerably less than 1. For example, the emittance of unoxidized aluminum is 0.06, while that of rough steel plate is 0.97. Most other materials fall between these two values. Emittance is a very uncertain characteristic of an object, and a major objective of a radiation pyrometer is to overcome this variable effect.

A *wide-band pyrometer* is designed to measure as much radiation from the object as possible. It has no filters and depends on an unobstructed view of the object. The presence of smoke or carbon dioxide will result in a low reading. The output must be calibrated for the emittance of the object. The calibration varies considerably with the emittance and the temperature of the object. For example, assume a wide-band pyrometer is calibrated to measure an object at 3000°F with an emittance of 0.9. If the emittance of the object changes to 0.7, the reading will be in error about 200°F.

A *narrow-band pyrometer* is designed to measure only energy in a narrow band of frequencies. The emittance usually does not vary as much over a narrow band as it does over the wide spectrum. The actual wavelength used is selected according to the application.

A *ratio pyrometer* measures the energy in two narrow bands that are very close to each other. The energy from one band is divided by the energy from the other band. The expectation is that the emittance will change by the same amount in each band. Taking the ratio should cancel the effect of equal changes in emittance. The two wavelengths are selected according to the application.

Desirable features of a radiation pyrometer include no physical contact with the object whose temperature is being measured, fast response, the ability to measure the temperature of small objects, and the ability to measure very high temperatures. Less desirable features include high cost, a nonlinear response approaching the fourth power of the temperature, variations in the emittance of the object cause erroneous readings, and a relatively wide temperature span is required.

10.2 FLOW RATE MEASUREMENT

Sensing Methods

The flow rate of liquids and gases is an important variable in industrial processes. The measurement of the flow rate indicates how much fluid is used or distributed in a process. Flow rate is frequently used as a controlled variable to help maintain the economy and efficiency of a given process.

The average flow rate is usually expressed in terms of the volume of liquid transferred in 1 s or 1 min.

$$\text{Average flow rate} = q_{\text{avg}} = \frac{\text{change in volume}}{\text{change in time}} = \frac{\Delta V}{\Delta t}$$

The instantaneous flow rate is determined by the limit of the average flow rate as Δt is reduced to zero. In mathematics, this limit is called the *derivative* of V with respect to t and is represented by the symbol dV/dt.

$$\text{Instantaneous flow rate} = q = \lim_{\Delta t \to 0} \frac{\Delta V}{\Delta t} = \frac{dV}{dt}$$

The flow rate in a pipe can also be expressed in terms of the average fluid velocity, v_{avg}, and the cross-sectional area of the pipe, A.

$$q_{\text{avg}} = A v_{\text{avg}}$$

The SI unit of flow rate is cubic meter/second.

Sometimes it is preferable to express the flow rate in terms of the mass of fluid transferred per unit time. This is usually referred to as the *mass flow rate*. The mass flow rate (W) is obtained by multiplying the flow rate (q) by the fluid density (ρ).

$$\text{Mass flow rate} = W = \rho q$$

The SI unit of mass flow rate is kilogram/second.

Differential Pressure Flow Meters

Differential pressure flow meters operate on the principle that a restriction placed in a flow line produces a pressure drop proportional to the flow rate squared. A differential pressure transmitter is used to measure the pressure drop (h) produced by the restriction. The flow rate (q) is proportional to the square root of the measured pressure drop.

$$q = K\sqrt{h} \tag{10.1}$$

The restriction most often used for flow measurement is the orifice plate—a plate with a small hole, which is illustrated in Figure 10.6a. The orifice is installed in the flow line in such a way that all the flowing fluid must pass through the small hole (see Figure 10.6b).

Special passages transfer the fluid pressure on each side of the orifice to opposite sides of the diaphragm unit in a differential pressure transmitter. The diaphragm arrangement converts the pressure difference across the orifice into a force on one end of a force beam. A force transducer on the other end of the beam produces an exact counterbalancing force. A displacement detector senses any motion resulting from an imbalance of the forces on the force arm. The amplifier converts this displacement signal into an adjustment of the current input to the force transducer that restores the balanced condition. The counterbalancing force produced by the force transducer is proportional to both the pressure drop and the input current (I). Thus the current (I) is directly proportional to the pressure drop across the orifice (h). This same electric current is used as the output signal of the differential pressure transducer.

In Figure 10.6 the orifice is the primary element, and the differential pressure transmitter is the secondary element. The orifice converts the flow rate into a differential pressure signal, and the transmitter converts the differential pressure signal into a proportional electric current signal. A typical calibration curve is illustrated in Figure 10.6c.

Desirable features of the orifice flow meter include the fact that it is simple and easy to fabricate, has no moving parts, that a single differential pressure transmitter can be used without regard to pipe size or flow rate, and that it is a widely accepted standard. A less desirable feature is that an orifice does not work well with slurries.

Turbine Flow Meters

A *turbine flow meter* is illustrated in Figure 10.7. A small permanent magnet is embedded in one of the turbine blades. The magnetic sensing coil generates a pulse each

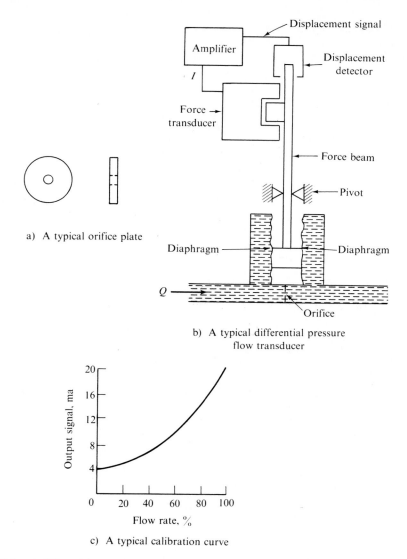

a) A typical orifice plate

b) A typical differential pressure
flow transducer

c) A typical calibration curve

Figure 10.6 A differential pressure flow transmitter is the most widely used method of flow measurement.

time the magnet passes by. The number of pulses is related to the volume of liquid passing through the meter by the following equation: $V = KN$, where V is the total volume of liquid, K the volume of liquid per pulse, and N the number of pulses. The average flow rate, q_{avg}, is equal to the total volume V divided by the time interval Δt.

$$q_{avg} = \frac{V}{\Delta t} = K \frac{N}{\Delta t}$$

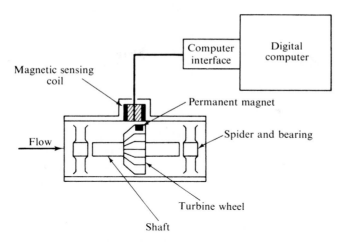

Figure 10.7   A turbine flow meter produces an accurate linear, digital flow signal. Turbine meters are used in the petrochemical and other industries for a broad range of applications.

But $N/\Delta t$ is the number of pulses per unit time (i.e., the pulse frequency f). Thus

$$q = Kf \tag{10.2}$$

The pulse output of the turbine flow meter is ideally suited for digital counting and control techniques. Digital blending control systems make use of turbine flow meters to provide accurate control of the blending of two or more liquids. Turbine flow meters are also used to provide flow rate measurements for input to a digital computer, as shown in Figure 10.7.

Example 10.2

A turbine flow meter has a K value of 12.2 cm^3 per pulse. Determine the volume of liquid transferred for each of these pulse counts: (a) 220; (b) 1200; (c) 470. Also determine the flow rate, if each of the pulse counts above occurs during a period of 140 s.

Solution

$$V = KN$$

a. For 220 pulses in 140 s.

$$V = KN = (12.2)(220) = 2684 \text{ cm}^3$$

$$Q = \frac{V}{\Delta t} = \frac{2684}{140} = 19.2 \text{ cm}^3/\text{s}$$

b. For 1200 pulses in 140 s,

$$V = (12.2)(1200) = 14{,}640 \text{ cm}^3$$

$$Q = \frac{14{,}640}{140} = 104.6 \text{ cm}^3/\text{s}$$

c. For 470 pulses in 140 s,

$$V = (12.2)(470) = 5734 \text{ cm}^3$$

$$Q = \frac{5734}{140} = 41 \text{ cm}^3/\text{s}$$

Vortex Shedding Flow Meters

A *vortex shedding flow meter* uses an unstreamlined obstruction in the flow stream to cause pulsations in the flow. The pulsations are produced when vortices (or eddies) are alternately formed and then shed on one side of the obstruction and then on the other side of the obstruction. The resulting pulsations are sensed by a piezoelectric crystal. The frequency of the pulses is directly proportional to the fluid velocity, thus forming the basis of a volumetric flow meter. Figure 10.8 illustrates a vortex shedding flow meter.

The frequency (f) of the vortex shedding is proportional to the average fluid velocity (v_{avg}) and inversely proportional to the width of the obstruction (w). The expression fw/v_{avg} is called the *Strouhal number*. The Strouhal number is constant over many ranges of Reynolds number. The relationship between the mass flow rate (W) and the vortex frequency (f) is given by the following equation:

$$W = \frac{\rho w A f}{\text{St}} \tag{10.3}$$

where A = cross-sectional area of the pipe, meter^2

f = frequency of the vortex shedding, hertz

W = mass flow rate, kilogram/second

w = width of the obstruction, meter

ρ = density of the fluid, kilogram/meter^3

St = Strouhal number

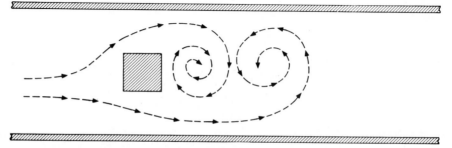

Figure 10.8   In a vortex flow meter, vortices are alternately formed and shed on one side of an obstruction and then on the other side. The resulting pulsations are sensed by a piezoelectrical crystal, and the frequency of the pulses is directly proportional to the volumetric flow rate.

Desirable features of the vortex shedding meter include a linear digital output signal, good accuracy over a wide range of flow, no moving or wearing parts, and low installed cost. Less desirable features include decreasing rangeability with increasing viscosity, and practical considerations limit the size to a diameter range of 1 to 8 in.

Magnetic Flow Meters

The *magnetic flow meter* has no moving parts and offers no obstructions to the flowing liquid. It operates on the principle that a voltage is induced in a conductor moving in a magnetic field. A magnetic flow meter is illustrated in Figure 10.9. The saddle-shaped coils placed around the flow tube produce a magnetic field at right angles to the direction of flow. The flowing fluid is the conductor, and the flow of the fluid provides the movement of the conductor. The induced voltage is perpendicular to both the magnetic field and the direction of motion of the conductor. Two electrodes are used to detect the induced voltage, which is directly proportional to the liquid flow rate. The magnetic flow transmitter converts the induced ac voltage into a dc electric current signal suitable for use by an electronic controller.

Desirable features of magnetic flow meters include no obstruction to the fluid flow, no moving parts, low electric power requirements, excellent for slurries, and very low flow capabilities. Less desirable features include the fact that the fluid must have a minimum electrical conductivity, the large size and high cost of a magnetic flow meter, and the fact that periodic zero flow checks are required.

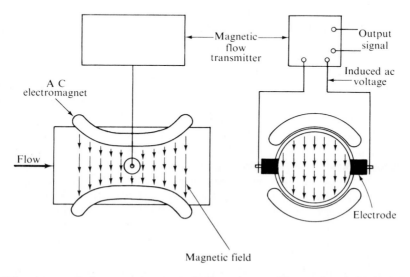

Figure 10.9   A magnetic flow meter has a completely unobstructed flow path, a decided advantage for slurries and food products.

10.3 PRESSURE MEASUREMENT

Sensing Methods

Pressure is defined as the force per unit area exerted by a liquid or gas on a surface. Liquid pressure is the source of the buoyant force that supports a floating object, such as a boat or a swimmer. Pressure is also the motivating force that causes liquids and gases to flow through a pipe.

An extremely wide range of pressures is measured and controlled in industrial processes—all the way from 0.1 Pa (about 0.001 mm of mercury) to above 100 MPa (about 10,000 psi). A great variety of primary elements has been developed to measure pressure over various parts of this extreme range. The great majority of these primary elements convert the measured pressure into a displacement or force. A signal conditioner then converts the force or displacement into a voltage, current, or air pressure signal suitable for use by a controller.

Pressure measurements are always made with respect to some reference pressure. Atmospheric pressure is the most common reference. The difference between the measured pressure and atmospheric pressure is called the *gage pressure*. Weather conditions and altitude both cause variations in the atmospheric pressure, and the gage pressure will vary accordingly. The standard atmospheric pressure is 101.3 kPa, or 14.7 psi. A pressure less than atmospheric is called a *vacuum*. A pressure of zero is called a *perfect vacuum*. A perfect vacuum is sometimes used as the pressure reference. The difference between a measured pressure and a perfect vacuum is called the *absolute pressure*. An arbitrary pressure is also used as the reference pressure. The difference between a measured pressure and an arbitrary reference pressure is called the *differential pressure*. A differential pressure sensor was discussed in the section on flow rate measurement.

Strain Gage Pressure Sensors

A strain gage is based on the fact that stretching a metal wire changes its resistance. The change in resistance bears an almost linear relationship to the change in length. The strain gage uses the change in resistance to measure extremely small changes in displacement. Figure 10.10 is a schematic diagram of a strain gage pressure sensor. The body acts as a mechanical transducer that converts the pressure into a displacement that increases the length of upper strain wires R_1 and R_3 and decreases the length of lower strain wires R_2 and R_4. A Wheatstone bridge converts the change in resistance of the upper and lower strain wires into an electric voltage proportional to the pressure. The voltage output of the Wheatstone bridge is then amplified to produce a usable signal. The primary element is the body and strain wire assembly. The signal conditioner consists of the Wheatstone bridge and amplifier. Strain gage pressure sensors are used in the high pressure range (from 100 to 10,000 psi).

Strain gages are divided into two types: bonded and unbonded. Figure 10.10 is an example of an unbonded type of strain gage. The displacement is transferred to the strain wire by a mechanical linkage. Bonded strain gages are cemented directly onto the body of the transducer, as shown in Figure 9.19.

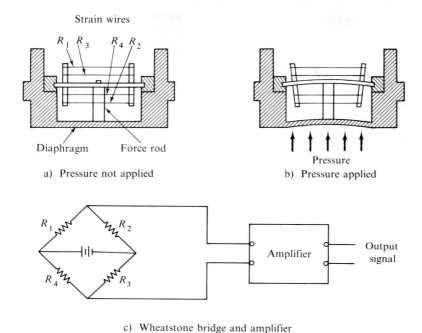

Strain wires

R_1 R_3 R_4 R_2

Diaphragm   Force rod

a) Pressure not applied

Pressure

b) Pressure applied

R_1 R_2

Amplifier

Output
signal

R_4 R_3

c) Wheatstone bridge and amplifier

Figure 10.10   Strain gage pressure sensor.

Deflection-Type Pressure Sensors

Deflection-type pressure sensors consist of a primary element, a secondary element, and a signal conditioner. The primary element converts the measured pressure signal into a proportional displacement. The secondary element converts the displacement into a change of an electrical element. The signal conditioner converts the change of the electrical element into a signal suitable for use by a controller, computer, or indicating device.

Figure 10.11 illustrates the common primary elements used in deflection-type pressure transducers. The *Bourdon element* is a flattened tube that is shaped into an incomplete circle, spiral, or helix. The tube tends to straighten out as the internal pressure increases, providing a displacement proportional to the pressure. Bourdon elements of bronze, steel, or stainless steel are available to cover the range from 0 to 10,000 psi or greater.

The *bellows* element is a thin-walled metal cylinder with corrugated sides. The shape allows the element to extend as the internal pressure is balanced against a calibrated spring. The bellows displacement is proportional to the measured pressure. Bellows elements are widely used to measure pressures up to 100 psi. They are also used to make vacuum and absolute pressure measurements.

The diaphragm element may be flat or corrugated. The diaphragm allows sufficient movement to balance the pressure against a calibrated spring or force transducer.

Figure 10.11   Deflection-type
pressure elements.

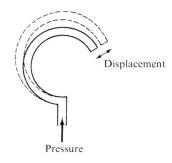

a) A circular Bourdon pressure element

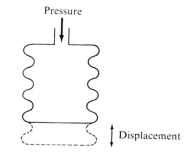

b) A bellows pressure element

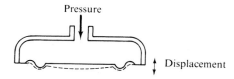

c) A diaphragm pressure element

   The secondary element converts the primary element displacement into an electrical change that the signal conditioner can use to produce a usable signal. The conversion is accomplished by using the displacement to adjust one of the three electric
circuit elements: resistance, capacitance, or inductance. The signal conditioner then
produces an electric signal based on the value of the variable element. Examples of
the three types of secondary elements are illustrated in Figure 10.12.
   A variable resistance pressure sensor is illustrated in Figure 10.12a. The calibrated
spring is displaced by an amount proportional to the pressure in the bellows. The
sliding contact causes a change in the resistance between the two leads connected to
the transmitter. The transmitter, in turn, produces an electrical signal based on the
resistance value.

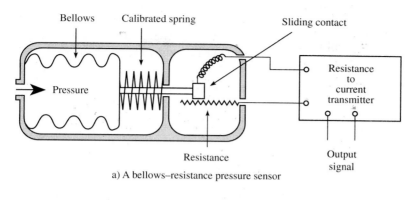

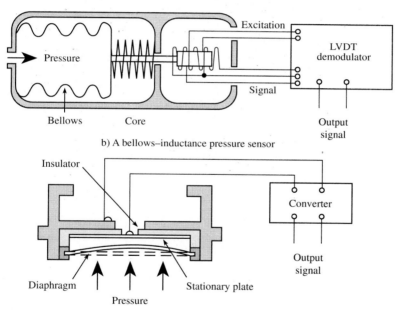

a) A bellows–resistance pressure sensor

b) A bellows–inductance pressure sensor

c) A diaphragm–capacitance pressure sensor

Figure 10.12 Examples of deflection-type pressure sensors.

A variable inductance pressure sensor is illustrated in Figure 10.12b. The LVDT displacement transducer and demodulator produces a linear dc voltage signal proportional to the displacement of the core from a central null position. As the core moves in one direction from null, a positive voltage is produced. Movement in the other direction produces a negative voltage. A major advantage of the LVDT is the fact that it does not touch the internal bore of the transformer. This eliminates problems of mechanical wear, errors due to friction, and electrical noise due to a rubbing action.

A variable capacitance pressure transducer is illustrated in Figure 10.12c. The diaphragm and the stationary plate form the two plates of the capacitor. The displacement of the diaphragm reduces the distance between the two plates, thereby increasing the capacitance. The signal conditioner produces an electrical signal based on the capacitance value of the primary element.

Example 10.3

A bellows pressure element similar to Figure 10.12b has the following values.

$$\text{Effective area of bellows} = 12.9 \text{ cm}^2$$
$$\text{Spring rate of the spring} = 80 \text{ N/cm}$$
$$\text{Spring rate of the bellows} = 6 \text{ N/cm}$$

What is the pressure range of the sensor if the motion of the bellows is limited to 1.5 cm?

Solution

The total spring rate is $80 + 6 = 86$ N/cm. The force required to deflect the spring a distance of 1.5 cm is $(1.5 \text{ cm}) \times (86 \text{ N/cm}) = 129$ N. The pressure required to produce this force is $(129 \text{ N})/(12.9 \text{ cm}^2) = 10 \text{ N/cm}^2$. The range is 0 to 10 N/cm^2, or 0 to 100 kN/m^2.

10.4 LIQUID LEVEL MEASUREMENT

Sensing Methods

The measurement of the level or weight of material stored in a vessel is frequently encountered in industrial processes. Liquid level measurement may be accomplished directly by following the liquid surface, or indirectly by measuring some variable related to the liquid level. The direct methods include sight glasses and various floats with external indicators. Although simple and reliable, direct methods are not easily modified to provide a control signal. Consequently, indirect methods provide most level control signals.

Many indirect methods employ some means of measuring the static pressure at some point in the liquid. These methods are based on the fact that the static pressure is proportional to the liquid density times the height of liquid above the point of measurement.

$$p = \rho g h \tag{10.4}$$

where p = static pressure, pascal

ρ = liquid density, kilogram/cubic meter

h = height of liquid above the measurement point, meter

g = 9.81 meter/second^2 (acceleration due to gravity)

Thus any static pressure measurement can be calibrated as a liquid level measurement. If the vessel is closed at the top, the differential pressure between the bottom and the top of the vessel must be used as the level measurement.

The following are examples of some of the other indirect methods used to measure liquid level.

1. The displacement float method is based on the fact that the buoyant force on a stationary float is proportional to the liquid level around the float.
2. The capacitance probe method is based on the fact that the capacitance between a stationary probe and the vessel wall depends on the liquid level around the probe.
3. The gamma-ray system is based on the fact that the number of gamma rays that penetrate a layer of liquid depends on the thickness of the layer.

Displacement Float Level Sensors

A displacement float level sensor is illustrated in Figure 10.13. The float applies a downward force on the force beam equal to the weight of the float minus the buoyant force of the liquid around the float. The force on the beam is given by the following equation:

$$f = Mg - \rho g A h \tag{10.5}$$

where f = net force, newton

M = mass of the float, kilogram

g = 9.81 m/s² (gravity)

ρ = liquid density, kilogram/cubic meter

A = horizontal cross-sectional area of the float, square meter

h = length of the float below the liquid surface, meter

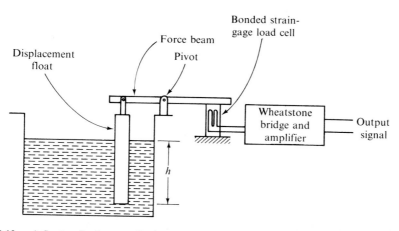

Figure 10.13 A float and a force or displacement sensor use the buoyant force on the float as a measure of the liquid level.

Equation (10.5) shows that the force, f, bears a linear relationship to the liquid level.

The load cell applies a balancing force on the force beam that is proportional to f and, consequently, bears a linear relationship with the liquid level. The load cell is a strain gage force transducer that varies its resistance in proportion to the applied force. A signal conditioner converts the load cell resistance into a usable electrical signal.

Static Pressure Level Sensors

Static pressure level sensors use the static pressure at some point in the liquid as a measure of the level. They are based on the fact that the static pressure is proportional to the height of the liquid above the point of measurement. The relationship is given by the equation

$$p = \rho g h \tag{10.4}$$

where p = static pressure, pascal

ρ = liquid density, kilogram/cubic meter

h = height of liquid above the measurement point, meter

g = 9.81 m/s^2

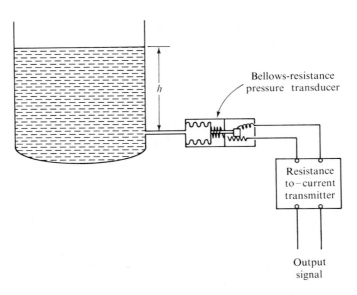

Figure 10.14   A static pressure sensor uses the pressure near the bottom of the tank as a measure of the liquid level.

If the top of the tank is open to atmospheric pressure, an ordinary pressure gage may be used to measure the pressure at some point in the liquid. A variety of methods is used to measure the static pressure. One method is illustrated in Figure 10.14, where a bellows resistance pressure sensor and transmitter is used to measure the level. The output of the transmitter is a 4- to 20-mA current signal corresponding to a level range from 0 to 100%.

If the top of the tank is not vented to the atmosphere, the static pressure is increased by the pressure in the tank at the liquid surface. The height of the liquid above the point of measurement is proportional to the difference between the static pressure and the pressure at the top of the tank. A differential pressure measurement is required. Figure 10.15 illustrates the use of a differential pressure transducer to measure the level in a closed tank.

Capacitance Probe Level Sensors

A capacitance probe level sensor is illustrated in Figure 10.16. The insulated metal probe is one side of the capacitor and the tank wall is the other side. The capacitance varies as the level around the probe varies. The transmitter uses a capacitance bridge to measure the change in capacitance. Solid and liquid levels up to 70 m can be measured with a capacitance probe.

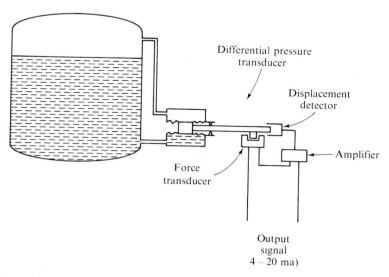

Figure 10.15   A differential pressure sensor can be used to measure liquid level in a closed tank that is under high pressure.

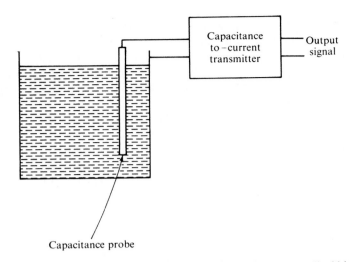

Figure 10.16   A capacitance probe level sensor can measure solid (granular) or liquid levels up to 70 m.

Example 10.4

The displacement float level sensor in Figure 10.13 has the following data.

    Mass of the float, $M = 2.0$ kg

    Cross-sectional area of float, $A = 20$ cm^2

    Length of the float, $L = 2.5$ m

    Liquid in the vessel, kerosene

Determine the minimum and maximum values of the force, f, applied to the force beam by the float.

Solution

From Table 3 in Appendix A, the density (ρ) of kerosene is 800 kg/m^3. The force, f, is given by Equation (10.5).

$$f = Mg - \rho g A h$$
$$= (2)(9.81) - (800)(9.81)\left(\frac{20}{10^4}\right)h$$
$$= 19.62 - 15.7h \text{ N}$$

The minimum force occurs when $h = L$:

$$f_{min} = 19.62 - (15.7)(2.5)$$
$$= -19.63 \text{ N}$$

The maximum force occurs when $h = 0$:

$$f_{max} = 19.62 - (15.7)(0)$$
$$= 19.62 \text{ N}$$

The force applied by the float on the force beam ranges from 19.62 N when the vessel is empty to -19.63 N when the vessel is full.

Example 10.5

Water is to be stored in an open vessel. The static pressure measurement point is located 2 m below the top of the tank. A pressure transducer is selected to measure the liquid level. Determine the range of the pressure transducer for 100% output when the tank is full.

Solution

Equation (10.4) gives the desired relationship. The density of water is obtained from Table 3 in Appendix A.

$$\rho = 1000 \text{ kg/m}^3$$
$$p = \rho g h$$
$$p_{max} = (1000)(9.81)(2)$$
$$= 1.962 \times 10^4 \text{ Pa}$$

or

$$p_{max} = 19.62 \text{ kPa}$$

GLOSSARY

Absolute pressure: The difference between a measured pressure and an absolute vacuum (0 pressure). (10.3)

Bellows: A thin-walled cylinder with corrugated sides used to measure pressure. (10.3)

Bimetallic thermostat: Two strips of different metals bonded together to form a temperature-activated switch. (10.1)

Blackbody: An ideal body used to model the heat radiated from objects. A blackbody has an emittance of 1. (10.1)

Bourdon element: A flattened tube bent into an incomplete circle, spiral, or helix that is used to measure pressure. (10.3)

Differential pressure: The difference between a measured pressure and a reference pressure. (10.3)

Differential pressure flow meter: A flow meter that operates on the principle that a restriction in a flowing fluid produces a pressure drop that is proportional to the flow rate squared. (10.2)

Emittance: The ratio of the radiant energy emitted from a body to the radiation emitted from an ideal blackbody. (10.1)

Filled thermal system: A temperature sensor that uses a bulb filled with a liquid, gas, vapor, or mercury. Thermal expansion of the fluid in the bulb produces a motion that is a measure of the temperature of the fluid. (10.1)

Gage pressure: The difference between a measured pressure and atmospheric pressure. (10.3)

Magnetic flow meter: A flow meter that uses the voltage induced when a conductor moves in a magnetic field (the flowing fluid is the conductor). (10.2)

Perfect vacuum: A pressure of zero. (10.3)

Radiation pyrometer: A temperature sensor that measures the temperature of an object by sensing the thermal radiation emanating from the object. (10.1)

RTD: Abbreviation of resistance temperature detector; a temperature sensor that uses the change in resistance of a conductor due to a change in temperature of the conductor. (10.1)

Strouhal number: A constant used in the flow rate equation of a vortex shedding flow meter. (10.2)

Thermistor: A temperature sensor using a semiconductor that has a large change in resistance with changes in temperature. (10.1)

Thermocouple: A temperature sensor that uses the fact that two dissimilar wires connected at each end generate a voltage that is a measure of the difference in temperature between the two ends. (10.1)

Turbine flow meter: A flow meter that uses the rotation of a turbine blade to measure fluid flow rate. (10.2)

Vacuum: A pressure that is less than atmospheric pressure. (10.3)

Vortex shedding flow meter: A flow meter that uses pulsations caused by an unstreamlined obstruction in the flow stream to measure flow rate. (10.2)

EXERCISES

10.1 Name the four types of fluids used in filled thermal system temperature sensors.

10.2 The following data were obtained in a calibration test of a class III FTS temperature transmitter similar to Figure 10.3.

Temperature (°C)	0	38	81	120	162	199
Output Signal (psi)	3.01	5.34	7.86	10.14	12.78	15.01

Construct a calibration graph by plotting the data points. Draw a line through the endpoints and estimate the terminal-based nonlinearity of the transmitter in degrees celsius and percentage of full scale.

10.3 Check the equation developed in Example 10.1 at 75°C.

10.4 The resistance of nickel wire at 20°C is given by the following equation:

$$R = \frac{\rho L}{A}$$

where R = resistance at 20°C, ohm

ρ = resistivity of nickel = 47.0

A = area of the wire, circular mil

[circular mil = (diameter in mils)^2]

L = length of the wire, foot

A nickel resistance thermometer element is to have a resistance of 100 Ω at 20°C. Determine the length of wire required if the diameter of the wire is 0.004 in. (4 mils).

10.5 The EMF produced by a thermocouple may be approximated by the following equation:

$$E = E_0 + a_1 T + a_2 T^2$$

where E = thermocouple EMF at T°C, volt

E_0 = thermocouple EMF at 0°C, volt

T = temperature, °C

a_1, a_2 = constants

Determine the values of E_0, a_1, and a_2 for the iron–constantan thermocouple described in Table 10.2. Use the EMF values at 0, 100, and 200°C to find E_0, a_1, and a_2. Check the accuracy of the equation at 50°C. (*Hint:* See Example 10.1.)

10.6 Construct a graph of the temperature versus the output signal for the iron–constantan thermocouple in Table 10.2. Determine the terminal-based nonlinearity of the graph (i.e., the maximum difference between the actual curve and a straight line connecting the two endpoints).

10.7 A differential pressure flow meter is used as the sensor in a liquid flow control system. A technician obtained the following data in a series of calibration tests. Each test consisted of measuring the time required to fill a 0.5-gal container. The tests were conducted at controller settings of 25, 50, 75, and 100%.

Controller Setting (%)	Time to Fill a 0.5-gal Container (min)
25	6.63
50	4.78
75	3.95
100	3.45

Determine the average flow rate in gal/min for each controller setting by dividing 0.5 gal by the time in minutes. Then construct a calibration curve from your calculated results.

10.8 The flow control system in Exercise 10.7 is described by the equation

$$q = K\sqrt{sp}$$

where q = flow rate, gallon/minute

K = constant

sp = controller setting, percent

Calculate the value of K for each of the four calibration test results in Exercise 10.7. Find the average of the four K values and use this value to express the relationship between the flow rate, q, and the square root of the controller setting, sp.

10.9 A turbine flow meter has a K value of 34.1 cm^3 per pulse. Determine the volume of liquid transferred for each of the following pulse counts.

a. 8200 **b.** 32,060 **c.** 680

10.10 Determine the flow rate for each pulse count in Exercise 10.9 if the pulse counts occur during a 210-s time interval.

10.11 A bellows pressure element similar to Figure 10.12a has the following values.

Effective area of the bellows = 21 cm^2

Spring rate of the spring = 200 N/cm

Spring rate of the bellows = 10 N/cm

What is the pressure range of the transducer if the motion of the bellows is limited to 2 cm?

10.12 A bellows pressure element similar to Figure 10.12b has the following values.

Effective area of the bellows = 4 cm^2

Spring rate of the spring = 400 N/cm

Spring rate of the bellows = 25 N/cm

What stroke of the bellows is required for a range of 0 to 50 N/cm^2?

10.13 A displacement float level transducer has the following data.

Mass of the float, M = 6.5 kg

Area of the float, A = 50 cm^2

Length of the float, L = 4.0 m

Liquid in the vessel, water

Determine the minimum and maximum values of the force, F, applied to the force beam by the float.

10.14 Gasoline is to be stored in a vented vessel. The static-pressure measurement point is 8 m below the top of the vessel. A pressure sensor is to be used as the level sensor. Determine the range of the pressure sensor for 100% output when the tank is filled.

10.15 Complete the following for the iron–constantan thermocouple transmitter described in Table 10.2:

a. Determine the terminal-based linearity of the transmitter output.

b. Design a four-step, piecewise linear function that will linearize the transmitter output.

PART THREE

Manipulation

CHAPTER 11

Switches, Actuators, Valves, and Heaters

OBJECTIVES

A controller has two interfaces with the process it controls; one is the input to the controller, the other is the output from the controller. Sensors and signal conditioners handle the input side. Various types of switching elements, actuators, control valves, heaters, and electric motors handle the output side.

The purpose of this chapter is to give you an entry-level ability to discuss, select, and specify switching elements, pneumatic actuators, hydraulic actuators, process control valves, and heaters. After completing this chapter, you will be able to

1. Interpret standard symbols for mechanical switches, solid-state switching components, relays, and solenoid valves

2. Describe pushbutton switches, limit switches, level switches, pressure switches, and temperature switches

3. Describe mechanical relays, contactors, motor starters, and time-delay relays

4. Discuss and sketch simple circuits that illustrate the operation of the following solid-state switching devices: transistor, SCR, triac, UJT, and diac

5. Describe pneumatic cylinders and motors

6. Describe hydraulic cylinders and motors

7. Use example spec sheets to select a pneumatic or hydraulic actuator

8. Describe process control valves—include the following in your discussion: direct and reverse acting, inherent versus installed characteristics, quick opening, linear and equal percentage

9. Determine the size of control valve required to meet specs that include the maximum flow rate, properties of the fluid, available pressure drop, and the safety factor

10. Compute the heat flow rate required to melt and raise the temperature of a given amount of a substance in a given elapsed time

11.1 MECHANICAL SWITCHING COMPONENTS

Switches are devices that make or break the connection in an electric circuit. The switching action may be accomplished mechanically by an actuator, electromechanically by a solenoid, or electronically by a solid-state device. All three methods of accomplishing the switching action are used in control systems. Switches turn on electric motors and heating elements, sense the presence of an object, regulate the speed of an electric motor, actuate solenoid valves that control pneumatic or hydraulic cylinders, and initiate actions in sequential control systems.

A sequential control system performs a set of operations in a prescribed manner. The automatic washing machine is a familiar example of a sequential control system. The operations of filling the tub, washing the clothes, draining the tub, rinsing the clothes, and spin drying the clothes are controlled by switches.

Mechanical Switches

Mechanical *switches* use one or more pairs of contacts to make or break an electric circuit. The contacts may be normally open (NO) or normally closed (NC). A normally open switch will close the circuit path between the two terminals when the switch is actuated and will open the circuit path when the switch is deactuated. A normally closed switch will open the circuit path when the switch is actuated and will close the circuit path when the switch is deactuated. The more common actuating mechanisms include pushbuttons, toggles, levers, plungers, and rotary knobs.

The switching action may be momentary-action or maintained-action. In a *momentary-action* switch, the operator pushes the button, moves the toggle, or rotates the knob to change the position of the contacts. When the operator releases the switch, the contacts return to the normal condition. When an operator actuates a *maintained-action* switch, the contacts remain in the new position after the operator releases the actuator. In most momentary-action switches, the actuator returns to its original position when released; and in most maintained-action switches, the actuator remains in the new position when released. However, in some maintained-action switches the actuator returns to the original position, even though the contacts remain in the new position.

Mechanically actuated switches may be operated manually by an operator, or automatically by fluid pressure, liquid level, temperature, flow, thermal overload, a cam, or the presence of an object. Figure 11.1 shows standard wiring diagram symbols for various types of switches. The limit switches in Figure 11.1 are actuated by a cam or some other object that engages the switch actuator. Pressure-actuated switches open or close their contacts at a given pressure. Liquid level-actuated switches open or close their contacts at a given liquid level. Temperature-actuated switches open or close their contacts at a given temperature. Flow-actuated switches open or close their contacts when a given flow rate is sensed. The overload switches are circuit breakers that open a normally closed contact when an overload condition occurs. Overload switches are intended to protect motors and other equipment from damage caused by an overload condition.

Pushbutton Switches
Momentary-action
Maintained-action
NO
NC
Limit Switch
Pressure Switch
NO
NC
NO
NC
Liquid Level Switch
Temperature Switch
NO
NC
NO
NC
Flow Switch
Thermal Overload
Magnetic Overload
NO
NC

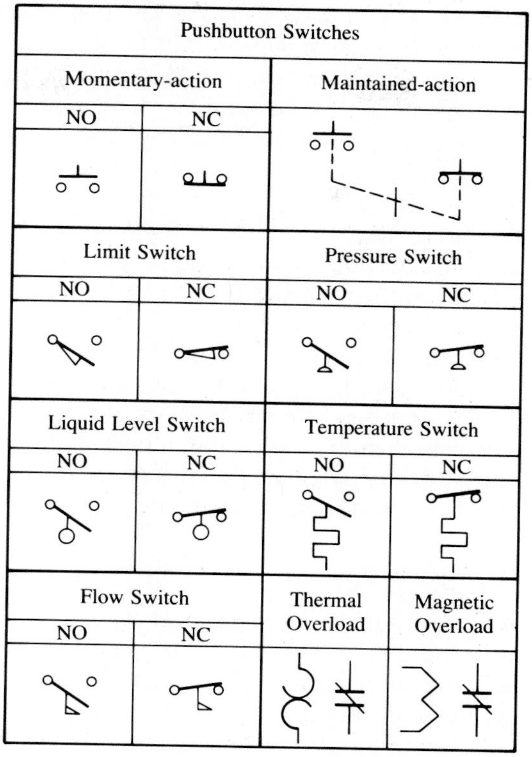

Figure 11.1 Wiring diagram symbols for mechanical switches.

Relays

A *relay* is a set of switches that are actuated when electric current passes through a coil of wire. The electric current passing through the coil of wire generates a magnetic field about the core of the coil. This magnetic field pulls a movable arm that forces the contact to open or close. The *pull-in current* is the minimum coil current that causes the arm to move from its OFF position to its ON position. The *drop-out current* is the maximum coil current that will allow the arm to move from its ON position to its OFF position. Figure 11.2 illustrates two *control relays*, one with two

1CR
1CR
1CR
3CR
3CR
3CR
3CR
3CR

Figure 11.2 Control relays use the coil designation (e.g., 1CR, 2CR) to identify the contacts that are actuated by the coil.

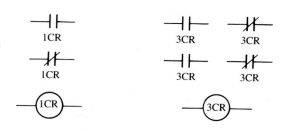

switches, the other with four switches. The relay coils are represented schematically by the circles with the designation 1CR and 3CR. The CR signifies a control relay, and the numbers 1 and 3 are used to distinguish between the two control relays. Each relay in a drawing must have a unique designation.

The relay is actuated (or energized) by completing the circuit branch that contains the relay coil. The relay coil designation is used to identify the contacts that are actuated by a particular relay coil. Normally open contacts are designated by a pair of parallel lines. Normally closed contacts are designated by parallel lines with a diagonal line connecting the two parallel lines. A normally open contact will close the circuit path when the relay coil is energized and will open the circuit path when the relay is deenergized. A normally closed contact will open the circuit path when the relay coil is energized and will close the circuit path when the relay coil is deenergized.

Notice that relay 1CR has one normally open contact and one normally closed contact. Relay 3CR has two normally open and two normally closed contacts. This method of using the relay coil designation to identify the relay contacts is necessary because the contacts may occur anywhere in an electric circuit diagram. The designation identifies which relay coil actuates each set of contacts.

Time-Delay Relays

Time-delay relays are control relays that have provisions for a delayed switching action (see Figure 11.3). The delay in switching is usually adjustable, and it may take place when the coil is energized or when the coil is deenergized. An arrow is used to identify the switching direction in which the time delay takes place. In relay 1TR, the delay occurs when the coil is energized. Contact 1TR delays before it closes. When coil 1TR is deenergized, contact 1TR opens immediately. In relay 2TR, the delay also occurs when the coil is energized. The 2TR contact delays before it opens and closes immediately when the coil is deenergized. In relays 3TR and 4TR, the delay occurs when the coil is deenergized. Contact 3TR closes immediately when coil 3TR is

Time Delay Relay			
Time delay when the coil is energized		Time delay when the coil is deenergized	
NO	NC	NO	NC
1TR	2TR	3TR	4TR

Figure 11.3 In time-delay relays, the arrow points in the direction in which the delayed action occurs.

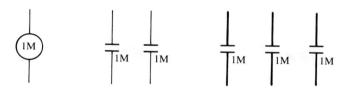

Figure 11.4   Motor starters and contactors have three large contacts that are used to switch large amounts of electric power. The small auxiliary contacts are used in the control circuit.

energized and delays before it opens when the coil is deenergized. Contact 4TR opens immediately when coil 4TR is energized, and delays before it closes when coil 4TR is deenergized.

In a time-delay relay, the delay occurs in the direction in which the arrow is pointing.

Contactors and Motor Starters

Relay with heavy-duty contacts are used to switch circuits that use large amounts of electric power. When the circuit load is an electric motor, the relay is called a *motor starter*; otherwise, it is called a *contactor*. An example of a relay for switching a three-phase system is shown in Figure 11.4. The three heavy contacts are used to switch the three lines supplying electric power to the load. The two light contacts are used in the control circuit.

Figure 11.5 illustrates a circuit for starting and stopping a 480-V ac three-phase electric motor. The two momentary-action pushbutton switches in the 115-V ac circuit are the start–stop station for the motor. When the start button is pressed, coil 1M

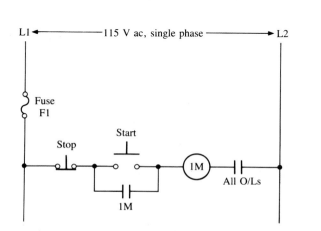
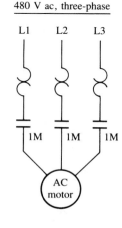

Figure 11.5   Control circuit for starting a large ac motor.

is energized, closing all four 1M contacts. The three large contacts connect the three 480-V ac lines to the motor. The small contact in parallel with the start switch is used to hold the circuit closed after the start button is released. The small contact is called a "holding contact" because it "holds" coil 1M in the energized condition after the operator releases the start button. When the operator presses the stop button, the circuit breaks and all four 1M contacts open. The circuit remains deenergized after the stop button is released because the 1M holding contact is open.

Table 11.1 Partial List of Solid-State Components

Name of Solid-state Component	Graphical Symbol	Volt-ampere Characteristic	Typical Applications
Diode			Rectifier block bypass
NPN transistor			Amplifiers, switches, oscillators
PNP transistor			Amplifiers, switches, oscillators
Unijunction transistor (UJT)			Timers, oscillators, SCR trigger
Programmable unijunction transistor (PUT)			Timers, oscillators, SCR trigger
Silicon-controlled rectifier (SCR)			Power switches, inverters, frequency converters
Triac			Switches, relays
Diac			Triac and SCR trigger

11.2 SOLID-STATE SWITCHING COMPONENTS

The number and variety of solid-state components used in control is sufficient to fill several books. Solid-state components are found in relays, timers, contactors, motor starters, temperature controllers, a variety of switching circuits, ac-to-dc converters, dc-to-ac inverters, battery chargers, variable-speed drives, cycloconverters, and rectifiers—just to name a few. The purpose of this section is to present a few typical examples that will illustrate how solid-state components are used in control. Table 11.1 gives the symbols and volt-ampere characteristics of the solid-state components most frequently used in control.

Silicon-Controlled Rectifiers

One of the major uses of solid-state components is in switching circuits of all sizes. Although transistors are used in some switching circuits, power switching is the domain of the *silicon-controlled rectifier (SCR)* and the triac. The SCR has a number of useful characteristics. First, it is a rectifier—an SCR will conduct current only in one direction. Second, the SCR is a latching switch—the SCR can be turned on by a short pulse of control current into the gate and it remains on as long as current is flowing from the anode to the cathode. Third, the SCR has a very high gain—the anode current is about 3000 times as large as the control current. This means that a small amount of power in the turn-on circuit is used to control a large amount of power in the main circuit. Figure 11.6 shows the volt–ampere graph of an SCR.

When the anode-to-cathode voltage, V_{AC}, is negative, the SCR is said to be *reverse biased*. With a reverse bias, the anode current is negligible until the breakdown occurs—typically 50 to 2000 V. When the anode-to-cathode voltage is made positive, the anode current is still negligible. However, when V_{AC} is positive and a small voltage is applied between the gate and cathode, the SCR switches on. When the SCR

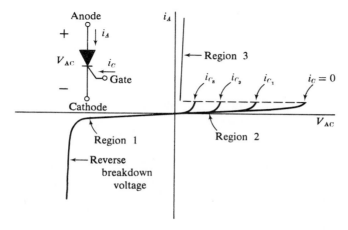

Figure 11.6   The volt–ampere characteristics of an SCR can be divided into three operating regions, as shown above.

is on, it remains on even if the triggering voltage is removed. The SCR has three operating regions.

Region 1. V_{AC} is negative and i_A is negligible.

Region 2. V_{AC} is positive and i_A is negligible.

Region 3. V_{AC} is positive and i_A is large; the SCR is on.

There are four ways that an SCR can make the transition from region 2 to region 3. The first way is to increase the anode voltage until the forward breakdown voltage is reached and the device turns on. The second way is by applying a positive voltage pulse to the gate input to "trigger" the SCR on. A third method is by applying light to the gate/cathode junction. The light energy turns the SCR on. This is called a LASCR (Light-Actuated Silicon Controlled Rectifier). The fourth method is to rapidly increase the anode-to-cathode voltage, V_{AC}. The rapid increase in voltage will also turn the SCR on. The second method, a triggering voltage pulse at the gate, is usually used in control applications.

Once the SCR is turned on, it will remain on. There are three ways that an SCR can be turned off.

1. The anode current can be reduced below a minimum value called the *holding current*. The holding current is typically about 1% of the rated current. This causes a transition from region 3 to region 2. The SCR is still forward biased, but it is off instead of on.
2. If the anode voltage is reversed (i.e., $V_{AC} < 0$), the SCR will go from region 3 to region 1. This method is used in ac circuits where the voltage reverses each half-cycle. It is also used in inverters and frequency converters where a charged capacitor is switched into the SCR circuit to force a reversal of the anode-to-cathode voltage.
3. In small current applications, the SCR can also be turned off by supplying a negative gate current to increase the holding current. When the increased holding current exceeds the load current, the SCR switches into region 2.

The second method, reversing V_{AC}, is the method most often used in control circuits.

One application of the SCR is the inverter circuit illustrated in Figure 11.7. An inverter is a device that converts the voltage from a dc source into an ac voltage. The dc source might be a bank of storage batteries used to store electrical energy collected by solar cells or a wind generator. The ac voltage is needed to operate ac equipment such as fluorescent lights, appliances, and electric power tools. The circuit operation consists of the SCRs alternately switching the dc current from one half of the primary winding to the other half. The effect is equivalent to an alternating current in a single primary winding so that the secondary winding delivers an alternating current to the load. The term *commutation* is used to name the transfer of current from one SCR to the other SCR. Reliable commutation is one of the major problems in inverters. In Figure 11.7, additional commutation circuitry is required to turn off the SCRs by reverse biasing (method 2) each time the other SCR is turned on by a triggering impulse.

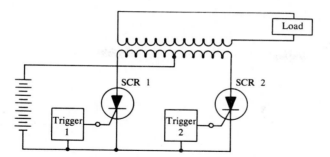

Figure 11.7 This center-tapped primary, SCR-controlled inverter converts a dc voltage into an ac voltage.

Triacs

The *triac* was developed as a means of providing improved controls for ac power. The major difference between the triac and the SCR is that the triac can conduct in both directions, whereas the SCR can conduct in only one direction. The volt–ampere characteristic of the triac is shown in Table 11.1. A positive or negative gate current of sufficient amplitude will trigger the triac on when V_{21} is either positive or negative. A typical triac circuit is shown in Figure 11.8. The trigger circuit produces pulses that turn the triac on. The first trigger pulse turns the triac on during the positive half-cycle. The triac remains on until the ac voltage reverses and turns the triac off. The next triggering pulse turns the triac on during the negative half-cycle. The triac again remains on until the ac voltage reverses and turns the triac off. The trigger circuit determines when the trigger pulse will turn the triac on. This in turn determines how much current is delivered to the load.

Unijunction Transistors

A number of devices are used to produce the trigger pulse for SCRs and triacs. The *unijunction transistor* (UJT) is one device used in the triggering circuit. A basic

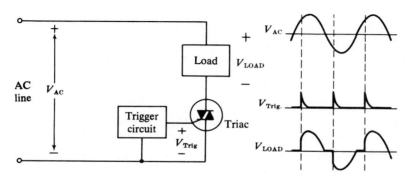

Figure 11.8 Basic triac circuit for control of an ac load.

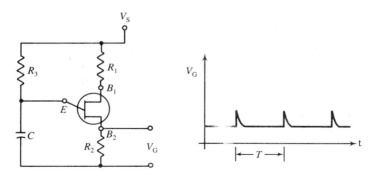

Figure 11.9   UJT trigger circuit.

unijunction trigger circuit and its waveforms are shown in Figure 11.9. In this circuit, the capacitor is charged by the current through R_3 until the emitter voltage reaches V_p (see Table 11.1) and the UJT turns on. This discharges the capacitor through R_2, producing the trigger spike in V_G. When the emitter voltage reaches 2 V, the UJT turns off and the cycle is repeated. The values of R_3 and C determine the time between the triggering pulses.

Figure 11.10 shows an example of a UJT used as an SCR trigger circuit. The circuit operates as follows:

1. The ac input is applied to the load in series with the SCR. The load voltage remains at zero until the SCR is switched on by a gate pulse.
2. Voltage e_{in} is also applied to the zener diode clipper circuit (R_S and V_Z). The zener diode breaks down at V_Z volts and thus limits or clips the positive peaks of e_{in} (see the v_{xy} waveform in the figure). During the negative half-cycle of e_{in}, the zener is forward biased and maintains a near 0 voltage between x and y.
3. The clipped sine wave, v_{xy}, is the supply voltage for the UJT circuit. During the positive half-cycle of e_{in} while v_{xy} is at $+V_Z$ volts, the capacitor charges through R until it reaches V_p of the UJT. When it does, the UJT turns "on" and discharges C, producing a positive pulse across R_1 (see v_{B1} waveform). This pulse is fed to the gate of the SCR and turns "on" the SCR.
4. Once the SCR is on, the load voltage becomes approximately equal to e_{in} for the duration of the positive half-cycle. (See the v_{load} waveform.)
5. During the negative half-cycle, the SCR stays off and v_{load} stays at zero.
6. The amount of power delivered to the load is controlled by varying the RC time constant, which causes C to charge slower or faster, thereby triggering the UJT and SCR later or earlier in the positive half-cycle of e_{in}. In other words, the RC time constant controls the SCR trigger time and therefore the load power. As RC is increased, the trigger time increases and the load decreases.*

---

* R. Tocci, *Fundamentals of Electronic Devices*, Fourth Edition (Columbus, Ohio: Merrill/Macmillan, 1991), p. 592.

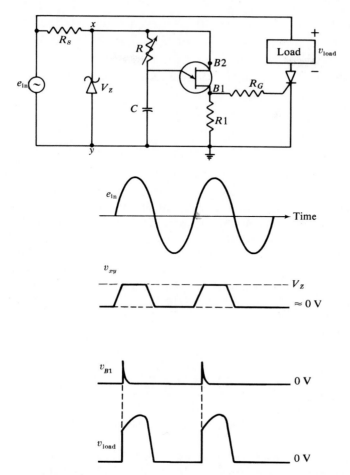

Figure 11.10   A UJT is used to trigger an SCR in this ac power control circuit. [From R. Tocci, *Fundamentals of Electronic Devices*, Fourth Edition (Columbus, Ohio: Merrill/Macmillan, 1991), p. 592.]

Diodes

Diodes are used as rectifiers in dc power supplies and as one-way "valves" to block or bypass undesired electric currents. The volt-ampere characteristic of the diode is shown in Figure 11.11. The diode has three operating regions depending on the anode-to-cathode voltage (V). If the anode-to-cathode voltage is positive, the diode is said to be forward biased; if the voltage is negative, the diode is reverse biased. In operating region 1, the diode is forward biased, and it conducts electric current with very low resistance. In region 2, the diode is reverse biased, and it blocks electric current with very high resistance. In region 3, the reverse bias has increased to the point of breakdown, and the diode conducts electric current. For most applications, operation in region 3 is avoided. An exception to this rule is the zener diode, which is operated in region 3 as a voltage regulator.

Figure 11.11 Volt–ampere characteristic of a diode.

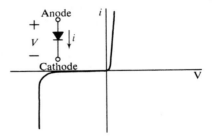

A full-wave dc power supply is illustrated in Figure 11.12. The diodes are used as one-way switches that pass current only when they are forward biased. The center-tapped transformer configuration is such that the two diodes are forward biased by opposite polarities of the ac input voltage (V_{in}). When V_{in} is positive, diode D_1 is forward biased, and when V_{in} is negative, diode D_2 is forward biased. Therefore, diode D_1 conducts current during each positive half-cycle of the input voltage, and diode D_2 conducts current during each negative half-cycle. The output of each diode is connected to output terminal a, producing the full-wave rectified output voltage shown in Figure 11.12. A capacitor is usually placed across terminals a and b to produce a more constant output voltage.

Transistors

Transistors are used in control systems as switches, amplifiers, and oscillators. There are two types of transistors, the NPN type and the PNP type. The graphic symbols and volt–ampere characteristics of both types are included in Table 11.1. The operating characteristics of NPN and PNP transistors are the same except that the directions of the voltages and currents are opposite. The following discussion is confined to the NPN transistor.

An NPN transistor has three terminals called the collector, base, and emitter, as shown in Figure 11.13. Transistors are essentially controlled-current amplifiers. A

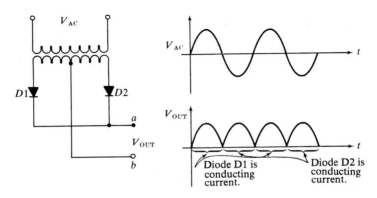

Figure 11.12 Full-wave dc power supply.

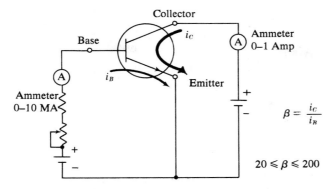

Figure 11.13   Test circuit for an NPN transistor.

relatively small current entering the base is used to control a much larger current entering through the collector. Both currents leave through the emitter. The ratio of the collector current (i_C) to the base current (i_B) is called the β (beta) of the transistor ($\beta = i_C/i_B$). The β of a transistor ranges from 20 to 200, and it varies from one transistor to another, even if they are of the same type. Aging and changes in temperature may also cause the β to change.

A transistor may be used in one of three different amplifier configurations, depending on which terminal is common to the input and the output. The three configurations and a summary of their characteristics are shown in Figure 11.14.

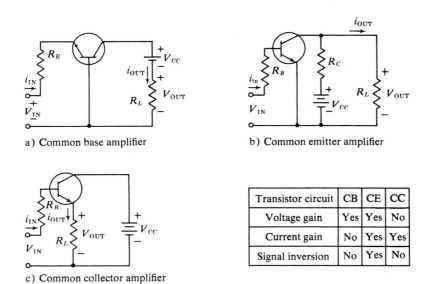

a) Common base amplifier

b) Common emitter amplifier

c) Common collector amplifier

Transistor circuit	CB	CE	CC
Voltage gain	Yes	Yes	No
Current gain	No	Yes	Yes
Signal inversion	No	Yes	No

Figure 11.14   Configurations and characteristics of transistor amplifiers.

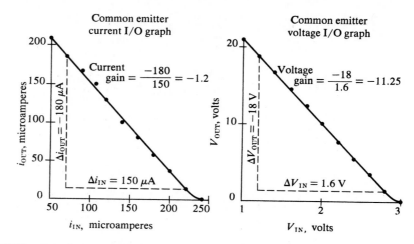

Figure 11.15   Input/output graphs of a common-emitter amplifier.

The input/output graphs of a typical common-emitter amplifier are shown in Figure 11.15. Notice the signal inversion in both graphs (i.e., as the input voltage or current increases, the output voltage or current decreases). The voltage gain is determined as follows. Select two points on the straight portion of the graph that are spaced as far apart as possible. Measure the change in the input voltage (ΔV_{in}) and the change in the output voltage (ΔV_{out}) between the two points. The voltage gain is the change in output voltage divided by the change in input voltage, $\Delta V_{out}/\Delta V_{in}$. The current gain is determined in a similar manner.

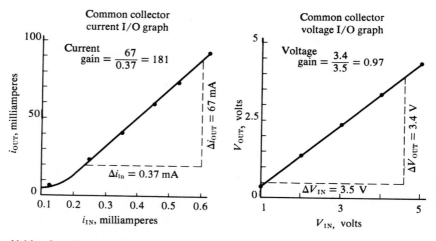

Figure 11.16   Input/output graphs of a common-collector amplifier.

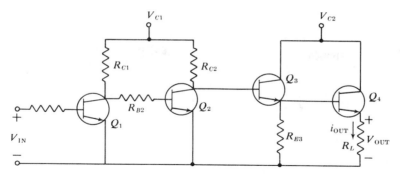

Figure 11.17   Four-stage transistor amplifier.

In the example illustrated in Figure 11.15, the voltage gain is -11.25 and the current gain is -1.2. The negative sign is due to the signal inversion. This signal inversion is a characteristic of the common-emitter amplifier. The inversion is eliminated when two common-emitter amplifiers are connected in series, because the inverted output of the first stage is inverted again by the second stage. The overall voltage gain of a two-stage amplifier is the product of the voltage gains of the two stages. The same is true for the current gain. For example, if stage 1 has a voltage gain of -12 and stage 2 has a voltage gain of -10, the overall voltage gain is $(-12)(-10) = 120$.

The input/output graphs of a typical common-collector amplifier are shown in Figure 11.16. The voltage gain of this amplifier is 0.97, and the current gain is 181. Notice the absence of the signal inversion that characterized the common-emitter amplifier. This direct relationship between input and output signals and the absence of voltage gain are characteristics of the common-collector amplifier. The common-collector amplifier is a current amplifier.

The common-emitter and common-collector amplifiers may be combined to form multistage amplifiers with increased voltage and current gains. A four-stage amplifier is shown in Figure 11.17. Transistors Q_1 and Q_2 form common emitter amplifiers for stages 1 and 2. Transistors Q_3 and Q_4 form common-collector amplifiers for stages 3 and 4.

Example 11.1

A schematic diagram of the industrial process described in Example 3.19 is shown in Figure 11.18. In Example 3.19, a control program was developed for the ICST-1 microcomputer to control the required sequence of operations. Two-position solenoid valves are used to control the water and syrup flows into the tank. A third solenoid valve is used to control the flow out of the tank. The solids feeder is driven by a synchronous motor, and a universal motor is used for the agitator drive. The operator's console completes the main components of the process.

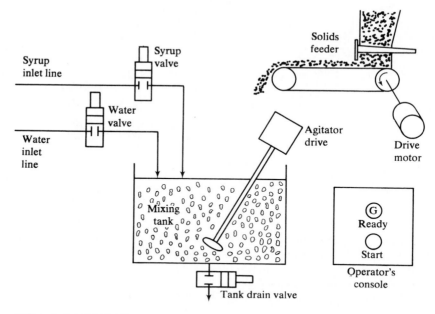

Figure 11.18   Industrial blending process.

The main contacts for starting the solids feeder drive motor are shown in Figure 11.19. The control circuit for energizing the solids feeder starter coil is shown in Figure 11.20. Triac 4 in Figure 11.20 is triggered by a dc signal controlled by bit 2 of the output port of the ICST-1 microcomputer. Triac 4 is ON when bit 2 is a logic 1 and it is OFF when bit 2 is a logic 0. Transistor T3 acts as a solid-state switch to control the triggering of the triac. Similar triac circuits are used to control the solenoid valves and the ready light.

The agitator drive motor is controlled by triac 1 in Figure 11.20. Variable resistor R_2, capacitor C_2, and the diac make up the triggering circuit. Resistor R_2 and capacitor C_2 form a variable time constant circuit that triggers triac 1 when the voltage across C_2 reaches the breakover voltage of the diac. Resistor R_1 and capacitor

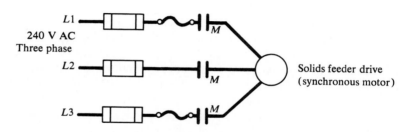

Figure 11.19   Power circuit for the solids feeder drive motor in the blending process in Figure 11.18.

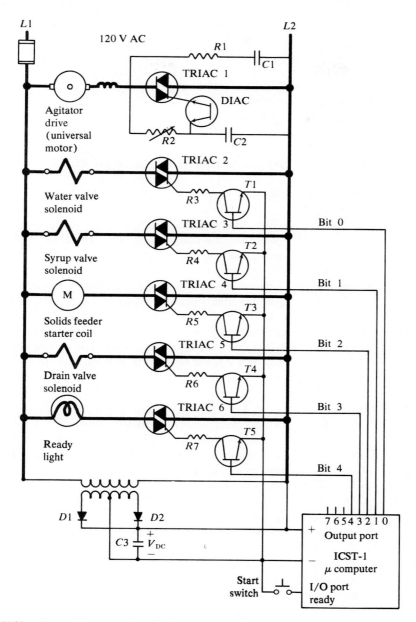

Figure 11.20   Control circuit for the blending process in Figure 11.18.

C_1 reduce the transient voltages developed during switching due to the inductive energy stored in the motor windings. The diac triggering circuit provides a smooth control of motor speed from an intermediate value up to full speed. Other triggering techniques are used when the full control range is required.

The control circuit shown in Figure 11.20 is simpler than the actual circuits used in an industrial process, but it illustrates some of the variety of applications of solid-state components in control systems.

11.3 HYDRAULIC AND PNEUMATIC VALVES AND ACTUATORS

Actuators are an important part of every control loop. They provide the power needed to position the final control element. Hydraulic and pneumatic actuators often provide the most economical and trouble-free method for positioning the final control element. *Hydraulic actuators* are used where slow, precise positioning is required, or where heavy loads must be moved. Hydraulic actuators operate at higher pressures than pneumatic actuators, so they can provide larger forces from a given-size cylinder or motor. *Pneumatic actuators* are used where lighter loads are encountered, where an air supply is available, and where fast response is required. Pneumatic actuators are less expensive than hydraulic actuators, and they are capable of much faster movement. The speed of a hydraulic actuator is limited by the maximum flow rate provided by the hydraulic pump. Pneumatic diaphragm operators are used in most control valves, a topic covered in Section 11.4. In this section we cover hydraulic and pneumatic valves, cylinders, and motors.

Solenoid Valves

Solenoid valves are used to control fluid flow in hydraulic or pneumatic systems. A *solenoid valve* consists of a movable spool fitted inside a housing with one or two solenoids capable of moving the spool to different positions relative to the housing. Both the spool and the housing have fluid passages. The passages in the housing are connected together or blocked in different ways, depending on the position of the spool in the housing. A valve with one solenoid has two spool positions, one position when the solenoid is deenergized, the other position when the solenoid is energized. A valve with two solenoids has three spool positions, one position when both solenoids are deenergized, a second position when solenoid *a* is energized, and a third when solenoid *b* is energized.

Each passage in the housing may be blocked, or it may be connected to the supply line, to the return line, or to a line from a cylinder. Each such connection is called a *way*. A solenoid valve is characterized by the number of positions the valve has and by the number of connecting lines (ways) it has. Thus a valve with one solenoid and two connecting lines is called a two-position, two-way valve. A valve with two solenoids and four connecting lines is called a three-position, four-way valve.

Symbols of common types of solenoid valves are illustrated in Figure 11.21. In the diagram, the spool is represented by two or three squares, one square for each position of the valve. Each connecting line or way is indicated by a line outside of the

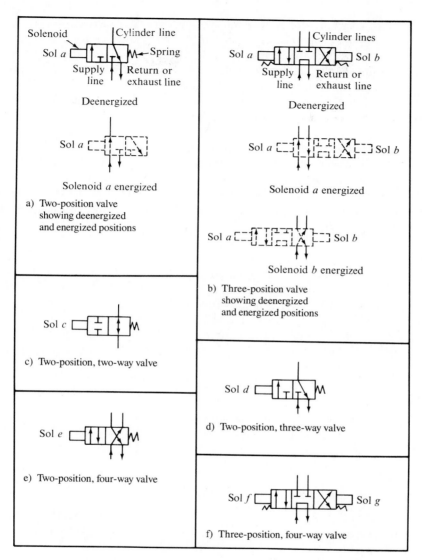

Figure 11.21 Drawing symbols for solenoid valves used for hydraulic and pneumatic systems.

squares that represent the spool. The passages in the spool are indicated by lines within the squares. Blocked lines indicate blocked passages. Arrows indicate the direction of fluid flow. The spool is always shown in the deenergized position, and the two, three, or four connecting lines (ways) are attached to the square that is in effect when the valve is deenergized.

Figure 11.21a illustrates the operation of a two-position valve. The solid-line diagram on top shows the valve in the deenergized position, the way it would be shown in a schematic diagram. Notice that the spool square next to the spring is lined up with the three connecting lines. In this position, the cylinder is connected to the return or exhaust line, and the supply line is blocked. The dashed-line diagram in the bottom of Figure 11.21a shows the valve in the energized position. In this position, the supply line is connected to the cylinder and the return line is blocked.

Figure 11.21b illustrates the operation of a three-position valve. The solid-line diagram on top shows the valve in the deenergized position. Notice that the two cylinder lines are blocked and the supply line is connected to the return line. The middle dashed-line diagram shows the valve in the position it will be in when solenoid *a* is energized. Here the supply line is connected to the left cylinder line, and the right cylinder line is connected to the return line. The lower dashed-line diagram shows the valve in the position it will be in when solenoid *b* is energized. In this position, the supply line is connected to the right cylinder line, and the left cylinder line is connected to the return line.

Cylinders

Hydraulic and *pneumatic cylinders* are used to produce a linear motion with a definite maximum distance of travel. Cylinders may be single acting or double acting, as illustrated in Figure 11.22. Single-acting cylinders allow fluid on only one side of the cylinder. The piston is returned to the starting position by gravity or by a return spring. Double-acting cylinders allow fluid on both sides of the cylinder, providing a power return of the piston. Table 11.2 lists the theoretical forces and speeds for typical hydraulic and pneumatic cylinders.

Motors

Hydraulic and *pneumatic motors* develop large torques to produce rotational motion in a load or final control element. The four most common types of motor are the

Figure 11.22   Hydraulic and pneumatic cylinders produce large forces for linear movement of a load or final control element.

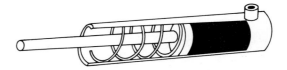

a) Single-acting spring-return cylinder

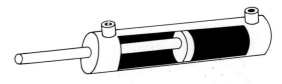

b) Double-acting cylinder

Table 11.2   Theoretical Force and Speed for Hydraulic and Pneumatic Cylinders

Cylinder Bore (in.)	Force (lb)		Speed (in./min)
	Hydraulic (3000 psi)	Pneumatic (100 psi)	Hydraulic (1 gal/min)
1.5	5,301	177	131
2.0	9,425	314	74
3.0	21,206	707	33
4.0	37,699	1257	18
5.0	58,905	1963	12
6.0	84,823	2827	8.2
8.0	150,796	5027	4.6
10.0	235,619	7854	2.9

piston motor, the gear motor, the rotary vane motor, and the turbine motor. Figure 11.23 illustrates these four types of motor.

The gear motors are used only for hydraulic fluids. The turbine motor is used only for pneumatic systems. The piston and rotary motors are used for both hydraulic and pneumatic systems.

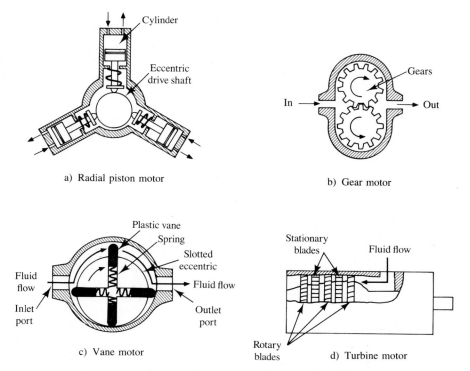

a) Radial piston motor

b) Gear motor

c) Vane motor

d) Turbine motor

Figure 11.23   The piston- and vane-type motors are used in both hydraulic and pneumatic systems. The gear motor is used only in hydraulic systems and the turbine motor is used only in pneumatic systems.

The *radial piston motor* has pistons that radiate out from an eccentric drive shaft. Fluid enters the cylinder and forces the piston against the eccentric thrust ring, causing the shaft to rotate. Radial piston motors are usually limited to speeds below 2000 rpm.

Gear motors are the most popular hydraulic motor. The entering fluid causes the two gears to rotate in the direction indicated. The gears carry the fluid from the inlet port to the outlet port. The output of the gear motor is delivered by a shaft attached to one of the gears. The hydraulic fluid moves the gears, the output gear moves the shaft, and the shaft moves the load. Gear motors can operate at speeds up to 3000 rpm.

The *vane motor* is basically the same for hydraulic and pneumatic systems, although the motors are not interchangeable. The vane motor consists of a slotted eccentric, each with a spring-loaded plastic vane. As fluid enters the motor, it applies force against the vanes, with more force against the vane that is extended. The unbalanced force on the extended vane causes the eccentric shaft to rotate. The fluid is trapped between two vanes and is carried from the inlet port to the outlet port. Hydraulic vane motors can operate at speeds up to 4000 rpm, while pneumatic vane motors can operate up to about 6500 rpm.

The *turbine motor* consists of several sets of rotating blades separated by rows of stationary blades. The rotating blades are attached to the shaft, which is also connected to the load. The fluid enters the motor and flows through the turbine blades, producing a force on the rotary blades that tends to rotate the shaft. Notice that the rotary blades are canted and the stationary blades are straight. The canted blades deflect the fluid, producing a torque that tends to rotate the shaft. The shaft transfers this torque to the load. The stationary blades straighten the fluid flow for the next set of rotary blades. Turbine motors have low starting torque but are capable of very high speeds.

Selection of a Cylinder

The first step in selecting a cylinder is to determine the force required to move the load. The second step is to determine the speed required by the application. Other considerations include the choice between hydraulic and pneumatic systems, the working fluid pressure, the flow rate (hydraulic systems), the valve size, and the oversize factor. The *oversize factor* is a number greater than 1 that is used to increase the size of the cylinder to provide the additional force necessary to accelerate the load. One supplier of hydraulic and pneumatic cylinders recommends selecting a cylinder with 25% extra force for slow-moving loads and 100% extra force for fast-moving loads.

Hydraulic cylinders are capable of much larger forces than pneumatic cylinders. However, pneumatic cylinders are capable of greater speeds than hydraulic cylinders. The speed of a hydraulic cylinder is determined by the flow rate of the fluid (see Table 11.2). The speed of an air cylinder cannot be calculated. Air cylinder speed depends on the force available to accelerate the load and the rate at which air can be vented ahead of the advancing piston. In fact, an adjustable valve on the cylinder vent port is one method of regulating the speed of a pneumatic cylinder.

The force produced by a cylinder is equal to the area of the cylinder times the working pressure in the cylinder.

$$\text{Force} = (\pi d^2/4)(\text{working pressure}) \tag{11.1}$$

Table 11.2 gives the force produced by working pressures of 100 psi and 3000 psi in cylinders with diameters ranging from 1.5 in. to 10 in. A very common design calculation is the computation of the *working pressure* required to produce a given force in a cylinder with a given diameter. The designer may use Equation (11.1) for this computation, but there is a simpler way based on Table 11.2 and the law of proportions. We now use Equation (11.1) to develop the equation for this simpler method. Here we let f_1 and WP_1 represent the force and working pressure from Table 11.2. We want the working pressure (WP_2) required to produce the given force (f_2). Using Equation (11.1), we get the following two equations:

$$f_1 = (\pi d^2/4)WP_1$$

and

$$f_2 = (\pi d^2/4)WP_2$$

Now we divide the second equation by the first:

$$\frac{f_2}{f_1} = \frac{(\pi d^2/4)WP_2}{(\pi d^2/4)WP_1} = \frac{WP_2}{WP_1}$$

Finally, multiply both sides of the equation by WP_1.

$$WP_2 = (WP_1)(f_2/f_1) \tag{11.2}$$

The speed of a hydraulic cylinder is equal to the flow rate in cubic inches per minute divided by the area of the cylinder in square inches. Equation (11.3) gives the speed in inches per minute for a flow rate of Q gallons per minute and a cylinder diameter of d inches (1 gallon = 231 cubic inches).

$$\text{Speed} = 231Q/(\pi d^2/4) \tag{11.3}$$

Table 11.2 also gives the speed of hydraulic cylinders for a flow rate of 1 gallon per minute. Another common design calculation is the computation of the flow rate required to produce a given speed in a cylinder with a given diameter. The designer may use Equation (11.3) for this computation, but again there is a simpler method, which we develop as follows. Let s_1 and Q_1 represent the speed in Table 11.2 ($Q_1 = 1$). We want the flow rate (Q_2) required to produce the given speed (s_2). Using a development similar to the one above, we obtain the following design equation:

$$Q_2 = (Q_1)[s_2/s_1] \tag{11.4}$$

The selection of the working pressure is interrelated with the size and cost of the cylinder and operating considerations such as leakage costs and pump costs. Trade-offs are involved. For example, a higher working pressure will permit a smaller cylinder and a lower flow rate to achieve a specified force and speed. The downside is that higher working pressures result in higher leakage rates and higher noise levels. Example 11.2 illustrates the process of selecting a cylinder for a given application.

Example 11.2

A manufacturing operation requires the movement of a workpiece a distance of 10 in. in 15 s. A force of 10,000 lb is required to move the workpiece. The available space limits the cylinder diameter to a maximum of 5 in. Leakage and noise considerations make it desirable to limit the working pressure to a maximum of 1000 psi. A hydraulic pump with capacity of 2.5 gal/min is available in the surplus equipment yard. Select a cylinder for this application.

Solution

The required force is 10,000 lb. The required speed is $10/0.25 = 40$ in./min. The speed is relatively slow, so a 25% oversize factor will be used.

$$\text{Required cylinder force} = 1.25 \times 10,000 = 12,500 \text{ lb}$$

The force and diameter requirements dictate a hydraulic cylinder. Three cylinder sizes will be considered: 3, 4, and 5 in.

For a 3-in.-diameter cylinder:

From Table 11.2, $WP_1 = 3000$-psi pressure and $f_1 = 21,206$ lb for a 3-in.-diameter cylinder. From Equation (11.2), the pressure required to produce a force of 12,500 lb is

$$WP = (3000)\left(\frac{12,500}{21,206}\right) = 1768 \text{ psi}$$

From Table 11.2, a flow rate of 1 gal/min produces a speed of 33 in./min in a 3-in.-diameter cylinder. From Equation (11.4), the flow rate required to produce a speed of 40 in./min is

$$Q = (1)\left(\frac{40}{33}\right) = 1.2 \text{ gal/min}$$

For a 4-in.-diameter cylinder:

From Table 11.2, 3000 psi produces a force of 37,699 lb in a 4-in. cylinder. The required working pressure is again given by Equation (11.2):

$$WP = (3000)\left(\frac{12,500}{37,699}\right) = 995 \text{ psi}$$

From Table 11.2, a flow rate of 1 gal/min produces a speed of 18 in./min in a 4-in. cylinder. The required flow rate is given by Equation (11.4):

$$Q = (1)\left(\frac{40}{18}\right) = 2.2 \text{ gal/min}$$

For a 5-in.-diameter cylinder the required working pressure is

$$WP = (3000)\left(\frac{12,500}{58,905}\right) = 637 \text{ psi}$$

The required flow rate is

$$Q = (1)\left(\frac{40}{12}\right) = 3.3 \text{ gal/min}$$

Conclusion: The 4-in.-cylinder is selected for this application. The required flow rate of 2.2 gal/min can be provided by the surplus pump. The working pressure of 995 psi is within the 1000-psi limit, and the 4-in.-diameter cylinder fits in the available space.

11.4   CONTROL VALVES

In many process control systems, the manipulated variable is the flow rate of a fluid, and the pneumatic *control valve* is the most common final control element. A typical pneumatic control valve is illustrated in Figure 11.24. The input to the valve is a

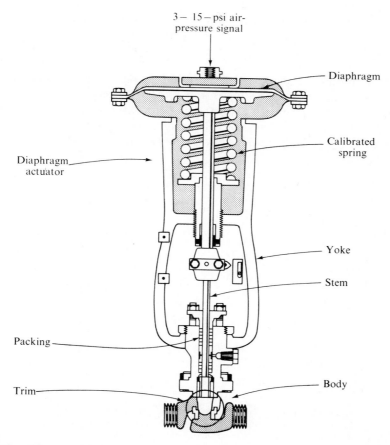

Figure 11.24   Pneumatic control valve.

3- to 15-psi air pressure signal, which is applied to the top of the diaphragm. The diaphragm actuator converts the air pressure into a displacement of the valve stem. The valve body and trim varies the area through which the flowing fluid must pass.

The input air pressure signal may come directly from a pneumatic controller; it may come from a pneumatic valve positioner; or it may come from an electropneumatic transducer. A valve positioner is a pneumatic amplifier that provides additional power to operate the valve. The electropneumatic transducer converts the milliampere signal from an electronic controller into the 3- to 15-psi air pressure signal required by the control valve.

The diaphragm actuator consists of a synthetic rubber diaphragm and a calibrated spring. The air pressure on the top of the diaphragm is opposed by the spring. Full travel of the valve stem is obtained with a change in the air pressure signal from 3 to 15 psi. The graph of the valve stem displacement versus the air pressure signal is within a few percent of a straight line. The actuator may be described as a *direct actuator* or a *reverse actuator*, depending on the location of the calibrated spring. The two types of actuator are illustrated in Figure 11.25.

A yoke attaches the diaphragm actuator to the valve body. The valve stem passes through a packing, which allows motion to the valve stem and prevents leakage of the fluid flowing around the stem. The trim consists of the valve plug and the valve seat or port. These two parts are often produced in matched pairs to provide tight shutoff.

The valve position (P) is the displacement of the plug from the fully seated position. The valve action is the direction in which the valve plug moves as the air pressure signal increases. If increasing air pressure closes the valve, the valve action is described as air-to-close. If increasing air pressure opens the valve, the action is described as air-to-open. The variations of valve action and actuator type are illustrated in Figure 11.26.

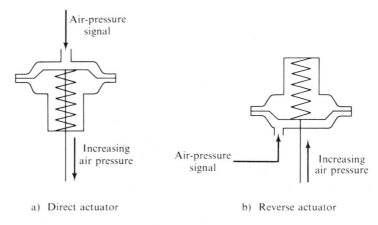

a) Direct actuator   b) Reverse actuator

Figure 11.25   Direct- and reverse-action valve actuators.

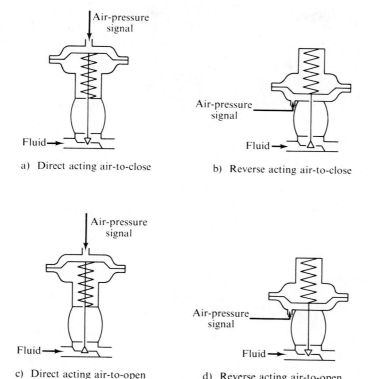

a) Direct acting air-to-close b) Reverse acting air-to-close

c) Direct acting air-to-open d) Reverse acting air-to-open

Figure 11.26 Four combinations of valve action (air-to-open or air-to-close) and actuator type (direct or reverse).

Valve Characteristics

The characteristic of a control valve is the relationship between the valve position and the flow rate through the valve. The *inherent characteristic* of a valve is obtained when there is a constant pressure drop across the valve for all valve positions. The *installed characteristics* are obtained when the valve is installed in a system and the pressure drop across the valve depends on the pressure drop in the remainder of the system. Thus the installed characteristics depend on both the inherent characteristics of the valve and the flow characteristics of the system in which the valve is installed.

The three most common valve characteristics—quick opening, linear, and equal percentage—are illustrated in Figure 11.27.

The *quick-opening characteristic* provides a large change in flow rate for a small change in valve position. This characteristic is used for on–off or two-position control systems in which the valve must move quickly from open to closed, or vice versa. A typical quick-opening characteristic is shown in Figure 11.27.

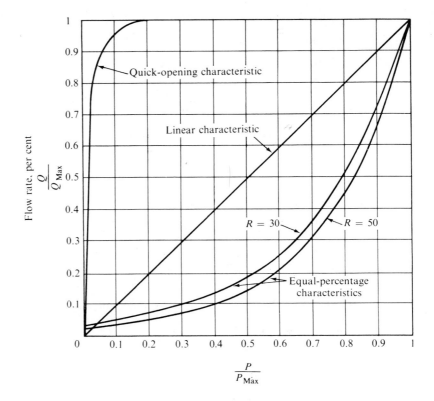

Figure 11.27   Three types of inherent characteristics for control valves.

A *linear characteristic* provides a linear relationship between the valve position (P) and the flow rate (Q). The linear characteristic is described by the following mathematical relationship.

$$\frac{Q}{Q_{max}} = \frac{P}{P_{max}} \tag{11.5}$$

where Q = valve flow rate

P = valve position (displacement of the plug from the closed position)

P_{max} = maximum valve position

Q_{max} = maximum flow rate

The flow rate and valve position are usually given in the English units of gallons per minute and inches. However, any convenient units may be used. The values of

Q_{max} and P_{max} are usually included in the valve manufacturer's literature. The linear characteristic is usually used in installed systems in which most of the system pressure drop is across the valve.

An *equal-percentage characteristic* provides equal percentage changes in flow rate for equal changes in valve position. An equal-percentage valve is designed to operate between a minimum flow rate (Q_{min}) and a maximum flow rate (Q_{max}). The ratio Q_{max}/Q_{min} is called the *rangeability* of the valve, R.

$$R = \frac{Q_{max}}{Q_{min}} \tag{11.6}$$

Most commercial control valves have a rangeability between 30 and 50. The simplest way to express the equal-percentage characteristic is in terms of the change in flow rate, corresponding to a small change in valve position.

$$\frac{Q_2 - Q_1}{Q_1} = K \frac{P_2 - P_1}{P_{max}} \tag{11.7}$$

where Q_1 = initial flow rate

P_1 = initial valve position

Q_2 = final flow rate

P_2 = final valve position

P_{max} = maximum valve position

The value of the constant K depends on the rangeability and the relative size of the change in valve position, $(P_2 - P_1)/P_{max}$. For example, if $R = 50$ and $(P_2 - P_1)/P_{max} = 0.01 = 1\%$, then $K = 4$. In other words, a 1% change in valve position will produce a 4% change in flow rate. If the flow rate is 3 gal/min when the valve is 10% open, a change in valve position to 11% will increase the flow rate by 4% of 3, or 0.12 gal/min. When the valve is 50% open, the flow rate will be 14.2 gal/min. A change in valve position to 51% open will increase the flow rate by 4% of 14.2, or 0.57 gal/min.

The exact mathematical relationship for the equal percentage valve is given by the equation

$$\frac{Q}{Q_{max}} = \left(\frac{Q_{max}}{Q_{min}}\right)^{\frac{P - P_{max}}{P_{max}}} = R^{\frac{P - P_{max}}{P_{max}}} \tag{11.8}$$

where Q = valve flow rate

Q_{max} = maximum flow rate

Q_{min} = minimum flow rate

R = rangeability of the valve

P = valve position

P_{max} = maximum valve position

Any convenient units of flow rate and position may be used in Equation (11.8).

The *installed characteristics* of a valve depend on how much of the total system pressure drop is across the valve. If 100% of the system pressure drop is across the valve, the installed characteristic is the same as the inherent characteristic. The installed characteristics of linear and equal-percentage valves are illustrated in Figure 11.28. As the percentage pressure drop across the valve decreases, the installed characteristic of the linear valve approaches the quick-opening characteristic, and the installed characteristic of the equal-percentage valve approaches the linear characteristic.

The proper selection of a control valve involves matching the valve characteristics to the characteristics of the process. When this is done, the control valve will contribute to the stability of the control system. Matching the valve characteristics to a particular system requires a complete dynamic analysis of the system. When a complete dynamic analysis is not justified, an equal-percentage characteristic is usually

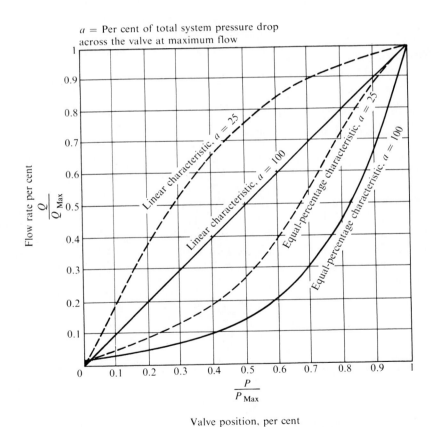

Figure 11.28 As the percent of total system pressure drop across the valve at maximum flow (a) decreases, the linear valve approaches a quick-opening characteristic, and the equal-percentage valve approaches a linear characteristic.

specified if less than half of the system pressure drop is across the control valve. If most of the system pressure drop is across the valve, a linear characteristic may be preferred.

Control Valve Sizing

Control valve sizing refers to the engineering procedure of determining the correct size of a control valve for a specific installation. A capacity index called the *valve flow coefficient* (C_v) has greatly reduced the difficulty in "sizing" a control valve. The valve flow coefficient is defined as the number of U.S. gallons of water per minute that will flow through a wide-open valve with a pressure drop of 1 psi. For example, a control valve with a C_v of 5 will pass 5 gal/min of water when the valve is wide open and the pressure drop is 1 psi.

The valve-sizing formulas (C_v formulas) are based on the basic equation of liquid flow:

$$Q = K\sqrt{\text{pressure drop}}$$

The following formulas may be used to determine the maximum flow rate for different flowing fluids. The approximate C_v values for standard valve sizes are given in Table 11.3.

C_v FORMULAS

$$\text{Liquids: } Q_L = C_v\sqrt{\frac{P_1 - P_2}{G_L}} \qquad (11.9)$$

$$\text{Gases: } Q_g = 960C_v\sqrt{\frac{(P_1 - P_2)(P_1 + P_2)}{G_g(T + 460)}} \qquad (11.10)$$

$$\text{Steam: } W = 90C_v\sqrt{\frac{P_1 - P_2}{V_1 + V_2}} \qquad (11.11)$$

where C_v = valve flow coefficient

G_g = gas specific gravity (density of the gas divided by the density of air with both gases at standard conditions)

G_L = liquid specific gravity (density of the liquid divided by the density of water)

W = steam flow, pound/hour

P_1 = valve inlet pressure, psia

P_2 = valve outlet pressure, psia

Q_g = gas flow rate, cubic feet per hour at 14.7 psia and 60°F

Q_L = liquid flow rate, gallon/minute

T = gas temperature, °F

V_1 = specific volume of the steam at the valve inlet, cubic foot/pound

V_2 = specific volume of the steam at the valve outlet, cubic foot/pound

---

Table 11.3 Approximate C_v Values for Common Valve Sizes

Valve Size	C_v	Valve Size	C_v
0.25	0.3	3	108
0.5	3	4	174
1	14	6	400
1.5	35	8	725
2	55	10	1100

Example 11.3

Determine the size of the control valve required to control the flow rate of a liquid with a specific gravity of 0.92. The maximum flow rate is 320 gal/min. The available pressure drop across the valve at maximum flow is 60 psi. Allow a safety factor of 25% (i.e., $Q_L = 1.25Q_{max}$).

Solution

Equation (11.9) may be used:

$$Q_L = C_v \sqrt{\frac{P_1 - P_2}{G_L}}$$

$$(1.25)(320) = C_v \sqrt{\frac{60}{0.92}}$$

$$400 = 8.08 C_v$$

$$C_v = 49.5 \text{ minimum}$$

Conclusion: A 2-in. valve is required (Table 11.3).

Example 11.4

Determine the size of the control valve required to control flow rate of a gas with a specific gravity of 0.85. The maximum flow rate is 20,000 standard cubic feet per hour (14.7 psia and 60°F). The gas temperature, inlet pressure, and outlet pressure are 500°F, 90 psia, and 70 psia, respectively. Allow a safety factor of 25%.

Solution

Equation (11.10) may be used.

$$Q_g = 960 C_v \sqrt{\frac{(P_1 - P_2)(P_1 + P_2)}{G_g(T + 460)}}$$

$$(1.25)(20,000) = 960 C_v \sqrt{\frac{(20)(160)}{(0.85)(960)}}$$

$$25,000 = 1901 C_v$$

$$C_v = 13.15$$

Conclusion: A 1-in. valve is required (Table 11.3).

Example 11.5

Determine the size of the control valve required to control the flow rate of steam to a process. The maximum flow rate is 600 lb/hr. The inlet and outlet steam conditions are given below.

$$P_1 = 40 \text{ psia}$$
$$V_1 = 10.5 \text{ ft}^3/\text{lb}$$
$$P_2 = 30 \text{ psia}$$
$$V_2 = 14.0 \text{ ft}^3/\text{lb}$$

Use a safety factor of 25%.

Solution

Equation (11.11) may be used.

$$W = 90C_v \sqrt{\frac{P_1 - P_2}{V_1 + V_2}}$$

$$(1.25)(600) = 90C_v \sqrt{\frac{10}{24.5}}$$

$$C_v = 13.0$$

Conclusion: A 1-in. valve is required (Table 11.3).

Control Valve Transfer Function

A pneumatic control valve is essentially a second-order system, usually underdamped or critically damped. The resonant frequency (ω_0) is usually between 1 and 10 rad/s. The transfer function is given by Equation (11.12).

CONTROL VALVE TRANSFER FUNCTION*

$$\frac{P}{I} = \frac{G}{1 + (2\zeta/\omega_0)s + (1/\omega_0^2)s^2}$$
(11.12)

where G = valve gain, inch/psi

I = input air pressure signal, psi

P = valve position, inch

s = frequency parameter, 1/second

ω_0 = resonant frequency, radian/second

ζ = damping ratio

---

* The second-order model presented here is adequate for small-signal changes. For large-signal changes, the model must include a limitation on the valve speed similar to the slew-rate limitation on an op amp.

11.5 ELECTRIC HEATING ELEMENTS

Electric heating elements convert electrical energy into heat energy. As energy converters, electric heaters are 100% efficient. All the electrical energy is converted into heat. However, some of the heat energy is lost to the surrounding environment, resulting in less than 100% efficiency of heat energy delivered to the product.

Electric heaters come in a variety of shapes and sizes to satisfy different heating requirements. The burner on an electric stove is an example of an electric heating element. Industrial heaters come in the following forms: cartridge, band, strip, cable, radiant, flexible, ceramic fiber, and various process heater assemblies. We will describe the first two forms in some detail. Further information can be obtained from manufacturers' catalogs.

A *cartridge heater* consists of a nickel-chromium wire wound on a supporting core and enclosed in a metal tube sheath (see Figure 11.29). The outside diameter of the tube ranges from 0.125 to 1 in. The length of the tube ranges from 1.25 to 12 in.

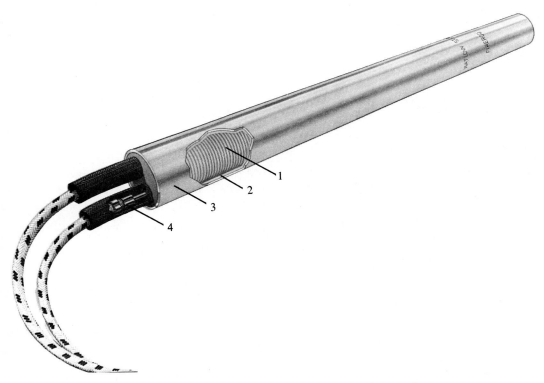

Figure 11.29   A cartridge heater has four main parts: (1) a coil of resistance wire wound on a supporting core; (2) magnesium oxide insulation that combines high dielectric strength with efficient, fast heat transfer; (3) a corrosion-resistant sheath; and (4) electrical leads. [From *Everything You Ever Wanted to Know About Electric Heaters and Control Systems* (St. Louis, Mo.: Watlow Electric Mfg. Co.), p. 1.]

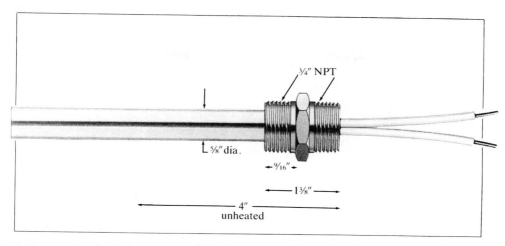

Figure 11.30 Electric immersion heaters are inserted into the heated fluid through a threaded hole in the wall of the container. Elements with a flange mounting or "over the side" mounting are also available. [From *Everything You Ever Wanted to Know About Electric Heaters and Control Systems* (St. Louis, Mo.: Watlow Electric Mfg. Co.), p. 13.]

for the 0.125-in.-diameter element. For the 1-in.-diameter element, the length ranges from 1.25 to 72 in. Typical wattage ratings are 25 to 50 W for the 0.125-in.-diameter by 1.25-in.-long element. A 1-in.-diameter by 36-in.-long element has a rating of 2500 W.

Cartridge heaters are immersed in the object to be heated. When used to heat solid objects, the cartridge is inserted into a hole in the object. The fit should be as tight as possible while allowing easy insertion and removal of the heating element. When used to heat liquids, the cartridge is immersed in the liquid to be heated. Immersion elements may be fitted with a threaded plug or a flange for mounting through the wall of the containing vessel (see Figure 11.30). Stock immersion heaters have a diameter of 0.625 in. Sizes vary from a 500-W unit (6.25-in. overall length) to an 18,000-W unit (35-in. overall length).

Band heaters are used to heat round objects or liquid product in round containers such as pipes, extruder barrels, drums, cylinders, and so on. The band heater wraps around the object to be heated much like a belt (see Figure 11.31). Stock band heaters range from a 100-W unit [1 in. inside diameter (ID) by 1 in. wide] to a 2300-W unit (12 in. ID by 2 in. wide). Units are available with diameters ranging from 0.875 to 44 in. and widths ranging from 0.625 to 15 in.

The first step in selecting an electrical heater is to determine the heat flow rate required to do the job. The heat energy requirements fall into one of the following three categories:

1. Heat required to raise the temperature of the product and any surrounding objects

Figure 11.31   Band heaters consist of five parts: (1) a clamping strap, (2) a high-temperature heating ribbon, (3) mica insulation, (4) a rust-resistant steel sheath, and (5) post terminals for electrical connections. [From *Everything You Ever Wanted to Know About Electric Heaters and Control Systems* (St. Louis, Mo.: Watlow Electric Mfg. Co.), p. 35.]

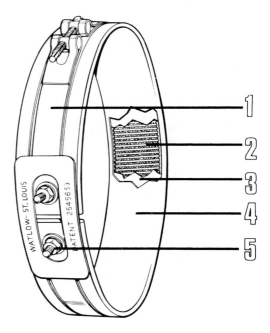

2. Heat required to change the state of the product (i.e., change the product from a solid to a liquid, or from a liquid to a gas)
3. Heat losses by conduction, convection, and radiation

The second step is to examine other considerations that will influence the selection. Examples include the following:

1. The operating environment (e.g., the size and shape of the container, temperature, humidity, etc.)
2. Life requirements
3. Operating costs
4. Safety factor

The third step is to select the type, size, and number of heating elements.

The heat flow rate required to raise the temperature of an object is determined by the thermal capacitance of the object and the elapsed time for the temperature rise. In this discussion, an amount of liquid product will be treated as an "object" to be heated. Thermal capacitance was defined in Chapter 4 as the amount of heat required to increase the temperature by 1°C. Thus the heat flow rate required to raise the temperature of an object is given by the following equation:

$$Q_1 = \frac{C_T(T_{\text{final}} - T_{\text{initial}})}{t_e} \tag{11.13}$$

$$C_T = mS_h = \rho V S_h \tag{4.39}$$

where Q_1 = heat flow rate required to raise the temperature, watt
  C_T = thermal capacitance, joule/kelvin
  T_{final} = final temperature, °C or kelvin
  $T_{initial}$ = initial temperature, °C or kelvin
  t_e = elapsed time, second
  m = mass of the object, kilogram
  ρ = density of the object, kilogram/meter^3
  S_h = heat capacity, joule/kilogram kelvin
  V = volume of the object, meter^3

The amount of heat required to convert 1 kilogram of a solid into a liquid is called the latent heat of fusion (H_f). Appendix A lists the melting point and the latent heat of fusion for several substances. The amount of heat required to convert 1 kilogram of a liquid into a gas is called the latent heat of vaporization (H_v). Water boils at 100°C and the latent heat of vaporization is 2.26 MJ/kg. The heat flow rate required to change the state of a given amount of product is given by the following equations:

$$Q_2 = \frac{mH_f}{t_e} \tag{11.14}$$

$$Q_3 = \frac{mH_v}{t_e} \tag{11.15}$$

where Q_2 = heat flow rate required to melt the product, watt
  Q_3 = heat flow rate required to vaporize the product, watt
  m = mass of the product, kilogram
  H_f = heat of fusion of the product, joule/kilogram
  H_v = heat of vaporization of the product, joule/kilogram
  t_e = elapsed time, second

The heat loss by conduction is determined by the program "THERMRES" of Chapter 4. Exact calculations of convection and radiation losses are quite difficult. The simplest approach is to use a combined surface loss factor that gives the heat loss for a unit area of surface at a given temprature. The heat loss rate is equal to the product of the surface area and the combined surface loss factor.*

* For further details, refer to *Everything You Ever Wanted to Know About Electric Heaters and Control Systems*, by Watlow Electric Mfg. Co., St. Louis, MO.

Example 11.6

Oil is to be heated from 10°C to 55°C and pumped at a rate of 2 L/min. Determine the theoretical heat flow rate required to heat the oil.

Solution

Equations (11.13) and (4.39) apply. An elapsed time of 1 min or 60 s will be used for the calculation.

$$t_e = 60 \text{ s}$$

The following properties of oil were obtained from Appendix A.

$$\rho = 880 \text{ kg/m}^3$$
$$S_h = 2180 \text{ J/kg·K}$$

The mass of oil that must be heated in 1 minute is

$$m = \left(\frac{2}{1000}\right)(880) = 1.76 \text{ kg}$$

The thermal capacitance of this amount of oil is

$$C_T = mS_h = (1.76)(2180) = 3837 \text{ J/K}$$

The heat flow rate required to heat the oil is

$$Q_1 = \frac{C_T(T_{\text{final}} - T_{\text{initial}})}{t_e}$$
$$= \frac{(3837)(55 - 10)}{60}$$
$$= 2878 \text{ W}$$

Example 11.7

A tank of paraffin is to be used for coating parts. The tank contains 85 kg of paraffin, which is at room temperature before the start of a production run. Determine the heat flow rate required to melt the paraffin and raise the temperature from room temperature (20°C) to the dipping temperature of 65°C. The desired heating time is 90 min. Assume the same value of heat capacity for solid and liquid paraffin.

Solution

Equations (11.13), (4.39), and (11.14) apply.

$$t_e = (90)(60) = 5400 \text{ s}$$

The following properties of paraffin were obtained from Appendix A.

$$\rho = 897 \text{ kg/m}^3$$
$$S_h = 2931 \text{ J/kg·K}$$
$$H_f = 147 \text{ kJ/kg}$$

The thermal capacitance of 85 kg of paraffin is

$$C_T = mS_h = (85)(2931) = 2.49 \times 10^5 \text{ J/K}$$

The heat flow rate required to raise the temperature of the paraffin is

$$
\begin{aligned}
Q_1 &= \frac{C_T(T_{\text{final}} - T_{\text{initial}})}{t_e} \\
&= \frac{(2.49\text{E}05)(65 - 20)}{5400} \\
&= 2076 \text{ W}
\end{aligned}
$$

The heat flow rate required to melt the paraffin is

$$
\begin{aligned}
Q_2 &= \frac{mH_f}{t_e} \\
&= \frac{(85)(147\text{E}03)}{5400} \\
&= 2314 \text{ W}
\end{aligned}
$$

The total heat flow required to melt the paraffin and raise the temperature is

$$
\begin{aligned}
Q_T &= Q_1 + Q_2 \\
&= 2076 + 2314 \\
&= 4390 \text{ W}
\end{aligned}
$$

GLOSSARY

Actuator: A control system component that provides the power needed to carry out the control action produced by the controller. (11.3)

Actuator, hydraulic: An actuator that uses an oil-based fluid under high pressure to carry out the control action. Hydraulic actuators are used where slow, precise positioning is required, or where heavy loads must be moved. (11.3)

Actuator, pneumatic: An actuator that uses air under moderate pressure to carry out the control action. Pneumatic actuators are used where relatively light loads are to be moved, where an air supply is available, and where fast response is required. (11.3)

Band heater: An electric heating element that wraps around circular containers such as pipes, extruder barrels, drums, cylinders, and so on. (11.5)

Cartridge heater: An electric heating element that consists of a nickel-chromium wire wound on a supporting core and enclosed in a metal tube sheath. (11.5)

Characteristic, equal-percentage: A control valve characteristic that provides equal percentage changes in flow rate for equal changes in valve position. (11.4)

Characteristic, inherent: The relationship between the position of a control valve and the flow rate through the valve when there is a constant pressure drop across the valve for all positions. (11.4)

Characteristic, installed: The relationship between the position of a control valve and the flow rate through the valve when the valve is installed in a system and the pressure drop across the valve depends on the pressure drop in the remainder of the system. (11.4)

Characteristic, linear: A control valve characteristic that provides a linear relationship between the valve position and the flow rate through the valve. (11.4)

Characteristic, quick-opening: A control valve characteristic that provides a large change in flow rate for a small change in valve position. (11.4)

Contactor: A type of relay that has heavy-duty contacts for switching large amounts of electric power and light-duty contacts for the control circuit. *See also* Motor starter. (11.1)

Control valve: A control system component used to control the flow of a fluid by regulating the size of an opening in the flow passage. A control valve consists of a valve plus an actuator for positioning the valve stem. (11.4)

Control valve sizing: The engineering procedure of determining the correct size of a control valve for a specific installation. (11.4)

Cylinder: An actuator used to produce linear motion with a definite distance of travel. A cylinder consists of a piston and shaft enclosed in a tube that is closed at both ends, except for the shaft which protrudes through one end. The tube has ports through which hydraulic fluid or air can be passed into and out of both ends. The piston is moved by forcing fluid in one side of the tube while allowing it to exit the other side. (11.3)

Cylinder, hydraulic: A cylinder that uses hydraulic fluid to produce a linear motion. (11.3)

Cylinder, pneumatic: A cylinder that uses pressurized air to produce a linear motion. (11.3)

Diode: A two-terminal, solid-state component that allows electric current to flow in one direction, but resists flow in the other direction. (11.2)

Hydraulic motor: An actuator that uses hydraulic fluid under high pressure to produce rotary motion in a load. Types of hydraulic motors include radial piston motors, gear motors and vane motors. (11.3)

Maintained-action switch: A type of switch in which the contacts remain in the position they are placed by a switching action until another switching action changes their position. (11.1)

Momentary-action switch: A type of switch in which the contacts are in the actuated position only while the operator holds the switch actuator in that position. As soon as the operator releases the actuator, the contacts return to their unactuated positions. (11.1)

Motor starter: A type of relay that has heavy-duty contacts for starting electric motors and light-duty contacts for the control circuit. *See also* Contactor. (11.1)

Oversize factor: A number greater than 1 which is used to increase the size of a cylinder to provide the additional force necessary to accelerate the load. An oversize factor of 1.25 is recommended for slow-moving loads and a factor of 2.00 is recommended for fast-moving loads. (11.3)

Pneumatic motor: An actuator that uses pressurized air to produce rotary motion in a load. Types of pneumatic motors include piston motors, vane motors, and turbine motors. (11.3)

Relay: A set of switches, a coil of wire, and supporting members arranged such that the switches are actuated when electric current passes through the coil of wire. (11.1)

Silicon-controlled rectifier (SCR): A three-terminal, solid-state switching component that requires a small amount of control power to control a large amount of electric power. An SCR is both a rectifier and a latching switch that can control currents as much as 3000 times as large as the control current. (11.2)

Solenoid valve: A movable spool fitted inside a housing with one or two solenoids capable of moving the spool to different positions relative to the housing. Both the spool and the housing have fluid passages. The passages in the housing are connected together or blocked in different ways, depending on the position of the spool relative to the housing. (11.3)

Switch: One or more pairs of contacts and supporting members that are used to make or break connections in an electric circuit. (11.1)

Time-delay relay: A relay with a timing mechanism that provides a delay between the time the actuating current is applied to the coil and the time the contacts move to their actuated position. (11.1)

Transistor: A three-terminal, solid-state switching component that is used as a switch, amplifier, and oscillator. A transistor is essentially a controlled current amplifier. (11.2)

Triac: A three-terminal, solid-state switching component that requires a small amount of control power to control a large amount of electric power. A triac is similar to an SCR except that the triac can conduct in both directions, whereas the SCR can conduct in only one direction. (11.2)

Unijunction transistor: A three-terminal, solid-state component used to produce the trigger pulse for controlling SCRs and triacs. (11.2)

Valve flow coefficient (C_v): A capacity index for control valves, which is defined as the number of U.S. gallons of water per minute that will flow through a wide-open valve with a pressure drop of 1 psi. (11.4)

Way: A passage in the housing of a solenoid valve. Each way can be connected to a supply line, a return line, or a line to a cylinder. (11.3)

Working pressure: The actual pressure required in a cylinder to move the load at the required speed. (11.3)

EXERCISES

11.1 A 115-V ac electric motor drives a sump pump. A normally open level switch turns the motor on and off directly (no relay is used). Draw a diagram for a 115-V ac circuit that includes the level switch, the pump motor, an overload heater, and the normally closed (NC) overload contact.

11.2 A 240-V ac three-phase electric motor drives a shallow well pump used for irrigation. The pump is connected to a 100-gal pressure tank. The pressure in the tank turns the pump on and off. When the pressure drops below 20 psi, the pump is turned on and remains on until the pressure reaches 50 psi. When the pressure reaches 50 psi, the pump is turned off and remains off until the pressure drops below 20 psi. A motor starter similar to Figure 11.5 is to be used to control the motor. Two NC pressure switches are to be used in place of the two pushbutton switches. Switch *a* is closed when the pressure is below 20 psi and open when the pressure is above 20 psi. Switch *b* is closed when the pressure is below 50 psi and open when the pressure is above 50 psi. Draw a diagram of the circuit for starting and stopping the pump motor. The 115-V ac control circuit should include the NC overload contact and the holding contact.

11.3 Describe the four ways that an SCR can be turned on (i.e., move from region 2 to region 3).

11.4 Describe the three ways that an SCR can be turned off and indicate which region the SCR is in when it is off.

11.5 Describe the major difference between the triac and the SCR and explain how the triac is equivalent to two SCRs connected in parallel but in opposite directions.

11.6 A half-wave dc power supply uses one diode instead of the two diodes used in a full-wave power supply. Sketch the output voltage waveform of the power supply in Figure 11.12 if diode D_2 is removed from the circuit.

11.7 The following voltage and current measurements were obtained from a common-emitter amplifier similar to Figure 11.14b. Draw voltage and current graphs similar to Figure 11.15 and determine the voltage gain and the current gain.

(1) $V_{in} = 0.70$ V, $V_{out} = 24.2$ V,
     $i_{in} = 0$ μA, $i_{out} = 0.95$ mA

(2) $V_{in} = 0.75$ V, $V_{out} = 20.0$ V,
     $i_{in} = 10.6$ μA, $i_{out} = 0.78$ mA

(3) $V_{in} = 0.80$ V, $V_{out} = 15.8$ V,
     $i_{in} = 21.0$ μA, $i_{out} = 0.61$ mA

(4) $V_{in} = 0.85$ V, $V_{out} = 11.6$ V,
     $i_{in} = 31.5$ μA, $i_{out} = 0.45$ mA

(5) $V_{in} = 0.90$ V, $V_{out} = 7.4$ V,
     $i_{in} = 42.2$ μA, $i_{out} = 0.27$ mA

(6) $V_{in} = 0.95$ V, $V_{out} = 3.4$ V,
     $i_{in} = 52.4$ μA, $i_{out} = 0.12$ mA

(7) $V_{in} = 1.00$ V, $V_{out} = 0$ V,
     $i_{in} = 62.8$ μA, $i_{out} = 0$ mA

11.8 A four-stage transistor amplifier similar to Figure 11.17 has the following voltage and current gains for each stage. Determine the overall voltage gain and current gain.

Stage 1: voltage gain = -8, current gain = -1.4

Stage 2: voltage gain = -10, current gain = -1.2

Stage 3: voltage gain = 0.95, current gain = 110

Stage 4: voltage gain = 0.96, current gain = 150

11.9 A manufacturing operation requires the movement of a workpiece a distance of 10 in. in 30 s. A force of 22,500 lb is required to move the workpiece. The available space limits the cylinder diameter to a maximum of 8 in. Leakage and noise considerations make it desirable to limit the working pressure to a maximum of 1000 psi. A hydraulic pump with a capacity of 2.5 gal/min is available in the surplus equipment yard. Select a cylinder for this application.

11.10 A manufacturing operation requires the movement of a workpiece a distance of 10 in. in 0.6 s. A force of 125 lb is required to move the workpiece. The available space limits the cylinder diameter to a maximum of 4 in. Heavy usage of the air supply sometimes limits the supply pressure to 80 psi. Select a cylinder for this application.

11.11 An equal-percentage valve has a rangeability of 20. Use Equation (11.4) to calculate the value of Q/Q_{max} for values P/P_{max} from 0 to 1 in increments of 0.1.

11.12 Determine the size of the control valve required to control the flow rate of water. The maximum flow rate is 850 gal/min, and the available pressure drop across the valve is 50 psi. Allow a safety factor of at least 30%.

11.13 Determine the size of the control valve required to control the flow rate of a gas with a specific gravity of 0.85. A maximum flow rate of 2500 standard cubic feet per hour is required. The gas temperature is 420°F, the inlet pressure is 40 psia, and the outlet pressure is 20 psia. Allow a minimum safety factor of 25%.

11.14 Determine the size of a control valve required to control the flow rate of steam to a process. The maximum flow rate is 320 lb/h. The inlet and outlet steam conditions are given below. Use a safety factor of 30%.

$$P_1 = 60 \text{ psia}$$
$$V_1 = 7.175 \text{ ft}^3/\text{lb}$$
$$P_2 = 40 \text{ psia}$$
$$V_2 = 10.72 \text{ ft}^3/\text{lb}$$

11.15 Water is to be heated from 5°C to 75°C and pumped at a rate of 4 L/min. Determine the theoretical heat flow rate required to heat the water.

11.16 A wave-soldering process uses a tank of 50–50 solder. The tank is 60 cm wide by 80 cm long, and the solder depth is 10 cm. The solder is at room temperature before the start of a production run. Determine the heat flow rate required to melt the solder and raise the temperature from room temperature (20°C) to the operating temperature of 250°C. The desired heating time is 2 hours. Assume the same value of heat capacity for solid and liquid solder.

CHAPTER 12

Electric Motors

OBJECTIVES

Electric motors are frequently used as the manipulative device in control systems. Stepper motors and servomotors provide the "muscle" for robotic arms and numerically controlled machine tools. Adjustable speed drives provide efficient control of pumps and blowers. Electric motors are used to sort mail, package food, make plastic sheet, and cut cereal pellets; to name a few of the host of applications.

The purpose of this chapter is to give you an entry-level ability to discuss, select, and specify various types of electric motors. After completing this chapter, you will be able to

1. Describe the major characteristics of ac induction motors, ac synchronous motors, and ac servomotors

2. Describe the major characteristics of each of the following dc motors: shunt wound, series wound, compound wound, permanent magnet, pancake, and brushless

3. Describe stepper motors and explain full-stepping, half-stepping, and micro-stepping

4. Discuss the applications, strengths, and weaknesses of ac motors, dc motors, and stepper motors

5. Discuss ac adjustable-speed drives—include the following in your discussion: rectifier, inverter, variable frequency, SCR, transistor, VVI, PWM, and sine-coded PWM

6. Discuss dc adjustable-speed drives—include the following in your discussion: power op amp, SCR, and computed velocity feedback

12.1 INTRODUCTION

Electric motors are often used as actuators in control systems. They are used to provide both rotational motion and linear motion. Electric motors move the arm of a robot, position the workpiece in a numerically controlled (NC) machine, pump fluids, move the belt on a conveyor, and drive the blades of a fan—to name a few of their many applications.

Classification of Electric Motors

Electric motors are classified in two ways: by function and by electrical configuration. Classification by function is based on how the motor is used. Examples include servomotors, gear motors, instrument motors, pump motors, and fan motors. Classification by electrical configuration separates motors into two major categories: dc motors and ac motors. A *dc motor* consists of two parts: a rotating cylinder called the *armature*, and two stationary, magnetic poles called *field poles* or simply the *field*. The armature is placed in the magnetic field between the field poles. An *ac motor* also consists of two parts: a rotating cylinder called the *rotor* and a stationary part called the *stator*. The stator is a thick-walled tube that surrounds the rotor. The stator has windings that produce a rotating magnetic field in the space occupied by the rotor. The following outline expands on this classification of electric motors.

Classification of Electric Motors

I. AC motors
   A. Single-phase motors
      1. Induction
         a. Squirrel cage
            (1) Split-phase
            (2) Capacitor-start
            (3) Two-value capacitor
         b. Wound rotor
            (1) Repulsion-start
            (2) Repulsion-induction
      2. Synchronous
         a. DC excited rotor
         b. Nonexcited rotor
            (1) Permanent-magnet rotor
            (2) Reluctance
            (3) Hysteresis
   B. Polyphase motors
      1. Induction
         a. Squirrel cage
         b. Wound rotor
      2. Synchronous
         a. DC excited rotor
         b. Self-excited, reluctance-type rotor

C. Universal motors
II. DC motors
   A. Wound field
      1. Series wound
      2. Shunt wound
      3. Compound wound
   B. Permanent-magnet field
      1. Magnet types
         a. Alnico magnets
         b. Ceramic magnets
         c. Rare earth–cobalt magnets
      2. Permanent-magnet stator, wound armature
         a. Wound, iron-core armature
         b. Printed circuit, nonmagnetic armature
         c. Moving coil, stationary iron core
      3. Permanent-magnet rotor (brushless)
         a. No brushes
         b. Wound stator
         c. Electronic commutation of stator winding

Force, Torque, and Induced Voltage

The operation of electric motors and generators is based on the following two facts:

1. A force is exerted on a conductor in a magnetic field whenever a current passes through the conductor.
2. A voltage is induced in a conductor whenever it is moved through a magnetic field.

These two facts are illustrated in Figure 12.1.

Figure 12.1 shows an electrical conductor that is rotating about an axis that is located in a magnetic field. The magnetic field is formed by the two magnetic pole pieces labeled "N" and "S" for north and south. The magnetic flux density (*B*) gives the strength of the magnetic field, and the thin lines in Figure 12.1a give the direction of the magnetic field. As the conductor rotates about the axis of rotation, it assumes positions *a*, *b*, *c*, and *d*, as shown in Figure 12.1a. The conductor is part of a loop in an electrical circuit. However, at this time we are concerned only with the portion of the conductor shown in Figure 12.1.

The force exerted on the current-carrying conductor is shown in Figure 12.1b. The direction of the force is perpendicular to both the direction of the current (*i*) and the direction of the magnetic field (*B*). In Figure 12.1b, the direction of the current is toward the reader, as indicated by the dot in each conductor (a plus sign would indicate a current direction away from the reader). The direction of the magnetic field is down, and the force is directed toward the right. If the current were reversed, the direction of the force would also be reversed. The magnitude of the force is given by the following relationship:

$$f = iBL/\sin \alpha \qquad\qquad (12.1)$$

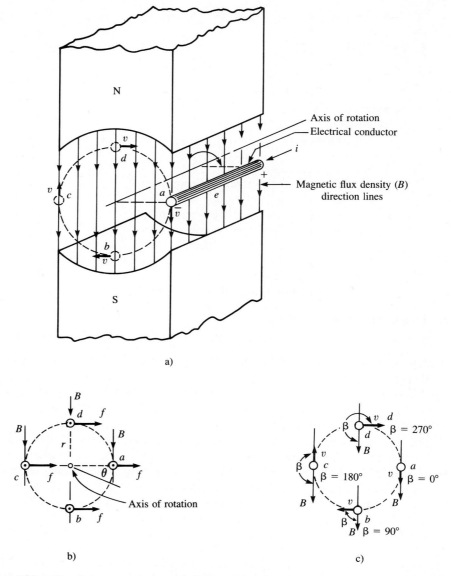

Figure 12.1 Two facts govern the operation of electric motors and generators. (1) A force is exerted on a current-carrying conductor in a transverse magnetic field. (2) A voltage is induced in a conductor moving through a transverse magnetic field.

where f = force, newton

i = current, ampere

L = length of the conductor, meter

B = flux density of the magnetic field, weber/square meter

α = angle between the direction of the current and the direction of the magnetic field, degree

In Figure 12.1, α is always equal to 90°, and the force is given by $f = iBL$.

The voltage induced in the moving conductor is labeled "e" in Figure 12.1a. Notice the polarity of the voltage as indicated by the plus and minus signs. If the conductor and the magnetic field are mutually perpendicular (as they are in Figure 12.1), the induced voltage is given by the following equation:

$$e = LvB \sin \beta \tag{12.2}$$

where e = induced voltage, volt

L = length of the conductor, meter

v = velocity of the conductor, meter/second

B = flux density, weber/square meter

β = angle between the direction of motion of the conductor and the direction of the magnetic field, degree

Figure 12.1c shows the values of β for the four positions labeled a, b, c, and d.

Notice in Figure 12.1b that in positions a and c, the direction of the force is through the axis of rotation, and no rotational torque is developed. However, in positions b or d, the force produces a torque that tends to rotate the conductor. If θ is the angular position of the conductor as shown in Figure 12.1b, the torque produced by force f is given by the following equation:

$$\text{Torque} = T = f_r \sin \theta \tag{12.3}$$

Table 12.1 gives the values of θ, $\sin \theta$, and torque (T) for the four positions of the conductor in Figure 12.1b.

The Motor-Generator Action

A simple dc motor/generator will be used to explain the operation of electric motors and generators. Figure 12.2 illustrates the essential parts of a dc machine that can operate as a dc motor or as a dc generator. Four coils of wire are mounted in slots on a cylinder of magnetic material called the *armature*. The armature is mounted on

Table 12.1 Values of θ, $\sin \theta$, and T (Figure 12.1b)

Position	θ (deg)	$\sin \theta$	Torque, T
a	0	0	0
b	90	1	f_r
c	180	0	0
d	270	−1	$-f_r$

Figure 12.2 Schematic diagram of a two-pole four-coil dc motor/generator.

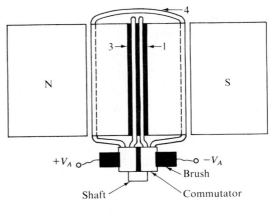

a) Top view

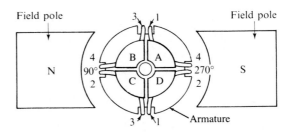

b) End view (brushes not shown)

bearings and is free to rotate in the magnetic field produced by the two *field poles*. The field poles may be permanent magnets or electromagnets, depending on the size of the motor (many smaller motors use permanent magnets). The ends of each coil are connected to adjacent segments of a segmented ring called the *commutator*. Electrical connection is made to the armature coils through carbon contacts called *brushes*.

The operation of the dc machine is based on the two facts presented in the preceding section and reworded as follows.

1. A force is exerted on a current-carrying conductor in a transverse magnetic field.
2. A voltage is induced in a conductor moving through a transverse magnetic field.

Both facts are present in electric motors, and both facts are present in electric generators.

When the dc machine is used as a motor, an external voltage source is connected across the brushes, producing a current in the armature conductors. This current produces a force on the conductors that tends to rotate the armature. Figure 12.3 shows a dc motor with eight conductors. The electrical diagram in Figure 12.3b shows

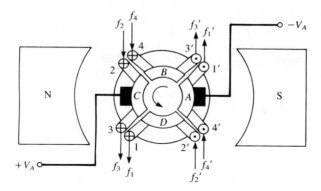

a) End view showing forces on conductors

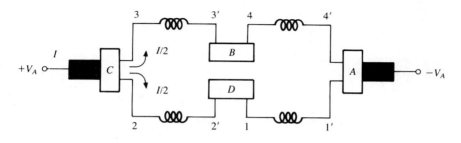

b) Electrical diagram

Figure 12.3   The forces on the conductors (end view) produce a torque that tends to rotate the armature in a counterclockwise direction. In the electrical diagram, commutator segment B connects coils 3
and 4 in series, while segment D connects coils 1 and 2 in series.

a current of $I/2$ in each conductor. This current produces a force on the conductor
which is given by $f = 0.5IBL$ [Equation (12.1) with $\alpha = 90°$]. Figure 12.3a shows the
direction of the force on each conductor. Notice that all eight forces have a moment
arm of $r/\sqrt{2}$ where r is the radius of the armature (note the 45° angle between the
direction of the force and the radius line). The collective torque of the eight forces is
equal to $8(0.5IBL)r/\sqrt{2} = 2.83IBLr$.

Figure 12.3b is a diagram of the electrical circuit for the dc motor as shown in
Figure 12.3a. The armature voltage, V_A, is applied to the armature windings through
the two brushes. The positive terminal of V_A is connected to the brush that is in
contact with commutator segment C. The negative terminal of V_A is connected to the
brush that is in contact with commutator segment A. Voltage V_A produces current I
through the armature windings. This current divides, and $I/2$ amperes passes through
each coil in the direction indicated in Figure 12.3.

As the motor rotates, the commutator reverses the current in each coil as it passes
from a north pole to a south pole, or vice versa. This reversal in the direction of the
current eliminates the reversal in the direction of the force that would otherwise occur.

The torque developed in a dc motor is a direct result of fact 1. Whenever the motor rotates, fact 2 results in the generation of a voltage in the coils as they move through the electric field. The magnitude of the voltage induced in each leg is given by Equation (12.2). The polarity of this induced voltage is opposite the polarity of the applied voltage, V_A. For this reason, the induced voltage is called the *back EMF* of the motor. We will use the symbol E_G for the back EMF. When the motor is in steady-state operation, the following equation applies to the armature circuit.

$$E_A = E_G + IR_A \qquad\qquad (12.4)$$

where E_A = armature voltage, volt

          E_G = voltage induced in the armature, volt

           I = armature current, ampere

        R_A = armature resistance, ohm

At full speed, the back EMF is almost as large as the armature voltage. A typical value of E_G is about $0.94E_A$, meaning that the rated motor current is developed by only $0.06E_A$. The most significant consequence of this fact is the very high current that occurs when the armature is stalled for any reason. A motor with a rated current of 60 A could have a locked rotor current of 1000 A. When this happens, a thermal overload switch opens to protect the motor windings.

When the dc machine in Figure 12.2 is used as a generator, the armature is rotated by an external prime mover. The motion of the coils through the magnetic field induces a voltage in the coil. The magnitude of the induced voltage is given by Equation (12.2). The polarity of the induced voltage reverses each time the coil passes from a north pole to a south pole, or vice versa. In a two-pole generator, the polarity reverses twice during each revolution. The function of the commutator is to reverse the connection between the brushes and the coil at the same time that the polarity of the induced voltage reverses. In other words, the commutator is a rectifier that converts an ac voltage into a dc voltage. In an ac generator, the commutator is replaced by two slip rings. Each slip ring has a brush used to bring out the ac voltage induced in the armature winding.

The generator described above supplies current to a load, and this current passes through the armature coils. The current in the armature winding causes a force to be exerted on the armature coils (fact 1). The magnitude of this force is given by Equation (12.1). This force produces a torque that opposes the motion of the armature. The prime mover must overcome this additional torque to maintain the armature speed.

Commutation

As the armature of a dc motor or generator rotates, the brushes make their way around the commutator, passing from segment to segment. When the brushes move from one segment to the next, both the direction of the current in the coil and the polarity of the induced voltage are reversed. This reversal of current and polarity in the coil is called *commutation*. Figure 12.4 illustrates the commutation of the dc

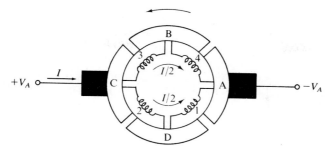

a) Before commutation

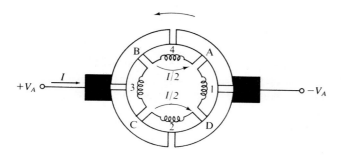

b) During commutation of coils 1 and 3

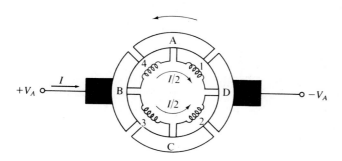

c) After commutation

Figure 12.4 The commutator provides a switching action. During commutation (b), coils 1 and 3 are shorted by the brushes. After commutation (c), the direction of the current in coils 1 and 3 is the reverse of what it was before commutation (a).

motor/generator shown in Figures 12.2 and 12.3. The four armature coils are drawn inside the commutator ring to make it easy to trace the circuit. In Figure 12.4a, the commutator is in the position shown in Figures 12.2 and 12.3. The circuit is essentially the same as the circuit shown in Figure 12.3b. In Figure 12.4b, the armature has rotated 45°, and coils 1 and 3 are undergoing commutation. Notice that coils 1 and 3 are shorted by the brushes in Figure 12.4b. The reversal of the current in coils 1 and 3 can be observed by comparing the direction before commutation (Figure 12.4a) with the direction after commutation (Figure 12.4c).

12.2 AC MOTORS

AC motors dominate the applications of electric motors that require a single operating speed. Single-phase motors are used for low-horsepower applications (from fractional horsepower up to about 20 hp). Polyphase motors overlap the low-horsepower range of single-phase motors and extend the range to much higher horsepower ratings.

DC motors dominate the variable-speed applications of electric motors. However, improvements in solid-state switching components and the application of large-scale integrated-circuit techniques to the complex ac drive circuits have made variable-speed ac drives more competitive with the "old standby" dc drives.

AC motors can be divided into two groups in two different ways, making four major categories of ac motors. One division is between single-phase and polyphase motors. The other division is between synchronous and induction motors. The four categories are: single-phase induction motors, polyphase induction motors, single-phase synchronous motors, and polyphase synchronous motors.

Single-Phase Induction Motors

The single-phase squirrel-cage induction motor is the most common type of motor. Squirrel-cage induction motors have no brushes to generate sparks or wear out. They are very reliable, have a low initial cost, and also have a low maintenance cost. An *induction motor* consists of a *stator* with one or two windings and a *rotor* that contains the current-carrying conductors upon which the force is exerted. The rotor winding of a squirrel-cage motor consists of copper or aluminum bars that fit into slots in the rotor. The bars are connected at each end by a closed continuous ring. The assembly of conductor bars and end rings resembles a squirrel cage and gives the motor its name.

Only the stator windings of an induction motor are externally excited. The winding on the rotor is shorted and receives its energy by electromagnetic induction. The induction motor is a rotating transformer in which the stator winding is the primary and the rotor winding is the secondary.

The magnetic field produced by the stator winding does not remain fixed as it does for a dc motor. Instead, the poles of the magnetic field alternate between two positions. In effect, the magnetic field is rotating about the rotor axis. For this reason, it is called a *rotating field*. The speed at which the field of an induction motor rotates is called the *synchronous speed.* If there could be zero torque on the rotor, the induction motor would run at its synchronous speed. As the torque on the rotor increases, the motor speed decreases. The difference between the synchronous speed and the actual rotor speed is called *slip.* Even at no load there is some slip due to the torque required to overcome friction and wind resistance. For a motor with a two-pole stator, the synchronous speed is equal to the line frequency, 60 revolutions per second or 3600 revolutions per minute. For a four-pole motor, the synchronous speed is equal to one-half the line frequency, 1800 rpm.

The basic single-phase induction motor has one stator winding called the *main winding.* This type of motor is not self-starting. When the rotor is stationary, an equal torque is produced in each direction. Therefore, the net torque is zero, and the rotor remains stationary. If the rotor is started by some starting device, the motor will continue to run in the direction it was started, even if the starting device is removed. A second stator winding called the *start winding* is the most common method of starting a single-phase induction motor. The start winding is rotated 90° from the main winding as shown in Figure 12.5. A speed-sensitive switch disconnects all or part of the start circuit when the rotor reaches a preset speed. The most common methods of using a start winding are split-phase, capacitor-start, and two-capacitor.

The *split-phase motor* has only a speed switch in series with the start winding. When the motor reaches a preset speed, the switch opens, disconnecting the start winding. The motor continues to run with only the main winding excited. Figure 12.5 illustrates a split-phase motor.

The *capacitor-start motor* has a capacitor and a speed switch in series with the start winding. The speed switch disconnects the start winding and the capacitor once the motor has started. The capacitor-start motor has a larger starting torque than the split-phase motor.

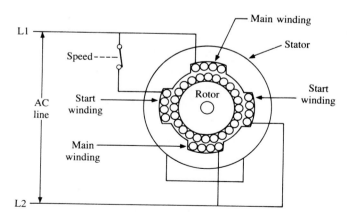

Figure 12.5 A split-phase induction motor uses a start winding to start the motor. When the rotor reaches a set speed, the speed switch opens and the motor runs with only the main winding energized by the ac line.

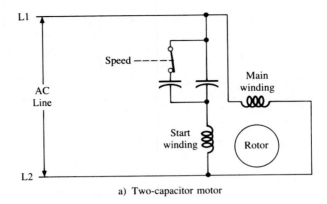

a) Two-capacitor motor

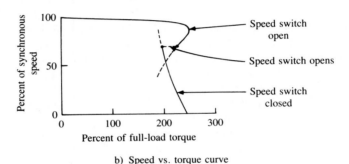

b) Speed vs. torque curve

Figure 12.6 A two-capacitor motor uses both capacitors to start the motor. When the rotor reaches about 70% of the operating speed, one of the capacitors is removed from the circuit. The other capacitor and the start winding remain in effect during normal operation of the motor.

The *two-capacitor motor* has two capacitors and a speed switch connected in a series/parallel combination as shown in Figure 12.6a. The start winding and one capacitor remain in the circuit after the motor has started. The speed switch removes the other capacitor from the circuit when the motor has started. The two-capacitor motor has a high starting torque and a low run current. Also, the capacitor improves the power factor, a definite advantage because induction motors are noted for their low power factor. A typical torque versus speed graph of a two-capacitor motor is shown in Figure 12.6b.

Polyphase Induction Motors

Most polyphase motors are three-phase motors. In this section we describe three-phase induction motors. There are two types of induction motors, the *squirrel-cage motor* and the *wound-rotor motor*. The rotor of a squirrel cage motor has the same conducting bars and end rings described for the single-phase induction motor. In a wound-rotor motor the conducting bars are connected in series to form three windings. The windings are joined on one end in a wye connection. The other each of each winding is connected to a slip ring, one slip ring for each winding. Three brushes are in contact with the three slip rings so that external resistances can be connected in series with the windings. The other ends of the resistors are also connected together in a wye connection. The purpose of the external resistors is to limit the rotor current during startup of loads with large inertia. When the external resistances are reduced to zero, the wound rotor motor is the same as a squirrel-cage motor.

A two-pole three-phase induction motor is shown in Figure 12.7. The stator has three windings that can be connected in either a delta connection or a wye connection. The three windings are located 120° apart and are connected to the three lines

Figure 12.7 Two-pole three-phase induction motor. The rotor windings may be connected to the three-phase ac supply in either a wye or a delta configuration.

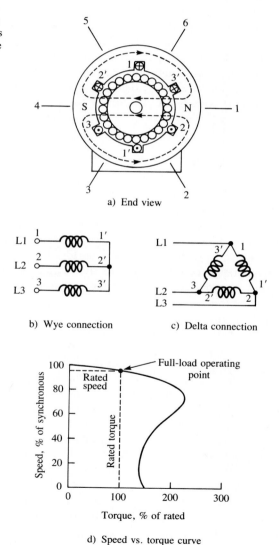

a) End view

b) Wye connection

c) Delta connection

d) Speed vs. torque curve

of a three-phase source. Figure 12.7 shows the position of the magnetic field and the location of the two poles at the beginning of a cycle of one of the three lines. During each sixth of a cycle, the poles move to a new location by rotating 60° in the clockwise direction. In one full cycle, the poles will have completed a 360° rotation. During that time, the north pole will have been in the six positions numbered 1 through 6 in Figure 12.7. The synchronous speed of the two-pole motor is 3600 rpm.

The rotation of the field poles in six steps is much smoother than the two-step rotation of the single-phase motor. One major advantage is that *three-phase induction motors are self-starting*. They do not need an auxiliary starting device. The direction of rotation is reversed by simply interchanging any two line connections. Interchanging two line connections also changes the phase sequence of the motor.

A four-pole three-phase motor has six windings, located at 60° intervals. Each coil uses stator slots that are 90° apart. The coils are paired with the coil on the opposite side of the stator. The paired coils are connected in series to form three branches. Each branch has two coils connected in series. The three branches are connected to a three-phase line in either a delta or a wye configuration. The resulting magnetic field forms four poles, with like poles located 180° apart. The four poles rotate in six steps, in the same manner as the two poles. The synchronous speed of the four-pole motor is 1800 rpm.

Synchronous Motors

Synchronous motors normally run at synchronous speed. They are used where precise constant speed is required. Synchronous motors are made in all size ranges from fractional horsepower to several thousand horsepower. They are available in both single phase (smaller sizes) and polyphase (larger sizes). The stator and stator windings of a synchronous motor are almost identical to the stator windings of the corresponding single-phase or polyphase induction motor. The rotor of a synchronous motor has fixed magnetic poles that lock into step with the rotating poles in the stator. The rotor may be nonexcited or direct-current excited, which gives us another way to classify synchronous motors.

Synchronous motors are not self-starting. This is true for both single-phase and polyphase synchronous motors. Two methods are used to start synchronous motors. One method uses a separate prime mover to start the synchronous motor and accelerate it to nearly synchronous speed. When the rotor locks into synchronous speed, the prime mover is removed. This method is sometimes called *prime mover starting*. The other method of starting a synchronous motor is called *induction motor starting*. A squirrel-cage winding is added to the rotor. The motor starts as an induction motor and accelerates to near-synchronous speed. When the rotor locks into synchronous speed, the current in the squirrel-cage conductor drops to zero, and the motor operates as a synchronous motor. The single-phase induction-start motors use one of the starting devices for single-phase induction motors, such as split-phase, capacitor-start, shaded-pole, and so on.

The *dc excited synchronous motor* has a winding for each pole on the rotor. (Synchronous motors may have two, four, six, or more poles.) The individual field pole windings are connected in series to form one large winding that is terminated in two slip rings. The winding is excited by an external dc source through brushes that contact the slip rings. The dc excited motor is sometimes referred to as the "true synchronous motor." Large polyphase synchronous motors are dc excited motors.

   Nonexcited synchronous motors include reluctance motors, hysteresis motors, and permanent-magnet motors. Reluctance-type synchronous motors use special construction to produce a variable reluctance in the rotor. This enables the rotor to establish the fixed poles without external excitation. Hysteresis motors develop fixed poles as the motor reaches synchronous speed. The shaded-pole hysteresis motor has wide application for clocks and timing devices. *Permanent-magnet motors* have permanent magnets embedded in a squirrel-cage rotor. This motor has become very popular because of its brushless construction, high power factor, and good efficiency.

AC Servomotors

AC servomotors are often used in control systems that require a low-power, variable-speed drive. The primary advantage of the ac motor over the dc motor is its ability to use the ac output of synchros, LVDTs, and other ac measuring means without demodulation of the error signal. An ac amplifier provides the gain for a proportional control mode. However, more elaborate control modes are difficult to implement with an ac signal. When additional control actions are required, the ac signal is usually demodulated, and the control action is inserted in the dc signal. The modified dc signal is then reconverted to an ac signal.

   The *ac servomotor* is a two-phase, reversible induction motor with special modifications for servo operation. The schematic diagram of an ac servo is shown in Figure 12.8. The motor consists of an induction rotor and two field coils located 90° apart. One field coil serves as a fixed reference field, the other as the control field.

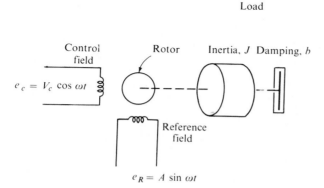

$$e_c = V_c \cos \omega t$$

$$e_R = A \sin \omega t$$

Figure 12.8   An ac servomotor is a two-phase reversible induction motor with special modifications for servo operation.

The amplified ac error signal is applied to the control field. The signal has a variable magnitude with a phase angle of either 0° or 180°. A constant ac voltage is applied to the reference field through a 90° phase-shift network. This signal has a constant magnitude and a phase angle of −90°. The two voltages are given below.

$$e_c = V_c \cos \omega t$$
$$e_R = A \cos(\omega t - 90°) = A \sin \omega t$$

where e_c = control field voltage

   e_R = reference field voltage

   ω = operating frequency

   V_c = variable amplitude of the control voltage

   A = constant amplitude of the reference voltage

The sign and magnitude of V_c is determined by the sign and magnitude of the error signal. A negative error signal results in a negative value of V_c. This is usually interpreted as a 180° phase shift in e_c.

The linearized operating characteristics of an ac servomotor are shown in Figure 12.9. The actual operating line will depend on the speed–torque characteristics of the process. Two typical process load lines are indicated by the dashed lines. The negative values of V_c in the third quadrant simply indicate that the motor reverses direction when V_c is negative.

The velocity and position transfer functions are given by Equations (12.5) and (12.6).

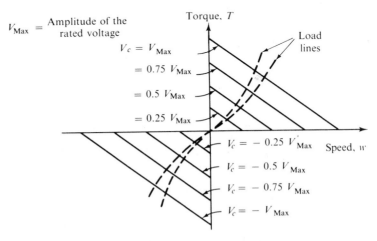

Figure 12.9   Linearized operating characteristics of an ac servomotor.

AC SERVOMOTOR TRANSFER FUNCTIONS

Velocity Transfer Function

$$\frac{\Omega}{V_c} = \frac{K}{1 + \tau s} \qquad\qquad \textbf{(12.5)}$$

Position Transfer Function

$$\frac{\theta}{V_c} = \frac{K}{s(1 + \tau s)} \qquad\qquad \textbf{(12.6)}$$

where V_c = control voltage amplitude, volt
   Ω = motor speed, radian/second
   θ = motor position, radian
   $K = K_1/(b + K_2)$
   $\tau = J/(b + K_2)$, second
   K_1 = stall torque/rated voltage, newton meter/volt
   K_2 = stall torque/no-load speed at rated voltage, newton meter/ (radian/second)
   J = moment of inertia of the load, kilogram meter^2
   b = damping resistance of the load, newton meter/(radian/second)

Example 12.1

Determine the velocity and position transfer functions of an ac servomotor with the following data.

Rated voltage: 120 V
Load inertia: $6 \times 10^{-6} = \text{k·m}^2$
Load damping: 2×10^{-5} n·m/(rad/s)
Stall torque: 0.04 N·m at 120 V
No-load speed: 4000 rpm at 120 V

Solution

The transfer functions are given by Equations (12.5) and (12.6).

1. Determine K_1 by dividing stall torque by rated voltage:

$$K_1 = \frac{0.04}{120} = 3.33\text{E}-04$$

2. Convert no-load speed to radians per second:

$$\text{No-load speed} = \frac{(4000)2\pi}{60} = 419 \text{ rad/s}$$

3. Determine K_2 by dividing stall torque by no-load speed:

$$K_2 = \frac{0.04}{419} = 9.55\text{E}-05$$

4. Solve for K and τ:

$$b + K_2 = (2\text{E}-05) + (9.55\text{E}-05) = 1.155\text{E}-04$$

$$K = \frac{K_1}{b + K_2} = \frac{3.33\text{E}-04}{1.155\text{E}-04} = 2.89$$

$$\tau = \frac{J}{b + K_2} = \frac{6\text{E}-06}{1.155\text{E}-04} = 0.052$$

5. Determine
   a. Velocity transfer function

$$\frac{\Omega}{V_c} = \frac{2.89}{1 + 0.052s}$$

   b. Position transfer function

$$\frac{\theta}{V_c} = \frac{2.89}{s(1 + 0.052s)}$$

12.3 DC MOTORS

DC motors are extremely versatile drives, capable of reversible operation over a wide range of speeds, with accurate control of the speed at all times. They can be controlled smoothly from zero speed to full speed in both directions. DC motors have a high torque-to-inertia ratio that gives them quick response to control signals. DC motors are available with horsepower ratings from 1/300 to over 700.

AC motors stall at torque loads about 2 to 2.5 times their rated torque and have a starting torque of about 1.5 times their rated torque. DC motors are capable of delivering over 3 times their rated torque for a short time.

DC motors can easily accomplish dynamic braking and regenerative braking of the load. *Braking* is accomplished by momentarily turning the motor into a dc generator driven by the inertia of the load. In *dynamic braking*, the voltage from the temporary generator is applied to a bank of resistors. The resistors draw current from the generator, causing a torque that tends to slow down the generator and load (fact 1). The resistors dissipate the energy as heat. In *regenerative braking*, the current from the generator is fed back into the dc supply, thus conserving the energy that is normally lost in bringing the load to a quick stop. If the dc supply is a battery, the generator current will charge the battery.

DC servomotors have lightweight low-inertia low-inductance armatures that can respond quickly to commands for a change in position or speed. Servomotors have very low electrical and mechanical time constants. Typical electrical time constants range from 0.1 to 6 ms, and mechanical time constants range from 2.3 ms to over 40 ms. Servomotors occur in a variety of configurations and features, including permanent-magnet field poles, wirewound iron-core armatures, ironless disk armatures (pancake motors), moving-coil armatures with a stationary iron core (shell motors), and brushless motors.

Wound Field DC Motors

Wound field dc motors are classified as series, shunt, compound, and separately excited, depending on how the field winding and the armature winding are connected. The four arrangements are illustrated in Figure 12.10.

The *dc series motor* (Figure 12.10a) has the highest starting torque and the greatest no-load speed of the four types of connections. Integral horsepower series motors are always direct coupled to the load. A belt drive is never used because a broken belt would result in a runaway motor condition. A dc series motor will continue to run in the same direction when the polarity of the line voltage is reversed. The universal motor is a special type of series motor that runs equally well on direct current or alternating current.

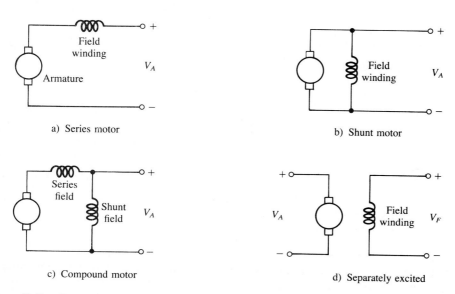

Figure 12.10   Four different methods of connecting a wound-field dc motor.

The *dc shunt motor* has a lower starting torque and a much lower no-load speed than the series motor. As one might expect, the speed and torque characteristics of the *compound motor* are between those of the series motor and the shunt motor.

The *separately excited motor* is a special case of the shunt motor, which allows separate control of the armature voltage and the field voltage. The speed of a separately excited motor can be increased by either decreasing the field voltage or increasing the armature voltage. Variable armature voltage with a fixed field voltage is the most popular method of controlling the speed of a dc motor. The term *armature-controlled dc motor* refers to this method of control.

Permanent-Magnet DC Motors

These motors use permanent magnets to provide the magnetic field instead of a field winding. *Permanent-magnet motors* are available in fractional and low integral horse-power ratings. The following are some of the advantages of permanent-magnet motors.

1. A simpler, more reliable motor because the field power supply is not required
2. Higher efficiency
3. Less heating, making it possible to completely enclose the motor
4. No possibility of overspeeding due to loss of field
5. A more linear torque-versus-speed curve

Alnico, ceramic, and rare-earth magnets are used in permanent-magnet (PM) motors. The use of rare-earth (samarium cobalt) magnets has led to significant increases in the torque-to-inertia ratio of PM motors. It has also allowed improved designs such as brushless pancake motors (see the section on brushless motors).

Moving-Coil Stationary-Core DC Motors

The iron-core armature of the dc servomotor is a major obstacle to increasing the torque-to-inertia ratio of the motor. A higher ratio means that more torque is available to accelerate the load. It also means a lower mechanical time constant. In many applications, the motor inertia is 60 to 70% of the total load inertia. Thus a reduction in the motor inertia has a significant effect on the speed of the control system. The *moving-coil stationary-core motor* is one approach to overcoming this obstacle. In this case the solution is to rotate only the armature winding, leaving the core stationary. The armature winding rotates in the annular space between the field poles on the outside and the iron core on the inside. A typical moving-coil, stationary-core motor has a torque-to-inertia ratio several times greater than an iron-core armature motor.

Disk-Armature (Pancake) Motors

The disk-armature dc motor is another approach to achieving a high torque-to-inertia ratio. These motors use a large-diameter, short-length armature of nonmagnetic material. The armature operates between permanent magnetic poles mounted on two stationary disks, one on either side of the armature. The magnetic field passes from the magnetic poles on one disk, through the armature, to the corresponding magnetic poles on the other disk. This type of construction makes use of the superior magnetic properties of rare-earth magnets. Disk armature motors are also called printed-circuit motors and *pancake* motors. The "printed circuit" name comes from the method used to place the conductors on the armature. The "pancake" name probably comes from the resemblance of the armature and the two stationary disks to a stack of pancakes.

Brushless DC Motors

The primary limitations of the dc servomotor are due to the armature winding and the brush/commutator assembly required to make electrical contact between the winding and the servo drive unit.* These limitations include replacement of worn brushes, arcing caused by commutation, voltage and current limitations, high rotor inertia, and a long path for dissipation of heat because most of the heat is generated in the armature. All of these limitations are eliminated by the brushless servomotor illustrated in Figure 12.11.

The rotor of the *brushless motor* is a permanent magnet, the winding is in the stator, and a solid-state circuit replaces the brush/commutator assembly. The heat from losses is almost entirely in the stator, with a short path for dissipation to ambient. The elimination of the winding and commutator reduces the inertia of the rotor and allows higher rotor speeds. The solid-state commutation circuit eliminates brush replacement and allows higher voltages and currents in the winding.

Two types of permanent magnets are used in the rotor: rare-earth magnets (samarium cobalt and neodymium iron boron) and ceramic magnets (ferrite). The ceramic magnet is low in cost and readily available, but has the poorest magnetic properties. The samarium cobalt magnet has excellent magnetic properties but is expensive and limited in supply. The neodymium iron boron magnet also has excellent magnetic properties and is readily available. Brushless motors with rare-earth permanent magnets have the lowest rotor inertia and the smallest motor size for a given torque rating.

The stator winding usually has three phases. The best brushless servo drives use a pulse-width-modulated (PWM) current amplifier to produce a three-phase sinusoidal current in the three stator windings. This type of drive has the smoothest operation at any speed or torque. Other drives use a simpler control circuit to produce a three-phase square-wave current, but the motor operation is not as smooth as it is with the sinusoidal drive.

---

* Reliance Motion Control's permission to use information from their Electro-Craft *Handbook* of brushless servo systems is gratefully acknowledged.

a) Brushless servo systems

Figure 12.11 The brushless servomotor has a permanent-magnet rotor, a three-phase stator winding, a rotor position sensor, and a solid-state circuit that controls the current in the three-phase stator winding. (Courtesy of Reliance Motion Control, Eden Prairie, Minn.)

Steady-State Characteristics

The steady-state operating characteristics of an armature-controlled dc motor are illustrated in Figure 12.12. The torque-versus-current graph (Figure 12.12a) shows a linear relationship between the armature current (i) and the motor torque (T). The slope of this line is called the torque constant (K_T). It indicates the change in torque (ΔT) produced by a change in current (Δi). The torque versus speed relationship is given by

$$T = K_T i - T_f \qquad (12.7)$$

$$K_T = \frac{\Delta T}{\Delta i} \qquad (12.8)$$

The intercept on the current axis, i_f, is the current required to overcome the friction torque, T_f.

The induced voltage-versus-speed graph (Figure 12.12b) shows a linear relationship between the armature speed (ω) and the voltage induced in the armature coil (e_i). The slope of this line is called the EMF constant (K_E). It indicates the change in induced voltage (Δe_i) produced by a change in armature speed ($\Delta\omega$). The induced

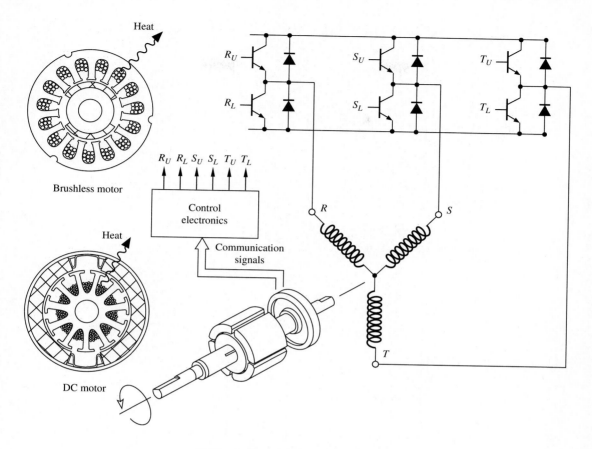

Heat

Brushless motor

R_U R_L S_U S_L T_U T_L

Control
electronics

Heat

Communication
signals

DC motor

b) Construction of a brushless servo motor

Figure 12.11   (continued)

voltage is given by

$$e_i = K_E \omega \qquad \text{(12.9)}$$

$$K_E = \frac{\Delta e_i}{\Delta \omega} \qquad \text{(12.10)}$$

The voltage-versus-speed graph (Figure 12.12c) is a graph of the armature voltage versus motor speed with constant torque. This also means a constant armature current (i). The armature voltage (e) is made up of two components: the induced voltage (e_i), and the voltage drop across the armature resistance (ir). The armature voltage (e) is given by

$$e = ir + e_i \qquad \text{(12.11)}$$

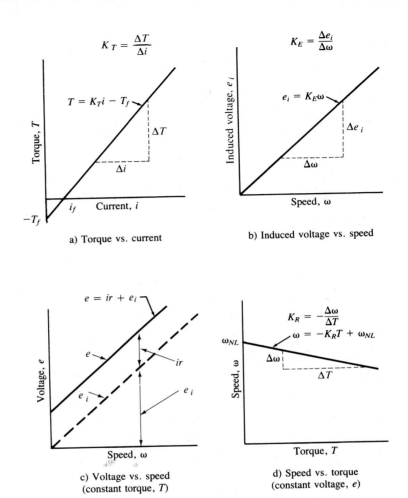

$$K_T = \frac{\Delta T}{\Delta i}$$

$$T = K_T i - T_f$$

a) Torque vs. current

$$K_E = \frac{\Delta e_i}{\Delta \omega}$$

$$e_i = K_E \omega$$

b) Induced voltage vs. speed

$$e = ir + e_i$$

c) Voltage vs. speed
(constant torque, T)

$$K_R = -\frac{\Delta \omega}{\Delta T}$$

$$\omega = -K_R T + \omega_{NL}$$

d) Speed vs. torque
(constant voltage, e)

Figure 12.12   Steady-state operating characteristics of the armature-controlled dc motor.

The speed-versus-torque graph (Figure 12.12d) is a graph of the motor torque versus motor speed with constant armature voltage. The equation for speed as a function of torque can be derived from Equations (12.7), (12.9), and (12.11).

Solve Equation (12.7) for i.

$$T = K_T i - T_f$$

$$i = \frac{(T + T_f)}{K_T}$$

Substitute $(T + T_f)/K_T$ for i in Equation (12.11):

$$e = \frac{(T + T_f)r}{K_T} + e_i$$

Substitute $K_E\omega$ for e_i [Equation (12.9)], rearrange the terms, and define two constants, K_R and ω_{NL}, to get

$$\omega = -K_R T + \omega_{NL} \tag{12.12}$$

$$K_R = \frac{r}{K_E K_T} \tag{12.13}$$

$$\omega_{NL} = \frac{eK_T - rT_f}{K_E K_T} \tag{12.14}$$

DC MOTOR STEADY-STATE CHARACTERISTICS

$$T = K_T i - T_f \tag{12.7}$$

$$e_i = K_E \omega \tag{12.9}$$

$$e = ir + e_i \tag{12.11}$$

$$\omega = -K_R T + \omega_{NL} \tag{12.12}$$

$$K_R = \frac{r}{K_E K_T} \tag{12.13}$$

$$\omega_{NL} = \frac{eK_T - rT_f}{K_E K_T} \tag{12.14}$$

$$p = \omega T \tag{12.15}$$

where e = armature voltage, volt

   e_i = induced voltage, volt

   i = armature current, ampere

   K_E = EMF constant, volt second/radian

   K_T = torque constant, newton meter/ampere

   p = power, watt

   r = armature resistance, ohm

   T = output torque, newton meter

   T_f = friction torque, newton meter

   ω = motor speed, radian/second

Example 12.2

An armature-controlled dc motor has the following ratings.

$$T_f = 0.012 \text{ N·m}$$
$$K_T = 0.06 \text{ N·m/A}$$
$$I_{max} = 2 \text{ A}$$
$$K_E = 0.06 \text{ V·s/rad}$$
$$\Omega_{max} = 500 \text{ rad/s}$$
$$r = 1.2 \text{ }\Omega$$

Determine the following.

    a. The maximum output torque, T_{max}.
    b. The maximum power output, P_{max}.
    c. The maximum armature voltage, E_{max}.
    d. The no-load motor speed when $e = E_{max}$.

Solution

    a. The maximum output torque is obtained from Equation (12.7), when $i = I_{max}$.

$$T_{max} = K_T I_{max} - T_f = (0.06)(2) - 0.012$$
$$= 0.108 \text{ N·m}$$

    b. The maximum power output is obtained from Equation (12.15) when ω and T are both a maximum.

$$P_{max} = \omega_{max} T_{max} = (500)(0.108)$$
$$= 54 \text{ W}$$

    c. The maximum armature voltage is obtained from Equations (12.9) and (12.10), when i and ω are both a maximum.

$$e_i = K_E \omega$$
$$e = ir + e_i = ir + K_E \omega$$
$$E_{max} = I_{max} r + K_E \omega_{max}$$
$$= (2)(1.2) + (0.06)(500)$$
$$= 32.4 \text{ V}$$

    d. The no-load motor speed is obtained from Equation (12.14) when $e = E_{max}$.

$$\omega_{NL} = \frac{E_{max} K_T - rT_f}{K_E K_T} = \frac{(32.4)(0.06) - (1.2)(0.012)}{(0.06)(0.06)}$$
$$= 536 \text{ rad/s}$$

Example 12.3

The motor in Example 12.2 is operated at 300 rad/s with a load torque of 0.05 N·m. Determine the following.

    a. The armature voltage.
    b. The armature speed if the torque increases to 0.075 N·m and the armature voltage is not changed.

Solution

    a. From Equation (12.7),

$$T = K_T i - T_f$$
$$i = \frac{T + T_f}{K_T} = \frac{0.05 + 0.012}{0.06}$$
$$= 1.03 \text{ A}$$

    From Equation (12.9),

$$e_i = K_E \omega = (0.06)(300) = 18 \text{ V}$$

From Equation (12.11),

$$e = ir + e_i = (1.03)(1.2) + 18$$
$$= 19.24 \text{ V}$$

b. From Equation (12.13),

$$K_R = \frac{r}{K_E K_T} = \frac{1.2}{(0.06)(0.06)} = 333.3$$

From Equation (12.12),

$$\omega_1 = -K_R T_1 + \omega_{NL} \Rightarrow \omega_{NL} = \omega_1 + K_R T_1$$
$$\omega_2 = -K_R T_2 + \omega_{NL} = -K_R T_2 + \omega_1 + K_R T_1$$
$$= \omega_1 - K_R(T_2 - T_1)$$
$$= 300 - (333.3)(0.075 - 0.050)$$
$$= 291.7 \text{ rad/s}$$

DC Motor Transfer Functions

The schematic diagram and the block diagram of an armature-controlled dc motor are illustrated in Figure 12.13. A dc voltage is applied to the field winding (or the field is

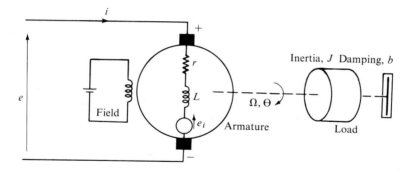

a) Schematic diagram

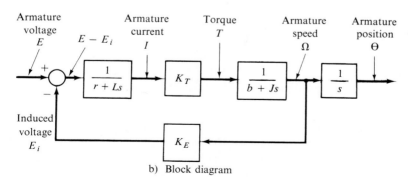

b) Block diagram

Figure 12.13   Armature-controlled dc motor.

provided by permanent-magnet field poles). A variable voltage (e) is applied to the armature windings. The armature is represented by a resistor, an inductor, and an induced voltage source connected in series. The armature current (i) is defined by the following time-domain equation:

$$e = L\frac{di}{dt} + ir + e_i$$

The corresponding frequency-domain equation is

$$E = LsI + Ir + E_i$$

or

$$I = \left(\frac{1}{r + Ls}\right)(E - E_i) \tag{12.16}$$

Equation (12.16) is represented on the block diagram by the summing junction, and the block between the summing junction and the current (I). The remaining system equations (each represented by a block in Figure 12.13) are given below.

$$T = K_T I$$

$$\Omega = \left(\frac{1}{b + Js}\right)T$$

$$E_i = K_E\Omega$$

$$\Theta = \frac{\Omega}{s}$$

The transfer functions are obtained by solving the system equations algebraically for the desired output-to-input ratio.

ARMATURE-CONTROLLED DC MOTOR TRANSFER FUNCTIONS

Velocity Transfer Function

$$\frac{\Omega}{E} = \frac{K_T}{rb + K_E K_T + rb(\tau_m + \tau_e)s + rb\tau_m\tau_e s^2} \tag{12.17}$$

Positional Transfer Function

$$\frac{\Theta}{E} = \frac{1}{s}\left(\frac{\Omega}{E}\right) \tag{12.18}$$

where b = damping resistance, newton meter second/radian

E = armature voltage, volt

J = moment of inertia of the load, kilogram meter^2

K_E = EMF constant, volt second/radian

K_T = torque constant, newton meter/ampere

L = armature inductance, henry

> r = armature resistance, ohm
> s = frequency parameter, 1/second
> $\tau_e = L/r$ = electrical time constant, second
> $\tau_m = J/b$ = mechanical time constant, second
> Θ = armature position, radian
> Ω = motor speed, radian/second

Example 12.4

Determine the velocity and position transfer functions of the motor in Example 12.3. The following additional values are required.

$$J = 6.2 \times 10^{-4} \text{ kg·m}^2$$
$$b = 1 \times 10^{-4} \text{ N·m·s/rad}$$
$$L = 0.020 \text{ H}$$

Solution

From Equation (12.18),

$$\tau_m = \frac{J}{b} = \frac{6.2\text{E}-04}{1\text{E}-04} = 6.2 \text{ s}$$

$$\tau_e = \frac{L}{r} = \frac{0.02}{1.2} = 0.0167 \text{ s}$$

$$rb = (1.2)(1\text{E}-04) = 1.2\text{E}-04$$

$$rb\tau_m\tau_e = 1.24\text{E}-05$$

$$rb(\tau_m + \tau_e) = (1.2\text{E}-04)(6.2 + 0.0167) = 7.46\text{E}-04$$

$$rb + K_E K_T = 1.2\text{E}-04 + (0.06)(0.06) = 0.00372$$

For velocity control,

$$\frac{\Omega}{E} = \frac{0.06}{0.00372 + 7.46 \times 10^{-4}s + 1.24 \times 10^{-5}s^2}$$

or

$$\frac{\Omega}{E} = \frac{16.13}{1 + 0.201s + 0.00333s^2}$$

For position control,

$$\frac{\Theta}{E} = \frac{16.13}{s + 0.201s^2 + 0.00333s^3}$$

12.4 STEPPER MOTORS

A *stepper motor* transforms electrical pulses into equal increments of rotary shaft motion called *steps*. A one-to-one correspondence exists between the electrical pulses and the motor steps. A drive circuit converts the electrical pulses into a sequence of command signals that actually accomplish the rotation of the motor. Stepper motors are capable of rotation in either direction, depending on the sequence of command signals from the drive circuit.

Stepper motors are used in computer peripherals, x–y plotters, scientific instruments, robots, and machine tools. Advantages of stepper motors include compatibility with digital signals from a computer, simplicity of open-loop control with capability for closed-loop control, accuracy within 5% of the last step taken (regardless of the total number of steps), and reduced maintenance due to the absence of brushes.

The construction of a stepper motor is very similar to a brushless dc motor. It has a wound stator and a nonexcited rotor. Stepper motors are classified as variable reluctance, permanent magnet, or hybrid, depending on the type of rotor. They are also classified as two-phase, three-phase, or four-phase, depending on the number of windings on the stator (two, three, or four windings). The number of teeth or poles on the rotor and the number of poles on the stator determine the size of the step (called the *step angle*). The step angle is equal to 360° divided by the number of steps per revolution. A motor that has 200 steps per revolution has a step angle of 1.8°. On the other extreme, a motor that has four steps per revolution has a step angle of 90°.

The need for precision positioning has resulted in steady improvement in the resolution of stepper motors. The earliest steppers had a resolution of only four steps per revolution (90° step angle). More recent motors have resolutions as high as 200 steps per revolution. However, further increases in resolution appear to be limited by mechanical and physical constraints. This limitation has been circumvented by electronic methods of reducing the step size. Half-stepping and microstepping are techniques of electronically dividing each step into two half-steps or from 10 to 125 microsteps. See the sections on half-stepping and microstepping for further explanations of these techniques.

Variable-reluctance (VR) steppers have a soft-iron rotor with a number of teeth, giving it the appearance of a gear. The stator also has teeth in addition to a number of wound poles. A two-phase VR motor has two windings, with the coils on alternate poles connected in series to form the two windings. When electric current is applied to a coil in the stator, magnetic flux is generated that causes teeth in the rotor to line up with teeth in the stator. When the current is switched to the other winding, the rotor moves a distance of one step angle, and a new set of teeth line up. Variable-reluctance stepper motors are used in many computer peripherals.

Permanent-magnet (PM) stepper motors have one or more pole pairs in the rotor. Early PM steppers had only one pole pair, and resolution depended entirely on the number of poles in the stator. The use of rare-earth magnets has greatly improved the design and performance of PM stepper motors. One such design features a thin disk rotor made of rare earth–cobalt magnetic material. The disk is magnetized with up to 50 pairs of alternating north and south poles. The excellent magnetic properties of the rare-earth magnets and the absence of iron in the rotor make this an excellent rotor for microstepping applications.

Hybrid stepper motors have a variable-reluctance rotor with a permanent magnet in its magnetic path. The magnet is usually in the rotor. Hybrid steppers have high torque, high inertia, and small step angles. A typical hybrid stepper has 50 teeth in the rotor and 50 teeth in the stator. The stator has eight poles with two center-tapped windings. When the center taps are used, the motor is considered to be a four-phase motor. When the center taps are removed or left open-circuited, the motor is considered a two-phase motor.

Full-Step Operation

As the name implies, *full-step operation* of a stepper motor consists of a movement of one full step for each input pulse. Full-step operation is the simplest of the three methods of stepping. An understanding of full-step operation is also a prerequisite to understanding two-step and microstep operation. Figure 12.14a shows the model of a stepper motor that will be used to explain the three methods of stepping.

The model stepper motor is a two-phase motor with four wound poles in the stator. The top and bottom coils are connected in series to form phase A. The two side coils are connected in series to form phase B. The input currents are applied to the terminals marked A and B on the top and right side. The rotor is a permanent-magnet type with 10 alternating north and south poles. The resolution of the motor is equal to the product of the number of poles in the rotor (10) times the number of pole pairs (2) in the stator. Thus the model stepper motor has 20 steps per revolution and a step angle of 18°. The top rotor pole in Figure 12.14a is marked with an arrow. The arrow will enable us to follow the movement of this "marked" pole as the rotor makes several full steps.

In Figure 12.14a, the motor is in the start position. A current of I_{max} amperes enters the phase A winding at terminal A. No current enters the phase B winding. The direction of current I_A is such that the top stator pole is a south pole (labeled "S") and the bottom stator pole is a north pole (labeled "N"). The absence of current in phase B results in a demagnetized condition for the two side poles (labeled "0"). The alignment of the rotor poles is such that the magnetic attraction between unlike poles will hold the rotor in the position shown.

In Figure 12.14b, a current of I_{max} enters the phase B winding at terminal B, and no current enters the phase A winding. Notice the 0 labels on the top and bottom stator poles, the S label on the right-side stator pole, and the N label on the left-side stator pole. The magnetic attraction between opposite poles and the repulsion between like poles has caused the rotor to rotate one step in the clockwise direction. Once again, the magnetic forces will hold the rotor in the new position shown in Figure 12.14b. Two more steps are shown in Figure 12.14c and d. Table 12.2 lists the angle θ and the two phase currents for five full steps.

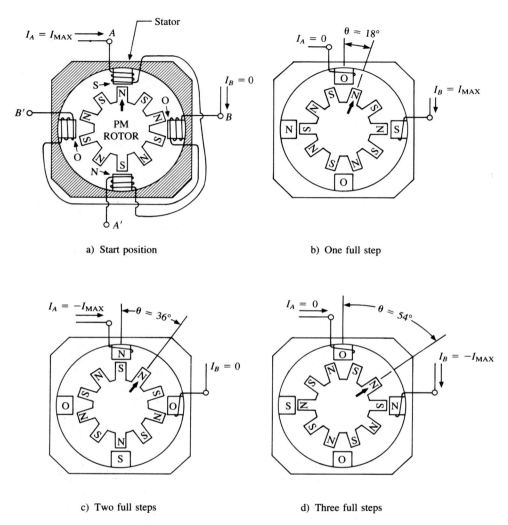

a) Start position

b) One full step

c) Two full steps

d) Three full steps

Figure 12.14 This model of a stepper motor has four teeth/pole on the stator and 10 teeth/pole on the PM rotor, which gives it 20 full steps per revolution. Each full step is 18° (i.e., 360°/20 = 18°).

Table 12.2 Sequence of Phase Currents for Five Full Steps of the Model Stepper Motor (Figure 12.14)

Step	θ (deg)	I_A/I_{max}	I_B/I_{max}
0	0	1	0
1	18	0	1
2	36	−1	0
3	54	0	−1
4	72	1	0
5	90	0	1

Half-Step Operation

Half-steps are accomplished by applying partial currents to both phase windings to position the rotor halfway between two full step positions. Figure 12.15 illustrates three half-steps of the model stepper motor. The start conditions in Figure 12.15a are identical to the conditions in Figure 12.14a. The first half-step is accomplished by reducing the phase A current to $0.707I_{max}$ amperes and increasing the phase B current to $0.707I_{max}$ amperes (see Figure 12.15b). Notice that all four stator coils are magnetized. The top and right side are south poles while the bottom and left side

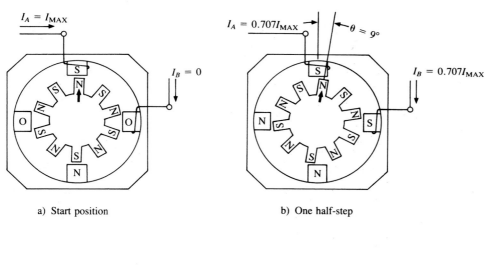

a) Start position   b) One half-step

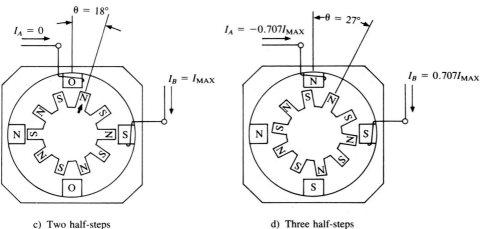

c) Two half-steps   d) Three half-steps

Figure 12.15   Half-stepping increases the resolution of the model stepper motor to 40 half-steps per revolution. In half-stepping, the driver first turns on phase A, then both A and B, then B only, then both A and B (with reversed polarity on A).

Table 12.3 Sequence of Phase Currents for 10 Half-Steps of the Model Stepper Motor (Figure 12.14)

Half-Step	θ (deg)	I_A/I_{max}	I_B/I_{max}
0	0	1	0
1	9	0.707	0.707
2	18	0	1
3	27	−0.707	0.707
4	36	−1	0
5	45	−0.707	−0.707
6	54	0	−1
7	63	0.707	−0.707
8	72	1	0
9	81	0.707	0.707
10	90	0	1

are north poles. The magnetic forces are such that the rotor is held in the position shown in Figure 12.15b. Notice that the position of the marked rotor pole, θ, is 9°, exactly half of a full 18° step. Two more half-steps are shown in Figures 12.15c and d. Table 12.3 lists the angle θ and the two phase currents for 10 half-steps.

Microstep Operation

In half-step operation, we saw how the rotor could be positioned halfway between two full-step positions by supplying current to both phase windings. Microstepping simply extends this technique to more than one midposition by using different values of current in each phase. The microstep sizes that are most commonly used are $\frac{1}{10}$, $\frac{1}{16}$, $\frac{1}{32}$, and $\frac{1}{125}$ of a full step. An obvious advantage of microstepping is the much finer resolution it provides. For example, when 125 microsteps are used in a stepper that has 200 full steps per revolution, the resolution is $200(125) = 25,000$ microsteps per revolution.

The values of the phase currents required for a microstep are given by

$$I_A = \cos\left(\frac{90n}{s}\right)I_{max} \qquad (12.19)$$

$$I_B = \sin\left(\frac{90n}{s}\right)I_{max} \qquad (12.20)$$

where I_{max} = maximum value of phase current, ampere
I_A = current in phase A winding, ampere
I_B = current in phase B winding, ampere
n = number of microsteps from start position
s = number of microsteps in a full step

Table 12.4 Sequence of Phase Currents for 10 Microsteps of the Model Stepper Motor (Figure 12.14)[a]

Micro-step	θ (deg)	I_A/I_{max}	I_B/I_{max}
0	0	1.000	0.000
1	1.8	0.988	0.156
2	3.6	0.951	0.309
3	5.4	0.891	0.454
4	7.2	0.809	0.588
5	9.0	0.707	0.707
6	10.8	0.588	0.809
7	12.6	0.454	0.891
8	14.4	0.309	0.951
9	16.2	0.156	0.988
10	18.0	0.000	1.000

[a] Microstep size = 0.1(full step size).

The values of these currents may be stored in a ROM memory chip to be read when needed by the driver circuit. Table 12.4 lists the angle θ and the two phase currents for ten $\frac{1}{10}$ microsteps for the model stepper motor (Figure 12.14).

12.5 AC ADJUSTABLE-SPEED DRIVES

AC adjustable-speed drives (both single phase and three-phase) consist of two major parts, a converter and an inverter. The *converter* converts the ac input power to a dc voltage. The *inverter* changes the dc voltage back into an ac voltage of any desired frequency from about 3 to 60 (or 120) Hz. The output of the inverter will drive an ac synchronous motor or an ac induction motor at a speed determined by the frequency of the inverter output. For example, the speed of an 1800-rpm synchronous motor can be adjusted from 90 to 1800 rpm by a 3- to 60-Hz inverter. Figure 12.16a shows the block diagram of a three-phase inverter and the ideal waveforms of the output. The actual voltage produced by the inverter is not sinusoidal but approximates a sine wave. Two methods of approximating a sine wave, the variable-voltage inverter and the pulse-width modulated inverter, are explained later in this section.

Adjustable-frequency ac drives are used to control pumps, fans, and conveyors. Many applications with single-speed drives were converted to adjustable speed with the intent of reducing energy consumption or increasing manufacturing flexibility. In many of these conversions, the same ac motor was used, with the ac drive providing the variable frequency required for adjustable speed. In pumps and fans, variable speed control is more efficient than throttling to obtain lower flow rates. Energy savings of 40 to 50% have been achieved by replacing throttling controls with variable-speed controls. Another advantage of variable ac drives is the low cost and rugged construction of ac motors. AC induction motors, in particular, are much more reliable and maintenance free than dc motors. However, variable controls for ac motors are

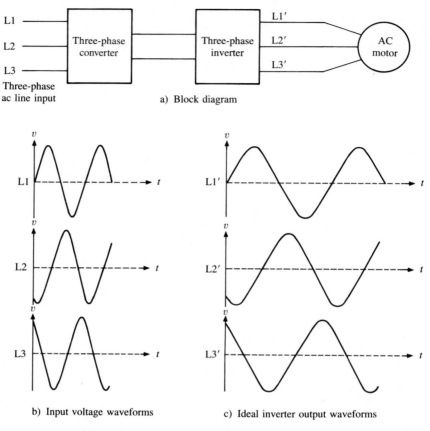

b) Input voltage waveforms

c) Ideal inverter output waveforms

Figure 12.16 A variable-speed ac drive uses a converter to convert the three-phase 60-Hz input voltage into a dc voltage source for the inverter. The inverter converts the dc voltage into a three-phase voltage with a frequency that can be varied from about 3 Hz to 60 (or 120) Hz.

much more complex than equivalent controls for dc motors. This complexity has hindered the replacement of dc motors in robots and motion controls with the rugged and maintenance-free ac motors.

Variable-Voltage Inverter

The purpose of the *variable-voltage inverter* (*VVI*) is to manufacture a sinusoidal waveform whose frequency can be varied at will over some frequency range. In the VVI method, a converter converts the ac input voltage into a variable dc voltage. The VVI inverter then inverts the dc voltage into a voltage whose amplitude and frequency are variable. The voltage is a simple approximation to a sine wave. Figure 12.17 shows the output waveforms of a three-phase variable-voltage inverter. This simple approximation of a sine wave is easy to produce, but it also results in high

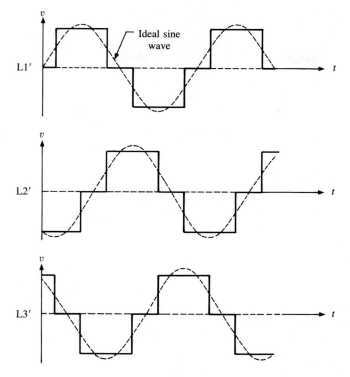

Figure 12.17   The three-phase output of a voltage-variable inverter (VVI) produces a simple approximation of a three-phase sinusoidal voltage. At low speeds, a VVI-driven motor produces a jerky motion called cogging, due to the crude approximation of a sine wave.

current spikes. Another problem occurs at motor speeds below 20% of full speed. The crude approximation of the sine wave causes a jerky movement of the motor called *cogging.*

SCRs were the first solid-state switching devices used in VVI drives. SCR-based VVI drives capable of driving ac motors of hundreds of horsepower have been available for many years. The latching characteristic of an SCR (an advantage in some applications) is a problem in ac inverters. Once an SCR is turned on, it remains on until the current is reduced to zero. This is no problem in a converter because the ac line becomes negative at the end of each positive cycle and turns the SCR off. In an inverter, there is nothing to turn the SCR off, and the control circuit must provide the means of doing so. This requires complex and expensive commutating circuitry that doubles or even triples the number of SCRs required in the inverter. The commutating circuit also requires additional capacitors and inductors, further increasing the cost of the circuit.

As higher-power transistors became available, designers began replacing SCRs with transistors. Transistors can be turned on and off at will with a considerably simpler circuit and with less power than required by an SCR. Transistors also switch much faster than SCRs. This gives the transistor two more advantages over the SCR. First, the fast switching time, usually nanoseconds against microseconds for an SCR, allows very little time for heating to occur. Consequently, the transistor accomplishes the switching more efficiently than the SCR. Second, fast switching gives the transistor-based ac drive much faster response for critical control applications.

The transistor's major problem is lower voltage and current ratings than SCRs. As transistors with higher ratings became available, transistor ac drives began replacing the slower, more expensive SCR drives.

Pulse-Width-Modulated Inverters

Pulse-width modulation (*PWM*) is another method of approximating a sine wave. The PWM inverter produces a much better approximation of a sine wave than that obtained with a VVI. Pulse-width modulation is not a new idea. It has been used in communication for many years to transmit information on a sequence of pulses called a carrier. The pulses have a constant amplitude and a variable width. A pulse-width modulator varies the width of the pulse according to an information signal.

The PWM inverter uses a sequence of pulses to approximate a sine wave with a variable amplitude and a variable frequency. Figure 12.18 illustrates how a sine wave can be approximated by pulse-width modulation. After appropriate filtering, the output of a PWM inverter is a fairly good approximation of a sine wave. One PWM method uses the magnitude of the sine function to determine the width of each pulse. This method is called *sine-coded PWM*. The values of the sine function are stored in a RAM memory unit for use by the controller in manufacturing a sine wave with the desired amplitude and frequency. Another advantage of PWM is that it uses a constant-amplitude dc input voltage. This means that the rectifier can be a simple diode bridge circuit. The VVI requires a variable dc voltage, which is usually provided by an SCR rectifier. The following comparison of size and cost of a 20-hp ac controller illustrates the progress in ac drives over a 10-year period.

Year	Controller Type	Relative Cost (%)	Package Size (in.)
1978	SCR—VVI	100	20 × 30 × 86
1982	Transistor—VVI	40	15 × 26 × 29
1983	Transistor—PWM	20	14 × 16 × 25

Source: Control Engineering, February 1988, p. 73.

12.6 DC MOTOR AMPLIFIERS AND DRIVES

Amplifiers

Permanent-magnet brush-type dc servomotors are sometimes controlled with a power operational amplifier. Power op amps obey all the rules of the op-amp family and

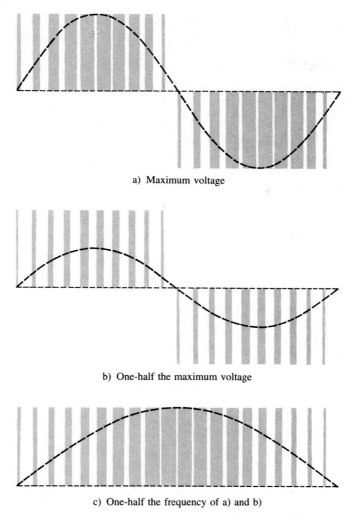

a) Maximum voltage

b) One-half the maximum voltage

c) One-half the frequency of a) and b)

Figure 12.18 A pulse-width-modulated inverter (PWM) produces a reasonably good approximation of a sine wave. The inverter chops the constant dc input voltage into a sequence of pulses (1 to 4 kHz) and modulates the width of the pulses to produce a sine wave. As shown here, pulse modulation can vary both the amplitude and the frequency of the inverter output voltage.

are also designed to operate at higher voltage and current levels than the lower-power members of the family. Power op amps operate at voltages above 44 V and can deliver currents above 0.1 A. Low-voltage power op amps (below 100 V) consist of an IC op-amp front end plus a high-current output stage housed in a heat-dissipating package. High-voltage power op amps (above 100 V) have a more complex circuit using many individual transistors. Construction of the high-voltage units is more labor intensive and hence more expensive. They do tend to have better accuracy due to improved control of individual components and laser trimming of the input stage. Figure 12.19 illustrates the use of a power op amp in a dc speed control system.

The transfer function of the position control system can be obtained from the block diagram in Figure 12.19. The objective is to obtain an equation for the ratio of output, Ω, over input, SP (i.e., Ω/SP). Each block in the transfer function defines an equation of the form

$$\text{Output} = (\text{input})(\text{transfer function of the block})$$

The transfer function of a block is written inside the block. For example, the output of the power op amp, E, is given by the equation

$$E = (SP - E_G)(P) \tag{12.21}$$

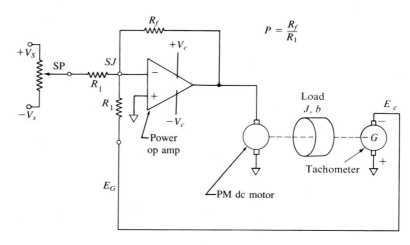

a) Schematic diagram

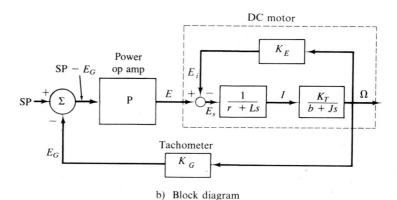

b) Block diagram

Figure 12.19 A power op amp is used to control the velocity of a small permanent-magnet brush-type dc motor.

The dc motor is a little more complicated, due to the internal feedback loop. However, with a little algebra, we can obtain an equation for the output of the motor, Ω. Proceed as follows.

1. Determine an equation for E_i.

$$E_i = K_E\Omega \tag{12.22}$$

2. Get the output of the internal summing junction.

$$E_s = E - E_i \tag{12.23}$$

3. Get an equation for I.

$$I = E_s\left(\frac{1}{r + sL}\right) \tag{12.24}$$

4. Get an equation for Ω.

$$\Omega = I\left(\frac{K_T}{b + Js}\right) \tag{12.25}$$

5. Use Equations (12.22) through (12.25) to solve for Ω in terms of E with both E_i and I eliminated from the equation. Equation (12.26) is the final equation for the dc motor.

$$\Omega[(r + Ls)(b + Js) + K_EK_T] = K_TE \tag{12.26}$$

The last component, the tachometer, is easy.

$$E_G = K_G\Omega \tag{12.27}$$

Finally, use Equations (12.21), (12.26), and (12.27) to obtain the transfer function as given by

$$\frac{\Omega}{SP} = \frac{K_TP}{(K_EK_T + K_GK_TP + rb) + (Jr + Lb)s + JLs^2} \tag{12.28}$$

DC Adjustable-Speed Drives

DC adjustable-speed drives use a rectifier to produce a controlled dc voltage that can be smoothly adjusted from 0 to 100% of the rated voltage. Some dc drives also provide a reverse energy path for regenerative braking. Figure 12.20 illustrates an SCR-based dc adjustable-speed drive with a built-in, computed velocity signal.

In Figure 12.20a, two SCRs and two diodes are connected to form a full-wave bridge rectifier. When the SCRs are turned off, the bridge acts as an open switch. When the SCRs are turned on, the bridge acts as a full-wave rectifier. The output of the bridge is a pulsating voltage which is applied to the armature of a dc motor (see the upper right corner of Figure 12.20a).

The magnetic amplifier trigger circuit produces the triggering pulses that turn the SCR on. The SCR is said to be 100% on if the trigger occurs at the start of each ac half-cycle. With the SCR 100% on, the bridge acts as a full-wave rectifier. The SCR is 75% on if the trigger occurs 45° after the start of each ac half-cycle. With the

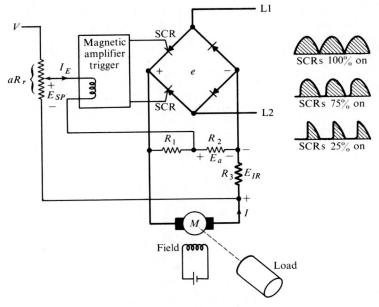

a) Schematic diagram

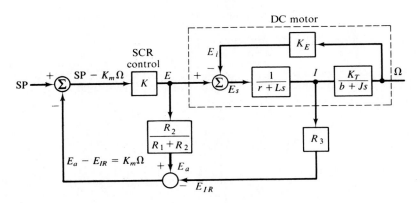

b) Block diagram

Figure 12.20 This SCR-based dc adjustable-speed drive has a full-wave bridge rectifier and a computed-velocity feedback signal.

SCR 75% on, the first 25% of each half-cycle is removed from the output waveform, as shown in Figure 12.20a. The SCR is 25% on if the trigger occurs 135° after the start of each half-cycle. With the SCR 25% on, the first 75% of each half-cycle is removed from the output. The inductance of the armature coil and the inertia of the motor and load help smooth out the pulsations in the power caused by the pulses in the armature voltage.

An interesting feature of the system illustrated in Figure 12.20 is the fact that a tachometer is not used to measure the velocity of the load. Instead, the velocity is calculated from measurements of the armature voltage (E) and the armature current (I). The calculation is based on the following frequency domain version of the combination of Equations (12.9) and (12.11):

$$E - Ir = K_E \Omega \tag{12.29}$$

In Figure 12.20, E_a is the voltage drop across resistor R_2 and E_{IR} is the voltage drop across resistor R_3. Voltage E_a can be obtained by applying the voltage-divider rule to resistors R_1 and R_2.

$$E_a = \left(\frac{R_2}{R_1 + R_2}\right) E \tag{12.30}$$

Voltage E_{IR} is obtained by applying Ohm's law to R_3.

$$E_{IR} = IR_3 \tag{12.31}$$

In the design of the internal feedback, R_3 is defined as

$$R_3 = \left(\frac{R_2}{R_1 + R_2}\right) r \tag{12.32}$$

where r represents the internal resistance of the motor armature. Equations (12.30) through (12.32) can be used to form the following equation for $E_a - E_{IR}$:

$$E_a - E_{IR} = \left(\frac{R_2}{R_1 + R_2}\right)(E - Ir) \tag{12.33}$$

A combination of Equations (12.29) and (12.33) produces the following result:

$$E_a - E_{IR} = K_m \Omega \tag{12.34}$$

where $K_m = \dfrac{K_E R_2}{R_1 + R_2}$, volt/radian per second

Ω = armature speed, radian/second

E_a = voltage across resistor R_2, volt

E_{IR} = voltage across resistor R_3, volt

R_1 and R_2 are arbitrarily selected resistors

R_3 is defined by Equation (12.32)

Equation (12.34) states that the voltage difference ($E_a - E_{IR}$) is proportional to the armature speed Ω and can be used as a measurement of Ω.

GLOSSARY

AC motor: An electric motor that is powered by an ac voltage. (12.2)

AC motor, induction: An ac motor that has no electrical connection to the rotor. The transformer action induces a current in current-carrying conductors in the rotor. Induction motors are very reliable and have a low initial cost and a low maintenance cost. (12.2)

AC motor, servo: A two-phase, reversible induction motor with special modifications for servo control. (12.2)

AC motor, squirrel-cage: A type of induction motor that has conducting bars and end rings on the rotor to provide an electrical path for the induced current. The name comes from the resemblance between the bars/end ring assembly and a squirrel cage. (12.2)

AC motor, synchronous: An electric motor that normally runs at synchronous speed. The stator has windings similar to an induction motor. The rotor has fixed poles that lock into step with the rotating poles in the stator. (12.2)

AC motor, wound-rotor: A type of induction motor in which the conduction bars are connected in series to form three windings. The windings are joined on one end to form a wye connection. The other end of each winding is connected to a slip ring. (12.2)

Armature: The cylindrical, rotating part of a dc motor. (12.1)

Back EMF: A voltage induced in the coils of a motor by the generator effect. This induced voltage opposes the external voltage applied to the coil and thus reduces the motor current. (12.1)

Braking: A method of slowing down a dc motor by causing the motor to act as a dc generator, thus producing a force that tends to slow down the armature. (12.3)

Braking, dynamic: A method of braking a dc motor in which the voltage produced by the generator action is applied to a bank of resistors. The resistors draw current from the generator, causing a force that tends to slow down the armature. (12.3)

Braking, regenerative: A method of braking a dc motor in which the current from the generator is fed back into the dc supply. If the dc supply is a battery, the current will charge the battery. (12.3)

Cogging: A jerky motion that occurs when an ac motor is driven at slow speeds by a variable-voltage inverter. The jerky motion is caused by the crude approximation of the sine wave. (12.5)

Commutation: The reversal of current and polarity in the coil of a dc motor as the brushes move from one segment to the next. (12.1)

Commutator: A segmented ring on the armature of a dc motor. Each segment is connected to one end of one of the armature coils. Electrical connection is made to the commutator segments (and hence the armature coils) through carbon contacts called brushes. (12.1)

DC motor: An electric motor that is powered by a dc voltage. (12.3)

DC motor, brushless: A dc motor that has a permanent-magnet armature and a wound field. There is no need for brushes and a commutator. Instead, the commutation is performed electronically on the stator winding. (12.3)

DC motor, compound: A dc motor in which part of the field winding is connected in series with the armature winding, and the remainder is connected in parallel with the armature-series field windings. Compound motors have a starting torque and no-load speed that lie between the series motor and the shunt motor. (12.3)

DC motor, moving coil: A dc motor in which the armature winding rotates in an annular space between the field pole on the outside and the iron core on the inside. The result is a motor with a torque-to-inertia ratio several times greater than an iron-core armature motor. (12.3)

DC motor, pancake: A dc motor with a disk-shaped armature that achieves a high torque-to-inertia ratio. (12.3)

DC motor, permanent-magnet: A dc motor in which the field is provided by permanent magnets. The speed of the permanent-magnet motor can be controlled by the armature voltage, making this a popular motor for servo control systems. (12.3)

DC motor, separately excited: A dc motor in which the field winding and the armature winding are excited by separate sources. The speed of a separately excited motor can be increased by either decreasing the field voltage or increasing the armature voltage. (12.3)

DC motor, series: A dc motor in which the field winding and the armature winding are connected in series. Series motors produce the highest starting torque and the greatest no-load speed of the four types of dc motors. (12.3)

DC motor, shunt: A dc motor in which the field winding and the armature winding are connected in parallel. Shunt motors have a lower starting torque and a much lower no-load speed than a series motor. (12.3)

Field poles: The stationary part of a dc motor. (12.1)

Full-step operation: A movement of a stepper motor of one full step for each input pulse. *See also* Half-step and Microstep operation. (12.4)

Half-step operation: A movement of a stepper motor of one-half step for each input pulse. *See also* Full-step and Microstep operation. (12.4)

Inverter: An electronic device that converts a dc voltage into an ac voltage. The frequency of the output may be 60 Hz, or it may be variable over a specific range of frequencies such as 3 to 60 Hz. (12.5)

Inverter, pulse-width-modulated (PWM):   A variable-frequency inverter that uses pulse-width modulation of a sequence of pulses to produce a fairly good approximation of a sine wave. (12.5)

Inverter, variable-voltage (VVI):   A variable-frequency inverter that uses variable voltage pulses to form a stairstep approximation of a sinusoidal voltage. (12.5)

Microstep operation: A movement of a stepper motor of a fraction of a step for each input pulse. The fraction is usually $\frac{1}{10}$, $\frac{1}{16}$, $\frac{1}{32}$, or $\frac{1}{125}$. *See also* Full-step and Half-step operation. (12.4)

Rotor: The cylindrical rotating part of an ac motor. (12.2)

Slip: The difference between the synchronous speed and the actual rotor speed of an ac motor. (12.2)

Stator: The stationary part of an ac motor. (12.2)

Stepper motor: An electric motor that transforms electrical pulses into equal increments of rotary shaft motion called steps. (12.4)

Synchronous speed: The speed at which the field of an induction motor rotates. (12.2)

EXERCISES

12.1 Figure 12.3 shows the forces on the conductors and the electrical diagram that correspond to the "before commutation" diagram in Figure 12.4a.

a. Construct diagrams similar to Figure 12.3a and b for the "during commutation" diagram in Figure 12.4b.

b. Construct diagrams similar to Figure 12.3a and b for the "after commutation" diagram in Figure 12.4c.

12.2 Imagine you are explaining how electric motors and generators work to a friend who has little technical background. Write a paragraph that explains the two facts that are the basis of the operation of motors and generators.

12.3 Write a paragraph that explains the function of the brushes and commutator in a dc motor and in a dc generator.

12.4 Explain what happens to the voltage at the output terminals of a dc generator if the commutator is replaced by two slip rings, one for each end of the armature winding. Assume that one brush is in contact with each slip ring.

12.5 A three-phase induction motor has a rated torque of 15 ft-lb and is connected to a load that has a starting torque of 30 ft-lb. The motor speed versus torque curve is given in Figure 12.7d. Explain what will happen when the motor is turned on.

12.6 Determine the velocity and position transfer functions of an ac servomotor with the following data.

Rated voltage: 120 V

Load inertia: 15×10^{-6} kg m^2

Load damping: 6×10^{-6} N·m·s/rad

Stall torque: 0.12 N·m

No-load speed: 5000 rpm

12.7 Explain dynamic breaking and regenerative braking.

12.8 An armature-controlled dc motor has the following ratings.

$$T_f = 0.03 \text{ N·m}$$
$$K_T = 0.15 \text{ N·m/A}$$
$$I_{max} = 2.4 \text{ A}$$
$$K_E = 0.15 \text{ V·s/rad}$$
$$\omega_{max} = 5000 \text{ rpm}$$
$$r = 0.8 \ \Omega$$

Determine the following:

a. The maximum output torque.

b. The maximum power output.

c. The maximum armature voltage.

d. The no-load motor speed when $e = E_{max}$.

12.9 The motor in Exercise 12.8 is operated at 3000 rpm with a load torque of 0.15 N·m. Determine the following:

a. The armature voltage.

b. The armature speed if the torque increases to 0.30 N·m and the armature voltage is not changed.

12.10 Determine the velocity and position transfer functions of the motor in Exercise 12.8. The following additional values are required:

$$J = 10.2 \times 10^{-4} \, kg \cdot m^2$$
$$b = 1.2 \times 10^{-4} \, N \cdot m \cdot s/rad$$
$$L = 0.045 \, H$$

12.11 Extend Table 12.2 to include full steps from 6 through 20.

12.12 Extend Table 12.3 to include half-steps from 11 through 40.

12.13 Extend Table 12.4 to include microsteps from 11 through 40.

12.14 Explain the voltage-variable inverter and the pulse-width-modulated inverter.

12.15 The SCR control in Figure 12.20 is to be used to control a dc motor with an armature resistance (r) of 0.4 Ω. The feedback voltage (E_a) must be 20% of the armature voltage (E) for proper operation of the control circuit. A total value of 10 kΩ for R_1 and R_2 is required. Determine the resistance values required for R_1, R_2, and R_3.

12.16 Show that Equation (12.26) can be transformed into the transfer function given by Equation (12.17).

12.17 Use Equation (12.34) to simplify the block diagram in Figure 12.20b. In doing this, a single block with a transfer function equal to K_m replaces the two blocks with transfer functions equal to R_3 and $R_2/(R_1 + R_2)$. Derive the transfer function of the control system from your new block diagram.

PART FOUR

Control

CHAPTER 13

Control of Discrete Processes

OBJECTIVES

A discrete process consists of a series of distinct operations with a definite condition for initiating each operation. When the series of operations has a beginning, an end, and a definite controlled form, the process is called a sequential process or a batch process. Discrete process operations can be grouped into two categories, those which are initiated by time and those which are initiated by an event. We call these categories time-driven operations and event-driven operations. The simplest discrete process consists of a single output that has only two possible values (e.g., on and off, high and low, hot and cold, etc.). More complicated discrete processes consist of a number of operations, and each operation may consist of a number of distinct parts called steps.

The purpose of this chapter is to give you an entry-level ability to discuss, select, specify, and design discrete process control systems. After completing this chapter, you will be able to

1. Given a description of a time-driven process, complete a statement list and a process timing diagram

2. Given a description of a time/event-driven process, complete the following:

   a. A statement list

   b. A sequential function chart

   c. A state chart

   d. Define the transition conditions

   e. Define the output functions

   f. Construct a ladder diagram of the controller

   g. Construct an output ladder diagram

   h. Document the design

3. Explain the ladder diagram of a sequential control system and prepare a statement list or a sequential function chart from the ladder diagram

4. Describe programmable controllers and use the terms in the glossary of this chapter in discussions about programmable controllers

13.1 INTRODUCTION

Discrete processes occur in a number of different places. Home appliances use discrete processes to wash dishes, to dry clothes, and to cook a frozen dinner. Manufacturing industries use a sequence of operations to produce discrete parts and assemble them into finished products. The chemical industry uses a batch process in a chemical reactor to produce a definite amount of a product. The food industry uses batch processes for operations such as sterilization, freeze drying, and extraction.

Some discrete processes are nonsequential in nature, consisting primarily of event-driven operations. These processes are controlled by programmable logic controllers (PLC) or by panels of mechanical relays. However, most discrete processes are sequential, consisting of a combination of event-driven operations and time-driven operations. Depending on where they are used, sequential, discrete processes are called sequential processes or *batch processes*. In the following discussion, the term *sequential process* will include both sequential and batch processes.

A *sequential process* consists of a sequence of one or more operations (called steps) that must be performed in a defined order. The completion of this sequence of steps creates a definite amount of finished product. The product may be a discrete part, such as an automobile; an amount of liquid, such as mouthwash; or an amount of solid material, such as sugar. The sequence of operations must be repeated to produce more of the product. Sequential processes have a *set of directions* that define each step. (In food processing the set of directions is called a recipe; in computer programming it is called an algorithm). Most sequential processes assemble specific parts or ingredients and process them according to the set of directions. The processing may be drilling, punching, painting, cooking, blending, reacting, or other operations that modify the product.

The steps in a sequential process can often be grouped into a few general operations, such as preparation, loading, processing, unloading, and post preparation. Each general operation consists of many individual steps. Each step is a single event, such as opening a valve, starting a motor, putting a controller on automatic control, and so on. The general operations are sometimes called phases. In this book they will be called *operations*, and the individual parts of an operation will be called *steps*.

In a strictly sequential process, each step must be completed before the next step can be initiated. However, many actual processes include *parallel operations* that are not part of any other operation. For example, high- and low-level switches may start and stop a pump to maintain the level of liquid in a tank between two limits. The supply of liquid in the tank may be part of a sequential process, but the control of the level is independent of the sequential control of the process.

Auxiliary operations are another type of operation that is not included in the main sequential process. The auxiliary operation appears as one or two steps in the main process. However, it consists of a number of steps that are not included in the main process. The auxiliary operation is usually initiated and terminated by steps in the main process and runs concurrently with the main process. An example of an auxiliary operation is a continuous PID controller used to control the temperature in a batch cooker. The main process includes steps that switch the controller from manual to automatic and from automatic to manual. The main process may also include steps that adjust the setpoint of the controller to different values to follow a prescribed time-versus-temperature curve.

Methods of describing a sequential process include statement lists, flowcharts, ladder diagrams, sequential function charts, state charts, process timing diagrams, and a mathematical (Boolean) language. The *statement list* is an English-language list of actions that must be carried out for each step. A *flowchart* uses blocks to represent each step and lines to show the path from step to step. A *ladder diagram* is an electrical diagram showing the connections between various contacts, relay coils, solenoids, motors, and so on. A *state chart* is a truth table that shows the outputs produced by each step and is usually accompanied by a diagram or chart that shows the transfer paths from state to state. The *sequential function chart* uses blocks to represent each step and lines to represent the transitions from step to step. A *process timing diagram* is a graph of the outputs plotted versus time. The mathematical language is similar to a computer program. These methods will be used and explained at appropriate points in the remainder of this chapter.

13.2 TIME-DRIVEN SEQUENTIAL PROCESSES

In a *time-driven* sequential process, each step is initiated at a given time, or after a given time interval. The statement list and the process timing diagram are two methods used to describe a time-driven process. You may recognize the following statement list from a popular product found in every supermarket.

Microwave Instructions

STEP 1. Remove tray from carton and peel back film from one end to vent.

STEP 2. Place tray in microwave oven.

STEP 3. Cook 10 minutes on medium power (50%) or defrost.

STEP 4. Rotate $\frac{1}{2}$ turn.

STEP 5. Cook 4 minutes on full power.

STEP 6. Let stand for 1 minute.

STEP 7. Remove film and serve.

Figure 13.1 shows an instrumentation and piping diagram for a batch blending process. This is another example of a time-driven sequential process. Two liquid ingredients are blended and heated until they are warm and thoroughly mixed. Then a solid material is added and the blending continues until the solid is completely dissolved in the liquid mixture. Finally, the blended product is pumped to a bottling line. The mixture is heated during the blending operation to facilitate the dissolving of the solid material.

The two flow controllers, FQC-100 and FQC-101, deliver measured quantities of the two liquid ingredients. Upon receiving a start signal from the sequential

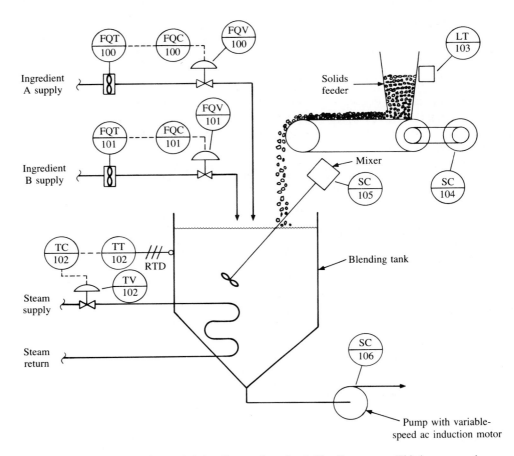

Figure 13.1   Instrumentation and piping diagram for a batch blending process. This is an example of a time-driven sequential process.

controller, FQC-100 and FQC-101 automatically deliver measured amounts of ingredients A and B to the mixing tank. When done, the controllers automatically shut off. The temperature controller, TC-102, controls the temperature of the contents of the blending tank. When the controller is in manual control, control valve TV-102 is turned off. When in automatic control, controller TC-102 manipulates the steam flow rate to maintain the temperature of the blending tank contents at the controller setpoint. The belt feeder is driven by an ac induction motor with an ac adjustable-speed drive, SC-104. The speed of the belt drive is adjusted by the operator to deliver the entire amount of solid material in slightly less than 5 minutes. The mixer and the pump are also driven by induction motors with ac adjustable-speed drives, SC-105 and SC-106. The following is a statement list of the process.

Statement List of the Batch Blending Process (Figure 13.1)

STEP 1 (4 minutes). Prepare the solid material and fill the solids feeder hopper with a measured amount of material.

STEP 2 (30 seconds). Press the START button.

STEP 3 (3 minutes). Fill the blending tank with a measured quantity of ingredients A and B.

STEP 4 (4 minutes). Heat and mix the contents of the blending tank (i.e., ingredients A and B).

STEP 5 (5 minutes). Gradually deliver solid material from the hopper to the blending tank. Continue to heat and mix the contents of the blending tank.

STEP 6 (1.5 minutes). Continue to heat and mix the contents of the blending tank.

STEP 7 (4.5 minutes). Pump the contents of the blending tank to the bottling line.

Figure 13.2 shows the process timing diagram for the batch blending process. Notice that the process is divided into five operations: preparation, filling, heating, blending, and draining. The filling and blending operations each have two steps. The other operations each have only one step. Batch processes often have several steps for each operation.

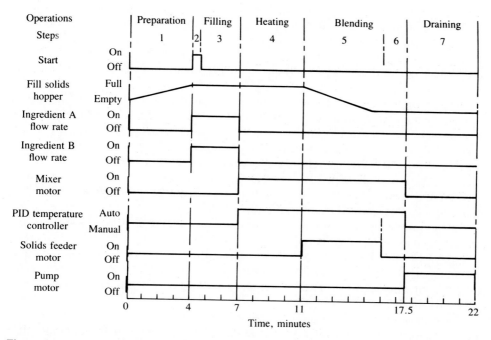

Figure 13.2   Process timing diagram for the batch blending process in Figure 13.1.

13.3 EVENT-DRIVEN SEQUENTIAL PROCESSES

In an *event-driven process*, each step is initiated by the occurrence of an event. The event may be a single action, such as an operator pressing a pushbutton, the closing of a limit switch, the opening of a pressure switch, or some other action that causes a switch to open or close. The event could also be a combination of several actions. For example, the event may consist of the simultaneous occurrence of several actions, with the contacts for each action connected in series. The event may also be the occurrence of any one of several actions, with the contacts for each action connected in parallel.

> The action-causing event in a sequential process is determined by an interconnection of one or more contacts. The event has occurred whether there is a closed path through this interconnection of contacts.

The *ladder diagram* is a very popular graphical method of describing an event-driven process. It was developed to represent systems consisting of switches, relays, solenoids, motor starters, and other switching components used to control industrial machinery. The ladder diagram is very easy to learn and very logical in its interpretation. Only one rule is required to understand a ladder diagram: When there is a closed path through a relay coil from one side to the other side, the coil will close its associated contacts. The ladder diagram got its name from its general appearance— it looks like a ladder. A ladder diagram consists of two vertical lines and any number of horizontal circuits that connect the two vertical lines. The horizontal circuits are called the *rungs* of the ladder diagram. The ladder diagram in Figure 13.3 has three rungs.

Each rung in a ladder diagram defines one operation (or step) in the process. A rung has two parts: an *input* section and an *output* element. In the top rung in Figure 13.3, the series connection of the two limit switches (LS1 and LS2) is the input section, and relay coil CR1 is the output element. Both limit switches must be closed to energize coil CR1. When CR1 is energized, all normally open (NO) contacts labeled CR1 will close, and all normally closed (NC) contacts labeled CR1 will open. These contacts may appear in any rung in the ladder diagram. The name "CR1" is used to associate a contact with the coil that operates the contact.

In the middle rung, the parallel combination of the level switch (LV1) and the pressure switch (PS1) is the input section, and solenoid A is the output element. If either LV1 or PS1 closes, solenoid A will be energized. In the bottom rung, the input section is the parallel/series combination of contacts A, B, and CR1, and solenoid B is the output element. Solenoid B is energized when the following expression is true: (A OR B) AND NOT CR1.

Boolean equations are another method of representing the rungs in a ladder diagram. The following Boolean equations represent the three rungs in the ladder diagram in Figure 13.3.

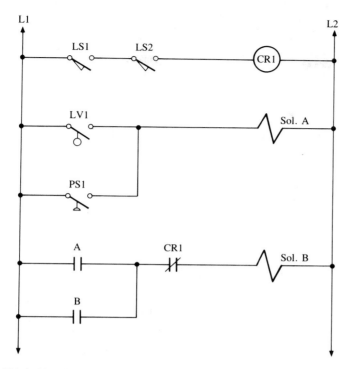

Figure 13.3   This ladder diagram has three rungs, each controlling one operation (or step) in an event-driven sequential process.

$$CR1 = LS1 \text{ AND } LS2$$
$$Sol. A = LV1 \text{ OR } PS1$$
$$Sol. B = (A \text{ OR } B) \text{ AND NOT } CR1$$

The hydraulic hoist in Figure 13.4 is an example of an event-driven control system. The schematic diagram shows the physical configuration, and the ladder diagram defines the sequential control system. The ladder diagram consists of six rungs between line L1 and line L2. Each rung consists of two or more switches or contacts on the left and a single operational element on the right. The operational element is turned on whenever there is a closed path through the rung from L1 to L2 (i.e., whenever all switches and contacts in the rung are closed).

The STOP symbol represents a *normally closed* pushbutton that is open only when someone is pressing the STOP button. The START, UP, and DOWN symbols represent *normally open* pushbutton contacts that are closed only when someone is pushing the appropriate button. The circle labeled 1M is the coil of a relay used to start the

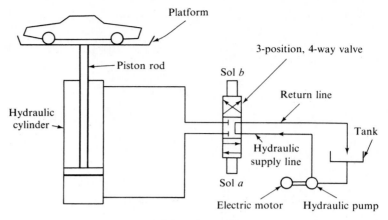

a) Schematic diagram

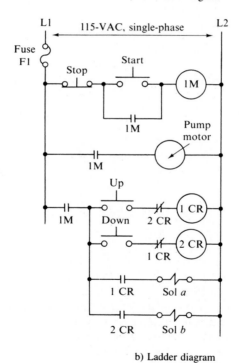

b) Ladder diagram

Figure 13.4   Event-driven sequential control system for a hydraulic hoist.

pump motor. The three contacts labeled 1M are open when coil 1M is off (deenergized), and they are closed when coil 1M is on (energized). The circles labeled 1CR and 2CR are the coils of two control relays. Each control relay has one normally open contact and one normally closed contact. The normally open contacts are open when the coil is deenergized and closed when the coil is energized. The normally closed contacts are just the opposite, closed when the coil is deenergized and open when the coil is energized.

The devices labeled "Sol a" and "Sol b" in the ladder diagram are the two coils of the three-position, four-way solenoid valve in the schematic diagram. The valve symbol consists of three position blocks located between the two coils. Each position block shows its method of connecting the four ways (the two lines on each side of the valve). The valve is shown in the position it occupies when both coils are deenergized. When "Sol a" is energized, the valve moves up one block, placing the bottom position block in line with the four ways. When "Sol b" is energized, the valve moves down, placing the top position block in line with the four ways. The ladder diagram includes an interlock feature that prevents simultaneous energizing of both solenoids.

The hydraulic hoist system is turned on by pressing the START button. This completes the path through coil 1M, so coil 1M is energized and all three contacts labeled 1M are closed. When the operator releases the START button, the 1M contact directly below the START button maintains the closed path and coil 1M remains energized. A contact used in this way is called a holding contact. The second 1M contact turns on the pump motor, and the third 1M contact activates the bottom half of L1. The operator can turn the hoist off at any time by pressing the STOP button. This will deenergize coil 1M and open the three 1M contacts. When the operator releases the STOP button, coil 1M remains off because both the START contact and the holding contact are open. This pushbutton and holding contact arrangement is a standard method of starting electric motors.

When the hoist is turned on, the bottom half of L1 is activated, and the operator can use the UP and DOWN pushbuttons to operate the hoist. When the operator presses the UP button, relay 1CR is energized, the normally open 1CR contact closes, and "Sol a" is energized. In the schematic diagram, the solenoid valve moves up because "Sol a" is energized. This connects the pump supply line to the bottom end of the cylinder and the top end of the cylinder to the return line. The pump forces hydraulic fluid into the bottom end of the cylinder and the piston moves up, raising the platform and its load. When the operator releases the UP button, the valve returns to its normal position, blocking both sides of the cylinder and locking the cylinder in its current position. When the operator presses the DOWN button, the platform moves down in a similar manner.

In summary: The platform goes up when the UP button is pressed, down when the DOWN button is pressed, and remains stationary when neither button is pressed. If the UP button is pressed when the cylinder is at the top, a relief valve (not shown) provides a direct path from the pump outlet to the tank. The same is true if the DOWN button is pressed when the cylinder is at the bottom. In addition, the system is interlocked so that both solenoids cannot be energized at the same time.

Automated Drilling Machine Example

The automated drilling machine in Figure 13.5 will be used as an example for a study of event-driven sequential processes. The drilling machine consists of an electric drill mounted on a movable platform. A hydraulic cylinder moves the platform and drill unit up and down between a *Drill Reset* position (up) and a *Hole Drilled* position (down). Limit switch LS1 is actuated when the drill platform is in the Reset position. Limit switch LS2 is actuated when the drill platform is in the Drilled position. Notice that the term *actuated* is used instead of the term *closed*. The reason is that limit switches have both normally open and normally closed contacts. Actuating a limit switch closes the normally open contacts and opens the normally closed contacts.

A stationary table supports the part (called the workpiece) in which a hole is to be drilled. The table has a hole in line with the drill bit, so the workpiece can be drilled clean through if desired. A single-action spring-return pneumatic cylinder clamps the workpiece against a bracket that is bolted to the workpiece table. Limit switch LS3 is actuated when the clamping cylinder is in the *Unclamped* position. Limit switch LS4 is actuated when the clamping cylinder is in the *Clamped* position.

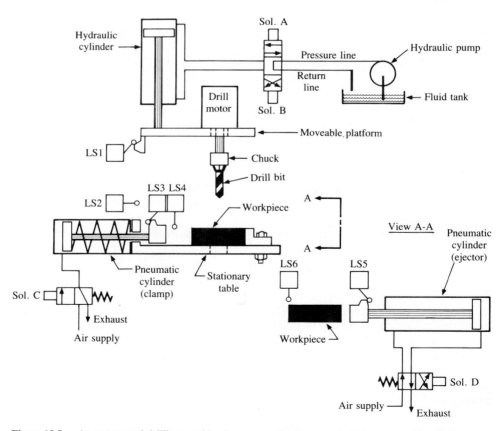

Figure 13.5   An automated drilling machine is an example of an event-driven sequential process.

A double-acting pneumatic cylinder ejects the workpiece after the hole is drilled. The ejector cylinder is oriented 90° from the clamping cylinder, so the workpiece slides along the clamping bracket as it is ejected. Limit switch LS5 is actuated when the cylinder is in the *Ejector Retracted* position. Limit switch LS6 is actuated when the cylinder is in the *Ejected* position.

Three solenoid valves are used to control the cylinders. A hydraulic three-position four-way valve controls the hydraulic cylinder. When the pump is running and solenoid A is energized, the cylinder will move the platform down. When solenoid B is energized, the cylinder will move the platform up. When neither solenoid is energized, the valve returns the fluid back to the supply tank and the platform is stationary. A two-position three-way pneumatic valve controls the clamping cylinder. When solenoid C is deenergized, the cylinder is vented and the spring returns it to the Unclamped position. When solenoid C is energized, the air supply is ported to the cylinder and it moves to the Clamped position. A two-position four-way valve controls the ejector cylinder. When solenoid D is deenergized, the ejector is forced into the Retracted position. When solenoid D is energized, the workpiece will be ejected as the cylinder moves to the Ejected position.

Sequential Function Chart

The first step in the design of a sequential control system for the drilling machine is to prepare a diagram or chart that describes the operations in the process. The sequential function chart shown in Figure 13.6 is an example of such a diagram. It uses a box to represent each step in the process. A double box is used for the first step in the cycle. The progression is from top to bottom, in order of the numbers of the steps. Arrows are not required to show the direction of the flow and are not used. A rectangular box on the right is used to describe the operation performed in each step. The condition for advancing to the next step is written next to the horizontal lines that cross the transfer path between two steps. For example, in the first step, the system waits for an operator to press the START button. The condition for advancing to step 2 is for an operator to press the START button. When this is done, we say that START is true and has a value of 1. When the START button is not pressed, we say that START is false and has a value of zero.

The sequential function chart has a provision for simultaneous sequences on parallel paths. A pair of horizontal lines just above the parallel paths indicates simultaneous activation of all the paths. Another pair of horizontal lines near the bottom indicates the simultaneous completion of all the paths. Each pair includes a wait step just above the lines, indicating simultaneous completion. Between the simultaneous start and the simultaneous completion, each path completes its operations independently of all the other paths. Parallel operations of this type are common in manufacturing workstations.

State Chart

A state chart is a truth table showing the condition of each output at each step in the cycle. The condition of the output is indicated by an × if the output is on or

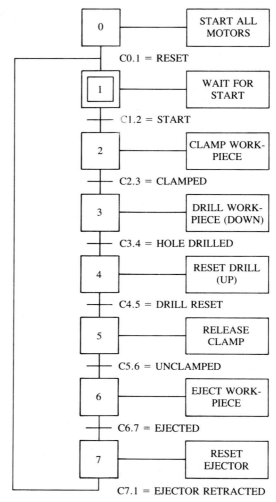

Figure 13.6 Function sequence chart for the drilling machine in Figure 13.5.

		START ALL MOTORS
0		

C0.1 = RESET

| 1 | | WAIT FOR START |

C1.2 = START

| 2 | | CLAMP WORK-PIECE |

C2.3 = CLAMPED

| 3 | | DRILL WORK-PIECE (DOWN) |

C3.4 = HOLE DRILLED

| 4 | | RESET DRILL (UP) |

C4.5 = DRILL RESET

| 5 | | RELEASE CLAMP |

C5.6 = UNCLAMPED

| 6 | | EJECT WORK-PIECE |

C6.7 = EJECTED

| 7 | | RESET EJECTOR |

C7.1 = EJECTOR RETRACTED

Table 13.1 State Chart for the Drilling Machine in Figure 13.5

Step No.	Sol. A	Sol. B	Sol. C	Sol. D	Motors
0					
1					×
2			×		×
3	×		×		×
4		×	×		×
5					×
6				×	×
7					×

nothing if the output is off. (Sometimes a 1 is used in place of the × and a 0 in place of nothing.) Table 13.1 is an example of a state chart.

Process Timing Diagrams

A process timing diagram similar to Figure 13.2 can be used to describe an event-driven process. The major difference is that the timing diagram for an event-driven process uses an arbitrary time scale while the timing diagram for a time-driven process uses a real-time scale. The timing diagram is a very useful tool for analyzing and designing an event-driven sequential process. It gives the designer a clear picture of the sequence in which the various inputs and outputs occur. In particular, it clearly shows the inputs and outputs whose values change during a particular step. The designer may need this information to determine whether or not a holding contact is necessary to complete an action initiated by a condition that changes before the action is completed. Figure 13.7 is the timing diagram for the drilling machine in Figure

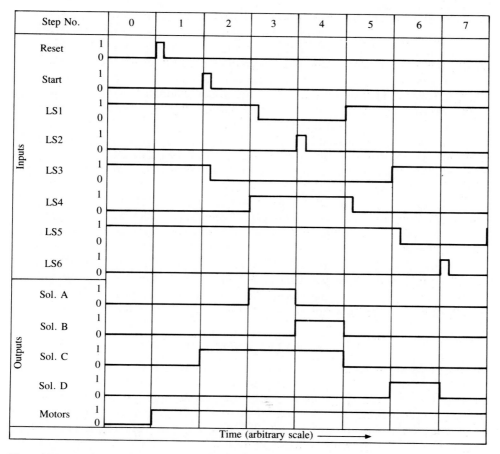

Figure 13.7 Process timing diagram for the drilling machine in Figure 13.5.

13.5. The 0 level in the timing diagram indicates that the device is deactivated or deenergized. The 1 level indicates that the device is activated or energized.

Sequential Circuit Design

This section presents a step-by-step procedure for the design of an event-driven sequential circuit. The purpose is to illustrate the design procedure and to obtain a finished ladder diagram that is easy to follow. The design does not attempt to minimize the number of relays and contacts that are used. With microcomputers and PLCs providing the control function, the number of relays and contacts is not a significant cost factor. It is usually more cost-effective if the design and its documentation are easy to understand, maintain, and modify than it is to reduce the number of components in the controller. The next example does, however, illustrate a design that uses fewer components.

The basic concept of the design is the use of a control relay for each step in the process. When the process is in a given step, the control relay associated with that step is energized and all other relays are deenergized. We say that each control relay represents a "state" of the controller. There is a one-to-one correspondence between steps in the process and states of the controller, and the step numbers are used to identify the states. When the operator presses the RESET button (or the control system is automatically reset), relay CR1 is energized, and all other relays are deenergized. The controller is in STEP 1, waiting for the START command. When the operator presses START, the controller moves to STEP 2 by energizing CR2 and deenergizing CR1. When STEP 2 is completed, the controller waits for the occurrence of the condition for transition from STEP 2 to STEP 3. When the transition condition occurs, the controller moves to STEP 3 by energizing CR3 and deenergizing CR2. This process of moving from step to step continues until the controller is back at STEP 1 and the cycle has been completed.

Each state in the controller consists of one rung in the controller ladder diagram. The general form and one specific example of these "state" rungs are shown in Figure 13.8. Each rung begins on the left with a normally closed contact from the control relay for the next state. This NC contact opens when the advance to the next state is made. Its purpose is to turn off the previous state. Otherwise, the controller would end up with all states energized at the end of the cycle. Next comes a parallel section. The bottom branch of the parallel section contains the holding contact. The top branch establishes the transition condition. The transition function is of the following form:

$$\text{STATE } j = \text{STATE } i \text{ AND CONDITION } i.j$$

The AND represents the logical and operator. CONDITION i.j is the condition required to move from STATE i to STATE j. In essence the transition function states that arrival at STATE j depends on two things. First, the controller must be in STATE i (i.e., the previous state). Second, the condition for the transition from STATE i to STATE j must be true. If these two conditions are satisfied, the controller will move to STATE j, and relay CRj will be energized.

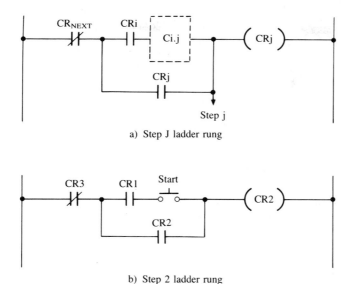

a) Step J ladder rung

b) Step 2 ladder rung

Figure 13.8 The rungs that establish the steps in a sequential control circuit have the general form shown above.

Figure 13.8b shows the ladder rung for STATE 2. The condition for moving to STATE 2 is (1) the controller is in STATE 1, and (2) the START button is pressed. We will use C1.2 to represent the condition for transfer from STATE 1 to STATE 2. It is obvious that C1.2 is described by the following expression.

$$C1.2 = START$$

The design process consists of the following steps:

1. *Define the process.* A piping and instrumentation drawing or a schematic diagram is an excellent method of graphically defining the process.
2. *Define the step (states).* The sequential function chart and the state chart are two methods of documenting the steps in the process.
3. *Define the input and output conditions.* The input and output conditions should be defined for each step in the process. Any values that change during a step should also be noted. The timing diagram is an excellent method of graphically defining the input and output conditions.
4. *Define the transition conditions.* The transition conditions are the Ci.j terms that define the condition for a transition from one state to the next state. They determine the transition portion of the circuit in the ladder rung for each state (or step). The timing diagram is an excellent tool for developing the transition conditions. The following equation defines the condition for transition from

STEP 2 to STEP 3:

$$C2.3 = CLAMPED = LS2$$

5. *Define the output functions.* The output functions are Boolean equations that define each of the output elements. The following output function defines solenoid A to be energized only during STEP 3.

$$Sol. A = STEP 3$$

6. *Construct the controller ladder diagram.* The controller ladder diagram has a rung similar to those in Figure 13.8 for each step in the process.
7. *Construct the output ladder diagram.* The output ladder diagram consists of a rung for each output element in the process. The output function defines the input section of each rung.
8. *Document the design.* The design documentation is a collection of the charts and diagrams described in design steps 1 through 7.

The first three steps of the design process have already been done. Figure 13.5 and the accompanying discussion covered step 1. The sequential function chart in Figure 13.6 and the state chart in Table 13.1 are the culmination of design step 2. The timing diagram in Figure 13.7 covers step 3. The remaining design steps follow.

Design Step 4: Define the transition conditions. The timing diagram in Figure 13.7 is most useful in determining each transition condition. Look for an input signal that changes exactly at the boundary between the present state and the next state. The signals are usually from some type of switch. Usually, there is only one signal that changes at the boundary, and the choice is obvious. Occasionally, there will be more than one input that changes at the boundary, and an arbitrary choice must be made. The following transition conditions were determined from an examination of Figure 13.7.

$$C0.1 = RESET$$
$$C1.2 = START$$
$$C2.3 = LS4$$
$$C3.4 = LS2$$
$$C4.5 = LS1$$
$$C5.6 = LS3$$
$$C6.7 = LS6$$

The function of the limit switches is obvious from the above set of transition conditions. It is also obvious that all six switches are required.

Design Step 5: Define the output functions. The output functions are obtained from either the timing diagram or the state chart. Examine each output and list all the steps during which a particular output is energized. The output function for that output is the logical OR of all the steps for which it is energized. The following output conditions were determined from an examination of Figure 13.7.

$$\text{Sol. A} = \text{STEP } 3$$
$$\text{Sol. B} = \text{STEP } 4$$
$$\text{Sol. C} = \text{STEP } 2 \text{ or } \text{STEP } 3 \text{ or } \text{STEP } 4$$
$$\text{Sol. D} = \text{STEP } 6$$
$$\text{Motors} = 1$$

The last output requires additional explanation. The motors are started by a START MOTORS pushbutton during STEP 0. From STEP 0, a RESET signal puts the controller into STATE 1 (or STEP 1). The control circuit is not involved in the starting or stopping of the two electric motors. Figure 13.9 shows the main power and motor starting circuit. The 110-V output of the transformer provides the electrical power for the control circuit (lines 2 and 6 in Figure 13.9) and for the output circuit (lines 1 and 7 in Figure 13.9).

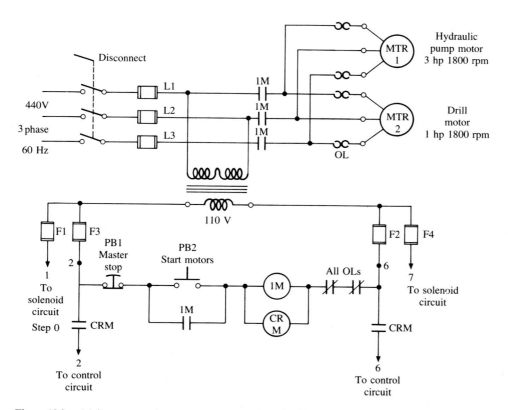

Figure 13.9 Main power and motor control circuit for the drilling machine in Figure 13.5.

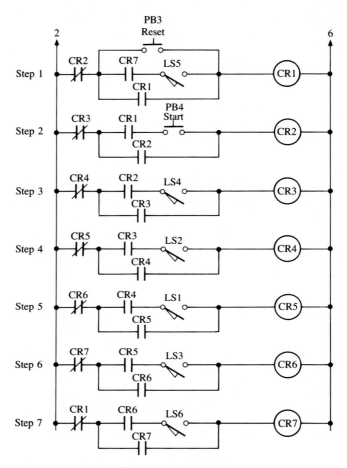

Figure 13.10 Control circuit for the drilling machine in Figure 13.5.

Design Steps 6, 7, and 8. The results of design steps 6 and 7 are shown in Figure 13.10 (the control circuit) and Figure 13.11 (the output circuit). Design step 8 consists of the following:

The schematic diagram (Figure 13.5)

The sequential function chart (Figure 13.6)

The state chart (Table 13.1) (optional)

The timing diagram (Figure 13.7)

The transition conditions

The output functions

The main power and motor control diagram (Figure 13.9)

The control system ladder diagram (Figure 13.10)

The output ladder diagram (Figure 13.11)

Figure 13.11 Solenoid circuit for the drilling machine in Figure 13.5.

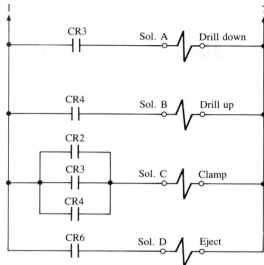

An Alternative Design

Figure 13.12 shows an alternative design for the drilling machine controller. This design reduces the number of relays from seven to five (four if CR4 and TD1 can be combined into a single time-delay relay). However, the design of the controller is more empirical than the previous design. An empirical design is similar to working a puzzle. The designer keeps trying different combinations until a design that works is found. In testing different ideas, the designer develops a "feel" for the process that helps to develop a satisfactory design.

In the alternative design, the process is divided into two major operations: drilling and ejection. The drilling operation consists of three steps. New step 1 is a combination of steps 1 and 2 from the original design. New step 2 is old step 3, and new step 3 is old step 4. The ejection operation is divided into two steps: the first step in ejection is old step 5; the second step is a combination of old steps 6 and 7. Actually, old step 7 is ignored and old step 6 is accomplished by a time-delay relay.

In an empirical design such as Figure 13.12, it is especially important to examine carefully a timing diagram of the control relays. Figure 13.13 shows such a diagram. A good way to check the design is to imagine that you are the controller, and walk through the operations for a complete cycle. The following is an example of a walk-through of the design in Figure 13.12 using the timing diagram in Figure 13.13.

The main power and motor control circuit is the same as in the original design. We begin in STEP 0 with the motors running and the ejector retracted (LS5 actuated). When the operator presses the START switch (Figure 13.12), CR1 is energized and the CR1 holding contacts (in parallel with the START switch) close, and holding relay CR1 is energized. The controller is in the drilling operation. A second CR1 contact energizes solenoid C, which causes the single acting air cylinder to clamp the workpiece.

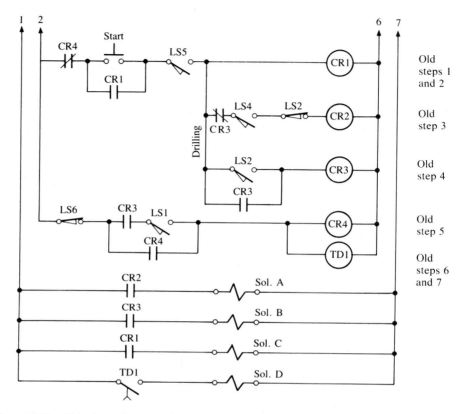

Figure 13.12 This alternative control circuit for the drilling machine in Figure 13.5 uses fewer components than the circuit in Figure 13.10, but the design of the alternative circuit is more empirical and may have some unexpected problems.

When the workpiece is clamped, LS4 is actuated and relay CR2 is energized. A CR2 contact closes to energize solenoid A, causing the hydraulic cylinder to move the drill platform down to drill a hole in the workpiece.

When the hole is drilled, switch LS2 is actuated. This opens the normally closed LS2 contact in the CR2 rung and closes the normally open LS2 contact in the CR3 rung. The result of these actions is that CR2 is deenergized and CR3 is energized. The CR3 holding contact is necessary because LS2 does not remain closed. When CR3 is energized, a CR3 contact closes to energize solenoid B. This causes the hydraulic cylinder to move the platform up, and LS2 soon becomes deactuated. The CR3 holding contact keeps the platform moving up until it reaches the drill reset position.

When the platform reaches the drill reset position, LS1 is actuated and relay CR4 is energized. The normally closed CR4 contact in the top rung opens, and coils CR1 and CR3 are deenergized. The controller has moved from the drilling operation to the ejection operation.

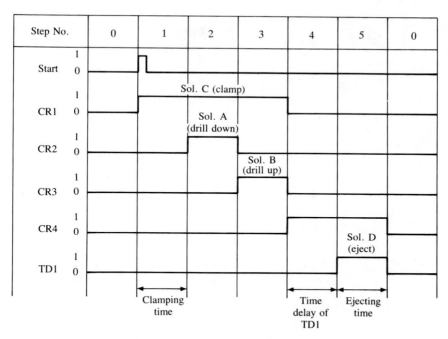

Figure 13.13   Timing diagram for the control circuit in Figure 13.12.

The ejection operation begins when CR4 is energized and time-delay relay TD1 begins a time delay. After the time delay, delayed TD1 contact closes and energizes solenoid D. The time delay allows the clamp to retract before the ejector removes the workpiece.

When the workpiece is ejected, LS6 is actuated, deactivating both CR4 and TD1. This ends the ejection operation. The controller is ready for another cycle as soon as the ejector is retracted and LS5 is actuated.

13.4   TIME/EVENT-DRIVEN SEQUENTIAL PROCESSES

Many discrete processes have both time-driven operations and event-driven operations. Batch processes, for example, are usually time/event-driven sequential processes. The following statement list describes a reaction and blending process that will serve as an example of a time/event-driven sequential process.

Statement List for a Reaction and Blending Process

STEP 1: prepare (nominal time = 3 minutes)

    a. Operator prepares ingredient A

    b. Operator prepares ingredient B

    c. Wait for START PB to close

STEP 2: charge (nominal time = 3 minutes)

    a. Initiated when operator presses START PB

    b. Ingredient A flow rate = 80% of FS

    c. Ingredient B flow rate = 65% of FS

    d. Mixer turned ON—stays ON until beginning of drain operation

    e. Level increases from 0 to 70%

STEP 3: reaction (exact time = 8 minutes)

    a. Initiated when level = 70%

    b. Ingredient A flow rate = 0

    c. Ingredient B flow rate = 0

    d. Temperature controller set to AUTO—stays in auto until end of blend operation

STEP 4: blend (exact time = 6 minutes)

    a. Initiated 8 minutes after start of reaction operation

    b. Ingredient C flow rate = 10 lb/min until total flow, $Q = 40$ lb

    c. Level increases to 95%

STEP 5: cool (nominal time = 3 minutes)

    a. Initiated 6 minutes after start of blend operation

    b. Temperature controller set to MANUAL

    c. Cooling water ON

STEP 6: drain (normal time = 3 minutes)

    a. Initiated when $T = 60°$

    b. Cooling water and mixer OFF

    c. Pump ON

    d. Terminated when level = 0%

Figure 13.14 is the process timing diagram for the reaction and blending process. The transition times in the diagram are nominal times. The actual transition times may occur before or after the nominal times. A large dot marks the transition signal or condition. There are four event-driven transitions and two time-driven transitions. The START PB, two level switches, and a temperature switch provide the event-driven transitions. Two time-delay relays provide the time-driven transitions. The following is a summary of these switching devices.

START PB = normally open pushbutton switch

LS1 = level switch set at a level of 70%

LS2 = level switch set at a level of 0%

TS1 = temperature switch set at 60°C

TD1 = time-delay relay set for a delay of 8 minutes

TD2 = time-delay relay set for a delay of 6 minutes

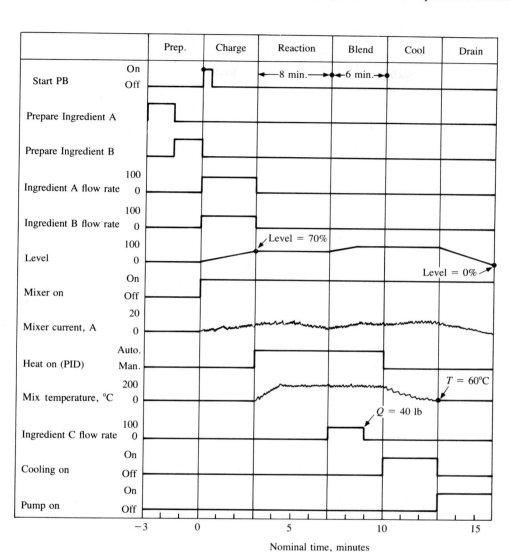

Figure 13.14 Process timing diagram for the example reaction and blending process. This process has both time-driven operations and event-driven operations.

All switches have a normally open contact and a normally closed contact. The delays in TD1 and TD2 occur when the coil is energized; they also have a normally open contact and a normally closed contact.

Ingredients A and B are controlled by PI flow controllers set at 80% of full scale and 65% of full scale, respectively. The charge operation consists of filling the blending tank up to the 70% full mark. When the 70% level is reached, the PI controllers are turned off and the flow of ingredients stops.

The reaction operation has a duration of exactly 8 minutes, determined by time-delay relay TD1. The PID temperature controller raises the temperature of the contents of the blending tank to 200°C and holds it there until the end of the blending operation. The blending operation has a duration of exactly 6 minutes, determined by time-delay relay TD2. The ingredient C flow controller determines both the flow rate of ingredient C and the total amount of ingredient C that is delivered to the blending tank. The flow rate of ingredient C is 10 lb/min; the total quantity delivered is 40 lb. The ingredient C controller turns itself off as soon as it delivers 40 lb to the blending tank.

The cool operation sends cool water through a coil to cool the contents of the blending tank. The cool operation ends as soon as the temperature in the blending tank reaches 60°C.

13.5 PROGRAMMABLE CONTROLLERS

A *programmable controller* (*PLC*) is a digital, electronic device designed to control machines and processes by implementing functions such as logic, sequence control, timing, counting, and arithmetic operations. A programmable controller is a digital computer—it has a processor, a memory unit, a control unit, and an input/output unit. However, it differs from a general-purpose computer in several ways. First, a programmable controller is designed to operate in harsh industrial environments. Second, it can be programmed without special programming skills. Third, a programmable controller can be maintained by factory maintenance technicians. Most programmable controllers can perform the following operations.

Relay logic	Latches
Timing	Counting
Shift register	Communication
Addition	Subtraction
Comparison	BCD operations
Binary conversion	PID controller
Data manipulation	Analog interface
AC interface	DC interface
Peripheral interface	

These operations make the programmable controller very useful for industrial control applications. Figure 13.15 shows the block diagram of a programmable controller.

Programmable controllers perform event-based operations that were originally intended to eliminate the high cost associated with inflexible, relay-controlled systems used by the automobile industry. The Hydramatic Division of the General Motors Corporation specified the design criteria for the first programmable controller in 1968. They specified a solid-state system with computer flexibility that would survive in the industrial environment, that could be easily programmed and maintained by plant engineers and technicians, and would be reusable. The intent of the specification was to reduce the time and cost involved as the car models changed. When relay logic control panels were used, hundreds of hard-wired relay panels were junked each

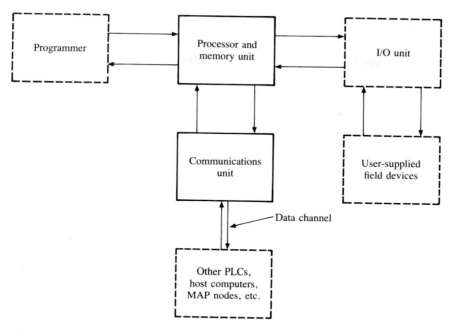

Figure 13.15   Block diagram of a programmable controller.

time a car model was changed. From this beginning in 1968, programmable controllers have developed into much more than just an inexpensive replacement for relay logic panels. The newer programmable controllers can perform arithmetic operations, analyze data, communicate with other programmable controllers and host computers, provide PID control of a process variable, and control discrete sequential processes. Some advantages of programmable controllers over relay control panels include:

Greater reliability

Smaller space requirements

Can be programmed for new applications

Can perform more functions

Easier to maintain

Less expensive

Programmable controllers are programmed with the aid of a device called a *programmer.* The program specifies the operations the PLC is to perform to carry out the desired control functions. A programmer is a device that allows the operator to enter a new program, examine the program in memory, change the program in memory, monitor the status of inputs or outputs, display the contents of registers, and display timer and counter values. In addition, a password can be entered to protect the program in memory from unauthorized changes. Programming devices include hand-held programmers, CRT terminals, and personal computers (with special software). The hand-held programmer resembles a pocket calculator. It has an LCD for displaying instructions, addresses, timer and counter values, data, and so on. It has a keypad for entering instructions, addresses, and data. A *CRT terminal* looks like a personal computer, but it is specifically designed for programming a programmable controller. The IBM PC has become very popular for a variety of control functions, including programming programmable controllers. The CRT and IBM PC programmers allow the operator to type in the program using symbols that are very similar to the contact and coil symbols used in ladder diagrams.

Most programmable controllers are programmed using a symbolic language that is very similar to the ladder diagrams used for relay logic controllers. This method of programming is sometimes called *contact symbology*, but more often it is simply called *ladder diagram programming*. As programmable controllers acquired more functions, ladder diagram programming became more cumbersome. Higher-level languages were developed to supplement ladder diagram programming. In the future, programmable controllers are expected to utilize more fully the computer capabilities of their processors. Languages such as C, Pascal, and BASIC will all have a place in programmable controllers.

Power flow is an important concept in the programming of a programmable controller. Power always flows from left to right, up or down, but never from right to left. In a relay panel, power flow from right to left through a closed contact is called a *sneak path*. Sneak paths are not allowed in programming a PLC. Not allowing sneak paths simplifies the programming, but does require some adjustment when converting a relay panel to a PLC program. The sneak paths must be converted into equivalent ladder diagram programs that do not have a sneak path.

The trend in programmable controllers has been to smaller controllers that are distributed near the process rather than large centralized controllers. Instead of one large programmable controller with thousands of inputs and outputs, users now tend to specify several small PLCs installed close to the process. A supervisory PLC communicates with the individual PLCs on a LAN to coordinate their activities. One reason for this trend is the tremendously complex program required to control a centralized system with one large PLC. Breaking the process into smaller, more manageable portions simplifies the program and also allows the use of smaller, less expensive PLCs.

Figure 13.16 shows a training and industrial work cell in which a CIM control station is networked to two CNC-controlled machines and three PLC-controlled robots. The work cell consists of the following parts:

1. A computer-controlled conveyor
2. A CNC-controlled lathe
3. A PLC-controlled transverse robot to load and unload the lathe
4. A CNC-controlled milling machine
5. A PLC-controlled pick-and-place robot to load and unload the mill
6. A PLC-controlled SCARA robot for drilling or small assembly on the conveyor
7. A computer-controlled storage rack (ASRS) Automatic Warehouse
8. A master computer (the CIM control station) to coordinate the entire operation

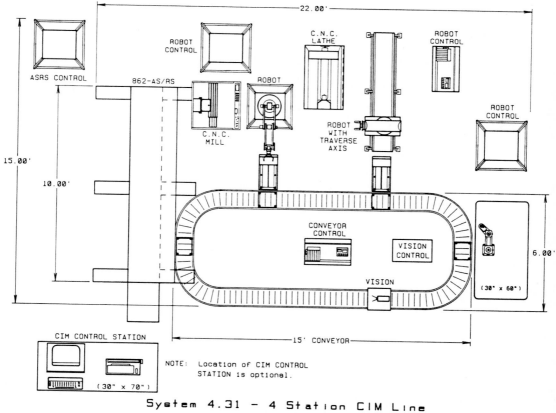

ASRS CONTROL

862-AS/RS

ROBOT CONTROL

C.N.C. MILL

ROBOT

C.N.C. LATHE

ROBOT WITH TRAVERSE AXIS

ROBOT CONTROL

ROBOT CONTROL

CONVEYOR CONTROL

VISION CONTROL

VISION

(30" x 60")

22.00'

15.00'

10.00'

6.00'

15' CONVEYOR

CIM CONTROL STATION

(30" x 70")

NOTE: Location of CIM CONTROL STATION is optional.

System 4.31 - 4 Station CIM Line
Basic Dimensions

Scale: 3/8" = 1'-0"

Figure 13.16 A training and industrial work cell contains two CNC-controlled machines and three PLC-controlled robots, all networked to a CIM control station (Courtesy of Amatrol, Inc.)

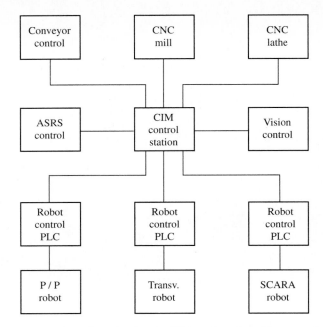

Figure 13.17 A star network configuration for the CIM work station. All communication is between the CIM control station and the individual units. Communication from one unit to another must pass through the CIM control station.

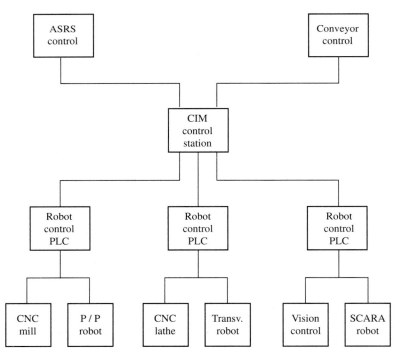

Figure 13.18 A semi-star network configuration for the CIM work station. The CIM control station communicates with the three PLCs, and each PLC communicates with the machine and robot under its control.

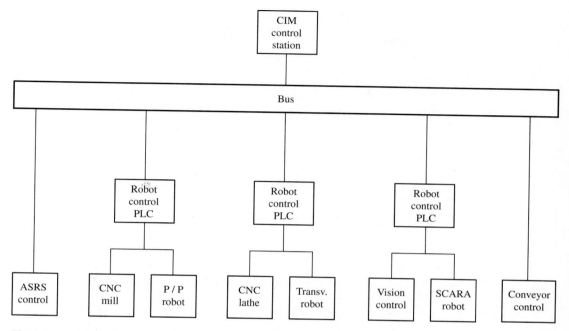

Figure 13.19   A bus network configuration for the CIM work station. The PLCs can communicate with each other without going through the CIM control station.

Three network configurations can be used with the work cell. Figure 13.17 shows a star configuration, Figure 13.18 shows a semi-star configuration, and Figure 13.19 shows a bus configuration. The star configuration is the least flexible and the easiest to program. The bus configuration is the most flexible and the hardest to program.

GLOSSARY

AC input interface: An input circuit that converts various ac signals from user devices to logic levels that can be used by a programmable controller. (13.5*)

AC output interface: An output circuit that converts logic levels from a programmable controller into an output signal capable of controlling an ac load. (13.5*)

Analog input interface: An input circuit that converts an analog input signal into a digital signal that can be used by a programmable controller. (13.5*)

Analog output interface: An output circuit that converts a digital number from a programmable controller into an analog signal for use by a user's analog device. (13.5*)

Backplane: A printed circuit board located in the back of a card cage. It has sockets that receive printed circuit card modules. (13.5*)

---

* Additional programmable controller terms that are not covered in the text.

Batch process: *See* Discrete process; Sequential process. (13.1)

Boolean equations: A mathematical method of representing the rungs in a ladder diagram. (13.3)

Contact histogram: A display of the ON and OFF times for any selected contact in a programmable controller. (13.5*)

Contact symbology diagram: A graphic expression of the logic to be implemented by a programmable controller. Also referred to as a ladder diagram. (13.5)

CRT terminal: A device that includes a CRT display and a keyboard that is used to write and examine programs in a programmable controller. Personal computers with special software can also be used for the same purpose as a CRT terminal. (13.5)

Data highway: A single-cable serial transmission line that provides communication among programmable controllers, data terminals, and computers. (13.5*)

Data link: The transmission cables and associated equipment that handle the transmission of information. (13.5*)

Discrete process: A series of distinct operations with a definite condition for initiating each operation. If the series of operations has a beginning, an end, and a definite controlled form, the process is called a sequential process or a batch process. (13.1)

Drop line: A line that connects a data highway to a programmable controller or other terminal device. (13.5*)

Event-driven process: A sequential process in which each step is initiated by the occurrence of an event. (13.3)

Fault: A malfunction that interferes with the normal operation of a programmable controller. (13.5*)

Flowchart: A diagram that uses blocks to represent the steps in a sequential process and lines to show the path from step to step. (13.1)

I/O interface: The circuit that goes between a programmable controller and the user devices. (13.5*)

I/O scan time: The time required for a programmable controller to monitor all the inputs and update all the outputs. (13.5*)

Ladder diagram: An electrical diagram that shows the connections between various contacts, coils, solenoids, motors, etc. in a sequential process. The ladder diagram got its name from its general appearance—it looks like a ladder. (13.1)

Module: A circuit that mates with a standard socket in a rack that usually has a number of identical sockets. Any of a number of interchangeable modules can be used in any of the sockets to adapt the rack to a number of different applications. I/O modules and an I/O rack are often used to form the I/O unit of a programmable controller. (13.5*)

Operation: A major step in a sequential process such as preparation, loading, processing, etc. An operation consists of one or more parts called steps. (13.1)

Operations, auxiliary: An operation that is initiated and terminated by steps in a sequential process but operates concurrently with the main process. A PID controller is an example of an auxiliary operation. (13.1)

Operations, parallel: Operations in a sequential process that occur at the same time. (13.1)

Output functions: Boolean equations that define each output element in a sequential process. (13.3)

PID module: An optional module for a programmable controller that provides a PID controller for control of a variable in the process. (13.5*)

Process timing diagram: A graph of the outputs of a sequential process plotted versus time. (13.1)

Programmable controller (PLC): A digital, electronic device designed to control machines and processes by implementing functions such as logic, sequence control, timing, counting, and arithmetic operations. (13.5)

Programmer: A device used to enter, examine, and change programs stored in a programmable controller. (13.5)

Sequential function chart: A diagram that uses blocks to represent each step in a sequential process and lines to represent the transition from step to step. (13.1)

Sequential process: A sequence of operations that must be performed in a defined order to make a definite amount of finished product or to perform a definite amount of work. (13.1)

Sneak path: Power flow in a relay panel that travels from right to left through a closed contact. (13.5)

State chart: A truth table that shows the condition of each output in a sequential process at each step in the cycle. The state chart is usually accompanied by a diagram or chart that shows the transfer paths from state to state. (13.1)

Statement list: An English-language list of the actions that must be carried out in each step of a sequential process. (13.1)

Steps: The individual parts of a sequential operation. (13.1)

Time-driven process: A sequential process in which each step is initiated at a given time or after a given time interval. (13.2)

Transition conditions: The condition for a transition to occur from one state to another state. (13.3)

EXERCISES

13.1 Construct a process timing diagram from the microwave instructions in Section 13.2. Estimate times for STEPS 1, 2, 4, and 7.

13.2 Write Boolean equations for the ladder rungs in Figure 13.7.

13.3 Write a Boolean equation for the ladder rung in Figure 13.8. Use the term NOT OL to account for all the overload contacts.

13.4 Write the Boolean equations for the ladder diagram in Figure 13.9. There will be seven equations, one for each rung in the ladder.

13.5 Convert the ladder diagram in Figure 13.11 into a set of Boolean equations.

13.6 Convert the following set of Boolean equations into a ladder diagram. The letters A through F represent normally open contacts.

$$CR1 = A \text{ AND } B \text{ AND } C$$
$$CR2 = D \text{ OR } E \text{ OR } F$$
$$CR3 = \text{NOT } CR1 \text{ AND NOT } CR2 \text{ AND } (LS1 \text{ OR } PS1)$$
$$SOLA = CR2$$

13.7 Convert the following set of Boolean equations into a ladder diagram.

1M OR CR1 = NOT STOP AND (MSTART OR 1M) AND NOT OL

CR2 = CR1 AND NOT CR3 AND (START OR CR2)

CR3 = CR1 AND NOT CR4 AND (LS2 OR CR3)

CR4 = CR1 AND NOT CR2 AND (LV1 OR CR4)

13.8 Sketch the circuit diagram of a sequential control system with the following features.

(1) Motor starter 1M is energized by pressing a momentary contact start switch.

(2) Motor starter 1M can be deenergized only by pressing a momentary contact stop switch.

13.9 Modify the circuit diagram of Exercise 13.8 to include the following additional features.

(1) Limit switch 1L must be closed before the start switch can energize the motor starter 1M.

(2) Either opening limit switch 1L or pressing the stop switch can deenergize motor starter 1M.

13.10 Modify the circuit diagram of Exercise 13.8 to include the following additional features.

(1) Limit switch 1L must close before the start switch can energize the motor starter 1M.

(2) Opening limit switch 1L cannot deenergize the motor starter. The motor starter 1M can be deenergized only by pressing a momentary contact stop switch.

13.11 Construct a timing diagram of the control relays in Figure 13.10.

13.12 Complete the following for the reaction and blending process described by the statement list in Section 13.4 and the process timing diagram in Figure 13.14.

a. Draw an instrumentation and piping diagram of the process. Assign a number for each control loop in the process. Use the following switching devices described in the text: START PB, LS1, LS2, TS1. Add any additional components you consider necessary or advantageous.

b. Construct a sequential function chart.

c. Construct a state chart.

d. Determine the transition conditions.

e. Determine the output conditions.

f. Construct the sequential controller ladder diagram.

g. Construct the output ladder diagram.

13.13 Assemble the documents prepared in Exercise 13.12 into a design document for the reaction and blending process. Add copies of the statement list and the process timing diagram to complete your design document.

CHAPTER 14

Control of Continuous Processes

OBJECTIVES

A continuous process has uninterrupted inputs and outputs. The value of at least one input is changed in a manner that tends to maintain the controlled variable equal to the setpoint. The output of a continuous-process controller is determined by one or more modes of control. The most common control modes are the two-position, floating, proportional, integral, and derivative modes. Usually, the proportional mode is combined with the integral and/or derivative modes to form two- or three-mode controllers. Continuous process controllers can be grouped into two categories, those in which the setpoint is constant for long periods of time and those in which the setpoint is constantly changing. Control systems in the first category are called regulator systems, and those in the second category are called servo systems. However, control system analysis and design methods work equally well on systems in either category.

The purpose of this chapter is to give you an entry-level ability to discuss, select, and specify continuous process controllers. After completing this chapter, you will be able to

1. Describe the two-position, floating, proportional, derivative, and integral control modes and sketch the time graph of the response of each mode, given a time graph of the error

2. Describe the conditions for which each of the following control modes would be a good choice: two-position, floating, proportional, proportional plus derivative (PD), proportional plus integral (PI), and proportional plus integral plus derivative (PID)

3. Identify the time-domain equation and the transfer function for each of the following control modes: proportional, derivative, integral, PI, PD, and PID

4. Sketch the Bode diagram of each of the following control modes: proportional, derivative, integral, PI, PD, and PID

5. Sketch diagrams of each of the following analog controllers: P, PI, PD, and PID

6. Explain data sampling and discuss the control algorithms used in digital controllers

7. Describe cascade control and give an example of a cascade control system

8. Describe feedforward control and give an example of a feedforward control system

9. Discuss adaptive control and explain two approaches to adaptive self-tuning controllers

10. Discuss multivariable control and give an example of a multivariable control system

14.1 INTRODUCTION

The block diagram of a closed-loop control system was introduced in Chapter 1 and is reproduced in Figure 14.1. Control is achieved by performing the following three operations:

Measurement. Measure the value of the controlled variable.

Decision. Compute the error (desired value minus measured value) and use the error to form a control action.

Manipulation. Use the control action to manipulate some variable in the process in a way that will tend to reduce the error.

The controller accomplishes the decision step and is the subject of this chapter. The controller consists of an error detector and a control mode unit. The error detector computes the error by subtracting the measured value (C_m) from the setpoint (SP). The control mode unit uses the error signal to produce the control action (V).

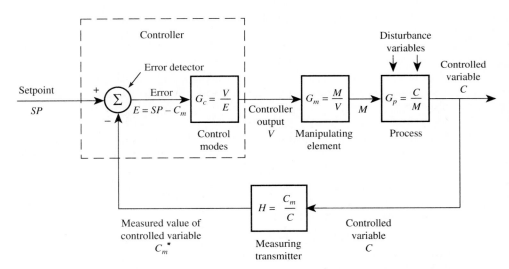

Figure 14.1 Block diagram of a closed-loop control system. The controller consists of an error detector and a control mode unit.

One of the most important characteristics of a controller is the way it uses the error to form the control action. The different ways the controller forms the control action are called *modes of control*. Common modes of control include (1) two-position, (2) floating, (3) proportional, (4) integral, and (5) derivative. Chapter 14 begins with a detailed study of these five modes of control. The methods used to describe control modes include input/output graphs, time-domain equations, frequency-domain equations, transfer functions, and Bode diagrams. Table 14.1 lists some variables that are used throughout this chapter.

The controller can be implemented by pneumatic circuits, analog electronic circuits, or digital electronic circuits. Pneumatic controllers use a pneumatic equivalent of the operational amplifier to generate the control action. Electronic analog controllers use a resistive circuit to compute the error and an operational amplifer to generate the control action. Digital controllers use a microprocessor and a control algorithm to generate the control action. Section 14.3 covers electronic analog controllers, and Section 14.4 deals with microprocessor-based digital controllers.

The chapter concludes with an overview of advanced control methods in Section 14.5. Advanced control refers to various methods of going beyond the single-loop single-variable feedback control system with three modes of control. Cascade control involves two controllers in which the output of the primary controller is the input to the secondary controller. Feedforward uses a model of the process to make changes in the controller output in response to measured changes in a major load variable without waiting for the error to occur. Adaptive control changes controller parameters to "adapt" to changes in the process. Expert system control uses artificial intelligence to incorporate the knowledge of control experts into the adaptive control system. Multivariable control uses measurements of several process load variables and may involve the manipulation of more than one process variable.

Table 14.1   Variables Used to Define Control Modes

Symbol	Description	Units
D	Derivative action time constant	seconds
e	Time-domain error	% of F.S.[a]
E	Frequency-domain error	% of F.S.
I	Integral action rate	1/s
P	Proportion gain	[b]
v	Time-domain controller output	% of F.S.
v_0	Controller output when error $= 0$	% of F.S.
V	Frequency-domain output[c]	% of F.S.
α	Derivative limiter coefficient	(none)

[a] F.S. = full scale.

[b] Gain is dimensionless when error and output are expressed in percent of full scale.

[c] Frequency-domain equations are simplified by the assumption that v_0 is zero.

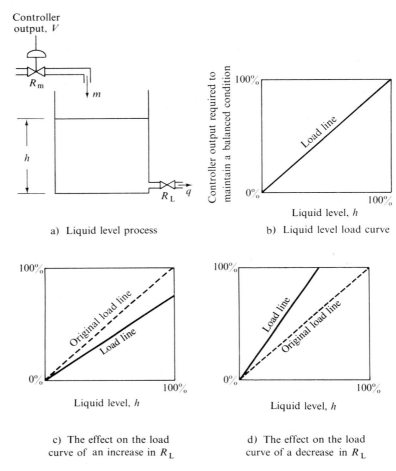

a) Liquid level process

b) Liquid level load curve

c) The effect on the load
curve of an increase in R_L

d) The effect on the load
curve of a decrease in R_L

Figure 14.2 The load line of a liquid level process depends on the outlet flow resistance, R_L. A nominal value of R_L produces the original load line (b). A value of R_L larger than the nominal moves the load line down (c), and a smaller value moves the load line up (d).

A brief review of the effect of *load changes* will be quite helpful in developing an understanding of the function of the controller. The first-order lag process illustrated in Figure 14.2 will be used to develop the concept of a process load curve. The liquid process in Figure 14.2a consists of a tank with liquid level (h), input valve resistance (R_m), and output resistance (R_L). The input valve is the final control element, and the input flow rate (m) is the manipulated variable. The liquid level (h) is the controlled variable. The output resistance (R_L) is the disturbance variable. Load changes are created whenever R_L changes.

The process is maintained in a balanced condition by adjusting the input flow rate (m) until it balances the output flow rate (q). Assuming laminar flow, the output flow rate (q) is given by

$$q = \left(\frac{g\rho}{R_L}\right)h \tag{14.1}$$

where g = acceleration by gravity = 9.81 meter/second^2
        ρ = liquid density, kilogram/cubic meter
       R_L = liquid resistance, pascal second/cubic meter
       h = liquid level, meter

If R_L is constant, q is proportional to h. For each level (h), there will be a specific output flow rate (q), and hence a specific input flow rate (m) necessary to maintain a balanced condition. Each input flow rate (m) requires a specific position of the final control element (R_m), which, in turn, requires a specific controller output signal (v). Eliminating the middle variables, each level requires a different controller output signal to maintain a balanced condition in the process. This concept is the basis of the process load curve in Figure 14.2b.

Now let us examine what happens to the load curve if R_L changes. For example, if R_L increases for some reason, the output flow rate (q) will decrease [see Equation (14.1)]. This will be reflected in a decrease in the controller output required to maintain a balanced condition, as illustrated by the load curve in Figure 14.2c. On the other hand, if R_L decreases, the output flow rate (q) will increase [Equation (9.1)]. This will be reflected in an increase in the controller output, as indicated in Figure 14.2d. A load change is any condition that changes the location of the load line. After a load change, the controller must adjust its output to the value determined by the new load line. This change is accomplished by the action of the control modes in the controller.

14.2   MODES OF CONTROL

Two-Position Control Mode

The *two-position control mode* is the simplest and least expensive mode of control. The controller output has only two possible values, depending on the sign of the error. If the two positions are fully open and fully closed, the controller is called an *on–off controller*. Most two-position controllers have a neutral zone to prevent chattering. The neutral zone is a range of values around zero in which no control action takes place. The error must pass through the neutral zone before any control action takes place. Figure 14.3 shows the input/output relationship of a two-position controller.

The two-position control mode supplies pulses of energy to the process, which causes a cycling of the controlled variable. The amplitude of the cycling depends on three factors: the capacitance of the process, the dead-time lag of the process, and the size of the load change the process is capable of handling. The amplitude of the oscillation is decreased by either increasing the capacitance, decreasing the dead-time

Figure 14.3 The output of a two-position controller has one of two values, depending on the value of the error signal from the error detector. The neutral zone prevents chattering (i.e., oscillation of the output between the two values when the error fluctuates around zero).

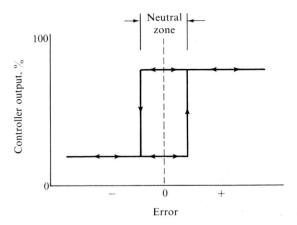

lag, or decreasing the size of the load change that the process can handle. Two-position control is suitable only for processes that have a large enough capacitance to counteract the combined effect of the dead-time lag and the load-change capability of the process. Two-position control is simple and inexpensive. Its use is preferred whenever the cycling can be reduced to an acceptable level.

A household heating system is an example of a two-position control system. The air in the house has a relatively large thermal capacitance, and the dead-time lag is small. The rate of heat input from the furnace is just sufficient to heat the house on the most severe winter day, and is small compared with the capacitance of the room. The room temperature cycles with an amplitude that is well within the acceptable limits of human comfort. This is an example of a good application of the two-position control mode. The two positions provide inputs equal to the maximum and minimum process loads (i.e., from no heat to just enough heat to handle the coldest day). A poor design is one that uses a furnace 10 times as large as required. The large rate of heat input from an oversized furnace results in a large amplitude of oscillation.

> The two-position control mode is used on processes with a capacitance large enough to reduce the cycling to an acceptable level. This implies a large capacitance process with a small dead-time lag and small load changes.

Example 14.1

The liquid-level process in Figure 14.2 is controlled by a two-position controller that opens and closes the inlet valve. The inlet flow rate (m) is 0 when the valve is closed and 0.004 m³/s when the valve is open. The oscillations in level are small enough that the outlet flow rate (q) is essentially constant at 0.002 m³/s. The tank has a cross-sectional area of 2.0 m², and the process dead-time lag is 10 s. The neutral zone of the controller is equivalent to a ± 0.005-m change in level. Determine the amplitude and period of the oscillation in level (h).

Solution

1. The rate of accumulation of liquid in the tank (a) is equal to the inflow minus the outflow.

$$a = m - q \qquad m^3/s$$

The rate of change of the level (dh/dt) is equal to the rate of accumulation divided by the area of the tank (A).

$$\frac{dh}{dt} = \frac{a}{A} = \frac{m - q}{2.0} \qquad m/s$$

When the valve is open,

$$\frac{dh}{dt} = \frac{0.004 - 0.002}{2.0} = 0.001 \ m/s$$

When the valve is closed,

$$\frac{dh}{dt} = \frac{0 - 0.002}{2.0} = -0.001 \ m/s$$

2. A graph of the oscillation is illustrated in Figure 14.4. At point a, the valve is closed. The level is changing at a rate of -0.001 m/s (decreasing), and the level has reached the lower limit of the neutral zone. After the 10-s dead-time lag, the controller opens the valve at point b. The level has decreased an additional amount equal to $(10 \ s) \times (-0.001 \ m/s) = -0.01 \ m$.

$$t_b - t_a = 10 \ s$$
$$h_b - h_a = -0.01 \ m$$

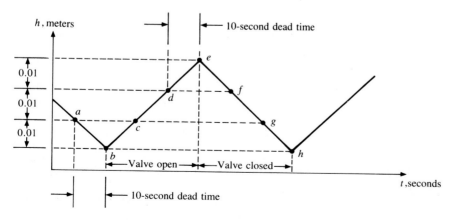

Figure 14.4   Oscillation of the level of the liquid tank in Example 14.1.

3. The valve is open from point b to point e, and the level is increasing at a rate of 0.001 m/s. It takes 10 s to reach point c.

$$t_c - t_b = 10 \text{ s}$$
$$h_c - h_b = 0.01 \text{ m}$$

The time required to move from point c to point d is equal to the change in level (0.01 m) divided by the rate of change of level (0.001 m/s).

$$t_d - t_c = \frac{0.01}{0.001} = 10 \text{ s}$$
$$h_d - h_c = 0.01 \text{ m}$$

At point d, the level has reached the upper level of the neutral zone. After the 10-s dead-time lag, the controller closes the valve at point e.

$$t_e - t_d = 10 \text{ s}$$
$$h_e - h_d = 0.01 \text{ m}$$

4. Since the rate of increase and the rate of decrease of the level are equal, the time from e to h is the same as the time from b to e.

$$t_h - t_e = t_e - t_b = 30 \text{ s}$$

5. The amplitude of oscillation is equal to $h_e - h_b$, and the period is equal to $t_h - t_b$.

$$\text{Amplitude} = (h_c - h_b) + (h_d - h_c) + (h_e - h_d)$$
$$= 0.01 + 0.01 + 0.01$$
$$= 0.03 \text{ m}$$
$$\text{Period} = (t_e - t_b) + (t_h - t_e)$$
$$= 30 + 30$$
$$= 60 \text{ s}$$

Floating Control Mode

The *floating control mode* is a special application of the two-position mode in which the final control element is stationary as long as the error remains within the neutral zone. When the error is outside the neutral zone, the final control element changes at a constant rate in a direction determined by the sign of the error. The final control element continues to change until the error returns to the neutral zone, or until the final control element reaches one of its extreme positions. The input/output curve of a floating controller is illustrated in Figure 14.5.

The floating control mode has a tendency to produce cycling of the controlled variable. The amplitude of the cycling depends on the dead-time lag of the process,

Figure 14.5   Input/output graph of the floating control mode.

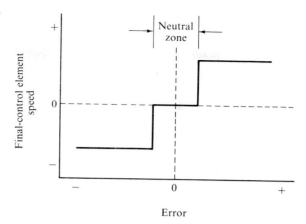

the capacitance of the process, and the speed at which the controller increases and decreases the final control element. The speed of the final control element determines the fastest load change that the controller can keep pace with, but not the size of the load change. The main advantage of floating control is its ability to handle load changes by gradually adjusting the final control element. As with two-position control, the amplitude is decreased by increasing the capacitance, decreasing the dead-time lag, or decreasing the speed of the final control element. Floating control is used when large load changes are anticipated, and the capacitance is large enough to counteract the effects of the dead-time lag and the speed of the final control element. Floating control is frequently used because it is inherent in the type of actuator used to drive the final control element (e.g., electric motors and hydraulic cylinders operated by on–off relays, or solenoid valves provide a floating control mode).

> The floating control mode is used on processes with large, slow-moving load changes and a capacitance large enough to reduce the cycling to an acceptable level. This implies a large capacitance process with a small dead-time lag. The floating control mode is inherent in some final control elements.

Proportional Control Mode

The *proportional control mode* produces a change in the controller output proportional to the error signal. There is a fixed linear relationship between the value of the controlled variable and the position of the final control element. A simple example of a proportional controller is shown in Figure 14.6. The controlled variable is the liquid level in the tank. The float is the measuring instrument, the valve is the manipulating element, and the lever provides the control action. Notice that there is a different valve position for each level. The desired level is h_0, and v_0 is the valve position corresponding to h_0 (v_0 is the position of the valve when the error is zero). The

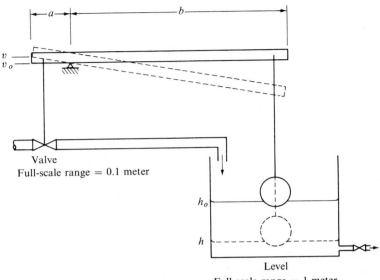

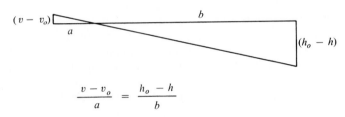

$$\frac{v - v_o}{a} = \frac{h_o - h}{b}$$

Figure 14.6 Simple proportional mode controller. The lever establishes a fixed linear relationship between the level (h) and the valve stem position (v).

valve position (v) is given by the following equation:

$$\frac{v - v_0}{a} = \frac{h_0 - h}{b}$$

$$\text{Error signal} = e = h_0 - h$$

$$v = \left(\frac{a}{b}\right)e + v_0 \tag{14.2}$$

where v = valve position, meter

v_0 = valve position with zero error, meter

e = error signal, meter

Equation (14.2) is one version of the defining equation of the proportional control mode.

The gain (P) of the proportional controller in Figure 14.6 is the change in valve position divided by the corresponding change in level. Both are expressed in percentage of the full-scale range.

$$\text{Percent change in valve position} = \frac{100(v - v_0)}{0.1}$$

$$= 1000(v - v_0)$$

$$\text{Percent change in level} = \frac{100(h_0 - h)}{1}$$

$$= 100(h_0 - h)$$

$$\text{Gain, } P = \frac{1000(v - v_0)}{100(h_0 - h)} = 10\left(\frac{v - v_0}{h_0 - h}\right) = 10\left(\frac{a}{b}\right)$$

Figure 14.7 includes input/output graphs of proportional controllers with gains of 0.5, 1, and 2. In general, an increase in the gain reduces the size of the error required to produce a 100% change in the valve position. In other words, a high gain requires

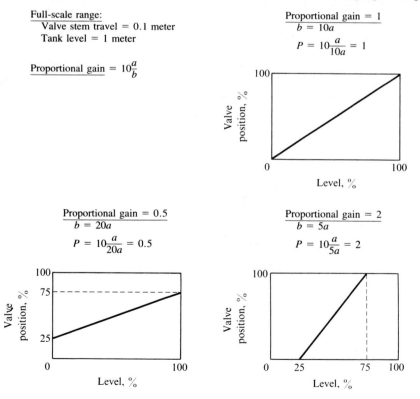

Full-scale range:
Valve stem travel = 0.1 meter
Tank level = 1 meter

Proportional gain $= 10\dfrac{a}{b}$

Proportional gain $= 1$
$b = 10a$

$P = 10\dfrac{a}{10a} = 1$

Proportional gain $= 0.5$
$b = 20a$

$P = 10\dfrac{a}{20a} = 0.5$

Proportional gain $= 2$
$b = 5a$

$P = 10\dfrac{a}{5a} = 2$

Figure 14.7   Input/output graphs for the controller in Figure 14.6 with gains of 1, 0.5, and 2. The gain is changed by moving the pivot point on the lever to change the ratio of distance a over distance b.

a small error to produce the change in valve position necessary to balance the process. Although this seems to imply that the gain should be as high as possible, unfortunately, increasing the gain increases the tendency for oscillation of the controlled variable. A compromise is necessary in which the gain is made as large as possible without producing unacceptable oscillations.

One problem with the proportional control mode is that it cannot completely eliminate the error caused by a load change. A residual error is always required to maintain the valve at some position other than v_0. This is obvious in Equation (14.2), and it is equally obvious in the simple system illustrated in Figure 14.6. A load change means that a different valve position is required to maintain a balanced condition in the process. With a proportional control mode, a change in level is the only way that the valve position can be changed. This change or *residual error* is called the *proportional offset*. The size of the offset is directly proportional to the size of the load changes and inversely proportional to the gain (P). The proportional control mode is used when the gain can be made large enough to reduce the proportional offset to an acceptable level for the largest expected load change.

The response of the proportional mode control action is instantaneous. There is no delay between a change in level and the corresponding change in valve position. The Bode diagram in Figure 14.8 is another way of looking at the response of the proportional control mode. Notice that the phase angle is $0°$ for all values of frequency. The absence of any phase lag is another indication of the speed of response of the proportional control mode. The gain is also constant for all values of frequency, with the decibel level determined by the value of the gain, P.

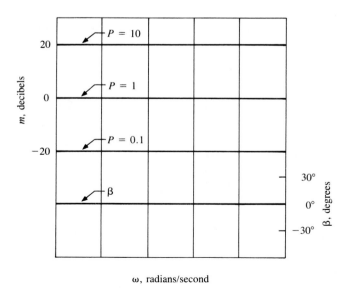

Figure 14.8 Bode diagram of proportional control modes with gains of 10, 1, and 0.1. The phase angle, β, is $0°$ and the magnitude, m, is constant for all values of radian frequency, ω.

> The proportional control mode is used on processes with a small capacitance and fast-moving load changes when the gain can be made large enough to reduce the off-set to an acceptable level. This implies a process with a capacitance that is too small to permit the use of two-position or floating control.

PROPORTIONAL CONTROL MODE

Time-Domain Equation

$$v = Pe + v_0 \tag{14.3}$$

Frequency-Domain Equation

$$V = PE \tag{14.4}$$

Transfer Function

$$\frac{V}{E} = P \tag{14.5}$$

(See Table 14.1 for definitions of the variables.)

   Note: In the frequency domain, v_0 is assumed to be zero to satisfy the condition of zero initial conditions for transfer functions.

Example 14.2

A proportional controller has a gain of 4. Determine the proportional offset required to maintain $v - v_0$ at 20%.

Solution

$$\text{Gain} = \frac{v - v_0}{E}$$

$$4 = \frac{20}{E}$$

$$E = 5\%$$

Integral Control Mode

The *integral control mode* changes the output of the controller by an amount proportional to the integral of the error signal. As long as there is an error, the integral control mode will change the output at a rate proportional to the size of the error. Figure 14.9 illustrates the relationship between the error signal and the controller output. Notice that the rate of change of the controller output is proportional to the error signal (the rate of change is equal to the slope of the graph).

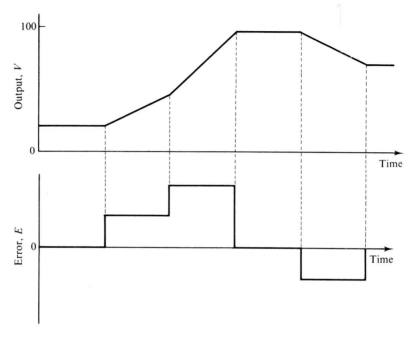

Figure 14.9 The integral control mode responds to an error signal by changing the controller output at a rate proportional to the size of the error.

The Bode diagram of the integral control mode is shown in Figure 14.10. The gain decreases at a rate of 20 dB per decade increase in frequency, and passes through 0 dB at a radian frequency equal to I, where I is the integral action rate. The phase angle is a constant $-90°$ for all frequency values. The integral control mode is almost always used with the proportional control mode.

INTEGRAL CONTROL MODE

Time-Domain Equation

$$v = I \int_0^t e \, dt + v_0 \tag{14.6}$$

Frequency-Domain Equation

$$V = \left(\frac{I}{s}\right) E \tag{14.7}$$

Transfer Function

$$\frac{V}{E} = \frac{I}{s} \tag{14.8}$$

(See Table 14.1 for definitions of the variables.)

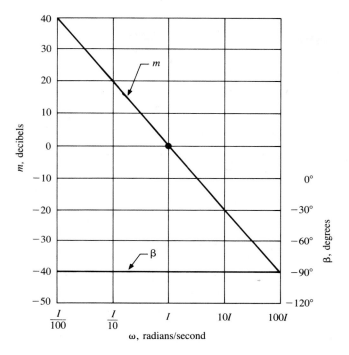

Figure 14.10   Bode diagram of an integral control mode. The gain decreases as frequency increases and the phase angle is a constant $-90°$.

PI Control Mode

The integral mode is frequently combined with the proportional mode to provide an automatic reset action that eliminates the proportional offset. The combination is referred to as the *proportional plus integral (PI) control mode*. The integral mode provides the reset action by constantly changing the controller output until the error is reduced to zero. Figure 14.11 illustrates the step response of a proportional plus integral controller. The proportional mode provides a change in the controller output that is proportional to the error signal. The integral mode provides an additional change in the output that is proportional to the integral of the error signal. The reciprocal of integral action rate (I) is the time required for the integral mode to match the change in output produced by the proportional mode.

One problem with the integral mode is that it increases the tendency for oscillation of the controlled variable. The gain of the proportional controller must be reduced when it is combined with the integral mode. This reduces the ability of the controller to respond to rapid load changes. If the process has a large dead-time lag, the error signal will not immediately reflect the actual error in the process. This delay often results in overcorrection by the integral mode—that is, the integral mode continues to change the controller output after the error is actually reduced to zero, because it is acting on an "old" signal.

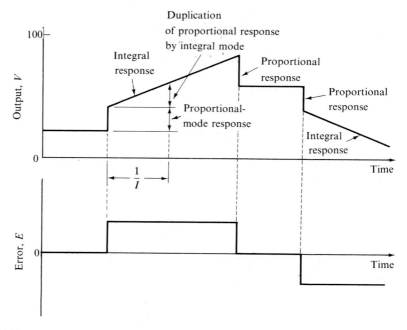

Figure 14.11   Step response of a proportional plus integral (PI) control mode.

The Bode diagram of a PI control mode is shown in Figure 14.12. The diagram is divided into two halves by the integral action break-point frequency, which is equal to the integral action rate.

$$\omega_i = I$$

On the left side of the diagram ($\omega < \omega_i$), the integral action dominates with the gain decreasing at 20 dB per decade and the phase angle equal to $-90°$. On the right side of the diagram ($\omega > \omega_i$), the proportional action dominates with a phase angle of $0°$ and a magnitude determined by the proportional gain, P. The region between $0.1\omega_i$ and $10\omega_i$ is a transition zone between the two sides of the diagram. In Figure 14.12, the proportional gain, P, is equal to 1, which gives a magnitude of 0 dB on the Bode diagram. The effect of a proportional gain, P, other than 1 is to raise or lower the gain curve without affecting the phase curve. A gain of $P = 10$, for example, would raise the entire gain curve 20 dB. A gain of 0.1 would lower the entire gain curve by 20 dB.

> The proportional plus integral control mode is used on processes with large load changes when the proportional mode alone is not capable of reducing the offset to an acceptable level. The integral mode provides a reset action that eliminates the proportional offset.

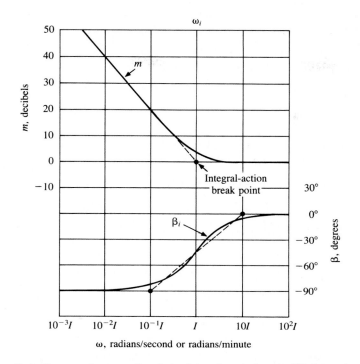

Figure 14.12   Bode diagram of a proportional plus integral control mode. The integral mode dominates at frequencies below ω_i. The proportional mode dominates at frequencies above ω_i.

PROPORTIONAL PLUS INTEGRAL (PI) CONTROL MODE

Time-Domain Equation

$$v = Pe + PI \int_0^t e\, dt + v_0 \tag{14.9}$$

Frequency-Domain Equation

$$V = PE + P\left(\frac{I}{s}\right)E \tag{14.10}$$

Transfer Function

$$\frac{V}{E} = P\left(\frac{I + s}{s}\right) \tag{14.11}$$

(See Table 14.1 for definitions of the variables.)

Example 14.3

A PI controller has a gain (P) of 2 and an integral action rate (I) of 0.02 s^{-1}. The value of v_0 is 32% (at $t = 0$). The graph of the error signal is given in Figure 14.13a. Determine the value of the controller output at the following times: $t =$ (a) 0, (b) 10, (c) 50, (d) 75, and (e) 100 s.

Solution

The controller output is given by Equation (14.9). Substituting the values of v_0, P, and I, the equation of the controller is

$$v = 2e + (2)(0.02) \int_0^t e\,dt + 32$$

The term $\int_0^t e\,dt$ is equal to the net area under the error curve between 0 and t seconds.

a. At $t = 0$ s, $e = 0$ and the net area $= 0$:

$$v = (2)(0) + \left(\frac{1}{25}\right)(0) + 32 = 32\%$$

b. At $t = 10$ s, $e = 0$ and net area $= 0$:

$$v = (2)(0) + \left(\frac{1}{25}\right)(0) + 32 = 32\%$$

c. At $t = 50$ s, $e = 10\%$ and net area $= (10)(50 - 20) = 300\%$:

$$v = (2)(10) + \left(\frac{1}{25}\right)(300) + 32 = 64\%$$

d. At $t = 75$ s, $e = 0$ and net area $= (10)(70 - 20) = 500$:

$$v = (2)(0) + \left(\frac{1}{25}\right)(500) + 32 = 52\%$$

e. At $t = 100$ s, $e = -10\%$ and
net area $= (10)(70 - 20) + (-10)(100 - 80) = 300$:

$$v = (2)(-10) + \left(\frac{1}{25}\right)(300) + 32 = 24\%$$

A graph of the controller output is shown in Figure 14.13b.

Derivative Control Mode

The *derivative control mode* changes the output of the controller proportional to the rate of change of the error signal. This change may be caused by a variation in the measured variable, the setpoint, or both. The derivative mode is an attempt to anticipate an error by observing how fast the error is changing, and using the rate of change to produce a control action that will reduce the expected error. The derivative mode contributes to the output of the controller only while the error is changing. For

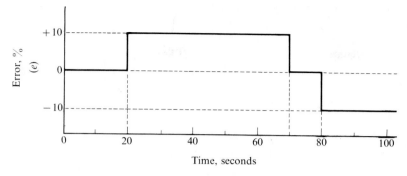

a) Graph of the error signal, *e*

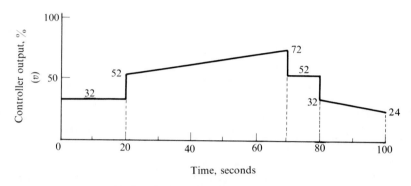

b) Graph of the controller output signal,

Figure 14.13   Error and controller output graphs for Example 14.3.

this reason, the derivative control mode is always used in combination with the proportional, or proportional plus integral control modes.

> The derivative control mode is never used alone. It is always used in combination with the proportional, or proportional plus integral modes.

The step and ramp responses of the ideal derivative control mode are given in Figure 14.14. At every instant, the output of the derivative control mode is proportional to the slope or rate of change of the error signal. The step response indicates the reason that the ideal derivative control mode is never used in practical controllers. The error curve has an infinite slope when the step change occurs. The ideal derivative mode must respond with an infinite change in the controller output. In practical controllers, the response of the derivative action to rapidly changing signals is limited. This greatly reduces the sensitivity of the controller to the unwanted noise spikes that frequently occur in practice.

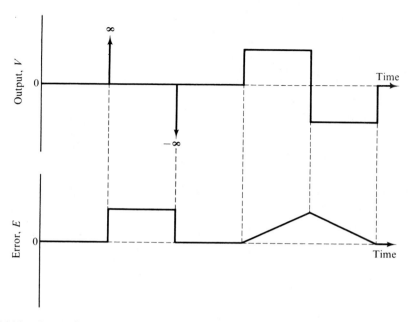

Figure 14.14   Step and ramp response of the ideal derivative control mode.

The Bode diagram of the ideal derivative mode (not shown) is the opposite of the integral mode diagram shown in Figure 14.10. The gain increases at a rate of 20 dB per decade increase in frequency, and passes through 0 dB at a radian frequency equal to $1/D$. The phase angle is a constant $+90°$ for all frequency values. The equations of the ideal derivative control mode are given below. The practical derivative mode is covered in the next section.

IDEAL DERIVATIVE CONTROL MODE

Time-Domain Equation

$$v = D\frac{de}{dt} \tag{14.12}$$

Frequency-Domain Equation

$$V = DsE \tag{14.13}$$

Transfer Function

$$\frac{V}{E} = Ds \tag{14.14}$$

(See Table 14.1 for definitions of the variables.)

PD Control Mode

The derivative control mode is sometimes used with the proportional mode to reduce the tendency for oscillations and allow a higher proportional gain setting. The combination of proportional and derivative modes is referred to as the *PD control mode*. The proportional mode provides a change in the controller output that is proportional to the error signal. The derivative mode provides an additional change in the controller output that is proportional to the rate of change of the error signal. The derivative mode anticipates the future value of the error signal and changes the controller output accordingly. This anticipatory action makes the derivative mode useful in controlling processes with sudden load changes. For this reason, the derivative mode is usually used with proportional, or proportional plus integral control when the sudden load changes produce excessive errors. The derivative mode control action opposes the change of a controlled variable, which helps damp out oscillations of the controlled variable.

> The proportional plus derivative control mode is used on processes with sudden load changes when the proportional mode alone is not capable of keeping the error within an acceptable level. The derivative mode provides an anticipatory action that reduces the maximum error produced by sudden load changes. It also allows a higher gain setting, which helps reduce the proportional offset.

Equation (14.15) is the time-domain equation of a practical proportional plus derivative control mode. The *Pe* term is the proportional mode action. The *PD de/dt* term is the ideal derivative mode action, and $\alpha D \, dv/dt$ is the term that limits the response produced by rapidly changing signals. Equation (14.16) is the frequency-domain equation, and Equation (14.17) is the transfer function.

PROPORTIONAL PLUS DERIVATIVE (PD) CONTROL MODE

Time-Domain Equation

$$v = Pe + PD \frac{de}{dt} - \alpha D \frac{dv}{dt} + v_0 \qquad (14.15)$$

Frequency-Domain Equation

$$V = PE + PDsE - \alpha DsV \qquad (14.16)$$

Transfer Function

$$\frac{V}{E} = P\left(\frac{1 + Ds}{1 + \alpha Ds}\right) \qquad (14.17)$$

$$0 < \alpha < 1$$

(See Table 14.1 for definitions of the variables.)

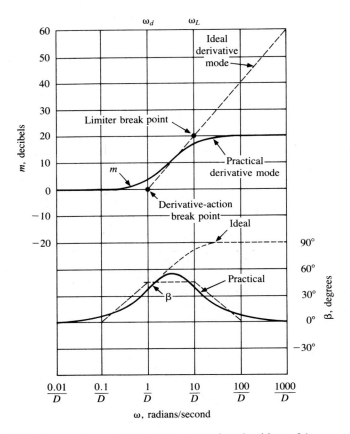

Figure 14.15 Bode diagram of a practical derivative control mode with $\alpha = 0.1$.

The Bode diagram of a PD control mode is shown in Figure 14.15. The proportional mode dominates the left side of the diagram (where $\omega < \omega_d = 1/D$). The proportional gain raises or lowers the entire gain curve, just as it did in the PI control mode. The derivative mode causes the gain curve to slope up at 20 dB per decade at the derivative-action break point. The derivative limiter causes the gain to return to horizontal at the limiter break point. The diagram clearly shows how the derivative mode amplifies high-frequency signals, and how the derivative limiter reduces the amplification of high-frequency signals. Notice also that the limiter causes the phase angle to return to 0° at the higher frequencies. In effect, the PD control mode provides a phase lead over a band of frequencies. Controller design involves the placement of this phase lead where it will do the most good.

Example 14.4

A PD controller has a gain (P) of 0.8, a derivative action time constant (D) of 1 s, an initial output (v_0) of 40%, and a derivative limiter coefficient (α) of 0.1. The graph

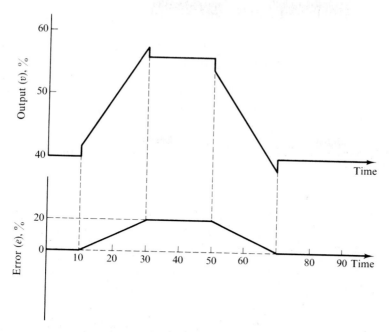

Figure 14.16   Error and controller output graphs for Example 14.4.

of the error signal is given in Figure 14.16. Determine the value of the controller output (v) at the following times: (a) $t = 0$ s, (b) $t = 10-$, an instant before $t = 10$ s, (c) $t = 10+$, an instant after $t = 10$ s, (d) $t = 20$ s, (e) $t = 40$ s, and (f) $t = 60$ s. Assume that the derivative limiter term is negligible.

Solution

The controller output is given by Equation (14.15). Using the values and assumptions given above, the equation for the controller is

$$v = 0.8e + 0.8 \frac{de}{dt} + 40$$

The term de/dt is equal to the slope of the curve at any given instant of time.

   a. At $t = 0$, $e = 0$, and the slope $= de/dt = 0$.

$$v = (0.8)(0) + (0.8)(0) + 40 = 40\%$$

   b. At $t = 10-$, an instant before $t = 10$ s, $e = 0$, and the slope $= de/dt = 0$.

$$v = (0.8)(0) + (0.8)(0) + 40 = 40\%$$

   c. At $t = 10+$, an instant after $t = 10$ s, $e = 0$, and the slope $= de/dt = 20/20 =$ 1%/s.

$$v = (0.8)(0) + (0.8)(1) + 40 = 40.8\%$$

(Notice the increase in v from 40.0% to 40.8% as t passes 10 seconds. We say there is a step change in v of 0.8% at $t = 10$ seconds.)

d. At $t = 20$ s, $e = 10\%$ and $de/dt = 20/20 = 1\%/\text{s}$.
$$v = (0.8)(10) + (0.8)(1) + 40 = 48.8\%$$
e. At $t = 40$ s, $e = 20\%$ and $de/dt = 0$.
$$v = (0.8)(20) + (0.8)(0) + 40 = 56\%$$
f. At $t = 60$ s, $e = 10\%$ and $de/dt = -20/20 = -1\%/\text{s}$.
$$v = (0.8)(10) + (0.8)(-1) + 40 = 47.2\%$$

The diagram of the output (v) is shown in Figure 14.16.

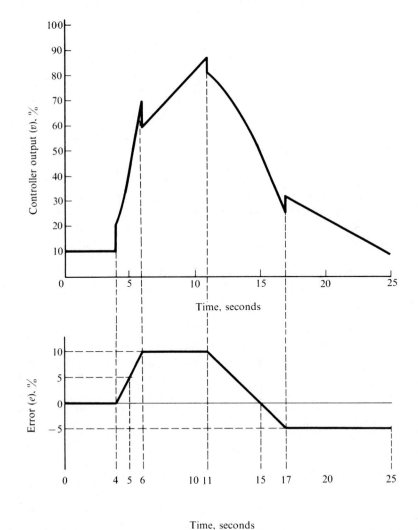

Figure 14.17 Error and controller output graphs for Example 14.5.

PID Control Mode

The *PID control mode* is a combination of the proportional, integral, and derivative control modes. A PID controller is also referred to as a three-mode controller. The integral mode is used to eliminate the proportional offset caused by large load changes. The derivative mode reduces the tendency toward oscillations and provides a control action that anticipates changes in the error signal. The derivative mode is especially useful when the process has sudden load changes.

The proportional-plus-integral-plus-derivative control mode is used on processes with sudden, large load changes when one or two mode control is not capable of keeping the error within acceptable limits. The derivative mode produces an anticipatory action that reduces the maximum error produced by sudden load changes. The integral mode provides a reset action that eliminates the proportional offset.

Equation (14.18) is the defining equation for an ideal three-mode controller.

$$v = Pe + PI \int_0^t e\, dt + PD\frac{de}{dt} + v_0 \qquad (14.18)$$

The practical PID controller includes the derivative limiter term introduced in the section on the PD control mode. Equations (14.19), (14.20), and (14.21) define the practical PID controller.

**PROPORTIONAL PLUS INTEGRAL PLUS
DERIVATIVE (PID) CONTROL MODE**

Time-Domain Equation

$$v = Pe + PI \int_0^t e\, dt + PD\frac{de}{dt} - \alpha D\frac{dv}{dt} + v_0 \qquad (14.19)$$

Frequency-Domain Equation

$$V = PE + P\left(\frac{I}{s}\right)E + PDsE - \alpha DsV \qquad (14.20)$$

Transfer Function

$$\frac{V}{E} = P\left(\frac{I + s + Ds^2}{s + \alpha Ds^2}\right) \qquad (14.21)$$

(See Table 14.1 for definitions of the variables.)

Example 14.5

A PID controller has the following parameters: $P = 4.3$, $I = (\frac{1}{7})\,\text{s}^{-1}$, $D = 0.5\,\text{s}$, $v_0 = 10\%$. The graph of the error signal is given in Figure 14.17. Determine the output of the controller at $t =$ (a) 5, (b) 10, (c) 15, and (d) 25 s. Assume that the derivative limiter term is negligible.

Solution

The controller output is given by Equation (14.19). Using the values and assumptions given above, the equation for the controller is

$$v = 4.3e + 0.614 \int_0^t e \, dt + 2.15 \left(\frac{de}{dt} \right) + 10$$

The term $\int_0^t e \, dt$ is equal to the net area under the error curve between 0 and t seconds. The term de/dt is equal to the slope of the curve at any given instant of time.

a. At $t = 5$ s, $e = 5\%$, net area $= (0.5)(1)(5) = 2.5\%/$s and slope $= 10/(6 - 4) = 5\%/$s:

$$v = (4.3)(5) + (0.614)(2.5) + (2.15)(5) + 10$$
$$= 44\%$$

b. At $t = 10$ s, $e = 10\%$, net area $= (0.5)(2)(10) + (4)(10) = 50$, and slope $= 0$:

$$v = (4.3)(10) + (0.614)(50) + (2.15)(0) + 10$$
$$= 84\%$$

c. At $t = 15$ s, $e = 0$, net area $= (0.5)(2)(10) + (5)(10) + (0.5)(4)(10) = 80$, and slope $= -15/(17 - 11) = -2.5$:

$$v = (4.3)(0) + (0.614)(80) + (2.15)(-2.5) + 10$$
$$= 54\%$$

d. At $t = 25$ s, $e = -5\%$, net area $= 80 - (0.5)(2)(5) - (8)(5) = 35$, and slope $= 0$:

$$v = (4.3)(-5) + (0.614)(35) + (2.15)(0) + 10$$
$$= 10\%$$

A graph of the controller output is included in Figure 14.17.

14.3 ELECTRONIC ANALOG CONTROLLERS

An electronic *analog controller* has two main parts: the error detector and the control mode unit. An example of an electrical error detector is illustrated in Figure 14.18. The output of the measuring transmitter is a 4- to 20-mA electric current signal. Each value of the current represents a unique value of the controlled variable (c). The 4-mA signal represents the minimum value of c, and the 20-mA signal represents the maximum value. The current signal is applied to a 62.5-Ω resistor, resulting in a 0.25- to 1.25-V signal across the resistor. The setpoint signal is produced by a potentiometer with a 0.25- to 1.25-V output range. The two voltage signals are connected in opposition so that the voltage between points a and b is equal to the setpoint signal minus the measured value signal.

$$e = sp - c_m$$

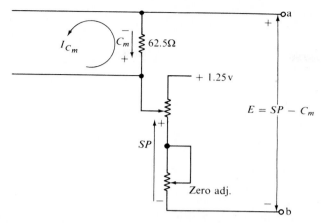

Figure 14.18   Typical error detector in an electronic analog controller.

The control mode unit is sometimes called "the controller," although it is actually one part of the unit that is usually called by that name. The electronic analog controller uses a single operational amplifier and some resistors and capacitors to form the control mode unit. The operational amplifier is used as a function generator, and the resistors and capacitors are arranged to implement the transfer function of the desired control mode or combination of modes.

The analog proportional controller uses three resistors to form an inverting amplifier (see Figure 14.19). The circuit has two inputs, the error (e) and the output

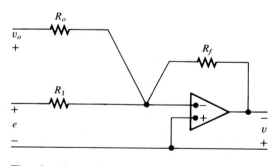

Time domain equation: $v = Pe + v_o$

Transfer function $= \dfrac{V}{E} = P$

$P = \dfrac{R_f}{R_1}$

$R_o = R_f$

Figure 14.19   An analog proportional controller is essentially an op-amp inverting amplifier.

Figure 14.20 Analog proportional plus integral (PI) controller.

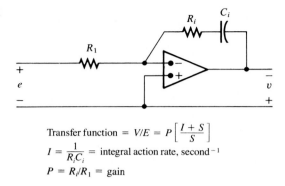

$$\text{Transfer function} = V/E = P\left[\frac{I+S}{S}\right]$$

$$I = \frac{1}{R_iC_i} = \text{integral action rate, second}^{-1}$$

$$P = R_i/R_1 = \text{gain}$$

offset (v_0). The proportional gain (P) is equal to the feedback resistor (R_f) divided by the error input resistor (R_1). The offset resistor (R_o) must be equal to the feedback resistor (R_f) to satisfy the time domain equation. The output lines can be reversed to make the output either positive or negative with respect to the sign of the error. Some applications of the controller will require a positive output for a positive error, and other applications will require a negative output for a positive error.

The proportional plus integral controller uses two resistors and a capacitor to implement the PI transfer function (see Figure 14.20). The capacitor (C_i) is placed in series with the feedback resistor (R_i). The gain (P) is equal to the feedback resistor (R_i) divided by the input resistor (R_1). The integral action rate is equal to the reciprocal of the product of the input resistor (R_i) and the capacitor (C_i).

The proportional plus derivative controller uses four resistors and a capacitor to implement the PD mode (Figure 14.21). The circuit is a proportional controller

Figure 14.21 Analog proportional plus derivative (PD) controller.

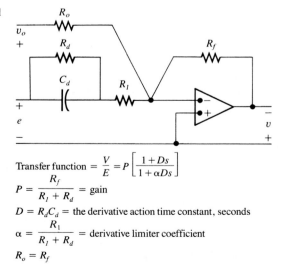

$$\text{Transfer function} = \frac{V}{E} = P\left[\frac{1+Ds}{1+\alpha Ds}\right]$$

$$P = \frac{R_f}{R_1 + R_d} = \text{gain}$$

$$D = R_dC_d = \text{the derivative action time constant, seconds}$$

$$\alpha = \frac{R_1}{R_1 + R_d} = \text{derivative limiter coefficient}$$

$$R_o = R_f$$

with a parallel combination of resistor (R_d) and capacitor (C_d) placed in series with the input resistor (R_1). The equations for the gain, derivative action time constant, and derivative limiter coefficient are given in Figure 14.21. A typical value of α is 0.1.

Two versions of the analog PID controller are shown in Figure 14.22. One version (Figure 14.22a) forms the derivative action on the input side and the integral action on the output side. The other version (Figure 14.22b) does just the opposite and forms the integral action on the input side and the derivative action on the output side. The equations for the controller parameters for each version are given in Figure 14.22.

The transfer function for the analog PID controller is a modified version of Equation (14.21). The modification is done for reasons of economy and consists of two first-order networks in series. The implementation of Equation (14.21) in its exact

Figure 14.22   Two versions of the analog proportional plus integral plus derivative (PID) controller.

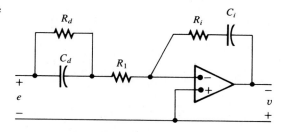

a) Derivative input and integral output

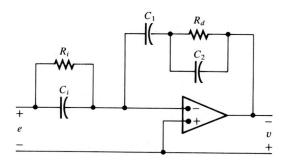

b) Integral input and derivative output

$$\text{Transfer function} = \frac{V}{E} = P\left[\frac{I+s}{s}\right]\left[\frac{1+Ds}{1+\alpha Ds}\right]$$

Figure 14.22a	Figure 14.22b
$P = \dfrac{R_i}{R_1 + R_d}$	$P = C_i/C_1$
$\alpha = \dfrac{R_1}{R_1 + R_d}$	$\alpha = \dfrac{C_2}{C_1 + C_2}$
$I = 1/(R_iC_i)$	$I = 1/(R_iC_i)$
$D = R_dC_d$	$D = R_d(C_1 + C_2)$

form requires three operational amplifiers. The derivative and integral terms must be formed in parallel and then summed with a summing amplifier. The modification consists of inserting an interaction term $(PIDe)$ in the time-domain equation, as shown below.

$$v = Pe + PIDe + PI \int_0^t e\,dt + PD\frac{de}{dt} - \alpha D\frac{dv}{dt} + v_0$$

A Laplace transformation of the equation above with $v_0 = 0$ gives the following frequency-domain equation:

$$V = PE + PIDE + P\left(\frac{I}{s}\right)E + PDsE - \alpha DsV$$

Solving for the ratio V/E gives the following transfer function:

$$\frac{V}{E} = \frac{P + PID + PI/s + PDs}{1 + \alpha Ds} \tag{14.22}$$

or

$$\frac{V}{E} = P\left(\frac{I + (1 + ID)s + Ds^2}{s + \alpha Ds^2}\right) = P\left(\frac{I + s}{s}\right)\left(\frac{1 + Ds}{1 + \alpha Ds}\right) \tag{14.23}$$

Example 14.6

Determine the values of R_1 and R_i for an electronic proportional plus integral controller with a gain (P) of 2 and an integral action rate (I) of $0.02\ \text{s}^{-1}$. Use a $0.1\text{-}\mu\text{F}$ capacitor for C_i. Determine the transfer function.

Solution

The equations are given in Figure 14.20.

1. $I = 1/(R_i C_i)$; therefore, $R_i = 1/(IC_i)$:

$$R_i = \frac{50}{10^{-5}} = 5\ \text{M}\Omega$$

2. $P = R_i/R_1$; therefore, $R_1 = R_i/P$:

$$R_1 = \frac{5 \times 10^6}{2} = 2.5\ \text{M}\Omega$$

The transfer function is

$$\frac{V}{E} = 2\left(\frac{1 + 50s}{50s}\right)$$

Example 14.7

Determine the value of R_1, R_i, R_d, and C_d for the analog PID controller in Figure 14.22a. The controller has a gain (P) of 4, an integral action rate (I) of $(\frac{1}{7})$ s^{-1}, a derivative action time constant (D) of 0.5 s, and a derivative limiter coefficient (α) of 0.1. Use a 10-μF capacitor for C_i. Also determine the transfer function.

Solution

The equations are given in Figure 14.22a.

1. $I = 1/(R_iC_i)$; therefore, $R_i = 1/(IC_i)$:

$$R_i = \frac{7}{10^{-5}} = 700 \text{ k}\Omega$$

2. $\alpha = R_1/(R_1 + R_d)$; therefore,

$$R_d = \left(\frac{1 - \alpha}{\alpha}\right)R_1 = \left(\frac{1 - 0.1}{0.1}\right)R_1$$
$$= 9R_1$$

3. $P = R_i/(R_1 + R_d) = R_i/(10R_1)$; therefore,

$$R_1 = \frac{R_i}{10P} = \frac{700 \text{ k}\Omega}{40} = 17.5 \text{ k}\Omega$$

4. $R_d = 9R_1 = (9)(17.5 \text{ k}\Omega) = 157.5 \text{ k}\Omega$
5. $D = R_dC_d$; therefore, $C_d = D/R_d$:

$$C_d = \frac{0.5}{157.5 \text{ k}\Omega} = 3.2 \text{ } \mu\text{F}$$

The transfer function is

$$\frac{V}{E} = 4\left[\frac{\frac{1}{7} + (1 + 0.5/7)s + 0.5s^2}{s + 0.05s^2}\right] = \frac{1 + 7.5s + 3.5s^2}{1.75s + 0.0875s^2}$$

14.4 DIGITAL CONTROLLERS

Microprocessor-based *digital controllers* are now very commonplace in industrial control systems. There are many reasons for the popularity of digital controllers. The power of the microprocessor provides advanced features such as adaptive self-tuning, multivariable control, and expert systems. The ability of the microprocessor to communicate over a field bus or local area network is another reason for the wide acceptance of the digital controller. Digital controllers used for closed-loop control generally implement the PI, PD, or PID control modes. In this section we examine the PID digital controller.

Sampling

A digital controller measures the controlled variable at specific times, which are separated by a time interval called the sampling time, Δt. Each sample (or measurement) of the controlled variable is converted to a binary number for input to a digital computer or microcomputer. The computer subtracts each sample of the measured variable from the setpoint to determine a set of error samples.

$$e_1 = sp - c_{m1} = \text{first error sample}$$
$$e_2 = sp - c_{m2} = \text{second error sample}$$
$$e_3 = sp - c_{m3} = \text{third error sample}$$
$$\vdots$$
$$e_n = sp - c_{mn} = \text{present error sample}$$

Control Algorithms

After computing each error sample, a digital PID controller follows a procedure called the PID algorithm to calculate the controller output based on the error samples: $e_1, e_2, e_3, \ldots, e_n$. The PID algorithm has two versions, the positional version and the incremental version.

The *positional PID algorithm* determines the valve position, v_n, based on the error signals. Equation (14.24) is a simplified version of the positional algorithm.

$$v_n = Pe_n + PI\,\Delta t \sum_{j=1}^{j=n} e_j + PD\frac{\Delta e_n}{\Delta t} \tag{14.24}$$

where v_n = present valve position, percent of full scale

P = controller gain

e_n = present error sample, percent of full scale

Δt = sample time, second

I = integral action rate, second^{-1}

D = derivative action time constant, second

$\Delta e_n = e_n - e_{n-1}$ = change in the error signal

A flow diagram of a positional PID algorithm is shown in Figure 14.23.

The *incremental PID algorithm* determines the change in the valve position, $\Delta v_n = v_n - v_{n-1}$, based on the error samples. The incremental algorithm can be determined by using Equation (14.24) to determine v_n and v_{n-1} and then subtracting to obtain Equation (14.25) as follows.

$$v_{n-1} = Pe_{n-1} + PI\,\Delta t \sum_{j=1}^{j=n-1} e_j + PD\frac{\Delta e_{n-1}}{\Delta t}$$

$$\Delta v_n = P\,\Delta e_n + PI\,\Delta t\,e_n + PD\left(\frac{\Delta e_n - \Delta e_{n-1}}{\Delta t}\right) \tag{14.25}$$

where $\Delta v_n = v_n - v_{n-1}$

$\Delta e_n = e_n - e_{n-1}$

$\Delta e_n - \Delta e_{n-1} = e_n - 2e_{n-1} + e_{n-2}$

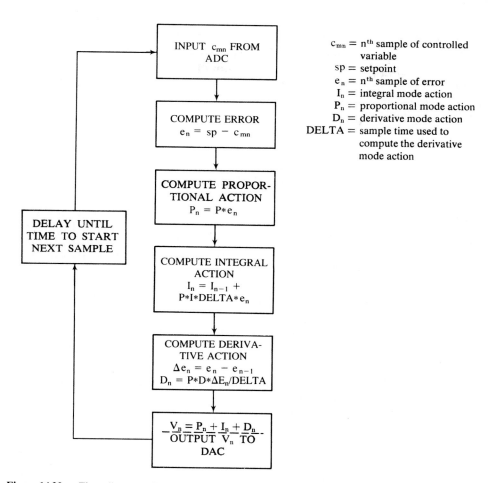

Figure 14.23   Flow diagram of a positional PID control algorithm.

The incremental algorithm is especially suited to incremental output devices such as stepper motors. The positional algorithm is more natural and has the advantage that the controller "remembers" the valve position. If the sample time, Δt, is much shorter than the integral action time constant, $T_i = 1/I$, the positional algorithm will produce a behavior similar to an analog controller.

The Integral Mode

The integral mode in Equation (14.24) presents computational problems that can produce unsatisfactory results. The integral mode is given by the following term in Equation (14.24).

$$\text{Integral term} = PI\,\Delta t \sum_{j=1}^{j=n} e_j \qquad \textbf{(14.26)}$$

For each sample, the integral mode must produce a change given by

$$\text{Integral mode change} = PI\,\Delta t\,e_j \qquad \textbf{(14.27)}$$

When the value of $PI\,\Delta t$ is less than 1, it is more convenient to work with the reciprocal of $PI\,\Delta t$, which could be stored in the computer as an integer. In this case, Equation (14.27) would be revised as follows:

$$I_{\text{DIV}} = \frac{1}{PI\,\Delta t}$$

$$\text{Integral mode change} = \frac{e_j}{I_{\text{DIV}}} \qquad \textbf{(14.28)}$$

If $PI\,\Delta t$ is very small, the computer may ignore relatively large errors because of insufficient resolution. For example, consider a digital controller with a 12-bit word length. The resolution of a 12-bit binary number is 1 part in 4096. To illustrate a point, let's assume that a 12-bit binary number is used to represent a range of errors from -2048 to $+2047$. If $P = 0.5$, $\Delta t = 1$ s, and $I = 0.002$ s^{-1}, then

$$PI\,\Delta t = (0.5)(1)(0.002) = 0.001$$
$$I_{\text{DIV}} = 1000$$

Any value of error greater than -1000 and less than $+1000$ (i.e., 48% of the full-scale range) would result in an integral mode change less than 1, which would be ignored. This small change would then be lost unless a special provision is made to include the change in the calculations for the next sample. The end result is a permanent offset error that the integral mode is unable to eliminate.

One solution to this integral offset problem is to increase the precision by increasing the word length in the computer. A 16-bit word length has a precision of 1 part in 65,536, which could represent a range of errors from $-32,768$ to $+32,768$. This would reduce the offset error to about 3% of the full-scale range.

Another solution is to add the unused portion of the sum of the error samples to the current error sample, e_n, before computing the integral mode change. In the preceding example, an error of 900 in each of two successive samples would not produce an integral mode change because each sample is less than 1000. However, if the first sample is retained, the sum of 1800 would produce an integral mode change of $1800/1000 = 1$ with a remainder of 800. The remainder of 800 would be retained to be added to the next error sample. Every time the accumulated remainder plus the current error is greater than 1000, another increment will be added to the integral mode change.

Derivative Mode

The derivative mode in Equation (14.24) also presents computational problems that can produce unsatisfactory results. A slowly changing signal, for example, results in a "jumpy" derivative mode action. Let's examine how this can occur and what can be done to smooth out the derivative action.

The derivative mode is given by the following term in Equation (14.24).

$$\text{Derivative term} = PD\left(\frac{e_n - e_{n-1}}{\Delta t}\right) \tag{14.29}$$

The term $(e_n - e_{n-1})/\Delta t$ is actually an *estimate* of the rate of change of the error, de/dt. Since Δt is fixed by the sampling rate, our attention will focus on the term $(e_n - e_{n-1})$, which we will represent as est_1. The derivative term produced by est_1 will be called D_1.

$$\text{est}_1 = e_n - e_{n-1} \tag{14.30}$$

$$D_1 = PD\frac{\text{est}_1}{\Delta t} \tag{14.31}$$

If $P = 6$, $\Delta t = 1$ s, and $D = 100$ s, then

$$D_1 = (6)\left(\frac{100}{1}\right)(\text{est}_1) = (600)(\text{est}_1)$$

Table 14.2 shows the derivative term that is produced by a controlled variable that decreases at the rate of 0.5 percent per second. Notice how the derivative term (D_1) jumps back and forth between 0 and 600, because the estimate, est_1, oscillates between 0 and 1.

What is needed is a better estimate of Δe. The theory of digital estimators is beyond the scope of this book. However, a simple example will show how a good estimator can smooth out the derivative term. The idea of an estimator is to use previous samples to improve the estimate. For our example, we use an estimator that uses the last four samples to estimate Δe. We will call this estimate est_2, and the derivative term it produces, D_2.

$$\text{est}_2 = (e_n + e_{n-1}) - (e_{n-2} + e_{n-3}) \tag{14.32}$$

$$D_2 = PD\frac{\text{est}_2}{2^2\,\Delta t} \tag{14.33}$$

Table 14.2   Derivative Action Produced by Two Δe Estimators[a]

n	c	c_m	e	est_1	D_1	est_2	D_2
1	9.5	9	0	0	0	0	0
2	9.0	9	0	0	0	0	0
3	8.5	8	1	1	600	1	150
4	8.0	8	1	0	0	2	300
5	7.5	7	2	1	600	2	300
6	7.0	7	2	0	0	2	300
7	6.5	6	3	1	600	2	300
8	6.0	6	3	0	0	2	300
9	5.5	5	4	1	600	2	300
10	5.0	5	4	0	0	2	300

[a] Setpoint, $sp = 9\%$.

If $P = 6$, $\Delta t = 1$ s, and $D = 100$ s, then

$$D_2 = 6\left(\frac{100}{4}\right)\text{est}_2 = 150\text{est}_2$$

Table 14.2 shows how our simple estimator has smoothed out the derivative term. The est_2 estimator has an effective sample period of 2 s. It used two samples to estimate e_n and two more samples to estimate e_{n-2}. The 2^2 term in the equation for D_2 accounts for the doubling of the sample period and the use of two samples to determine an average. The idea of the est_2 estimator can be extended to include more previous samples. An est_5 estimator would increase the effective sample time to 5 s and use five samples to estimate e_n and five samples to estimate e_{n-5}.

$$\text{est}_5 = (e_n + e_{n-1} + e_{n-2} + e_{n-3} + e_{n-4})$$
$$- (e_{n-5} + e_{n-6} + e_{n-7} + e_{n-8} + e_{n-9})$$

$$D_5 = PD\,\frac{\text{est}_5}{5^2\,\Delta t}$$

14.5 ADVANCED CONTROL

Advanced control refers to various methods of going beyond the single-loop single-variable feedback control system with three modes of control. Topics in this section include cascade control, feedforward control, adaptive self-tuning controllers, and multivariable control systems.

Cascade Control

Cascade control involves two controllers with the output of the primary controller providing the setpoint of the secondary controller. The level control loop in Figure 2.14 provides an excellent application of cascade control. The changes in level occur slowly due to the capacitance of the tank. In contrast, changes in flow occur very quickly. When a disturbance causes a change in the input flow rate, there is a considerable delay before the level changes enough to correct the disturbance. The disturbance often changes before the correction is made. The slow-moving correction of disturbances results in fluctuations of the level. Figure 14.24 shows how cascade control is used to improve the level control system.

In Figure 14.24, a flow transmitter and a secondary controller are used to form a flow control loop within the level control loop. The output of the level controller is the remote setpoint of the flow controller. The flow control loop responds quickly to flow disturbances, virtually eliminating the level fluctuations they caused in a simple level control loop. Industrial processes have many applications of cascade control.

Feedforward Control

Feedforward control uses a model of the process to make changes in the controller output in response to measured changes in a major load variable without waiting for the error to occur. The tubular heat exchanger control loop in Figure 2.3 is a

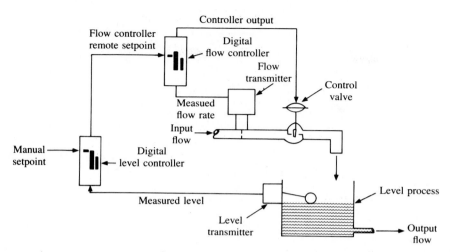

Figure 14.24 A cascade level control system uses a flow control loop inside the level control loop for improved response to flow disturbances.

prime candidate for feedforward control. The product flow rate is the major load on the process. An increase in the product flow rate requires an increase in the flow rate of the heating fluid to maintain the product temperature at the setpoint. Figure 14.25 shows the application of feedforward control to the tubular heat exchanger.

In Figure 14.25, a flow transmitter measures the product flow rate and sends the signal to a load compensator. The load compensator computes the correction necessary to adjust for the product flow rate. The output of the compensator is added to the output of the temperature controller. There is no delay in the compensator signal; the correction is made as soon as the change in product flow rate is measured. The term *feedforward* comes from the fact that the compensation signal travels in the same direction as the product. This is in contrast to the measured temperature signal, which travels in the opposite direction as the product; hence the name *feedback* for the primary loop.

If the feedforward compensation is perfect and there are no other disturbance variables in the process, the feedback loop could be eliminated. These ideal conditions never occur in practice, so feedforward control systems invariably include a feedback loop to make the final adjustments.

Adaptive Controllers

Adaptive controllers change the controller parameters to "adapt" to changes in the process. For example, a change in the product flow rate in the temperature control system in Figure 14.25 will change the dead time of the process. A change in the process dead time means a change in controller parameters is necessary to "tune" the controller to the process. An adaptive controller determines the values of P, I, and D necessary to adapt to the new process conditions and makes the necessary changes. Many different techniques are used to "adapt" the controller to changes in the process.

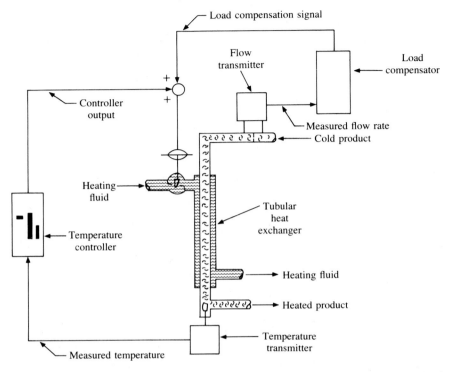

Figure 14.25 A feedforward signal compensates for changes in product flow rate in this temperature control system.

Self-tuning controllers fall into two general categories: those that use a model of the process as the basis of the tuning, and those that use pattern recognition and stored knowledge as the basis.

A typical model-based adaptive controller introduces a step change in the set-point and observes the resulting process response. The controller then forms a model of the process based on the response to the step change. This "self-learning" operation is repeated and the model and tuning parameters are adjusted until they match the actual process.

A pattern-recognition approach to adaptive control uses a graph of error versus time. The controller constantly examines the response to natural disturbances, looking for the presence or absence of peak heights, the time between peaks, and the proportional offset. Following a disturbance, the controller automatically computes P, I, and D based on the observed response pattern and knowledge stored in the controller's memory.

Multivariable Control

Multivariable control uses measurements of several process load variables and may involve the manipulation of more than one process variable. The computer control

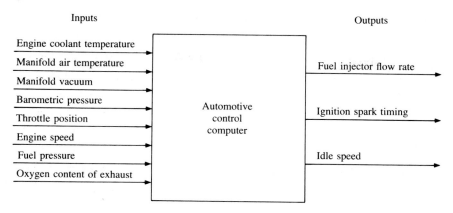

Figure 14.26 An automotive control computer is an example of a multivariable control system. The generic system shown here is intended only to illustrate multivariable control.

systems used to control the fuel injectors and spark timing of automobiles are excellent examples of multivariable control systems. Figure 14.26 illustrates a representative automotive control system.

The following is a general description of automotive control systems for the purpose of illustrating a multivariable control system. The discussion is intended to be generic, and no attempt was made to give a complete description of a particular system.

The purpose of the system shown in Figure 14.26 is to control the fuel injector flow rate, the ignition spark timing, and the idle speed. The inputs to the controller include engine coolant temperature, manifold air temperature, manifold vacuum, barometric pressure, throttle position, engine speed, fuel pressure, and the oxygen content in the exhaust gas. A single computer controls all three output variables. The control system has eight inputs and three outputs, making it a multivariable control system.

The major operating modes of the control system are:

1. *Starting.* The controller varies the amount of fuel sprayed into the intake manifold according to the engine coolant temperature. A cold engine receives more fuel than a warm engine. The ignition system generates the spark timing signals internally and ignores the timing signals from the computer.
2. *Normal running.* The computer uses four input signals to maintain a nearly ideal air/fuel ratio (about 14.7:1). The four input variables are manifold air temperature, manifold vacuum, fuel pressure, and the oxygen content. The computer also modifies the ignition timing based on engine speed, manifold vacuum, engine coolant temperature, and barometric pressure.
3. *Cold running.* The computer provides extra fuel when the engine coolant temperature is below a predetermined value.
4. *Acceleration.* The computer provides extra fuel during acceleration.

5. *Deceleration.* The computer reduces the amount of fuel during deceleration to reduce the pollution produced by the engine.
6. *Idle.* The idle speed is increased when the engine coolant temperature is below a predefined value. Idle speed is also increased when the battery voltage is low, when the transmission is shifted into drive or reverse, and when the air-conditioning compressor comes on.

GLOSSARY

Adaptive controller: A control system that changes the control mode settings in response to changes in the process. (14.5)

Analog controller: A controller producing a controller output that can have any value between 0% and 100% of the full-scale output. (14.3)

Cascade control: A system that uses two controllers, with the output of the primary controller providing the setpoint of the secondary controller. (14.5)

Control mode: Any of several different ways the controller forms the control action. Common control modes include two-position, floating, proportional, integral, and derivative. (14.1)

Control mode, derivative (D): The derivative control mode changes the output of the controller by an amount proportional to the rate of change of the error signal. (14.2)

Control mode, floating: A variation of the two-position mode in which the controller output is constant as long as the error remains within a small band around zero (called the neutral zone). When the error is outside the neutral zone, the controller output changes at a fixed rate until the error returns to the neutral zone or the output reaches one of its extreme positions (0% or 100%). (14.2)

Control mode, integral (I): The integral control mode changes the output of the controller by an amount proportional to the accumulation (integral) of the error signal. (14.2)

Control mode, PD: The combination of the proportional and derivative control modes. (14.2)

Control mode, PI: The combination of the proportional and integral control modes. (14.2)

Control mode, PID: The combination of the proportional, integral, and derivative control modes. (14.2)

Control mode, proportional (P): The proportional mode produces a change in the controller output that is proportional to the error signal. There is a fixed linear relationship between the error signal and the output of the controller. *See also* Proportional offset. (14.2)

Control mode, two-position: The two-position mode produces only two possible controller output values, depending on the sign of the error. If the two positions are 0% and 100%, the controller is called an on-off controller. (14.2)

Digital controller: A controller producing a controller output that can have only a definite number of discrete values between 0% and 100% of the full-scale output. (14.4)

Feedforward control: A control system that uses a model of the process to change the controller output in response to measured changes in a major load variable without waiting for the error to occur. (14.5)

Load change: Any condition in a control system that changes the location of the load line. (14.1)

Multivariable control: A control system that uses measurements of several process load variables to manipulate several other process variables in order to maintain a specified control objective. (14.5)

PID algorithm, incremental: A procedure used by a digital controller to calculate the change in the value of the controller output based on the error samples. (14.4)

PID algorithm, positional: A procedure used by a digital controller to calculate the value of the controller output based on the error samples. (14.4)

Proportional offset: The error required by a proportional control mode to hold the controller output at some value other than the value it has when the error is zero. *See also* Residual error. (14.2)

Residual error: Another name for the proportional offset. (14.2)

EXERCISES

14.1 Name five modes of control and give a brief description of each mode.

14.2 The two-position controller in Example 14.1 is modified so that the valve moves between two partially open positions instead of between on and off. The inlet flow rate (m) is 0.001 m^3/s when the valve is in the minimum flow position and 0.003 m^3/s when the valve is in the maximum flow position. The rest of the conditions are the same as in Example 14.1. Determine the amplitude and the period of oscillation of the level (h). Compare your results with the results in Example 14.1. What was given up to reduce the amplitude of the oscillation?

14.3 The solid flow rate control system shown in Figure 2.19 uses a gate to control the level of material on the belt. A single-speed reversible motor is used to drive the cam that positions the gate. If the solid feed rate is below a predetermined value, the controller drives the gate up at a constant rate. If the feed rate is above a second predetermined value, the controller drives the gate down at a constant rate. Between the two predetermined values, the gate is motionless. Identify the mode of control used in this system.

14.4 Sketch the input/output graph for each of the following proportional controller conditions.

a. $P = 8$, $v_0 = 40\%$ **b.** $P = 0.25$, $v_0 = 55\%$

c. $P = 0.5$, $v_0 = 25\%$

14.5 Determine the proportional offset required to maintain $v - v_0$ at 12% for proportional controllers with each of the following gain values.

a. $P = 0.2$ **b.** $P = 0.6$ **c.** $P = 1.2$

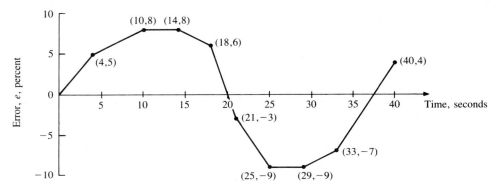

Figure 14.27 Error graph used in Exercises 14.6 to 14.8.

14.6 A PI controller has a gain of 0.5, an integral action rate of 0.0125 s^{-1}, and a value of v_0 of 25%. The graph of the error signal is given in Figure 14.27. Determine the value of the controller output at $t = 15$ s and $t = 30$ s.

14.7 A PD controller has the following parameter values: $P = 0.5$, $D = 12$ s, $\alpha = 0$, and $v_0 = 42\%$. The graph of the error signal is given in Figure 14.27. Determine the value of the controller output at $t = 15$ s and $t = 30$ s.

14.8 A PID controller has the following parameter values: $P = 0.5$, $I = 0.0125$ s^{-1}, $D = 12$ s, $\alpha = 0$, and $v_0 = 55\%$. The graph of the error signal is given in Figure 14.27. Determine the value of the controller output at **(a)** $t = 15$ s and **(b)** $t = 30$ s.

14.9 Use the program "Bode" to generate frequency response data for **(a)** the noninteracting PID controller [Equation (14.21)], and **(b)** the interacting PID controller [Equation (14.23)]. Plot Bode diagrams for both controllers. Comment on the difference between the two Bode diagrams. The controller has the following parameters: $P = 4.0$, $I = \frac{1}{7}$, $D = 0.5$, and $\alpha = 0.1$.

14.10 A certain process has a small capacitance. Sudden, moderate load changes are expected, and a small offset error can be tolerated. Recommend the control mode or combination of modes most suitable for controlling this process.

14.11 A process has a large capacitance and no dead-time lag. The anticipated load changes are relatively small. Recommend the control mode or combination of modes most suitable for controlling this process.

14.12 A dc motor is used to control the speed of a pump. A tachometer-generator is used as the speed sensor. The load changes are insignificant, and there is no dead-time lag in the process. Recommend the control mode or combination of modes most suitable for controlling the motor speed.

14.13 A liquid flow process is fast and the flow rate signal has many "noise spikes." Large load changes are quite common, and there is very little dead-time lag. Recommend the control mode or combination of modes most suitable for controlling the flow rate. Why is the derivative mode usually avoided in liquid flow controllers?

14.14 An electric heater is used to control the temperature of a plastic extruder. The process is a first-order lag with almost no dead-time lag. The time constant is very large. Under steady operation, the load changes are insignificant. Recommend the control mode or combination of modes most suitable for controlling the temperature of the extruder.

14.15 A dryer is a slow process with a very large dead-time lag. Sudden load changes are common, and proportional offset is undesirable. Recommend the control mode or combination of modes most suitable for controlling the dryer.

14.16 An analog proportional mode controller (Figure 14.19) uses a value of 100 kΩ for R_f. Determine the value of R_1 for each of the following controller gains.

    a. $P = 0.25$ **b.** $P = 1.75$ **c.** $P = 8.6$

14.17 Determine the values of R_1 and R_i for an analog PI controller (Figure 14.20) with $P = 6.4$ and $I = 0.00541$ s^{-1}. Use a 1000-μF capacitor for C_i.

14.18 Determine the value of R_1, R_d, and R_f for an analog PD controller (Figure 14.21) with $P = 3.6$, $D = 12$ s, and $\alpha = 0.1$. Use a 100-μF capacitor for C_d.

14.19 Determine the values of R_1, R_i, R_d, and C_i for an analog PID controller, as shown in Figure 14.22a. The controller parameters are $P = 0.65$, $I = 1/70$ s^{-1}, $D = 0.32$ s, $\alpha = 0.1$. Use a 1-μF capacitor for C_d.

14.20 Draw a PID flow diagram similar to Figure 14.23 with the following change: Replace the integral mode operation with the following operations:

(1) Add the nth sample error (e_n) to a new variable called the unused sum (S):

$$S = S + e_n$$

(2) Divide S by I_{DIV} to obtain a quotient (Q) and a remainder (R).

$$I_{DIV} = \frac{1}{PI\,\Delta t}$$
$$Q = S \text{ div } I_{DIV}$$
$$R = S \text{ mod } I_{DIV}$$

(3) Add Q to the past integral mode action (I_{n-1}) to form the current integral mode action (I_n).

$$I_n = I_{n-1} + Q$$

(4) Set the unused sum (S) equal to the remainder (Q):

$$S = Q$$

Explain the purpose of the operations above.

14.21 Draw a PID flow diagram similar to Figure 14.23 with the following change: Replace the derivative mode operation in the PID flow diagram (Figure 14.23) with a new derivative mode operation that uses an improved estimate of the rate of change of the error, de/dt. Use the estimate defined by Equation (14.32) and use Equation (14.33) to compute the derivative term.

14.22 A PI controller defined by Equation 14.11 has a gain that increases as frequency decreases. At very low frequencies, the gain becomes very large. In practical PI controllers, the maximum gain is limited. In the following transfer function of a PI controller, the term (b) in the denominator limits the gain to a maximum value of $1/b$.

$$\frac{V}{E} = P\left(\frac{I + s}{bI + s}\right)$$

Use the program "BODE" in Chapter 5 to generate two sets of frequency data for parameter sets 1 and 2 below. Plot both sets of data on a single Bode diagram and comment on the comparison of the effect of term (b).

Set 1: $P = 1, I = 0.1 \text{ s}^{-1}, b = 0$
Set 2: $P = 1, I = 0.1 \text{ s}^{-1}, b = 0.01$

PART FIVE

Analysis and Design

CHAPTER 15

Process Characteristics

OBJECTIVES

The objective of a control system is to maintain a desired value of a "controlled variable" in a process. To do this, the controller adjusts the value of another variable in the process, the "manipulated variable." The characteristics of a process are an expression of the relationship between the manipulated variable (input) and the controlled variable (output). The Bode diagram design method covered in Chapter 17 requires a knowledge of the characteristics of a process. Process characteristics can be expressed in several ways. In this chapter we use the following four methods: step response graph, time-domain equation, transfer function, and Bode diagram.

   The purpose of this chapter is to provide you with the means of determining the characteristics of the following five types of processes: **integral, first-order lag, dead time, second-order lag,** and **first-order lag plus dead time.** After completing this chapter, you will be able to

1. Sketch a typical step response graph and a typical Bode diagram for each of the following types of processes:

   a. Integral or ramp process

   b. First-order lag process

   c. Dead-time process

   d. Second-order lag process

   e. First-order lag plus dead time process

2. Identify the time-domain equation and the transfer function of each process listed in objective 1

3. Determine the integral action time constant of the following integral processes: liquid level and sheet loop

4. Determine the time constant of the following first-order lag processes: liquid level, electrical, thermal, gas pressure, and blending

5. Determine the resonant frequency, damping ratio, and steady-state gain of the following second-order lag processes: spring–mass–damping, series *RLC* circuit, noninteracting electrical, noninteracting liquid flow, interacting electrical, interacting liquid flow, and a dc motor-controlled load

6. Determine the dead-time lag of the following dead-time processes: solid flow and liquid flow

7. Determine the dead-time lag, the time constant, and sketch the step response graph of a first-order lag plus dead time blending process

15.1 INTRODUCTION

A process or component is characterized by the relationship between the input signal and the output signal. It is this input/output (I/O) relationship that determines the design requirements of the controller. If the I/O relationship of the process is completely defined, the designer can specify the optimum controller parameters. If the I/O relationship is poorly defined, the designer must provide a large adjustment of the controller parameters, so the optimum settings can be determined during startup of the system. This chapter provides the information necessary to determine the I/O relationship of the following types of processes:

1. The integral or ramp process (Section 15.2).
2. The first-order lag process (Section 15.3).
3. The second-order lag process (Section 15.4).
4. The dead-time process (Section 15.5).
5. The first-order lag plus dead-time lag process (Section 15.6).

The I/O relationship of a process may be defined by any or all of the following.

1. The time-domain equation
2. The transfer function
3. The step response graph
4. The frequency response graph

The *time-domain equation* defines the *size* versus *time* relationship between the input signal and the output signal. Time-domain equations are expressed in terms of the basic elements defined in Chapter 4 and frequently contain integral or derivative terms. The transfer function of a process is obtained by transforming the time-domain equation into a frequency-domain algebraic equation and then solving for the ratio of output over input. The *transfer function* defines the *gain and phase difference* versus *frequency* relationship between the input and output signals. The *step response graph* is the time graph of the output signal following a step change in the input signal from one value to another. The *frequency response graph* (or Bode diagram) is a dual plot of gain versus frequency and phase difference versus frequency.

The characteristics of a process depend on its basic elements (resistance, capacitance, inertance, or dead time), not on the type of system (thermal, electrical, mechanical, etc.). Two different systems may have the same process characteristics. That is, they may have the same time-domain equation, transfer function, step response, and Bode diagram. This means that we can extend our knowledge of one system to all other systems that have the same process characteristics. For example, electrical, liquid flow, gas flow, and thermal first-order lag processes are all characterized by a time constant that is equal to the product of resistance times capacitance ($\tau = RC$).

Each section begins with the time-domain equation, transfer function, step response graph, and Bode diagram that characterize one type of process. The equations are included in a box with all terms and their units. Each section concludes with various examples for use as guides in calculating the parameters of different systems with the same process characteristics.

15.2   THE INTEGRAL OR RAMP PROCESS

The *integral* or *ramp process* consists of a single capacitive element configured such that the outflow of material or energy is independent of the amount of material or energy stored in the capacitive element. The quantity of stored material or energy remains constant only if the inflow rate is exactly equal to the outflow rate. If the inflow rate is greater than the outflow rate, the quantity stored will increase at a rate proportional to the difference. If the inflow rate is less than the outflow rate, the quantity stored will decrease at a rate proportional to the difference. The input flow rate is the input signal to the integral process, but the output flow rate is not the output signal. The output signal of the integral process is a variable, such as liquid level, that is a measure of the amount of material or energy stored in the capacitive element.

The *step response* of an integral process is illustrated in Figure 15.1. Before the step change, the input flow rate is equal to the output flow rate and the level is maintained constant. After the step change, the input flow rate is greater than the output

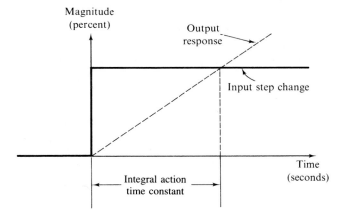

Figure 15.1   Response of an integral process to a step change in the input signal. Before the step change, the input signal (inflow minus outflow) was zero and the output signal (level) remained constant. After the step change, the inflow was greater than the outflow, and the level increased at a constant rate. The integral action time constant, T_i, is the time required for the output signal to change by an amount equal to the input step change.

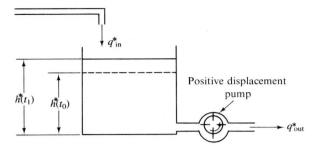

a) A liquid level integral process

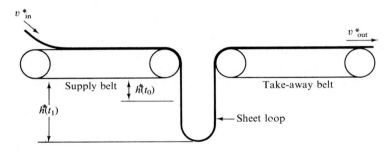

b) A sheet loop integral process

Figure 15.2 Examples of integral processes.

flow rate, and the level increases at a constant rate. The term ramp process is derived from the ramplike shape of the output response graph. The step response of an integral process is measured by the integral action time constant, which is the number of seconds (or minutes) required for the output to reach the same percentage change as the input (see Figure 15.1).

Two integral processes are illustrated in Figure 15.2. In the liquid level integral process, the input signal is the input flow rate; and the output signal is the level of liquid in the tank. In the sheet loop integral process, the input signal is the input sheet velocity; and the output signal is the distance from the belt to the bottom of the sheet loop. The input and output signals are expressed in percent of the full-scale range. In both integral processes, the change in the output signal during the time interval from t_0 to t_1 is equal to the integral of the difference between the input flow rate and the output flow rate divided by the *integral action time constant* (T_i). Equations (15.1) and (15.2) are the time-domain equation and transfer function of the liquid flow integral process. The equations for the sheet loop integral process are obtained from Equations (15.1) and (15.2) by replacing q_{in} and q_{out} by v_{in} and v_{out}. See the development of Equation 5.7 in Section 5.2 for further details concerning Equation (15.1). See also Example 5.12 for the development of Equation (15.2).

INTEGRAL PROCESS

Time-Domain Equation

$$h^*(t_1) - h^*(t_0) = \frac{1}{T_i} \int_{t_0}^{t_1} (q_{in}^* - q_{out}^*) \, dt \qquad \textbf{(15.1)}$$

Transfer Function

$$\frac{H^*(s)}{Q_{in}^*(s)} = \frac{1}{T_i s} \qquad \textbf{(15.2)}$$

where $h^*(t_0)$ = normalized output at time t_0, percent of FS_{out}

      $h^*(t_1)$ = normalized output at time t_1, percent of FS_{out}

      FS_{in} = full-scale range of the input

      FS_{out} = full-scale range of the output

      q_{in}^* = normalized input flow rate, percent of FS_{in}

      q_{out}^* = normalized output flow rate, percent of FS_{in}

      t = time, second

      T_i = integral action time constant, second

Liquid Flow Integral Process

$$T_i = A\left(\frac{FS_{out}}{FS_{in}}\right)$$

    A = cross-sectional area of the tank at the liquid surface, square meter

  FS_{in} = full-scale input range, m^3/s

 FS_{out} = full-scale output range, m

Sheet Loop Integral Process

Replace q_{in}^* and q_{out}^* by v_{in}^* and v_{out}^* in Equations (15.1) and (15.2).

      v_{in}^* = input sheet velocity, percent of FS_{in}

      v_{out}^* = output sheet velocity, percent of FS_{in}

$$T_i = 2\left(\frac{FS_{out}}{FS_{in}}\right)$$

The Bode diagram of an integral process (Figure 15.3) consists of straight-line gain and phase graphs. The gain line has a slope of -20 dB per decade increase in frequency. The phase line is horizontal at $-90°$. Notice that the integral process

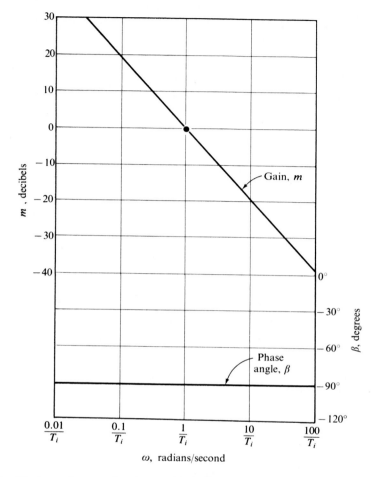

Figure 15.3 The integral process has the same Bode diagram as the integral control mode. For a liquid level integral process, $T_i = A(FS_{out})/(FS_{in})$. For the sheet loop integral process, $T_i = 2(FS_{out})/(FS_{in})$.

Bode diagram is identical to the Bode diagram of the integral control mode. An integral process has a built-in integral control mode action.

Example 15.1

A liquid level integral process has the following parameters.

    Tank height: 4 m

    Tank diameter: 1.5 m

    $FS_{in} = 0.01 \text{ m}^3/\text{s}$

$\mathrm{FS}_{\text{out}} = 4$ m

$h^*(t_0) = 22.5\%$ of FS (0.9 m)

$q^*_{\text{out}} = 60\%$ of FS (0.006 m^3/s)

$q^*_{\text{in}} = 80\%$ of FS (0.008 m^3/s)

Determine the time-domain equation, the transfer function, the integral action time constant, and the level at time $t_0 + 100$ s.

Solution

$$A = \text{area} = \frac{\pi d^2}{4} = \frac{\pi(1.5)^2}{4} = 1.77 \text{ m}^2$$

$$T_i = A \frac{\mathrm{FS}_{\text{out}}}{\mathrm{FS}_{\text{in}}} = \frac{(1.76)(4)}{0.01} = 703 \text{ s}$$

The time-domain equation is

$$h^*(t_1) = \frac{1}{703} \int_{t_0}^{t_1} (q^*_{\text{in}} - q^*_{\text{out}})\, dt + h^*(t_0)$$

The transfer function is

$$\frac{H^*(s)}{Q^*_{\text{in}}(s)} = \frac{1}{703\ s}$$

The change in level from t_0 to $t_0 + 100 = (80 - 60)(100)/703 = 2.83\%$ of FS.

$$h^*(t_0 + 100) = h^*(t_0) + \text{change in level}$$
$$= 22.5 + 2.83 = 25.33\% \text{ of FS}$$
$$= \frac{(25.33)(4)}{100} = 1.013 \text{ m}$$
$$= 25.33\% \text{ of FS} = 1.013 \text{ m}$$

15.3 THE FIRST-ORDER LAG PROCESS

The *first-order lag process* consists of a single capacitive element configured such that the outflow of material or energy is proportional to the amount of material or energy stored in the capacitive element. For each input flow rate, there is a corresponding amount of stored material or energy that will produce an output flow rate equal to the input. The first-order lag process is a self-regulating process because it automatically produces an output flow rate to match each input rate. In contrast, the integral process is a non-self-regulating process.

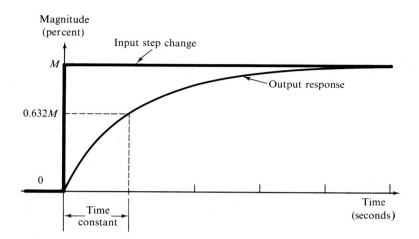

Figure 15.4   The step response of a first-order lag process is characterized by the time constant τ. From any point on the graph, the output will cover 63.2% of the remaining distance to the final value in one time constant.

The *step response* of the first-order lag process is shown in Figure 15.4. Before the step change, the input flow rate is equal to the output flow rate and the level is maintained constant. The step change consists of increasing the input flow rate. Let M represent the percentage increase in the input flow rate. The output flow rate is proportional to the level, which does not change immediately. The input is greater than the output, so the level will increase at a rate proportional to the difference. As the level increases, the difference between the input and the output decreases. This in turn reduces the rate at which the level increases. The result is the output response curve of Figure 15.4.

The step response of a first-order lag is measured by the time constant, which is the number of seconds (or minutes) required for the output to reach 63.2% of the total change. During each additional interval equal to the time constant, the output will reach 63.2% of the remaining change. In Figure 15.4, for example, the input step change is equal to M. During the first time constant interval, the output changed by an amount equal to $0.632M$, and the remaining change was $M - 0.632M = 0.368M$. During the second time interval, the output increased by $0.632(0.368M) = 0.232M$. Thus after two time constants, the output will reach $0.632M + 0.232M = 0.864M$. After five constants, the output will reach $0.993M$.

A liquid level, first-order lag, and three electrical first-order lag processes are illustrated in Figure 15.5. The *liquid level process* in Figure 15.5a was analyzed in Chapter 5 [see the development of Equation (5.6)]. The input of the liquid level process is the input flow rate (q_{in}), and the output is the level (h) of the liquid in the tank.

The *series RC circuit* in Figure 15.5b was also analyzed in Chapter 5 [see the development of Equation (5.10)]. In this circuit the input is the source voltage (e_{in}), and the output is the voltage across the capacitor (e_{out}).

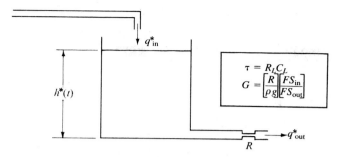

a) A liquid level, first-order lag process

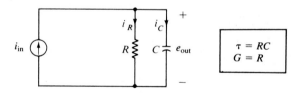

b) A series RC, first-order lag component

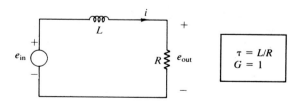

c) A parallel RC, first-order lag component

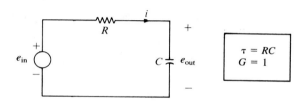

d) A series RL, first-order lag component

Figure 15.5   Liquid flow and electrical first-order lag processes.

In the *parallel RC circuit* in Figure 15.5c, the input is the source current (i_{in}), and the output is the voltage across the capacitor (e_{out}). The analysis of the parallel RC circuit begins by applying Kirchhoff's current law to the top node of the circuit.

$$i_{in} = i_R + i_C$$

But $i_R = e_{out}/R$ and $i_C = C\, de_{out}/dt$:

$$i_{in} = \frac{e_{out}}{R} + C\frac{de_{out}}{dt}$$

$$RC\frac{de_{out}}{dt} + e_{out} = Ri_{in} \tag{15.3}$$

Equation (15.3) is the equation of a first-order lag process with a time constant equal to RC and a gain equal to R.

The third electrical circuit is the *series RL circuit* shown in Figure 15.5d. In this circuit, the input is the source voltage (e_{in}), and the output is the voltage across the resistor (e_{out}). The analysis begins with Kirchhoff's voltage law.

$$e_{in} = e_L + e_R$$

But

$$e_L = L\frac{di}{dt}$$

and

$$i = \frac{e_{out}}{R}$$

Therefore,

$$\frac{di}{dt} = \frac{1}{R}\frac{de_{out}}{dt}$$

Substituting and rearranging gives the following equation for the RL circuit:

$$\frac{L}{R}\frac{de_{out}}{dt} + e_{out} = e_{in} \tag{15.4}$$

Equation (15.4) is the equation of a first-order lag process with a time constant equal to L/R and a steady-state gain equal to 1.

A *thermal, first-order lag process* is shown in Figure 15.6a. A jacketed kettle is used to heat a liquid. The mixer maintains a uniform temperature throughout the liquid. The input to the process is the jacket temperature (θ_j), and the output is the liquid temperature (θ_L). The increase in the liquid temperature ($\Delta\theta_L$) is equal to the amount of heat transferred to the liquid divided by the thermal capacitance (C_T) of the liquid in the tank. The amount of heat (Δq) transferred to the liquid depends on

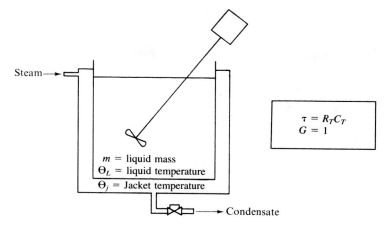

a) A thermal, first-order lag process

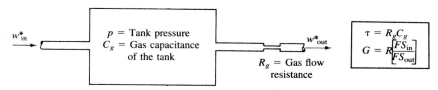

b) A gas pressure, first-order lag process

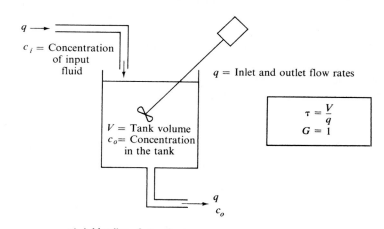

c) A blending, first-order lag process

Figure 15.6 Thermal, gas pressure, and blending first-order lag processes.

the thermal resistance (R_T) of the wall between the steam and the liquid, the temperature difference ($\theta_j - \theta_L$) between the jacket and the liquid, and the time interval (Δt).

$$\Delta\theta_L = \frac{\Delta q}{C_T} = \frac{1}{R_T C_T}(\theta_j - \theta_L)\Delta t$$

$$R_T C_T \frac{\Delta\theta_L}{\Delta t} + \theta_L = \theta_j$$

or as $\Delta t \to 0$,

$$R_T C_T \frac{d\theta_L}{dt} + \theta_L = \theta_j \qquad (15.5)$$

Equation (15.5) is the equation of a first-order lag process with a time constant equal to $R_T C_T$ and a steady-state gain equal to 1.

A *gas-pressure first-order lag process* is shown in Figure 15.6b. In this process the input is the mass flow rate of gas entering the tank (w_{in}), and the output is the pressure (p) of the gas in the tank. The gas process is described by the following equation in which R_g is the gas flow resistance of the outlet, and C_g is the capacitance of the tank.

$$R_g C_g \frac{dp}{dt} + p = R w_{in} \qquad (15.6)$$

Equation (15.6) is the equation of a first-order lag process with a time constant equal to $R_g C_g$ and a steady-state gain equal to 1.

A *blending, first-order lag process* is illustrated in Figure 15.6c. A constant flow rate of q cubic meters per second passes through the tank. The input of the process is the concentration (c_i) of component A in the incoming fluid. The output of the process is the concentration (c_o) of component A in the fluid in the tank (and the outgoing fluid). A material balance is used to determine the time-domain equation. In a time interval Δt, an amount of liquid equal to $q(\Delta t)$ is added at the inlet, and an equal amount is removed at the outlet. The amount of component A added to the tank in the incoming fluid is equal to $q(\Delta t)c_i$. The amount of component A removed in the outgoing fluid is $q(\Delta t)c_o$. The difference is the increase of component A in the tank, which is equal to $V(\Delta c_o)$.

Amount of buildup = amount inputed − amount removed

$$V(\Delta c_o) = q(\Delta t)c_i - q(\Delta t)c_o$$

$$\frac{V}{q}\frac{\Delta c_o}{\Delta t} + c_o = c_i$$

or as $\Delta t \to 0$,

$$\frac{V}{q}\frac{dc_o}{dt} + c_o = c_i \qquad (15.7)$$

Equation (15.7) is the equation of a first-order lag process with a time constant equal to V/q, and a steady-state gain equal to 1.

The *Bode diagram* of a first-order lag process with a steady-state gain of 1 is shown in Figure 15.7. The gain line has a break at a radian frequency equal to the

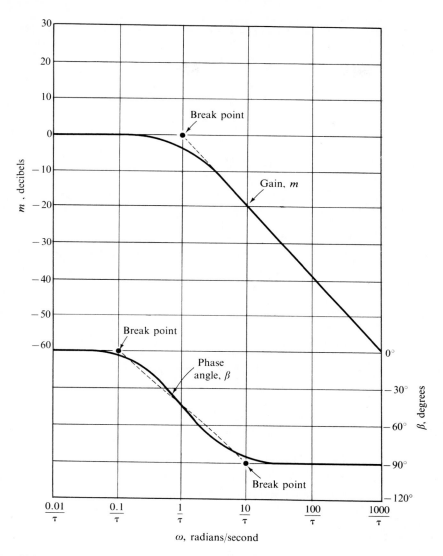

Figure 15.7   Bode diagram of a first-order lag process with a steady-state gain of 1.

reciprocal of the time constant. This frequency is called the break-point frequency, ω_b.

$$\omega_b = \frac{1}{\tau}$$ (15.8)

The phase line has two breaks, one on each side of the break-point frequency, ω_b. A gain different from one raises or lowers the gain line, but does not change the shape of the line. The phase line is unaffected by the gain of the process.

FIRST-ORDER LAG PROCESS

Time-Domain Equation

$$\tau \frac{dy}{dt} + y = Gx \qquad \textbf{(15.9)}$$

Transfer Function

$$\frac{Y}{X} = \frac{G}{1 + \tau s} \qquad \textbf{(15.10)}$$

where G = steady-state gain of the process

t = time, second

x = input to the process*

y = output of the process†

τ = time constant, second

* The input may be expressed with units or as a percentage of FS_{in}.
FS_{in} = full-scale range of the input.
† The output may be expressed with units or as a percentage of FS_{out}.
FS_{out} = full-scale range of the output.

Liquid Level First-Order Lag Process (Figure 15.5a)

Process input: q_{in}^* = input flow rate, percent of FS_{in}

(FS_{in} = input range, cubic meter/second)

Process output: h^* = level, percent of FS_{out}

(FS_{out} = output range, meter)

$$\tau = R_L C_L \qquad \textbf{(15.11)}$$

$$G = \left(\frac{R_L}{\rho g}\right)\left(\frac{FS_{in}}{FS_{out}}\right) \qquad \textbf{(15.12)}$$

R_L = liquid resistance, pascal second/cubic meter

C_L = liquid capacitance, cubic meter/pascal

ρ = liquid density, kilogram/cubic meter

g = 9.81 m/s²

Electrical First-Order Lag Processes (Figure 15.5)

Process input: e_{in} = input voltage (Figure 15.5b, d)

i_{in} = input current (Figure 15.5c)

Process output: e_{out} = output voltage

$$\tau = RC \qquad \text{(Figure 15.5b, c)} \qquad \textbf{(15.13)}$$

$$\tau = \frac{L}{R} \qquad \text{(Figure 15.5d)}$$

$$G = 1 \qquad \text{(Figure 15.5b, d)} \qquad \textbf{(15.14)}$$

$$G = R \qquad \text{(Figure 15.5c)} \qquad \textbf{(15.15)}$$

R = electrical resistance, ohm

C = electrical capacitance, farad

Thermal First-Order Lag Process (Figure 15.6a)

Process input: θ_j = jacket temperature, °C

Process output: θ_L = liquid temperature, °C

$$\tau = R_T C_T \qquad\qquad (15.16)$$

$$G = 1 \qquad\qquad (15.14)$$

R_T = thermal resistance, kelvin/watt

C_T = thermal capacitance, joule/kelvin

Gas Pressure First-Order Lag Process (Figure 15.6b)

Process input: w_{in}^* = input flow rate, percent of FS_{in}

(FS_{in} = input range, kilogram/second)

Process output: p^* = pressure in tank, percent of FS_{out}

(FS_{out} = output range, pascal)

$$\tau = R_g C_g \qquad\qquad (15.17)$$

$$G = R\left(\frac{FS_{in}}{FS_{out}}\right) \qquad\qquad (15.18)$$

R_g = gas flow resistance, pascal second/kilogram

C_g = gas flow capacitance, kilogram/pascal

Blending First-Order Lag Process (Figure 15.6c)

Process input: c_i = input concentration, percent

Process output: c_o = output concentration, percent

$$\tau = \frac{V}{q} \qquad\qquad (15.19)$$

$$G = 1 \qquad\qquad (15.14)$$

q = flow rate of the liquid, cubic meter/second

V = volume of the tank, cubic meter

Example 15.2

An oil tank similar to Figure 15.5a has a diameter of 1.25 m and a height of 2.8 m. The outlet at the bottom is smooth tubing with a length of 5 m and a diameter of 2.85 cm. The oil temperature is 15°C. The full-scale ranges are $FS_{in} = 4.0 \times 10^{-4}$ m³/s (24 L/min) and $FS_{out} = 2.8$ m. Determine each of the following.

a. The capacitance of the tank.
b. The resistance of the outlet.
c. The time constant of the process.
d. The gain of the process.
e. The time-domain equation.
f. The transfer function.

Solution

a. Equation (4.20) in Chapter 4 may be used to compute the tank capacitance.

$$C_L = \frac{A}{\rho g}$$

$$A = \frac{\pi D^2}{4} = \frac{\pi (1.25)^2}{4} = 1.23 \text{ m}^2$$

$$\rho = 880 \text{ kg/m}^3 \text{ (Appendix A)}$$

$$C = \frac{1.23}{880 \times 9.81}$$

$$= 1.4215\text{E} - 04 \text{ cubic meter/pascal}$$

b. The program "LIQRESIS" in Appendix F may be used to compute the liquid flow resistance. From Appendix A, the absolute viscosity of oil is 0.160 Pa·s, and the density is 880 kg/m³. Using a flow rate of 24 LPM, "LIQRESIS" gives the following results.

Reynolds number = 98

Flow is laminar

$R_L = 4.941\text{E} + 07 \text{ Pa·s/m}^3$

$p = 19.8 \text{ kPa}$

c. The time constant, $\tau = R_L C_L$

$$\tau = (4.941\text{E} + 07)(1.4215\text{E} - 04)$$
$$= 7024 \text{ s (or 117 min)}$$

d. The process gain,

$$G = \left(\frac{R_L}{\rho g}\right)\left(\frac{\text{FS}_{in}}{\text{FS}_{out}}\right)$$

$$= \left(\frac{4.941\text{E} + 07}{880 \times 9.81}\right)\left(\frac{4.0\text{E} - 04}{2.8}\right) = 0.8176$$

e. Equation (15.9) gives the time-domain equation.

$$7024 \frac{dh^*}{dt} + h^* = 0.818 q_{in}^*$$

f. Equation (15.10) gives the transfer function.

$$\frac{H^*(s)}{Q_{in}^*(s)} = \frac{0.818}{1 + 7024s}$$

Example 15.3

An electrical circuit similar to Figure 15.6b has an 8.2-kΩ resistance value and a 60-μF capacitance value. Determine the following.

a. The time constant.

b. The transfer function.

Solution

   a. Time constant,

$$\tau = RC$$
$$= (8.2\text{E}+03)(60\text{E}-06) = 0.492 \text{ s}$$

   b. The transfer function is

$$\frac{E_{\text{out}}(s)}{E_{\text{in}}(s)} = \frac{1}{1 + 0.492s}$$

Example 15.4

An oil-bath thermal process similar to Figure 15.6a has an inside diameter of 1 m and a height of 1.2 m. The inside film coefficient is 62 W/m²·K, and the outside film coefficient is 310 W/m²·K. The tank wall is a single layer of steel, 1.2 cm thick. Determine each of the following.

   a. The thermal resistance.
   b. The thermal capacitance.
   c. The time constant.
   d. The time-domain equation.
   e. The transfer function.

Solution

   a. Determine the thermal resistance, R_T.

From Appendix A, $K = 45$ W/m·K for steel
Thickness of the steel wall $= 0.012$ m

$$A = \frac{\pi D^2}{4} + \pi Dh = \pi D\left(\frac{D}{4} + h\right)$$

$$= \pi(1)\left(\frac{1}{4} + 1.2\right) = 4.5553 \text{ m}^2$$

From Equations (4.33) and (4.37) in Chapter 4

$$R_L = \left(\frac{1}{h_i} + \frac{x}{k} + \frac{1}{h_o}\right)\Big/ A$$

$$= \left(\frac{1}{62} + \frac{0.012}{45} + \frac{1}{310}\right)\Big/ 4.5553 = 4.307\text{E}-03 \text{ K/W}$$

   b. Equation (4.39) in Chapter 4 may be used to compute the thermal capacitance.

$$C = mS_h$$
$$S_h = 2180 \text{ J/kg·K (Appendix A)}$$
$$\rho = 880 \text{ kg/m}^3 \text{ (Appendix A)}$$
$$m = \frac{\rho h \pi D^2}{4} = \frac{(880)(1.2)(\pi)(1)^2}{4} = 829.4 \text{ kg}$$
$$C = (829.4)(2180) = 1.808\text{E}+06 \text{ J/K}$$

c. Time constant $= \tau = R_T C_T$.

$$\tau = (4.307\text{E}-03)(1.808\text{E}+06)$$
$$= 7788 \text{ s (or } 129.8 \text{ min)}$$

d. The time-domain equation is

$$7788 \frac{d\theta_L}{dt} + \theta_L = \theta_j$$

e. The transfer function is

$$\frac{\theta_L(s)}{\theta_j(s)} = \frac{1}{1 + 7788s}$$

Example 15.5

A carbon dioxide (CO_2) gas pressure process similar to Figure 15.6b has the following parameters.

Pressure vessel volume: 1.4 m³

Temperature: 530 K

Gas flow resistance: 2×10^5 Pa·s/kg

$FS_{in} = 2$ kg/s

$FS_{out} = 200$ kPa

Determine each of the following.

a. The capacitance of the pressure vessel.
b. The time constant.
c. The process gain.
d. The time-domain equation.
e. The transfer function.

Solution

a. Equation (4.31) in Chapter 4 may be used to determine the capacitance of the pressure vessel.

$$C = \frac{1.2\text{E}-04MV}{T}$$

$M = 44$ for CO_2 (Appendix A)

$V = 1.4$ m³

$T = 530$ K

$$C = \frac{(1.2\text{E}-04)(44)(1.4)}{530}$$

$$= 1.3947\text{E}-05 \text{ kg/Pa}$$

b. The time constant, $\tau = R_g C_g$

$$\tau = (2\text{E}+05)(1.3947\text{E}-05) = 2.79 \text{ s}$$

c. $G = R\left(\dfrac{FS_{in}}{FS_{out}}\right) = (2E+05)\left(\dfrac{2}{2E+05}\right)$

$\quad = 2$

d. The time-domain equation is

$$2.79\,\frac{dp^*}{dt} + p^* = 2w^*$$

e. The transfer function is

$$\frac{P^*(s)}{W^*(s)} = \frac{2}{1 + 2.79s}$$

Example 15.6

A blending tank similar to Figure 15.6c has a tank volume of 3.1 m³ and a flow rate of 0.0031 m³/s. Determine the following.

   a. The time constant.
   b. The time-domain equation.
   c. The transfer function.

Solution

   a. The time constant,

$$\tau = \frac{V}{q}$$

$$= \frac{3.1}{0.0031} = 1000 \text{ s}$$

   b. The time-domain equation is

$$1000\left(\frac{dc_o}{dt}\right) + c_o = c_i$$

   c. The transfer function is

$$\frac{C_o(s)}{C_i(s)} = \frac{1}{1 + 1000s}$$

15.4 THE SECOND-ORDER LAG PROCESS

A *second-order lag process* has two capacitance elements, one capacitance and one inertial element, or two inertial elements. (The inertial elements are mass, inductance, and inertance.) Three parameters characterize the response of a second-order system. The first parameter is the resonant frequency, denoted by ω_0. The second parameter is the amount of damping in the process, expressed by either the damping coefficient

(α), or the damping ratio (ζ). The damping ratio is simply the damping coefficient divided by the resonant frequency ($\zeta = \alpha/\omega_o$). The third parameter is the steady-state gain (G).

The step response of a second-order system is divided into three regions depending on the value of the damping ratio, ζ. If the damping ratio is greater than 1, the response is overdamped as shown in Figure 15.8a. If the damping ratio is less than 1, the response is underdamped as shown in Figure 15.8c. If the damping ratio is equal

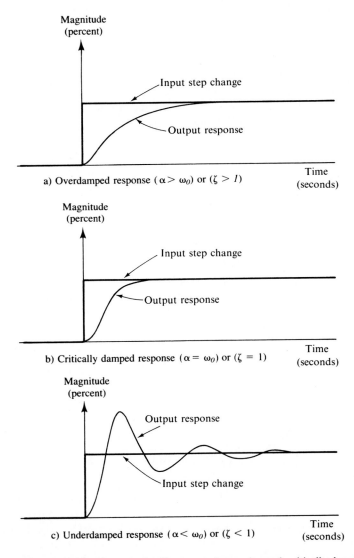

a) Overdamped response ($\alpha > \omega_0$) or ($\zeta > 1$)

b) Critically damped response ($\alpha = \omega_0$) or ($\zeta = 1$)

c) Underdamped response ($\alpha < \omega_0$) or ($\zeta < 1$)

Figure 15.8 The step response of a second-order process is overdamped, critically damped, or underdamped, depending on the damping ratio, $\zeta = \alpha/\omega_0$.

to 1, the response is critically damped as shown in Figure 15.8b. Notice the fast rise and the oscillatory nature of the underdamped response contrasted with the slow rise and the lack of oscillation of the overdamped response. The critically damped response has the fastest rise possible with no oscillation.

A *mechanical spring–mass–damping second-order process* is shown in Figure 15.9a. In this process the applied force (f) is the input signal, and the position of the moving body (h) is the output signal. Three forces act on the moving body: the force exerted by the spring (f_s), the force exerted by the damping piston (f_r), and the externally applied force (f). The sum of these three forces will cause the mass to accelerate according to Newton's law of motion.

$$f + f_s + f_r = M \frac{d^2 h}{dt^2}$$

The spring force (f_s) is equal to the distance the spring is compressed or extended, divided by the mechanical capacitance of the spring.

$$f_s = -\frac{h}{C_m}$$

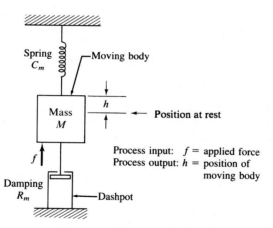

a) A spring-mass-damping second-order process

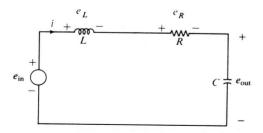

b) A series *RLC* second-order process

Figure 15.9 Mechanical and electrical second-order processes.

The negative sign in the equation for f_s indicates that the direction of the force is down when h is positive (up) and up when h is negative (down).

The damping force is equal to mechanical resistance of the dashpot multiplied by the velocity of the dashpot piston. The piston velocity is represented mathematically by the derivative of h with respect to time (i.e., dh/dt).

$$f_r = -R_m \frac{dh}{dt}$$

The negative sign in the equation for f_r indicates that the force is down when the piston velocity is positive (moving up) and the force is up when the piston velocity is negative (moving down).

Substituting the last two equations into the first equation, and rearranging the terms gives the following time-domain equation for a spring–mass–damping process.

$$MC_m \frac{d^2h}{dt^2} + R_m C_m \frac{dh}{dt} + h = C_m f \qquad (15.20)$$

where C_m = capacitance of the spring, meter/newton
f = externally applied force, newton
h = position of the moving body, meter
M = mass of the moving body, kilogram
R_m = dashpot resistance, newton second/meter

The circuit shown in Figure 15.9b is the electrical analog of the spring–mass–damping process. This can be easily seen by comparing Equation (15.20) with the following equation for the series RLC circuit.

$$LC \frac{d^2 e_{out}}{dt^2} + RC \frac{de_{out}}{dt} + e_{out} = e_{in} \qquad (15.21)$$

A *noninteracting second-order process* has two capacitive elements configured such that the second capacitive element has no effect on the first capacitive element. Equation (15.22) is the time-domain equation of a noninteracting process with input signal, x, and output signal, y.

$$\tau_1 \tau_2 \frac{d^2 y}{dt^2} + (\tau_1 + \tau_2) \frac{dy}{dt} + y = Gx \qquad (15.22)$$

where

$$\tau_1 = R_1 C_1 \qquad \text{and} \qquad \tau_2 = R_2 C_2$$

Figure 15.10 shows examples of liquid and electrical noninteracting second-order processes. The input signal of the liquid process is the input flow rate to the first tank (q_{in}), and the output signal is the level of the second tank (h). The input signal of the

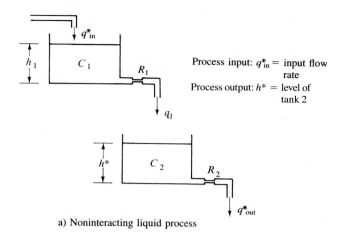

Process input: q_{in}^* = input flow rate

Process output: h^* = level of tank 2

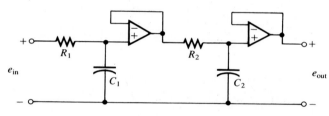

a) Noninteracting liquid process

b) Noninteracting electrical process

Figure 15.10 Two-capacitance noninteracting second-order processes.

electrical circuit is the input voltage (e_{in}), and the output signal is the output voltage (e_{out}). The time-domain equations and the transfer functions of the liquid and electrical noninteracting processes are in the boxed summary just before the exercises at the end of this section.

An *interacting second-order process* has two capacitive elements configured such that the second capacitive element does have an interactive effect on the first capacitive element. Equation (15.23) is the time-domain equation of an interacting, second-order process with input signal, x, and output signal, y.

$$A_2 \frac{d^2y}{dt^2} + A_1 \frac{dy}{dt} + y = Gx \qquad (15.23)$$

$$G = \frac{R_L}{R_1 + R_2 + R_L}$$

$$A_1 = G\left((\tau_1 + \tau_2) + \tau_1 \frac{R_2}{R_L} + \tau_2 \frac{R_1}{R_2}\right)$$

$$A_2 = G\tau_1\tau_2$$

Figure 15.11   Two-capacitance
interacting second-order processes.

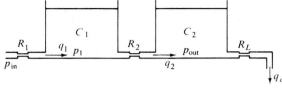

Process input: p_{in} = inlet pressure
Process output: p_{out} = pressure in tank 2

a) An interacting liquid process

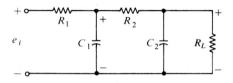

b) An interacting electrical process

Figure 15.11 shows examples of liquid and electrical interacting second-order processes. The input signal of the liquid process is the inlet pressure (p_{in}), and the output signal is the outlet pressure in the second tank (p_{out}). The input signal of the electrical circuit is the input voltage (e_{in}), and the output signal is the output voltage (e_{out}). The time-domain equations and the transfer functions of the liquid and electrical interacting, second-order processes are also in the boxed summary just before the exercises at the end of this section.

As a final example of a second-order process, we will examine the dc motor-driven load shown in Figure 15.12. The input signal is the voltage (e) applied to the

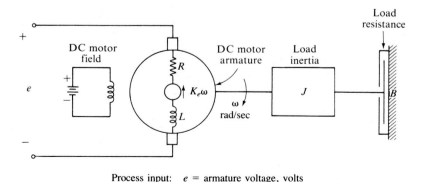

Process input: e = armature voltage, volts
Process output: ω = load speed, radian/second

Figure 15.12   A dc motor-driven load is a second-order process.

armature of the dc motor. The output signal is the rotational velocity of the load (ω). Three equations define this process. The first equation is the electrical equation of the armature circuit.

$$L\frac{di}{dt} + R_i + K_e\omega = e \tag{15.24}$$

The second equation is the current-versus-torque equation of the dc motor.

$$\text{Torque} = K_t i$$

The third equation is the mechanical equation of the load.

$$\text{Torque} = J\frac{d\omega}{dt} + B\omega$$

Combining the last two equations gives us the following:

$$i = \frac{J(d\omega/dt) + B\omega}{K_t}$$

The time-domain equation of the process is obtained by substituting the right-hand side of the last equation for i in the first equation. The result can be reduced to Equation (15.25), the time-domain equation of the dc motor-driven load.

$$A_2\frac{d^2\omega}{dt^2} + A_1\frac{d\omega}{dt} + \omega = Ge \tag{15.25}$$

where $\tau_m = \dfrac{J}{B}$ = mechanical time constant

$\tau_e = \dfrac{L}{R}$ = electrical time constant

$A_2 = \dfrac{\tau_m\tau_e RB}{K_e K_t + RB}$

$A_1 = \dfrac{(\tau_m + \tau_e)RB}{K_e K_t + RB}$

$G = \dfrac{K_t}{K_e K_t + RB}$

J = moment of inertia of the load, kilogram·meter²

B = damping resistance, newton·meter·second/radian

L = armature inductance, henry

R = armature resistance, ohm

K_t = torque constant, newton·meter/ampere

e = armature voltage, volt

ω = motor speed, radian/second

K_e = EMF constant, volt·second/radian

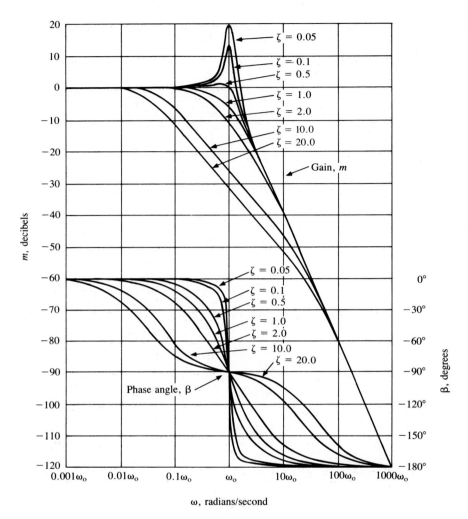

Figure 15.13 The Bode diagram of a second-order process varies considerably for different values of the damping ratio, $\zeta = \alpha/\omega_0$.

The *Bode diagram* of a second-order lag process is shown in Figure 15.13. The frequency scale is normalized in terms of the resonant frequency, and the gain of the process is 1. A range of values of damping ratio from 0.05 to 20 illustrates the pronounced effect that damping has on both the gain and phase angle, especially at the resonant frequency. Notice the amplification at the resonant frequency when the damping ratio is 0.1 and 0.05. The process with $\zeta = 0.05$ has an output at the resonant frequency that is 10 times as large as the input. The effect of the steady-state gain

(G) is to raise or lower the gain line without changing the shape of the line. The steady-state gain has no effect on the phase angle line.

SECOND-ORDER LAG PROCESS

Time-Domain Equation

$$A_2 \frac{d^2y}{dt^2} + A_1 \frac{dy}{dt} + y = Gx \qquad (15.26)$$

Transfer Function

$$\frac{Y(s)}{X(s)} = \frac{G}{1 + A_1 s + A_2 s^2} \qquad (15.27)$$

Parameters

$$\omega_0 = \sqrt{\frac{1}{A_2}} \qquad (15.28)$$

$$\alpha = \frac{A_1}{2A_2} \qquad (15.29)$$

$$\zeta = \frac{\alpha}{\omega_0} = \frac{A_1}{2\sqrt{A_2}} = A_1 \omega_0 / 2 \qquad (15.30)$$

$$A_2 = \frac{1}{\omega_0^2} \qquad (15.31)$$

$$A_1 = \frac{2\zeta}{\omega_0} = \frac{2\alpha}{\omega_0^2} \qquad (15.32)$$

where G = steady-state gain
   x = input signal
   y = output signal
   α = damping coefficient, second^{-1}
   ω_0 = resonant frequency, radian/second
   ζ = damping ratio

Mechanical Second-Order Process (Figure 15.9a)

   Process input: f = applied force, newton
   Process output: h = position of mass, meter
   $A_2 = MC_m$ (15.33)
   $A_1 = RC_m$ (15.34)
   $G = C_m$ (15.35)
   M = mass of the moving body, kilogram
   R = damping resistance, newton second/meter
   C_m = capacitance of the spring, meter/newton

Electrical Second-Order Process (Figure 15.9b)

$$\text{Process input: } e_{in} = \text{input voltage, volt}$$
$$\text{Process output: } e_{out} = \text{capacitor voltage, volt}$$

$$A_2 = LC \tag{15.36}$$
$$A_1 = RC \tag{15.37}$$
$$G = 1 \tag{15.38}$$

$$L = \text{electrical inductance, henry}$$
$$C = \text{electrical capacitance, farad}$$
$$R = \text{electrical resistance, ohm}$$

Noninteracting Second-Order Process (Figure 15.10)

$$\tau_1 = R_1 C_1 \tag{15.39}$$
$$\tau_2 = R_2 C_2 \tag{15.40}$$
$$A_2 = \tau_1 \tau_2 \tag{15.41}$$
$$A_1 = \tau_1 + \tau_2 \tag{15.42}$$

1. Liquid level noninteracting process

$$\text{Process input: } q_{in}^* = \text{input flow rate, percent of FS}_{in}$$
$$\text{Process output: } h^* = \text{level in tank 2, percent of FS}_{out}$$

$$G = \left(\frac{R_2}{\rho g}\right)\left(\frac{\text{FS}_{in}}{\text{FS}_{out}}\right) \tag{15.43}$$

$$R_1, R_2 = \text{liquid resistance, pascal second/meter}^3$$
$$C_1, C_2 = \text{liquid capacitance, meter}^3/\text{pascal}$$

2. Electrical noninteracting process

$$\text{Process input: } e_{in} = \text{input voltage, volt}$$
$$\text{Process output: } e_{out} = \text{voltage across } C_2, \text{ volt}$$

$$G = 1 \tag{15.44}$$

$$R_1, R_2 = \text{electrical resistance, ohm}$$
$$C_1, C_2 = \text{electrical capacitance, ohm}$$

Interacting Second-Order Process (Figure 15.11)

$$G = \frac{R_L}{R_1 + R_2 + R_L} \tag{15.45}$$

$$A_2 = \tau_1 \tau_2 G \tag{15.46}$$

$$A_1 = \left(\tau_1 + \tau_2 + \tau_1 \frac{R_2}{R_L} + \tau_2 \frac{R_1}{R_2}\right) G \tag{15.47}$$

1. Liquid level interacting process

   Process input: p_{in} = inlet pressure, pascal
   Process output: p_{out} = pressure in tank 2, pascal

2. Electrical interacting process

   Process input: e_{in} = input voltage, volt
   Process output: e_{out} = voltage across R_L, volt

DC Motor Second-Order Process (Figure 15.12)

Process input: e = armature voltage, volt
Process output: ω = rotational velocity of the load, radian/second

$$A_2 = \frac{\tau_m \tau_e RB}{RB + K_e K_t} \tag{15.48}$$

$$A_1 = \frac{(\tau_m + \tau_e)RB}{RB + K_e K_t} \tag{15.49}$$

$$G = \frac{K_t}{RB + K_e K_t} \tag{15.50}$$

where $\tau_m = \dfrac{J}{B}$, mechanical time constant, second

  $\tau_e = \dfrac{L}{R}$, electrical time constant, second

  B = damping resistance, newton meter second/radian
  e = armature voltage, volt
  J = moment of inertia of the load, kilogram meter^2
  K_e = EMF constant, volt·second/radian
  K_t = torque constant, newton·meter/ampere
  L = armature inductance, henry
  R = armature resistance, ohm
  ω = motor speed, radian/second

Example 15.7

A spring–mass–damping process consists of a 10-kg mass, a spring capacitance of 0.001 m/N, and a damping resistance of 20 N·s/m. Determine the following.

   a. The time-domain equation.
   b. The transfer function.
   c. The resonant frequency (ω_0).

    d. The damping ratio (ζ).

    e. Whether the process is overdamped, underdamped, or critically damped.

Solution

    a. The time domain is given by Equation (15.26) and Equations (15.33) through (15.35).

$$A_2 = MC_m = (10)(0.001) = 0.01$$
$$A_1 = RC_m = (20)(0.001) = 0.02$$
$$G = C_m = 0.001$$

$$0.01 \frac{d^2h}{dt^2} + 0.02 \frac{dh}{dt} + h = 0.001f$$

    b. Equation (15.27) gives the transfer function.

$$\frac{H(S)}{F(S)} = \frac{0.001}{1 + 0.02s + 0.01s^2}$$

    c. From Equation (15.28),

$$\omega_0 = \sqrt{\frac{1}{0.01}} = 10 \text{ rad/s}$$

    d. From Equation (15.30),

$$\zeta = \frac{0.02}{2\sqrt{0.01}} = 0.1$$

    e. The damping ratio is less than 1; therefore, the process is underdamped.

Example 15.8

An electrical series *RLC* circuit consists of a 0.022-H inductor, a 10-μF capacitor, and a 200-Ω resistance. Determine the following.

    a. The time-domain equation.

    b. The transfer function.

    c. The resonant frequency.

    d. The damping ratio.

    e. The type of damping.

Solution

    a. The time-domain equation is given by Equation (15.26) and Equations (15.36) through (15.38).

$$A_2 = LC = (0.022)(10E - 06) = 2.2E - 07$$
$$A_1 = RC = (200)(10E - 06) = 0.002$$
$$2.2 \times 10^{-7}\left(\frac{d^2 e_{out}}{dt^2}\right) + 0.002\left(\frac{de_{out}}{dt}\right) + e_{out} = e_{in}$$

b. The transfer function is given by Equation (15.27).

$$\frac{E_{out}(s)}{E_{in}(s)} = \frac{1}{1 + 0.002s + 2.2 \times 10^{-7}s^2}$$

c. From Equation (15.28),

$$\omega_0 = \sqrt{\frac{1}{2.2E - 07}} = 2.13 \times 10^3 \text{ rad/s}$$

d. From Equation (15.30),

$$\zeta = \frac{(2.0E - 03)(2.13E + 03)}{2} = 2.13$$

e. The damping ratio is greater than 1; therefore, the process is overdamped.

Example 15.9

A liquid, noninteracting, two-capacity system has time constants τ_1 and τ_2 of 520 s and 960 s. The liquid is water and the value of R_2 is 1.6×10^{-6} Pa·s/m^3. The full-scale ranges are $FS_{in} = 5.0 \times 10^{-3}$ m^3/s and $FS_{out} = 2.0$ m. Determine the following.

a. The time-domain equation.
b. The transfer function.

Solution

a. The time-domain equation is given by Equation (15.26) and Equations (15.41) through (15.43).

$$A_2 = \tau_1 \tau_2 = (520)(960) = 5E + 05 \text{ s}^2$$
$$A_1 = \tau_1 + \tau_2 = 520 + 960 = 1480 \text{ s}$$
$$\rho = 1000 \text{ kg/m}^3 \text{ (Appendix A)}$$
$$G = \left[\frac{1.6E + 06}{(1000)(9.81)}\right]\left(\frac{5.0E - 03}{2}\right) = 0.41$$
$$500,000\frac{d^2 h^*}{dt^2} + 1480\frac{dh^*}{dt} + h^* = 0.41 q_{in}^*$$

b. The transfer function is given by Equation (15.27).

$$\frac{H^*(s)}{Q_{in}^*(s)} = \frac{0.41}{1 + 1480s + 500,000s^2}$$

Example 15.10

An electrical, interacting, two-capacity process (Figure 15.11b) has the following component values.

$$R_1 = 1000 \ \Omega$$
$$R_2 = 800 \ \Omega$$
$$R_L = 400 \ \Omega$$
$$C_1 = 10 \ \mu F$$
$$C_2 = 50 \ \mu F$$

Determine the following.

    a. The time-domain equation.
    b. The transfer function.

Solution

    a. The time-domain equation is given by Equations (15.26), (15.39), (15.40), (15.45), (15.46), and (15.47).

$$\tau_1 = (1000)(10E-06) = 0.01 \text{ s}$$
$$\tau_2 = (800)(50E-06) = 0.04 \text{ s}$$
$$G = \frac{400}{1000 + 800 + 400} = 0.1818$$
$$A_2 = (0.01)(0.04)(0.182) = 7.27 \times 10^{-5}$$
$$A_1 = \left[0.01 + 0.04 + 0.01\left(\frac{800}{400}\right) + 0.04\left(\frac{1000}{800}\right) \right] 0.1818$$
$$= 0.0218$$

The time-domain equation is

$$7.27 \times 10^{-5} \frac{d^2 e_{\text{out}}}{dt^2} + 0.0218 \frac{de_{\text{out}}}{dt} + e_{\text{out}} = 0.182 e_{\text{in}}$$

    b. The transfer function is

$$\frac{E_{\text{out}}(s)}{E_{\text{in}}(s)} = \frac{0.182}{1 + 0.0218s + 7.27 \times 10^{-5}s^2}$$

Example 15.11

An armature-controlled dc motor has the following characteristics.

$$J = 6.2 \times 10^{-4} \text{ kg·m}^2$$
$$B = 1.0 \times 10^{-4} \text{ N·m·s/rad}$$

$$L = 0.020 \text{ H}$$
$$R = 1.2 \, \Omega$$
$$K_e = 0.043 \text{ V·s/rad}$$
$$K_t = 0.043 \text{ N·m/A}$$

Determine the following.

  a. The mechanical time constant.
  b. The electrical time constant.
  c. The time-domain equation.
  d. The transfer function.

Solution

  a. $\tau_m = J/B = 6.2\text{E}-04/1.0\text{E}-04 = 6.2$ s.
  b. $\tau_e = L/R = 0.020/1.2 = 0.0167$ s.
  c. The time-domain equation is given by Equations (15.26), (15.48), (15.49), and (15.50).

$$A_2 = \frac{(6.2)(0.0167)(1.2)(1.0\text{E}-04)}{(1.2)(1.0\text{E}-04) + (0.043)(0.043)} = 0.00630$$

$$A_1 = \frac{(6.2 + 0.0167)(1.2)(1.0\text{E}-04)}{(1.2)(1.0\text{E}-04) + (0.043)(0.043)} = 0.379$$

$$G = \frac{0.043}{(1.2)(1.0\text{E}-04) + (0.043)(0.043)} = 21.8$$

The time-domain equation is

$$0.00630 \frac{d^2\omega}{dt^2} + 0.379 \frac{d\omega}{dt} + \omega = 21.8e$$

  d. The transfer function is

$$\frac{\Omega(s)}{E(s)} = \frac{21.8}{1 + 0.379s + 0.00630s^2}$$

15.5 THE DEAD-TIME PROCESS

A *dead-time process* is one in which mass or energy is transported from one point to another. The output signal is identical to the input signal except for a time delay. The time delay is called the dead-time lag and is denoted by t_d. The dead-time lag is the time required for the signal to travel from the input location to the output location. Dead time was included as one of the common elements in Chapter 4 (see Sections 4.1, 4.2, 4.3, and 4.6 for further discussion and examples of dead-time elements).

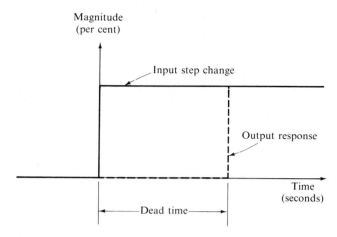

Figure 15.14   The step response of a dead-time process is identical to the input but is delayed by the amount of dead-time lag, t_d.

The step response of a dead-time process is shown in Figure 15.14. Before the step change, the output signal and the input signal are equal. The step change increases the input signal to a new value at time $t = 0$ s. The output signal remains at the original value until time $t = t_d$, when it also increases to the new value. The graph of the output signal is a duplication of the input graph moved to the right by t_d seconds.

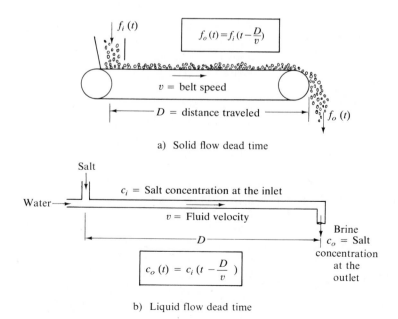

Figure 15.15   Solid flow and liquid flow dead-time processes.

The response of a dead-time process is characterized by the dead-time lag—the number of seconds (or minutes) that elapse between an input change and the corresponding output change. The time-domain equation of the dead-time process was developed in Section 4.1. Figure 15.15 illustrates two examples of dead-time processes.

The *Bode diagram* of a dead-time process is shown in Figure 15.16. The dead-time process delays a signal but does not change the size of the signal. The gain is 1 at all frequencies, and the Bode diagram gain line is on the 0-dB line. The phase angle

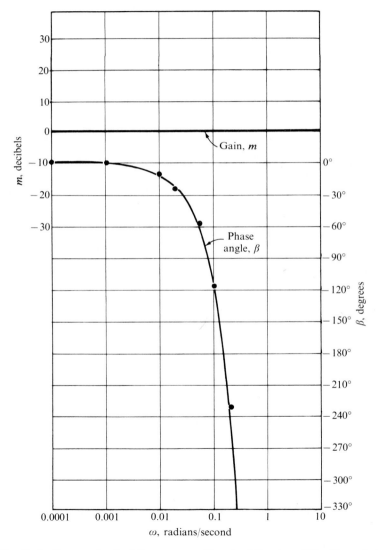

Figure 15.16   Bode diagram of a dead-time process with a dead-time lag of 20 s (Example 15.12). The gain is 1 at all frequencies, but the phase lag becomes very large as frequency increases.

(β) is given by the following equation:

$$\beta = -57.3\omega t_d \qquad (15.51)$$

DEAD-TIME PROCESS

Time-Domain Equation

$$f_o(t) = f_i(t - t_d) \qquad (15.52)$$

$$t_d = \frac{D}{v} \qquad (4.1)$$

Transfer Function

$$\frac{F_o(s)}{F_i(s)} = e^{-t_d s} \qquad (15.53)$$

where D = distance from input to output, meter

$f_i(s)$ = input signal

$f_o(s)$ = output signal

t = time, second or minute

t_d = dead-time lag, second or minute

v = velocity of signal travel, meter/second

Example 15.12

A dead-time process similar to Figure 15.15a consists of a 12-m-long belt conveyor with a belt velocity of 0.6 m/s. Determine the dead-time lag (t_d), the time-domain equation, the transfer function, and the Bode diagram.

Solution

$$t_d = \frac{D}{v} = \frac{12}{0.6} = 20 \text{ s}$$

The time-domain equation is given by Equation (15.52).

$$f_o(t) = f_i(t - 20)$$

The transfer function is given by Equation (15.53).

$$\frac{F_o(S)}{F_i(S)} = e^{-20s}$$

The Bode diagram gain line is a constant 0 dB. The phase line is given by Equation (15.51).

$$\beta = -57.3\omega t_d = -57.3\omega(20) = -1146\omega$$

The following table of values was used to construct the phase angle line in the Bode diagram in Figure 15.16.

ω (rad/s)	β (deg)
0.001	1.15
0.01	11.5
0.02	22.9
0.05	57.3
0.1	114.6
0.2	229.2
0.4	458.4

15.6 THE FIRST-ORDER LAG PLUS DEAD-TIME PROCESS

The first-order lag plus dead-time process is a series combination of a first-order lag process and a dead-time element. The step response, illustrated in Figure 15.17, is characterized by the first-order lag time constant (τ) and the dead-time lag (t_d). The first-order lag plus dead-time characteristic is frequently used as a first approximation of the model of more complex processes for the purpose of analysis and design. The values of t_d and τ can be determined from a step response test of the process as indicated in Figure 15.17. An example of a first-order lag plus dead-time salt brine process is shown in Figure 15.18. A mixing tank is used to blend salt and water continuously to form a brine solution. The inlet water flow rate is regulated to match the outlet flow of brine solution. The salt flow rate is regulated to maintain the desired salt concentration in the mixing tank. The salt flow rate, $f_i(t)$, is the input signal of

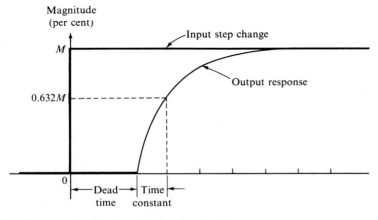

Figure 15.17   Step response of a first-order lag plus dead-time process.

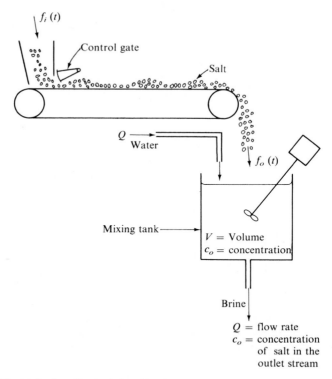

Figure 15.18   First-order lag plus dead-time blending process.

the process. The salt concentration in the tank, $c_o(t)$, is the output signal of the process. The belt conveyor is the dead-time element and the mixing tank provides the first-order lag characteristic. In this example, the dead-time element preceded the first-order lag element. However, the response is the same for processes in which the first-order lag characteristic precedes the dead-time element.

GLOSSARY

Dead-time process: A process in which the input appears at the output after a time delay of t_d seconds. The time delay, t_d, completely characterizes the dead-time process. This process usually involves transportation of mass or energy from one point to another. (15.5)

First-order lag process: A process in which the rate of change of the output is proportional to the difference between the input and the output. The first-order lag process is completely characterized by its time constant, τ, and its steady-state gain, G. First-order lag processes consist of a single resistance element and a single storage element. (15.3)

Frequency response graph: Two graphs that display the gain and the phase angle of a component plotted versus frequency. The two graphs share a common log frequency scale on the x-axis. The y-axis of one graph is the decibel gain of the component; the other is the phase angle of the component. (15.1)

Integral action time constant: The time required for the output of an integral process to change by an amount equal to the difference between the input and the nominal value that will maintain the output constant. (15.2)

Integral process: A process in which the output changes at a rate proportional to the difference between the input and a nominal value that will maintain the output constant. The integral process is characterized by the integral action time constant as defined above. (15.2)

Interacting second-order process: A second-order process configured such that the second storage element has an effect on the first storage element. (15.4)

Noninteracting second-order process: A second-order process configured such that the second storage element has no effect on the first storage element. (15.4)

Ramp process: Another name for the integral process. (15.2)

Second-order lag process: A process that has two storage elements and at least one resistive element. The second-order lag process is completely characterized by its resonant frequency (ω_0), damping ratio (ζ), and steady-state gain (G). The response of the second-order lag process is divided into three regions: underdamped, critically damped, and overdamped, depending on whether the damping ratio is less than 1, equal to 1, or greater than 1. (15.4)

Step response graph: A graph of the output of a process following a step change in the input to the process. (15.1)

Time-domain equation: An equation that defines the size-versus-time relationship between the input and the output of a process. (15.1)

Transfer function: The frequency domain ratio of the output of a process over its input with all initial conditions set to zero. The transfer function defines the gain and phase difference versus frequency relationship between the input and the output of the process. (15.1)

EXERCISES

15.1 A liquid level integral process (Figure 15.2a) has the following parameters and conditions.

$$\text{Tank height} = 5.2 \text{ m}$$
$$\text{Tank diameter} = 1.8 \text{ m}$$
$$FS_{in} = 0.02 \text{ m}^3/\text{s}$$
$$FS_{out} = 5 \text{ m}$$
$$h^*(t_o) = 1.7 \text{ m} = 34\% \text{ of FS}$$
$$q^*_{out} = 0.009 \text{ m}^3/\text{s} = 45\% \text{ of FS}$$
$$q^*_{in} = 0.011 \text{ m}^3/\text{s} = 55\% \text{ of FS}$$

Determine the time-domain equation, the transfer function, the integral action time constant, and the level at time $t_o + 50$ s.

15.2 Construct a Bode diagram of the integral process in Exercise 15.1. The input signal is expressed in percent of FS_{in}, and the output signal is expressed in percent of FS_{out}. Use the Bode diagram to determine the output amplitude produced by a sinusoidal input at each frequency given below. The input signal has an amplitude of 10% of FS_{in}. Express the output amplitude both as a percent of FS_{out}, and in meters. Indicate any output that is limited by FS_{out}.

Frequency (rad/s)

1.57×10^{-5}

1.57×10^{-4}

1.57×10^{-3}

1.57×10^{-2}

1.57×10^{-1}

1.57×10^{0}

15.3 An oil tank similar to Figure 15.5a has a diameter of 1.658 m and a height of 3.6 m. The outlet at the bottom is a 4-m-long smooth tube with a diameter of 2.5 cm. The oil temperature is 15°C. The full-scale ranges are: $FS_{in} = 2.5 \times 10^{-4}$ m³/s and $FS_{out} = 3.2$ m. Determine each of the following.

a. The capacitance of the tank.

b. The resistance of the outlet.

c. The time constant of the process.

d. The gain of the process.

e. The time-domain equation.

f. The transfer function.

15.4 An electrical circuit similar to Figure 15.5b has a capacitance value of 10 μF. Determine the resistance value that will result in a time constant equal to 0.083 s.

15.5 Construct the Bode diagram of the first-order electrical circuit in Example 15.3. Determine the amplitude and phase of the output signal corresponding to each of the following input signals.

a. $e_{in} = 0.25 \cos(0.2t + 45°)$ volts

b. $e_{in} = 0.5 \cos(2t + 45°)$ volts

c. $e_{in} = 1.5 \cos(20t + 45°)$ volts

15.6 An oil-bath thermal system similar to Figure 15.6a has an inside diameter of 0.75 m and a height of 0.6 m. The inside film coefficient is 58 W/m²·K, and the outside film coefficient is 350 W/m²·K. The tank wall is a single layer of aluminum 2.1 cm thick. Determine each of the following.

a. The thermal resistance.   **b.** The thermal capacitance.

c. The time constant.   **d.** The time-domain equation.

e. The transfer function.

15.7 Construct the Bode diagram of the oil-bath thermal process in Example 15.4.

15.8 A nitrogen gas process similar to Figure 15.6b has the following parameters:

Pressure vessel volume: 1.1 m³

Temperature: 540 K

Gas flow resistance: 1.8×10^5 Pa·s/kg

FS_{in}: 2.5 kg/s

FS_{out}: 100 kPa

Determine each of the following.

a. The capacitance of the pressure vessel.

b. The time constant.

c. The process gain.

d. The time-domain equation.

e. The transfer function.

15.9 A blending system similar to Figure 15.6c has a flow rate of 0.006 m³/s. Determine the tank volume that will result in a time constant of 10 min. Then determine the time-domain equation and transfer function using the 10-min time constant.

15.10 A spring–mass–damping system consists of a 25-kg mass, a spring capacitance of 6.9×10^{-4} m/N, and a damping resistance of 42 N·s/m. Determine the time-domain equation, the transfer function, the resonant frequency (ω_0), the damping ratio (ζ), and whether the process is overdamped, underdamped, or critically damped.

15.11 Construct the Bode diagram of the spring–mass–damping system in Example 15.7. Determine the amplitude and phase angle of the output produced by each of the following input signals.

a. $f = 10 \cos(t + 0°)$ newtons

b. $f = 10 \cos(10t + 0°)$ newtons

c. $f = 10 \cos(100t + 0°)$ newtons

15.12 An electrical series *RLC* circuit consists of a 0.04-H inductor, a 4-μF capacitor, and a 100-Ω resistor. Verify that the system is underdamped. The damping of the circuit may be increased by adding a second resistor in series with the 100-Ω resistor. Determine the value of the second resistor that will result in a critically damped circuit. Determine the time-domain equation and the transfer function of the critically damped circuit.

15.13 A liquid noninteracting two-capacity system has time constants τ_1 and τ_2 of 300 s and 1200 s. The liquid is water, and the value of R_2 is 2.2×10^6 Pa·s/m³. The full-scale ranges are $FS_{in} = 1.0 \times 10^{-2}$ m³/s and $FS_{out} = 4$ m. Determine the time-domain equation and transfer function.

15.14 An electrical interacting second-order circuit has the following component values:

$$R_1 = 100 \ \Omega$$
$$R_2 = 300 \ \Omega$$
$$R_L = 50 \ \Omega$$
$$C_1 = 0.1 \ \mu F$$
$$C_2 = 0.8 \ \mu F$$

Determine τ_1, τ_2, the time-domain equation, and the transfer function.

15.15 An armature-controlled dc motor has the following characteristics.

$$B = 2.0 \times 10^{-3} \ N \cdot m \cdot s/rad$$
$$J = 3.2 \times 10^{-3} \ kg \cdot m^2$$
$$K_e = 0.22 \ V \cdot s/rad$$
$$K_t = 0.22 \ N \cdot m/A$$
$$L = 0.075 \ H$$
$$R = 1.2 \ \Omega$$

Determine the mechanical time constant, the electrical time constant, the time-domain equation, the transfer function, the resonant frequency (ω_0), the damping ratio (ζ), and the type of damping (i.e., overdamped, critically damped, or underdamped).

15.16 Construct the Bode diagram of the dc motor second-order system in Example 15.11.

15.17 A dead-time process similar to Figure 15.15a consists of an 8.2-m-long belt conveyor with a belt velocity of 0.44 m/s. Determine the dead-time lag, the time-domain equation, and the transfer function.

15.18 Construct a Bode diagram for a dead-time lag process with a dead-time of 250 s.

15.19 A first-order lag plus dead-time process similar to Figure 15.18 has the following parameter values:

Distance traveled on the belt: 7.6 m

Belt speed: 1.1 m/s

Tank volume: 12.2 m³

Water flow rate: 0.01 m³/s

Determine the dead-time lag (t_d) and the first-order time constant (τ).

15.20 The response of a control system is very much like the response of a second-order system. A closed-loop system may be overdamped, critically damped, or underdamped. The damping coefficient (α) determines the type of damping present in a system. In Equation (15.26), the equation of the second-order system is expressed in terms of A_1 and A_2. The equation can also be expressed in terms of ω_0 and α, as given below.

$$\frac{1}{\omega_0^2} \frac{d^2y}{dt^2} + 2\left(\frac{\alpha}{\omega_0^2}\right)\frac{dy}{dt} + y = Gx$$

Notice that α is part of the coefficient of the first derivative of the output (dy/dt). The first derivative is the rate of change (or velocity) of the output. Consider a dc motor position control system that is underdamped. A speed sensor measures the rate of change of the output (i.e., dy/dt). Explain how the signal from the speed sensor could be used to increase the damping coefficient of the system.

15.21 A blending and heating system is illustrated in Figure 1.10. The thermal system is a first-order lag and the blending system is a first-order lag plus dead time. Determine the time constant and dead-time lag of the blending system and the time constant of the thermal system. Also determine the transfer function of the thermal system. The system parameters are as follows:

Production rate: 1.8×10^{-4} m^3/s

Mixing tank diameter: 0.55 m

Height of liquid in the mixing tank: 0.65 m

Inside thermal film coefficient: $h_i = 810$ W/m$^2 \cdot$K

Outside thermal film coefficient: $h_o = 1200$ W/m$^2 \cdot$K

Tank wall thickness: 0.7 cm

Tank wall material: steel

Density of liquid in the tank: 1008 kg/m^3

Specific heat of the liquid: 4060 J/kg$\cdot$K

Diameter of the mixing tank outlet pipe: 2 cm

Distance from tank outlet to the analyzer feed pipe inlet: 1.5 m

Concentration analyzer feed pipe flow velocity: 0.2 m/s

Length of the concentration analyzer feed pipe: 2 m

CHAPTER 16

Methods of Analysis

OBJECTIVES

The analysis of a control system centers on answering the question: Is the system stable? A stable control system responds to any reasonable input with an error that diminishes with time. The error of a stable system may or may not oscillate, and it may or may not diminish to zero. If the error oscillates, the amplitude of the oscillation diminishes exponentially with time. If the error does not diminish to zero, it eventually reaches a small, steady value. In contrast, an unstable control system responds to at least some reasonable inputs with an error that increases with time. The error usually oscillates. For continuous, linear systems, an error that oscillates with constant amplitude is considered to be unstable. However, some discontinuous systems are inherently oscillatory, and an oscillating error is considered stable as long as the amplitude does not increase with time.

A second concern of control system analysis is the question: How quickly does the error diminish? Control system engineers must answer both questions in the process of designing a control system that is stable and minimizes errors resulting from disturbances, changes in setpoint, or changes in load.

The purpose of this chapter is to introduce you to three methods the control engineer uses to analyze control systems: Bode diagrams, Nyquist diagrams, and root locus. After completing this chapter, you will be able to

1. Combine the Bode diagrams of several components connected in series to produce the overall Bode diagram

2. Use graphical or computer-aided methods to construct the open-loop Bode diagram of a control system

3. Use the Nichols chart or computer-aided methods to construct the closed-loop Bode diagram of a control system

4. Use computer-aided methods to construct the error ratio and the deviation ratio graphs of a control system

5. Determine if a control system is stable from the open-loop Bode diagram of the system

6. Determine the gain margin and the phase margin of a control system from the open-loop Bode diagram

7. Construct a Nyquist diagram of a control system and determine from the diagram whether the system is stable, marginally stable, or unstable

8. Construct a root-locus diagram of a control system and determine the values of gain, K, for which the system is unstable

9. Determine from a root-locus diagram the value of gain, K, that results in a given damping ratio

16.1 INTRODUCTION

In this chapter you will study a number of methods used to analyze control systems. The analysis of a control system involves the determination of the response of the system to various disturbances and the answer to the question: Is the system stable? A closed-loop control system consists of several components connected in series to form a closed loop. These components include the controller, manipulating element, process, and measuring transmitter (see Figure 1.7). Thus our study begins with the overall transfer function and frequency response of several components connected in series.

There are actually two overall transfer functions of a closed-loop control system. One transfer function is obtained with the feedback line disconnected at the error detector. We call this the *open-loop transfer function* of the control system. The second transfer function is obtained with the feedback line connected. We call this the *closed-loop transfer function* of the system. Both transfer functions are expressed as the ratio C_m/SP.

The *error ratio* and the *deviation ratio* give additional insight into the response of a closed-loop control system to disturbances of various frequencies. The response of a control system can be divided into three frequency zones. Zone 1 is the range of frequencies below the "frequency limit" of the control system. Zone 2 is the middle range of frequencies between the "frequency limit" and a second, higher-frequency value. Zone 3 is the range of frequencies above the second frequency value. In zone 1, feedback reduces the error caused by a disturbance or setpoint change. In zone 2, feedback actually increases the error caused by a disturbance or setpoint change. In zone 3, feedback has no effect on the error. The importance of knowing the frequency limit of a control system is obvious. A control system cannot handle disturbances or setpoint changes that have a frequency greater than its frequency limit.

The *frequency response* of a control system is usually displayed as a pair of graphs called Bode diagrams. One graph shows the overall gain (amplitude change from SP to C_m) versus frequency. The second graph shows the overall phase angle (phase change from SP to C_m) versus frequency. The same frequency scale is used on both graphs, and they are plotted on one diagram, thereby making it easy to read the gain and phase angle at any given frequency. As with transfer functions, two frequency responses are useful in control system analysis. The *open-loop frequency response* is obtained with the feedback line disconnected at the error detector. The *closed-loop frequency response* is obtained with the feedback line connected.

Both graphical and computed-aided methods are used to obtain the open-loop and closed-loop Bode diagrams of a control system. A BASIC program for computer-aided analysis and design is included on a disk that comes with the instructor's manual accompanying this book. Appendix F includes a listing of this program (named "DESIGN"). Program "DESIGN" produces both open-loop and closed-loop Bode diagrams of a control system. It also produces a Nyquist diagram and an error-ratio graph of the system. In Chapter 17, program "DESIGN" is used for computer-aided design of PID controllers and compensation networks. With program "DESIGN", a designer can try a number of different designs and observe the change in the frequency response on a graphic screen. The designer can use a "what if" approach to the analysis and design of a control system. However, a certain amount of hand plotting of graphical data will help students acquire the understanding and judgment required to make full use of powerful computer-aided methods such as program "DESIGN".

Stability is a major concern in the design of a control system. Control system designers use two safety margins, called the *gain margin* and *phase margin*, to assure the stability of the control system. The designer reads these two margins directly from the open-loop Bode diagram, and uses them as a guide while shaping the Bode diagram to obtain the optimum control system design.

The Bode diagram does not apply to some complex types of systems. For these systems, the designer may use the *Nyquist stability criterion*, which applies to a wider range of systems than the Bode diagram. The Nyquist criterion uses a graphic technique to determine the stability or instability of the control system. Although the Nyquist criterion answers the question "Is the system stable?", it is not as useful as Bode diagrams and root locus in the actual design of the controller.

The *root-locus* technique is another graphical method of determining the stability or instability of a control system. The root locus consists of a plot of the roots of a certain characteristic equation of the control system as the gain varies from zero to infinity. From the root-locus plot, the designer can observe the relationship between the system gain and the stability of the system. The root locus is particularly useful to the designer in determining the effect of gain changes on the stability and performance of the control system.

16.2 OVERALL BODE DIAGRAM OF SEVERAL COMPONENTS

A control system consists of several components connected in series to form a closed loop. The design and analysis of a control system require a knowledge of the transfer function and the frequency response of this group of components considered as a unit. In other words, the overall transfer function and frequency response must be known.

Consider the three components in Figure 16.1 with transfer functions TF_1, TF_2, and TF_3. Signal S_1 is the input signal of component 1. Signal S_2 is the output signal of component 1 and the input signal of component 2. Signal S_3 is the output signal of component 2 and the input signal of component 3. Signal S_4 is the output signal of component 3. Also, signal S_1 is the input signal, and S_4 is the output signal of the three components as a unit. By definition, the transfer function of a component

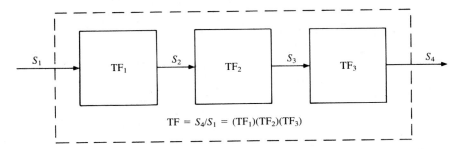

Figure 16.1   The transfer function of three components connected in series is equal to the product of the transfer functions of the individual components.

or group of components is the ratio of the output signal over the input signal. Thus

$$\text{Transfer function of component 1} = \text{TF}_1 = \frac{S_2}{S_1}$$

$$\text{Transfer function of component 2} = \text{TF}_2 = \frac{S_3}{S_2}$$

$$\text{Transfer function of component 3} = \text{TF}_3 = \frac{S_4}{S_3}$$

$$\text{Transfer function of the group of components} = \text{TF} = \frac{S_4}{S_1}$$

But

$$\frac{S_4}{S_1} = \left(\frac{S_2}{S_1}\right)\left(\frac{S_3}{S_2}\right)\left(\frac{S_4}{S_3}\right)$$

and

$$\text{TF} = (\text{TF}_1)(\text{TF}_2)(\text{TF}_3) \tag{16.1}$$

Equation (16.1) states that the overall transfer function of several components in series is equal to the product of the transfer functions of the individual components.

The frequency response of a component or group of components is obtained by substituting $j\omega$ for s in the transfer function. The transfer function can then be reduced to a complex number in the polar form. The magnitude of the complex number is the frequency response gain (g). The angle of the complex number is the frequency response phase angle (β). Let the frequency response be represented by the following terms.

$g_1\underline{/\beta_1}$ = frequency response of component 1

$g_2\underline{/\beta_2}$ = frequency response of component 2

$g_3\underline{/\beta_3}$ = frequency response of component 3

$g\underline{/\beta}$ = overall frequency response of the three components as a unit

Then, from Equation (16.1)

$$g\underline{/\beta} = (g_1\underline{/\beta_1})(g_2\underline{/\beta_2})(g_3\underline{/\beta_3}) \tag{16.2}$$

However, the product of three complex numbers in polar form is a complex number whose magnitude is the product of the magnitudes of the three complex numbers, and whose angle is the sum of the angles of the three complex numbers.

$$g\underline{/\beta} = g_1 g_2 g_3 \underline{/\beta_1 + \beta_2 + \beta_3} \tag{16.3}$$

Equation (16.3) states that the *overall gain* of several components in series is equal to the product of the gains of the individual components. It further states that the *overall phase angle* of several components in series is equal to the sum of the phase angles of the individual components. In other words, the gains are multiplied and the phase angles are added.

The Bode diagram uses a logarithmic gain scale so that multiplication of gain terms can be accomplished by graphically adding the logarithms of the individual gain terms; that is,

$$\log g = \log (g_1 \cdot g_2 \cdot g_3) = \log g_1 + \log g_2 + \log g_3$$

The decibel scale is a logarithmic scale, and the addition of decibel values is equivalent to multiplying the corresponding gain values.

$$m(\text{dB}) = m_1(\text{dB}) + m_2(\text{dB}) + m_3(\text{dB}) \tag{16.4}$$

The overall transfer function of several components in series is equal to the product of the transfer functions of the individual components.

The overall gain of several components in series is equal to the product of the gains of the individual components.

The overall phase angle of several components in series is equal to the sum of the phase angles of the individual components.

The overall frequency response of several components can be determined on a Bode diagram graphically by adding the decibel gains and by adding the phase angles of the individual components.

Example 16.1

A thermal process and a temperature measuring means are connected in series in a control system. Determine the overall transfer function from the following individual transfer functions:

$$\text{Thermal process transfer function} = \frac{1}{1 + 6420s}$$

$$\text{Measuring means transfer function} = \frac{1}{1 + 78s}$$

Solution

The overall transfer function is the product of the individual transfer functions:

$$\text{Overall transfer function} = \left(\frac{1}{1 + 6420s}\right)\left(\frac{1}{1 + 78s}\right)$$

$$= \frac{1}{(1 + 6420s)(1 + 78s)}$$

Example 16.2

Three components are connected in series. The transfer functions are given below. Determine the overall transfer function (TF).

$$\text{TF}_1 = \frac{1 + \tau_1 s}{1 + \tau_2 s}$$

$$\text{TF}_2 = \frac{1}{1 + A_1 s + A_2 s^2}$$

$$\text{TF}_3 = \frac{1}{1 + \tau_3 s}$$

Solution

The overall transfer function is the product of the individual transfer functions:

$$\text{TF} = (\text{TF}_1)(\text{TF}_2)(\text{TF}_3)$$

$$= \left(\frac{1 + \tau_1 s}{1 + \tau_2 s}\right)\left(\frac{1}{1 + A_1 s + A_2 s^2}\right)\left(\frac{1}{1 + \tau_3 s}\right)$$

$$= \frac{1 + \tau_1 s}{(1 + \tau_2 s)(1 + A_1 s + A_2 s^2)(1 + \tau_3 s)}$$

Example 16.3

The following gains and phase angles were measured at a frequency of 1 cycle per second. Determine the overall gain (g) and phase angle (β) if the components are connected in series.

	Gain	Phase Angle (deg)
Final control element	1.5	-5
Process	2.0	-170
Measuring means	0.9	-15

Solution

The gain (g) is the product of the individual gains and the phase angle (β) is the sum of the individual phase angles.

$$g = (1.5)(2.0)(0.9) = 2.7$$
$$\beta = (-5°) + (-170°) + (-15°) = -190°$$

16.3 OPEN-LOOP BODE DIAGRAMS

The *open-loop Bode diagram* expresses the *frequency response* of the control system when the measuring transmitter output is disconnected from the error detector. The Bode diagram data can be obtained from the transfer function of the "open-loop control system" shown in Figure 16.2. The first block represents the forward path components (controller, manipulating element, and process), and the symbol $G(s)$ is used to represent the transfer function of these components. The second block represents the feedback path component (measuring transmitter). Since the measurement (C_m) is not used, the output of the error detector is equal to the setpoint signal (i.e., $E = SP$).

The open-loop transfer function is obtained by substituting $SP[G(s)]$ for C in the equation for C_m (Figure 16.2) and then dividing the resulting equation by SP.

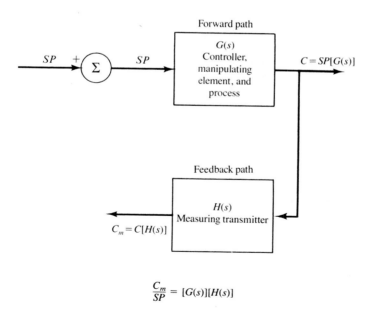

$$\frac{C_m}{SP} = [G(s)][H(s)]$$

Figure 16.2   The open-loop transfer function is obtained by disconnecting the feedback line from the error detector and solving for C_m/SP.

OPEN-LOOP RESPONSE

$$\text{Open-loop transfer function} = \frac{C_m}{SP} = G(s)H(s) \qquad (16.5)$$

The open-loop Bode diagram is constructed with the aid of a computer by multiplying the gains and adding the phase angles of the controller, manipulating element, process, and measuring transmitter.

The open-loop Bode diagram is constructed graphically by adding the decibel gain and phase graphs of the controller, manipulating element, process, and measuring transmitter.

16.4 CLOSED-LOOP BODE DIAGRAMS

The *closed-loop Bode diagram* expresses the *frequency response* of the control system when the measuring transmitter output is connected to the error detector. The Bode diagram data can be obtained from the transfer function of the closed-loop control system shown in Figure 16.3.

In Figure 16.3, the loop is closed, making the output of the error detector equal to the setpoint minus the measuring transmitter output (i.e., $E = SP - C_m$). The output of the forward path components (C) is equal to the input (E) times the transfer function $G(s)$.

$$C = E[G(s)] = [SP - C_m]G(s)$$

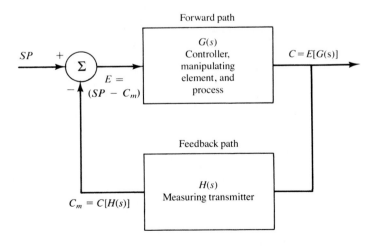

$$\frac{C_m}{SP} = \frac{[G(s)][H(s)]}{[G(s)][H(s)] + 1}$$

Figure 16.3 The closed-loop transfer function is obtained by connecting the feedback line to the error detector and solving for C_m/SP.

The output of the feedback path component (C_m) is equal to the input (C) times the transfer function $H(s)$.

$$C_m = C[H(s)] = [SP - C_m]G(s)H(s)$$

The preceding equation may be solved for the closed-loop transfer function C_m/SP.

$$C_m = SP[G(s)H(s)] - C_m[G(s)H(s)]$$

$$C_m + C_m[G(s)H(s)] = SP[G(s)H(s)]$$

$$C_m[1 + G(s)H(s)] = SP[G(s)H(s)]$$

$$\frac{C_m}{SP} = \frac{G(s)H(s)}{1 + G(s)H(s)} = \text{closed-loop transfer function}$$

where $G(s)H(s)$ is the open-loop transfer function. In other words, the closed-loop response is equal to the open-loop response divided by 1 plus the open-loop response.

Figure 16.4 shows the open-loop and closed-loop Bode diagrams of a control system that consists of a proportional controller and a first-order lag plus dead-time

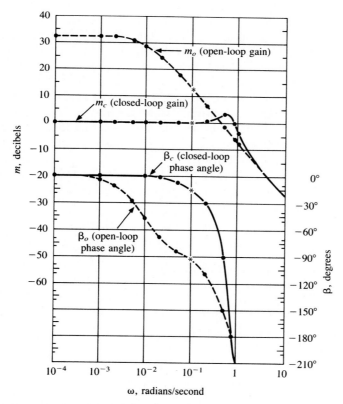

Figure 16.4   Open-loop and closed-loop Bode diagrams of a control system consisting of a proportional controller and a first-order lag plus dead-time process.

process. The controller gain is 40. The process time constant is 100 s, and the dead-time lag is 2 s. The open-loop transfer function is

$$G(s)H(s) = 40\left(\frac{e^{-2s}}{1 + 100s}\right)$$

CLOSED-LOOP RESPONSE

$$\text{Closed-loop transfer function} = \frac{C_m}{SP} = \frac{G(s)H(s)}{1 + G(s)H(s)} \qquad \textbf{(16.6)}$$

The closed-loop Bode diagram is constructed with the aid of a computer by dividing the open-loop response by 1 plus the open-loop response.

The closed-loop Bode diagram is constructed graphically by converting the open-loop gain and phase values to the corresponding closed-loop values. A Nichols chart is used to convert each set of open-loop gain and phase values into the corresponding closed-loop gain and phase values.

The closed-loop Bode diagram may be determined by using program "DESIGN" or by using a Nichols chart. Figure 16.5 illustrates the use of a Nichols chart to graphically obtain the closed-loop response from the open-loop response. The open-loop data points from Figure 16.4 are plotted on the Nichols chart using the rectangular, open-loop coordinates. (The open-loop phase angle is on the horizontal axis, and the open-loop decibel gain is on the vertical axis.) The closed-loop gain and phase angle are read from the same plotted points using the curved, closed-loop coordinates. (The closed-loop gain coordinates are the lines with the short dashes. The closed-loop phase-angle coordinates are the lines with the long and short dashes.)

Notice the points that are marked with an asterisk on Figures 16.4 and 16.5. We will use these marked points to trace the plotting of one open-loop point on the Nichols chart and the reading of the corresponding closed-loop gain and phase angle. The trace begins by reading the open-loop gain and phase angle.

From Figure 16.4 at $\omega = 0.1$ rad/s:

$$\text{Open-loop gain} = 12 \text{ dB}$$

$$\text{Open-loop phase angle} = -95°$$

This point is plotted in the Nichols chart (Figure 16.5) and is marked with an asterisk. The frequency, $\omega = 0.1$, is written beside the point. Notice that the marked point is on the 0-dB closed-loop gain line, so we read the closed-loop gain as 0 dB. Notice also that the marked point is midway between the $-10°$ and the $-20°$ closed-loop phase-angle lines, so we read the closed-loop phase angle as $-15°$.

From Figure 16.5 at $\omega = 0.1$ rad/s:

$$\text{Closed-loop gain} = 0 \text{ dB}$$

$$\text{Closed-loop phase angle} = -15°$$

These two closed-loop points are plotted in Figure 16.4 and are also marked with asterisks.

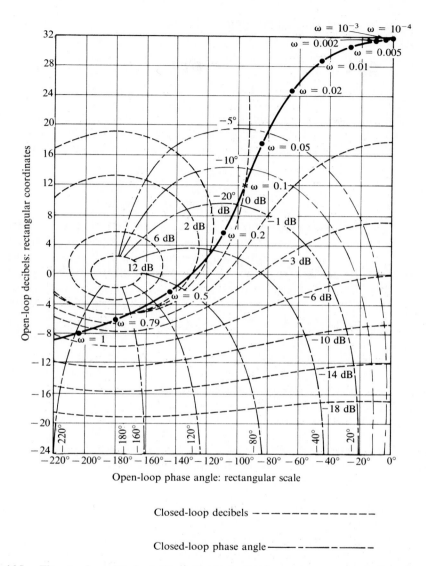

Figure 16.5 The open-loop data from Figure 16.4 is plotted in the Nichols chart using the rectangular open-loop coordinates. The closed-loop data are then read from the same plotted points using the curved closed-loop coordinates. [Based on information from H. M. James, N. B. Nichols, and R. S. Phillips, *Theory of Servomechanisms* (New York: McGraw-Hill Book Company, 1947), p. 179.]

16.5 ERROR RATIO AND DEVIATION RATIO

Every closed-loop control system has a maximum frequency limit. A control system cannot respond accurately to setpoint changes above its frequency limit. Disturbances with frequencies above the frequency limit are especially troublesome. The controller cannot reduce the error produced by the disturbance and may actually increase the error. The maximum frequency limit can be determined by comparing the error when feedback is used with the error when feedback is not used. A convenient means of comparison is the *error ratio*, the ratio of the closed-loop error over the open-loop error.

$$\text{Error ratio} = \frac{\text{closed-loop error magnitude}}{\text{open-loop error magnitude}}$$

A typical error ratio graph is illustrated in Figure 16.6. The graph is divided into three frequency *zones:* 1, 2, and 3. In zone 1, the error ratio is less than 1 (0 dB). This means that the closed-loop error is less than the open-loop error. The controller will reduce errors occurring in zone 1. In zone 2, the error ratio is greater than 1 (0 dB). The closed-loop error is actually greater than the open-loop error, and the controller is doing more harm than good. In zone 3, the error ratio is equal to 1 (0 dB). The

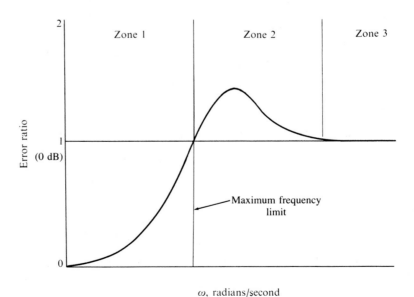

Figure 16.6   A graph of the error ratio may be divided into three distinct zones based on a comparison of the error with the loop closed and the error with the loop open. In zone 1, closed-loop control reduces the error. In zone 2, closed-loop control increases the error. In zone 3, closed-loop control neither increases nor decreases the error.

closed-loop error is equal to the open-loop error, and the presence of the controller neither increases or decreases the error. The *maximum frequency limit* is the frequency that divides zone 1 and zone 2.

The equation for the error ratio is derived from the open-loop and the closed-loop transfer functions. The closed-loop error is the error signal (E) in the closed-loop control system illustrated in Figure 16.3. The open-loop error is the setpoint signal (SP) minus the measured value signal (C_m) in the open-loop control system shown in Figure 16.2.

$$\text{Error ratio} = \frac{\text{closed-loop error magnitude}}{\text{open-loop error magnitude}}$$

$$\text{Closed-loop error} = (SP - C_m) \qquad \text{closed loop}$$

$$= SP - SP\left[\frac{C_m}{SP}\right] \qquad \text{closed loop}$$

$$= SP - SP\left[\frac{G(s)H(s)}{1 + G(s)H(s)}\right]$$

$$= SP\left[1 - \frac{G(s)H(s)}{1 + G(s)H(s)}\right]$$

$$= SP\left[\frac{1 + G(s)H(s) - G(s)H(s)}{1 + G(s)H(s)}\right]$$

$$= SP\left[\frac{1}{1 + G(s)H(s)}\right] \qquad (16.7)$$

$$\text{Open-loop error} = (SP - C_m) \qquad \text{open loop}$$

$$= SP - SP\left[\frac{C_m}{SP}\right] \qquad \text{open loop}$$

$$= SP - SP[G(s)H(s)]$$

$$= SP[1 - G(s)H(s)] \qquad (16.8)$$

The error ratio is equal to the magnitude of the right-hand side of Equation (16.7) divided by the magnitude of the right-hand side of Equation (16.8).

$$\text{Error ratio} = \left|\frac{1}{[1 + G(s)H(s)][1 - G(s)H(s)]}\right| \qquad (16.9)$$

The right-hand expression is a complex number. The parallel lines indicate that the error ratio is equal to the magnitude of this complex number, and the angle is ignored.

Every closed-loop control system has a zone 2: a range of frequencies for which the controller increases the error. The results in Table 16.1 clearly illustrate one frequency that is always in zone 2—the frequency for which the open-loop phase angle is $-180°$. This frequency is designated $\omega_{-180°}$. The open-loop gain at $\omega_{-180°}$ is designated by $g_{-180°}$.

$$g_{-180°}\underline{/-180°} = -g_{-180°} + j0 = -g_{-180°}$$

Table 16.1   Error Ratio Versus Gain When $\beta = -180°$

$g_{-180°}$	$1 + g_{-180°}$	$1 - g_{-180°}$	Error Ratio Ratio	Error Ratio dB
1.0	2.0	0.0	infinite	infinite
0.9	1.9	0.1	5.26	14.4
0.8	1.8	0.2	2.78	8.9
0.7	1.7	0.3	1.96	5.85
0.6	1.6	0.4	1.56	3.85
0.5	1.5	0.5	1.33	2.50
0.4	1.4	0.6	1.19	1.50
0.3	1.3	0.7	1.10	0.80
0.2	1.2	0.8	1.04	0.35
0.1	1.1	0.9	1.01	0.10
0.0	1.0	1.0	1.00	0.00

At $\omega_{-180°}$, the error ratio is given by

$$\text{Error ratio} = \frac{1}{(1 - g_{-180°})(1 + g_{-180°})}$$

The *deviation ratio** is equal to the ratio of the closed-loop error magnitude over the setpoint magnitude. It is an indication of how accurately a control system can follow a change in setpoint and is used to evaluate follow-up control systems. The deviation ratio is obtained directly from Equation (16.7).

$$\text{Closed-loop error} = SP\left[\frac{1}{1 + G(s)H(s)}\right] \qquad \textbf{(16.7)}$$

$$\text{Deviation ratio} = \left|\frac{\text{closed-loop error}}{\text{setpoint}}\right|$$

$$= \left|\frac{1}{1 + G(s)H(s)}\right| \qquad \textbf{(16.10)}$$

Figure 16.7 compares the graphs of the error ratio and the deviation ratio for the control system with the Bode diagrams shown in Figure 16.4. The control system consists of a proportional controller with a gain of 40 and a process with a time constant of 100 s and a dead time lag of 2 s.

---

* G. K. Tucker and D. M. Wills, *A Simplified Technique of Control System Engineering* (Philadelphia: Minneapolis-Honeywell Regulator Company, 1962), pp. 91–93.

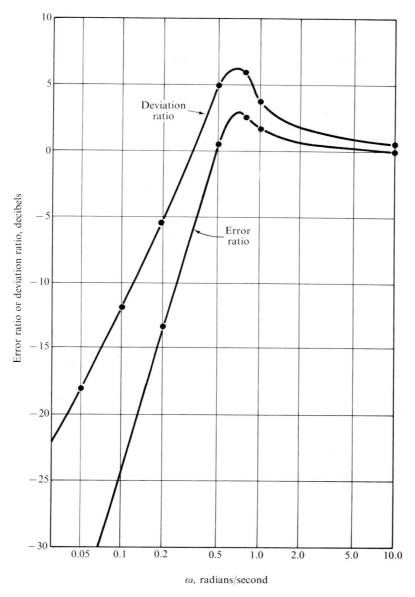

Figure 16.7 Graph of the error ratio and the deviation ratio for the control system described by the Bode diagrams in Figure 16.4.

ERROR RATIO

The error ratio gives the relative size of the error produced by a disturbance when the loop is closed compared with the error when the loop is open.

$$\text{Error ratio} = \frac{1}{(1 + g_o/\beta_o)(1 - g_o/\beta_o)} \tag{16.11}$$

where g_o = open-loop gain

   β_o = open-loop phase angle

DEVIATION RATIO

The deviation ratio gives the relative size of the closed-loop error produced by a change in setpoint compared with the size of the change in the setpoint.

$$\text{Deviation ratio} = \frac{1}{1 + g_o/\beta_o} \tag{16.12}$$

Three Zones of Control

   Zone 1: error/deviation ratio < 0 dB, good control

   Zone 2: error/deviation ratio > 0 dB, poor control

   Zone 3: error/deviation ratio $= 0$ dB, no control

16.6 COMPUTER-AIDED BODE PLOTS

The determinations of the open-loop Bode diagram, the closed-loop Bode diagram, and the error ratio of a control system are tedious, time-consuming chores that are well suited to computer analysis. Program "DESIGN" is a BASIC program for computer-aided analysis and design of closed-loop control systems. In Chapter 17, program "DESIGN" is used to design PID controllers and control system compensation networks. In this chapter, program "DESIGN" is used to produce Bode diagrams, error-ratio graphs, and Nyquist diagrams on the computer screen and a data table on the printer.

   Program "DESIGN" is included on a disk that comes with the instructor's manual. The program is not copy protected, and distribution of the program is permitted and encouraged. Appendix F includes a listing of program "DESIGN".

   Program "DESIGN" begins with the input of the transfer functions of the components that make up the control system. The system can have up to 10 polynomial transfer functions and up to 10 dead-time lags. All components are assumed to be connected in series to form a closed-loop system. Each transfer function is a polynomial of the following form:

$$\text{TF} = \frac{A(0) + A(1)s + A(2)s^2 + A(3)s^3}{B(0) + B(1)s + B(2)s^2 + B(3)s^3}$$

Upon completion of the input of the transfer functions, program "DESIGN" computes and stores analysis data for a frequency range from 1×10^{-6} to 5.6×10^5 radians/second. The analysis data includes the data necessary to plot the open-loop Bode graph, the closed-loop Bode graph, the error-ratio graph, and the Nyquist diagram of the control system.

When the analysis data has been stored, the program switches to the high-resolution graphics mode and puts the open-loop Bode diagram on the screen. The top line of the screen presents controller design parameters that are used in Chapter 17 and can be ignored in Chapter 16. The second line is the command line. The commands (Dmode), (Imode), and (Pmode) are also used only in Chapter 17. In Chapter 16, only the (Analysis), (Zoom), (Unzoom), and (Quit) commands are used. To execute a command, simply press the first letter of the command. For example, press the A-key to execute the (Analysis) command.

As the name implies, the (Zoom) command provides a close-up look at the graph on the screen. The (Unzoom) command returns the screen to its original scale. The (Analysis) command brings up a new command line with the commands (Closed-loop), (Nyquist), (Error-ratio) and (Open-loop). The (Closed-loop) command puts the closed-loop Bode diagram on the screen. The (Nyquist) command puts the Nyquist diagram on the screen, and the (Error-ratio) command puts the error-ratio graph on the screen.

Use the (Quit) command to exit the program. Before the program quits, it queries the user for the option to print a data table. A Yes response will result in a printed copy of the analysis data. For convenience, the analysis data table is terminated when the phase angle falls below -999 degrees.

Example 16.4 illustrates the use of program "DESIGN" to analyze a closed-loop control system.

Example 16.4

A control system consists of a proportional controller and a first-order lag plus dead-time process. The following values were obtained from a step response test of the process.

$$\text{Dead-time lag} = t_d = 2 \text{ s}$$
$$\text{Time constant} = \tau = 100 \text{ s}$$

Therefore,

$$H(s) = e^{-2s}\left(\frac{1}{1 + 100s}\right)$$

The controller gain is set at 40, so $G(s) = [40]$. Combining $G(s)$ and $H(s)$ gives the following open-loop transfer function:

$$\text{TF(open-loop)} = G(s)H(s) = \left(\frac{40}{1 + 100s}\right)e^{-2s}$$

Use program "DESIGN" to produce the following:

1. An open-loop Bode graph
2. A closed-loop Bode graph
3. An error-ratio graph
4. A printed Bode data table for open-loop, closed-loop, and error-ratio graphs

Solution

The following inputs were used in a run of program "DESIGN".

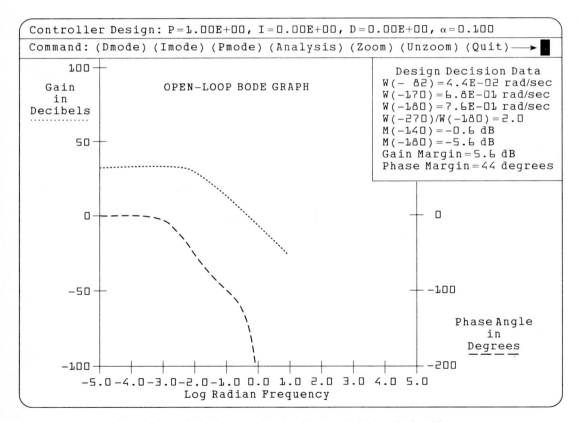

Figure 16.8 Open-loop Bode graph for the control system in Example 16.4 as displayed by program "DESIGN". The dotted line on the top is the open-loop gain; the dashed line on the bottom is the open-loop phase angle. The Controller Design line on the top and the Design Data table are used in Chapter 17, and can be ignored in Chapter 16.

```
Transfer function coefficients of component number 1:

        A(0) = 40,   A(1) =   0,   A(2) = 0,   A(3) = 0
        B(0) =  1,   B(1) = 100,   B(3) = 0,   B(3) = 0

Dead time delay number 1 = 2
```

The results of the run are as follows:

1. The open-loop Bode graph: Figure 16.8
2. The closed-loop Bode graph: Figure 16.9

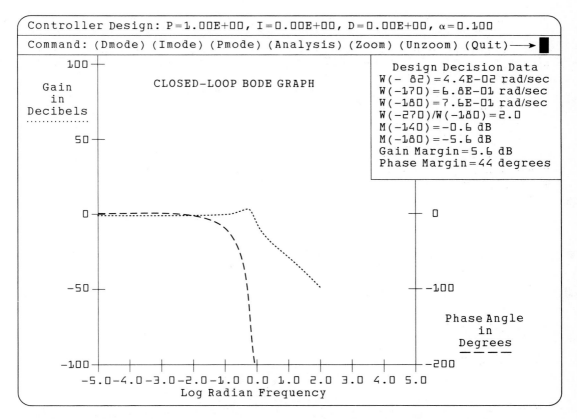

Figure 16.9   Closed-loop Bode graph for the control system in Example 16.4 as displayed by program "DESIGN". The dotted line is the closed-loop gain; the dashed line is the closed-loop phase angle.

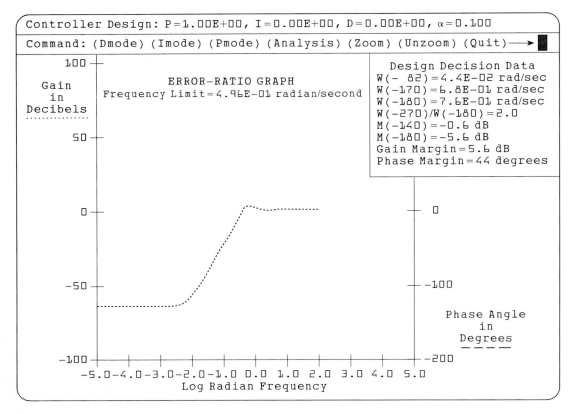

```
Controller Design: P=1.00E+00, I=0.00E+00, D=0.00E+00, α=0.100

Command: (Dmode) (Imode) (Pmode) (Analysis) (Zoom) (Unzoom) (Quit)——→ ■
```

ERROR-RATIO GRAPH
Frequency Limit=4.96E-01 radian/second

Gain
in
Decibels

Design Decision Data
W(- 82)=4.4E-02 rad/sec
W(-170)=6.8E-01 rad/sec
W(-180)=7.6E-01 rad/sec
W(-270)/W(-180)=2.0
M(-140)=-0.6 dB
M(-180)=-5.6 dB
Gain Margin=5.6 dB
Phase Margin=44 degrees

Phase Angle
in
Degrees

Log Radian Frequency

Figure 16.10   Error-ratio graph for the control system in Example 16.4 as displayed by program
"DESIGN".

3. The error-ratio graph: Figure 16.10
4. The Bode data: Table 16.2

16.7 STABILITY

The possibility of sustained oscillations always exists in a closed-loop control system. When a system goes into a continuous oscillation, it is said to be unstable. *Stability* refers to the ability of a control system to dampen out any oscillations that result from an upset. The analysis of a control system seeks the answer to the question: Is the control system stable? The design of a control system involves the determination of controller parameters that will minimize the error produced by a disturbance without making the system unstable.

Every closed-loop control system has a tendency to produce an oscillation in the controlled variable. This tendency for self-oscillation is caused by the presence

Table 16.2 Bode Data for Example 16.4

Frequency (rad/sec)	Open Loop Response Gain (Decibel)	Open Loop Response Angle (Degree)	Closed Loop Response Gain (Decibel)	Closed Loop Response Angle (Degree)	Error Ratio (Decibel)
1.0E−06	32.0	−0.0	−0.2	−0.0	−64.1
1.8E−06	32.0	−0.0	−0.2	−0.0	−64.1
3.2E−06	32.0	−0.0	−0.2	−0.0	−64.1
5.6E−06	32.0	−0.0	−0.2	−0.0	−64.1
1.0E−05	32.0	−0.1	−0.2	−0.0	−64.1
1.8E−05	32.0	−0.1	−0.2	−0.0	−64.1
3.2E−05	32.0	−0.2	−0.2	−0.0	−64.1
5.6E−05	32.0	−0.3	−0.2	−0.0	−64.1
1.0E−04	32.0	−0.6	−0.2	−0.0	−64.1
1.8E−04	32.0	−1.0	−0.2	−0.0	−64.1
3.2E−04	32.0	−1.8	−0.2	−0.0	−64.1
5.6E−04	32.0	−3.3	−0.2	−0.1	−64.0
1.0E−03	32.0	−5.8	−0.2	−0.1	−64.0
1.8E−03	31.9	−10.3	−0.2	−0.3	−63.8
3.2E−03	31.6	−17.9	−0.2	−0.5	−63.2
5.6E−03	30.8	−30.0	−0.2	−0.8	−61.7
1.0E−02	29.0	−46.1	−0.2	−1.4	−58.1
1.8E−02	25.8	−62.7	−0.2	−2.5	−51.7
3.2E−02	21.6	−76.1	−0.2	−4.5	−43.3
5.6E−02	16.9	−86.4	−0.2	−8.0	−34.0
1.0E−01	12.0	−95.7	−0.1	−14.4	−24.5
1.8E−01	7.0	−107.2	0.3	−26.1	−15.4
3.2E−01	2.0	−124.4	1.4	−49.7	−6.7
5.6E−01	−3.0	−153.4	3.4	−112.3	1.9
1.0E+00	−8.0	−204.0	−4.3	−218.4	0.9
1.8E+00	−13.0	−293.5	−13.9	−304.2	−0.3
3.2E+00	−18.0	−452.2	−18.0	−445.0	−0.1
5.6E+00	−23.0	−734.3	−23.5	−733.4	0.0

of the feedback signal and is directly related to $\omega_{-180°}$, the frequency at which the open-loop phase angle is equal to $-180°$. If the open-loop gain at $\omega_{-180°}$ is less than 1, the oscillations will diminish in size with each successive cycle. If the open-loop gain at $\omega_{-180°}$ is equal to 1, the oscillation will continue at a constant amplitude. If the open-loop gain at $\omega_{-180°}$ is greater than 1, the oscillation will increase in size with each successive cycle. If the oscillations diminish, the control system is said to be stable. If not, the system is said to be unstable. Therefore, *a stable control system is one in which the open-loop gain is less than one when the open-loop phase angle is* $-180°$.

Figure 16.11 illustrates a control system with a sustained oscillation. Graphs of the setpoint (SP), error (E), and controlled variable (C) are located directly above the corresponding signal lines. Figure 16.11 traces a single half-wave, sinusoidal pulse

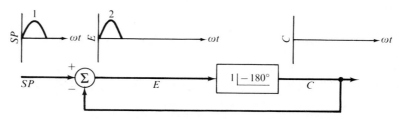

a) A single half-wave pulse (1) is introduced at the setpoint (*SP*). Since
$E = SP - C$ and $C = 0$, the error (2) is identical to the setpoint (1).

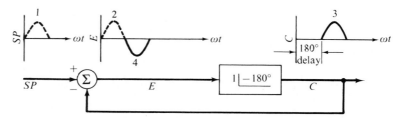

b) The system delays pulse 2 by 180°, forming pulse 3 at output (*C*).
The error detector inverts pulse 3, forming error pulse 4. Notice how
pulse 4 forms the next half of a sine wave.

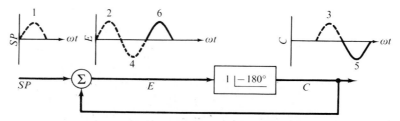

c) The system delays pulse 4 by 180°, forming pulse 5 at the output
(*C*). The error detector inverts pulse 5, forming error pulse 6. Notice
how pulse 6 continues the formation of a sine wave.

Figure 16.11   Sustained oscillations in a closed-loop system.

as it travels around the closed loop. The system has an open-loop gain of 1 and an
open-loop phase angle of −180°. In Figure 16.11a, the half-wave pulse is introduced
at the setpoint. This is the only signal that appears at the setpoint, and the setpoint
is zero for the remainder of the trace. The half-wave input pulse appears immediately
on the error signal graph, but does not appear on the controlled variable graph until
the error signal line has seen the entire 180° half-wave input pulse.

In Figure 16.11b, the half-wave pulse has moved through the process block and
now appears as pulse 3 on the controlled variable graph. There is no time delay
through the error detector, and pulse 4 appears on the error graph as the inversion
of pulse 3. Notice how pulse 4 completes one cycle of the sine wave on the error
graph. The dashed lines for pulses 1 and 2 indicate that those pulses occurred at an
earlier time than the current pulses (3 and 4).

In Figure 16.11c, the half-wave pulse has again moved through the process block to form pulse 5 on the controlled variable graph and inverted pulse 6 on the error graph. Notice how pulse 6 continues the formation of a closed-loop control system.

A control system will oscillate with a constant amplitude if the open-loop gain is 1 at the frequency for which the open-loop phase angle is −180°.

Figure 16.12 illustrates a control system with a diminishing oscillation. Figure 16.12 traces a single half-wave pulse as it travels around the closed loop, just as

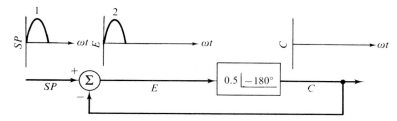

a) A single half-wave pulse (1) is introduced at the setpoint (SP). The error (2) is identical to the setpoint (1).

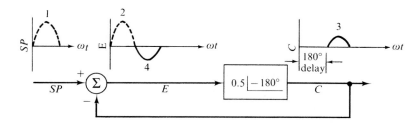

b) The system delays pulse 2 by 180° and halves the amplitude, forming pulse 3 at output (C). The error detector inverts pulse 3, forming error pulse 4. Notice how pulse 4 forms the next half of a diminishing sine wave.

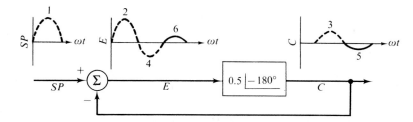

c) The system delays and halves pulse 4, forming pulse 5 at the output (C). The error detector inverts pulse 5, forming error pulse 6. Notice how pulse 6 continues the formation of a diminishing sine wave.

Figure 16.12   Diminishing oscillations in a closed-loop control system.

Figure 16.11 did. This time, however, the open-loop gain is 0.5 instead of 1.0. The open-loop phase angle remains at $-180°$. Notice how the gain of 0.5 reduces the amplitude of the sine wave as it is formed by the cycling pulse.

> A control system will oscillate with a diminishing amplitude if the open-loop gain is less than 1 at the frequency for which the open-loop phase angle is $-180°$.

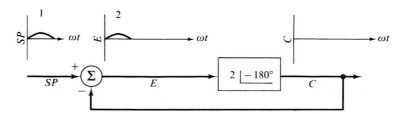

a) A single half-wave pulse (1) is introduced at the setpoint (*SP*). The error (2) is identical to the setpoint (1).

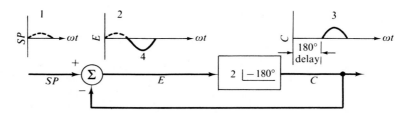

b) The system delays pulse 2 by 180° and doubles the amplitude, forming pulse 3 at the output (*C*). The error detector inverts pulse 3, forming error pulse 4. Notice how pulse 4 forms the next half of an increasing sine wave.

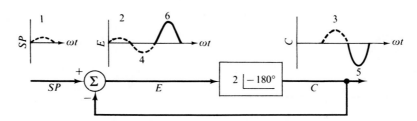

c) The system delays and doubles pulse 4, forming pulse 5 at the output (*C*). The error detector inverts pulse 5, forming error pulse 6. Notice how pulse 6 continues the formation of an increasing sine wave.

Figure 16.13 Increasing oscillations in a closed-loop control system.

Figure 16.13 illustrates a control system with an increasing oscillation. Figure 16.13 traces a single half-wave pulse as it travels around the closed loop. Here, the open-loop gain is 2 while the open-loop phase angle remains at $-180°$. Notice how the gain of 2 increases the amplitude of the sine wave as it is formed by the cycling pulse.

> A control system will oscillate with an increasing amplitude if the open-loop gain is greater than 1 at the frequency for which the open-loop phase angle is $-180°$.

A complete discussion of stability requires a level of mathematics that is beyond the scope of this book. However, we can expand on the intuitive concept of stability developed in the preceding discussion. Another view of stability can be obtained from an examination of the open- and closed-loop transfer functions given by Equations (16.5) and (16.6).

$$\text{Open-loop TF} = \frac{C_m}{SP} = G(s)H(s) \tag{16.5}$$

$$\text{Closed-loop TF} = \frac{C_m}{SP} = \frac{G(s)H(s)}{1 + G(s)H(s)} \tag{16.6}$$

A control system is clearly unstable if the denominator of the closed-loop transfer function is equal to zero. This condition is satisfied when the open-loop transfer function is equal to -1, which occurs when the open-loop gain is 1 and the open-loop phase angle is $-180°$. The same conclusion was reached in the discussion of Figure 16.11.

> A control system is unstable if the open-loop gain is equal to or greater than 1 at a frequency where the open-loop phase angle is $-180°$.

Our final view of stability involves the use of the inverse Laplace transformation to obtain the time-domain impulse response of a control system. We begin by assuming a polynomial, open-loop transfer function of the following general form.

$$G(s)H(s) = \frac{A_0 + A_1 s + A_2 s^2 + A_3 s^3}{B_0 + B_1 s + B_2 s^2 + B_3 s^3}$$

The denominator of the closed-loop transfer function is

$$1 + G(s)H(s) = 1 + \frac{A_0 + A_1 s + A_2 s^2 + A_3 s^3}{B_0 + B_1 s + B_2 s^2 + B_3 s^3}$$

$$= \frac{(A_0 + B_0) + (A_1 + B_1)s + (A_2 + B_2)s^2 + (A_3 + B_3)s^3}{B_0 + B_1 s + B_2 s^2 + B_3 s^3}$$

The closed-loop transfer function is

$$\frac{C_m}{SP} = \frac{G(s)H(s)}{1 + G(s)H(s)} = \frac{A_0 + A_1 s + A_2 s^2 + A_3 s^3}{(A_0 + B_0) + (A_1 + B_1)s + (A_2 + B_2)s^2 + (A_3 + B_3)s^3}$$

Assume that p_1, p_2, and p_3 are the roots of the denominator, and z_1, z_2, and z_3 are the roots of the numerator. The closed-loop transfer function can then be written as follows:

$$\frac{C_m}{SP} = \frac{(s - z_1)(s - z_2)(s - z_3)}{(s - p_1)(s - p_2)(s - p_3)}$$

Next, we apply a unit impulse at the setpoint. The unit impulse is chosen because the Laplace transformation of the unit impulse is 1, giving us the simplest possible function to work with.

$$C_m = \frac{C_m}{SP}[SP] = \frac{(s - z_1)(s - z_2)(s - z_3)}{(s - p_1)(s - p_2)(s - p_3)}[1]$$

$$C_m(s) = \frac{(s - z_1)(s - z_2)(s - z_3)}{(s - p_1)(s - p_2)(s - p_3)}$$

Finally, we do a partial fraction expansion to get the inverse Laplace transformation.

$$C_m(s) = \frac{K_1}{s - p_1} + \frac{K_2}{s - p_2} + \frac{K_3}{s - p_3}$$

By inverse Laplace transformation,

$$C_m(t) = K_1 e^{p_1 t} + K_2 e^{p_2 t} + K_3 e^{p_3 t}$$

The equation for $C_m(t)$ shows that p_1, p_2, and p_3 must all be negative for $C_m(t)$ to remain bounded as t increases. For example, if $p_1 = 2$, the term $K_1 e^{p_1 t} = K_1 e^{2t}$ will increase exponentially as t increases.

A control system is stable if all the roots of $1 + G(s)H(s) = 0$ have negative real parts.

16.8 GAIN AND PHASE MARGIN

The open-loop Bode diagram of a control system provides the information necessary to determine the stability of the system. It also gives the designer an idea of the stability safety margin; that is, how much the gain or phase-angle can change before the system will become unstable. The designer also uses the open-loop Bode diagram as a design tool—a topic covered in Chapter 17.

The stability of a closed-loop control system is defined by either of the following stability conditions.

Stability Conditions

1. A control system is stable only if the open-loop gain is less than 1 (or 0 dB) at the frequency for which the phase angle is $-180°$.
2. A control system is stable only if the phase angle is greater than $-180°$ (i.e., the phase lag is less than $180°$) at the frequency for which the gain is 1 (or 0 dB).

The two stability conditions suggest two margins of safety for the stability of a closed-loop control system. First, the gain at the $-180°$ frequency must be a safe amount less than 1 (or 0 dB). We call this the *gain margin*. Second, the phase angle at the 0 dB frequency must be a safe amount above $-180°$. We call this the *phase margin*. Standard practice is a gain margin of 6 dB and a phase margin of $40°$.

The designer considers a closed-loop control system to be *stable* only if it satisfies both the gain margin and the phase margin. In other words, a control system is considered to be stable only if the open-loop gain is less than 0.5 (or -6 dB) at the $-180°$ frequency, and the phase angle is greater than $-140°$ at the 0 dB frequency.

The designer considers a control system to be *marginally stable* if it satisfies the stability conditions but does not satisfy both the gain margin and the phase margin criteria.

---

1. A closed-loop control system is *stable* if it has a gain margin of at least 6 dB and a phase margin of at least $40°$.
2. A closed-loop control system is *marginally stable* if it satisfies the stability conditions, but its gain margin is less than 6 dB or its phase margin is less than $40°$, or both.
3. A closed-loop control system is *unstable* if its open-loop gain is 1 or greater at the $-180°$ frequency.

The designer can read the gain and phase margins directly from the open-loop Bode diagram of the control system. In doing this, the designer locates two frequencies on the Bode diagram. One is the frequency at which the open-loop gain is 0 dB. This frequency is called $\omega_{0\ dB}$. The other is the frequency at which the phase angle is $-180°$. This frequency is called $\omega_{-180°}$. The designer uses $\omega_{0\ dB}$ and $\omega_{-180°}$ to read the gain and phase margins as illustrated in Figure 16.14. The gain margin is obtained by reading the open-loop gain at $\omega_{-180°}$. We will call this gain value $m_{-180°}$. The gain margin is equal to $-m_{-180°}$. The phase margin is obtained by reading the open-loop phase angle at $\omega_{0\ dB}$. We will call this phase angle value $\beta_{0\ dB}$. The phase margin is equal to $180 + \beta_{0\ dB}$.

Gain and Phase Margins

Gain margin $= -m_{180°}$

Phase margin $= 180 + \beta_{0\ dB}$

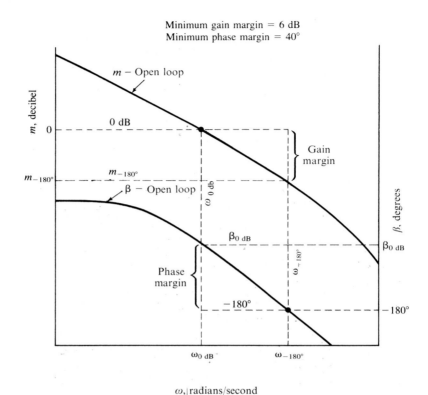

Minimum gain margin = 6 dB
Minimum phase margin = 40°

Figure 16.14 The control system designer uses $m_{-180°}$ (the gain at the $-180°$ frequency) and $\beta_{0\ dB}$ (the phase angle at the 0-dB frequency) to determine the gain margin and the phase margin. The gain margin is equal to $-m_{-180°}$, and the phase margin is equal to $180 + \beta_{0\ dB}$.

Example 16.5

Plot the open-loop Bode diagram for the control system in Example 16.4 and determine the gain and phase margins.

Solution

The easiest way to plot the open-loop Bode diagram is to use the Bode data table produced by program "DESIGN" (see Table 16.2). However, a certain amount of calculator computation and hand plotting of Bode data can enhance the learning process. With that in mind, we proceed to compute four Bode data points.

The first step in computing Bode data is to substitute $j\omega$ for s in the open-loop transfer function. Now for any given frequency (ω_i), the transfer function can be reduced to a single complex number in polar form. The magnitude and angle of this complex number are the open-loop gain and the open-loop phase angle at the given frequency (ω_i). From Example 16.4, we have the following open-loop transfer function:

$$\text{TF(open-loop)} = (40)\left(\frac{1}{1 + 100s}\right)e^{-2s}$$

Replacing s by $j\omega$, we obtain the following:

$$\text{Open-loop gain and phase angle} = (40)\left(\frac{1}{1 + j100\omega}\right)e^{-j2\omega}$$

(*Note:* The dead-time delay term, $e^{-j2\omega}$, defines a complex number with a magnitude of 1 and an angle of 2ω radians. We must multiply this angle by 57.3 to convert it to degrees.)

   a. $\omega = 0.001$ radian/second

$$\text{Gain and phase angle} = (40)\left(\frac{1}{1 + j0.1}\right)e^{-j0.002(57.3)}$$
$$= (40)(0.9950\,\underline{/-5.71°})(1\,\underline{/-0.11°})$$
$$= 39.8\,\underline{/-5.8°}$$

The decibel gain $= 20[\log_{10}(39.8)] = 32.0$ dB

     At $\omega = 0.001$ radian/second:

     open-loop gain $= 32.0$ dB

     open-loop phase angle $= -5.8$ degrees.

   b. $\omega = 0.01$ radian/second

$$\text{Gain and phase angle} = (40)\left(\frac{1}{1 + j1}\right)e^{-j0.02(57.3)}$$
$$= (40)(0.707\,\underline{/-45.0°})(1\,\underline{/-1.1°})$$
$$= 28.28\,\underline{/-46.1°}$$

The decibel gain $= 20[\log_{10}(28.28)] = 29.0$ dB

     At $\omega = 0.01$ radian/second:

     open-loop gain $= 29.0$ dB

     open-loop phase angle $= -46.1$ degrees.

   c. $\omega = 0.1$ radian/second

$$\text{Gain and phase angle} = (40)\left(\frac{1}{1 + j10}\right)e^{-j0.2(57.3)}$$
$$= (40)(0.0995\,\underline{/-84.3°})(1\,\underline{/-11.5°})$$
$$= 3.98\,\underline{/-95.7°}$$

The decibel gain $= 20[\log_{10}(3.98)] = 12.0$ dB

     At $\omega = 0.1$ radian/second:

     open-loop gain $= 12.0$ dB

     open-loop phase angle $= -95.7$ degrees.

d. $\omega = 1$ radian/second

$$\text{Gain and phase angle} = (40)\left(\frac{1}{1 + j100}\right)e^{-j2(57.3)}$$
$$= (40)(0.01\underline{/-89.4°})(1\underline{/-114.6°})$$
$$= 0.40\underline{/-204.0°}$$

The decibel gain $= 20[\log_{10}(0.40)] = -8.0$ dB

At $\omega = 1$ radian/second:

open-loop gain $= -8.0$ dB

open-loop phase angle $= -204.0$ degrees.

Examination of Table 16.2 reveals complete agreement between the calculator results and the program "DESIGN" results. Refer to Figures 15.7 and 15.16 for further insight concerning the frequency response of first-order lags and dead-time delays.

The Bode diagram is shown in Figure 16.15. The following approximate values are obtained from the Bode diagram.

$$m_{-180°} = -6 \text{ dB}$$
$$\beta_{0 \text{ dB}} = -135°$$

The gain margin $= -m_{-180°} = 6$ dB

The phase margin $= 180° + \beta_{0 \text{ dB}} = 180° + (-135°) = 45°$

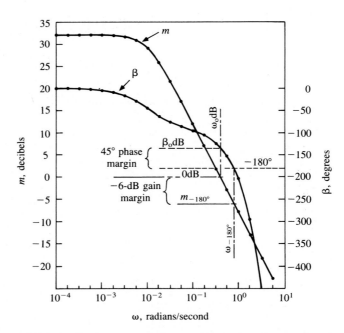

Figure 16.15 The open-loop Bode diagram is used to determine the gain margin and phase margin of a control system (see Example 16.5).

Exact values from program "DESIGN" are a gain margin of 5.6 dB and a phase margin of 44 degrees. Therefore, we conclude that the control system is, in the strictest sense, marginally stable. However, if the gain margin is rounded to the nearest integer, 6, and we accept the rounded value as the gain margin, we could conclude that the control system is stable.

16.9   NYQUIST STABILITY CRITERION

The *Nyquist stability criterion* is a graphic method of determining if the function $1 + G(s)H(s)$ has any positive roots. A Nyquist diagram is a plot, in a complex plane, of the open-loop gain and phase angle as the frequency, ω, is varied from 0 to infinity. In plotting a Nyquist diagram, the gain and phase angle are treated as the magnitude and angle of a complex number in polar form. Nyquist diagrams are presented as a polar plot; however, the gain and phase angle could be converted to rectangular form and plotted on rectangular coordinates with the same result. *In plotting a Nyquist diagram, decibel values must be converted to gain values.* The following equation can also be used to convert from decibel (*m*) to gain (*g*):

$$g = 10^{m/20} \tag{16.13}$$

If the control system under consideration is open-loop stable, the Nyquist criterion reduces to the observation of whether or not the Nyquist plot encloses the point $-1 + j0$. If it does, then $1 + G(s)H(s)$ has positive roots and the system is unstable. If the Nyquist plot does not enclose the point $-1 + j0$, then $1 + G(s)H(s)$ has only negative roots and the system is stable or marginally stable.

Figure 16.16 shows the Nyquist plot of a stable closed-loop control system with a close-up view showing the three points used to determine the stability of the system. The first point is the *stability point*, which has the rectangular coordinates $-1 + j0$ or polar coordinates $1\underline{/-180°}$. The second point is the *gain margin point* with coordinates $-0.5 + j0$ rectangular, or $0.5\underline{/-180°}$ polar. The third point is the *phase margin point* with polar coordinates $1\underline{/-140°}$. The following Nyquist criterion applies to all control systems that are open-loop stable (all systems in this book are open-loop stable).

Nyquist Stability Criterion

A closed-loop control system that is open-loop stable is

1. *Unstable* if the Nyquist plot encircles the stability point $(-1 + j0 = 1\underline{/-180°})$ (see Figure 16.16).

2. *Marginally stable* if the Nyquist plot does not encircle the stability point $(-1 + j0)$, but does encircle the gain margin point $(-0.5 + j0)$, or the phase margin point $(1\underline{/-140°})$, or both the gain margin and phase margin points.

3. *Stable* if the Nyquist plot does not encircle the gain margin point $(-1 + j0)$ and also does not encircle the phase margin point $(1\underline{/-140°})$.

Figure 16.17 shows Nyquist plots of stable, marginally stable, and unstable control systems. The marginally stable systems fall into one of three categories:

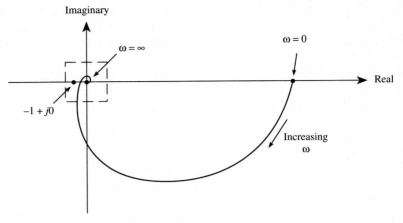

a) Nyquist diagram of a stable control system

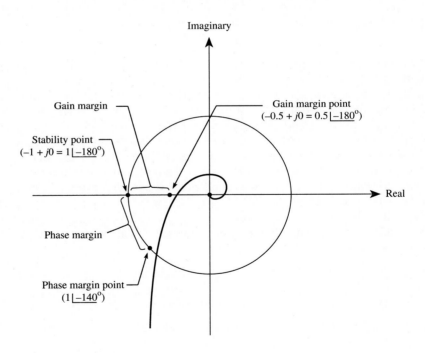

b) Close-up view of the boxed region in part a

Figure 16.16   The Nyquist diagram of a stable closed-loop control system and a close-up view show-ing the gain margin and phase margin. Observe how the gain and phase margins combine to keep the Nyquist plot a "safe" distance from the stability point.

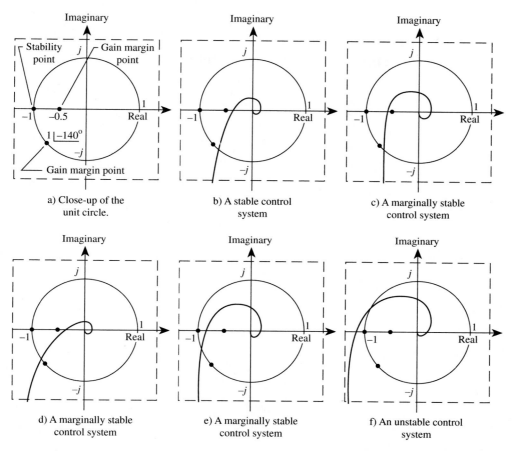

Figure 16.17   Close-up Nyquist plots of stable, marginally stable, and unstable control systems. The control system in (c) does not meet the gain margin criterion; the system in (d) does not meet the phase margin criterion; and the system in (e) does not meet either the gain or phase margin criterion.

systems with an acceptable phase margin but an unacceptable gain margin; systems with an acceptable gain margin but an unacceptable phase margin; and systems with unacceptable gain and phase margins.

Example 16.6

Plot a Nyquist diagram for the control system in Example 16.4 and determine if the system is stable or unstable.

Solution

The open-loop data from Example 16.4 will be used to construct the Nyquist diagram. The following table of values was taken from the data table in Example 16.4. Equation (16.13) was used to convert the decibel values to gain values.

ω (rad/s)	m (dB)	g	β (deg)
0.001	32.0	39.8	−5.8
0.010	29.0	28.2	−46.1
0.018	25.8	19.5	−63.0
0.032	21.5	11.9	−76.3
0.056	16.9	7.00	−86.3
0.10	12.0	3.98	−95.7
0.18	6.9	2.21	−107.4
0.32	1.9	1.24	−124.9
0.56	−2.9	0.72	−153.1
1.0	−8.0	0.40	−204.0
1.8	−13.1	0.22	−295.9
3.2	−18.1	0.12	−456.5
5.6	−22.9	0.07	−731.6

The Nyquist plot is shown in Figure 16.18. The plot does not enclose the point $-1 + j0$, so the control system is stable.

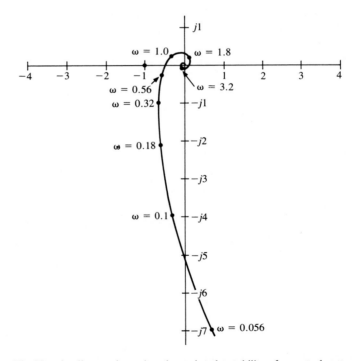

Figure 16.18 The Nyquist diagram is used to determine the stability of a control system (see Exercise 16.6).

16.10 ROOT LOCUS

One condition for stability of a closed-loop control system is that all roots of the following equation must have negative real parts:

$$1 + G(s)H(s) = 0 \qquad\qquad \textbf{(16.14)}$$

Equation (16.14) is referred to as the *closed-loop characteristic equation*. Recall that $G(s)H(s)$ is the open-loop transfer function and Equation (16.14) is the denominator of the closed-loop transfer function as defined by Equation (16.6).

The *root locus* is a graphical method of analysis in which all possible roots of the closed-loop characteristic equation are plotted as the controller gain (K) is varied from 0 to infinity. This graph of all possible roots gives the designer considerable insight into the relationship between gain (K) and the stability of the system. The root-locus method not only answers the question of stability, but also gives the designer information such as the damping ratio or damping constant of the system. The root-locus method was developed by W. R. Evans, who invented a device called a spirule that simplified the construction of root-locus plots. Computer software that produces printed root-locus plots is also available.

We begin with three simple examples to show how the roots of the characteristic equation change with changes in gain (K). We will use s_o to represent any value of s that satisfies the characteristic equation. Our first example has the following open-loop transfer function:

$$G(s)H(s) = \frac{K}{s + 2}$$

The characteristic equation is:

$$1 + \frac{K}{s_o + 2} = 0$$

Solving for s_o:

$$s_o = -(K + 2)$$

Thus the root locus begins at $s_o = -2$ when $K = 0$ and moves to the left on the real axis as K increases. Figure 16.19a shows the root locus of example 1.

Our second example has the following open-loop transfer function, characteristic equation, and solution:

$$G(s)H(s) = \frac{K}{(s + 2)^2}$$

$$1 + \frac{K}{(s_o + 2)^2} = 0$$

$$s_o = -2 \pm j\sqrt{K}$$

Thus the root locus begins at $s_o = -2$ when $K = 0$ and moves both up and down along a vertical line as K increases. Figure 16.19b shows the root locus of example 2.

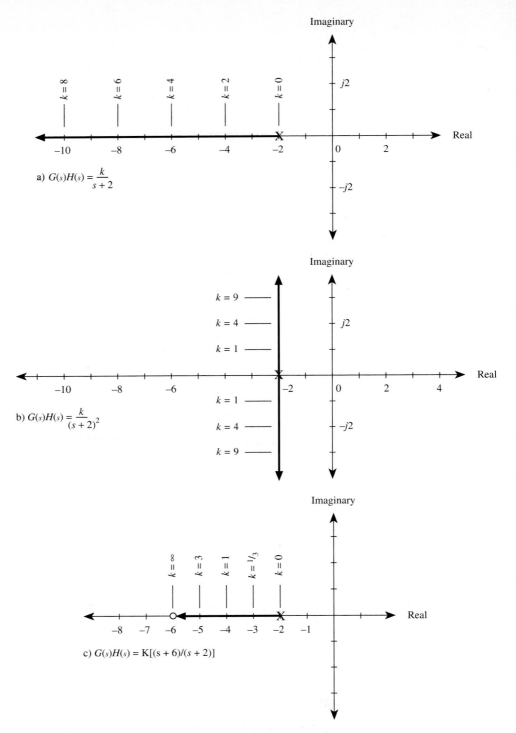

Figure 16.19 Root locus of $1 + G(s)H(s) = 0$ for three simple systems.

a) $G(s)H(s) = \dfrac{k}{s+2}$

b) $G(s)H(s) = \dfrac{k}{(s+2)^2}$

c) $G(s)H(s) = K[(s+6)/(s+2)]$

Our third example has the following open-loop transfer function, characteristic equation, and solution:

$$G(s)H(s) = \frac{K(s + 6)}{(s + 2)}$$

$$1 + \frac{K(s_o + 6)}{(s_o + 2)} = 0$$

$$s_o + 2 + K(s_o + 6) = 0$$

$$s_o(K + 1) + 6K + 2 = 0$$

$$s_o = -\left(\frac{6K + 2}{K + 1}\right)$$

The root locus is the plot of s_o as K varies from 0 to infinity. Our first observation is that s_o will always be a negative real number, and the magnitude of s_o will increase as K increases. We really only need the two endpoints of the locus, the values of s_o when $K = 0$ and $K = \infty$.

When $K = 0$,

$$s_o = -\left(\frac{6K + 2}{K + 1}\right)\Bigg|_{K=0} = -2$$

When $K = \infty$,

$$s_o = -\left(\frac{6K + 2}{K + 1}\right)\Bigg|_{K=\infty}$$

The last expression will be easier to evaluate if we divide the numerator and denominator of the fraction by K.

$$s_o = -\left(\frac{6 + 2/K}{1 + 1/K}\right)\Bigg|_{K=\infty} = -6$$

Figure 16.19c shows the root-locus diagram of this system. The following table of additional values of s_o shows how s_o varies as K increases.

K	$\frac{1}{3}$	1	3	39	399	3999
s_o	-3	-4	-5	-5.9	-5.99	-5.999

Example 16.7

Determine the root locus of a control system with the following open-loop transfer function.

$$G(s)H(s) = \frac{K}{s(s + 10)}$$

Solution

$$1 + G(s_o)H(s_o) = 1 + \frac{K}{s_o(s_o + 10)} = \frac{s_o(s_o + 10) + K}{s_o(s_o + 10)}$$

The characteristic equation is

$$s_o(s_o + 10) + K = s_o^2 + 10s_o + K = 0$$

The roots are given by the quadratic formula:

$$s_1 = -5 + \sqrt{25 - K}$$
$$s_2 = -5 - \sqrt{25 - K}$$

The following table of values shows how s varies as K increases.

K	s_1	s_2
0	0	-10
9	-1	-9
16	-2	-8
21	-3	-7
24	-4	-6
25	-5	-5
50	$-5 + j5$	$-5 - j5$
125	$-5 + j10$	$-5 - j10$

Figure 16.20 shows the root-locus diagram of this system.

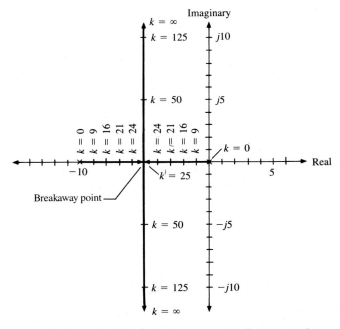

Figure 16.20   Root locus of $1 + G(s)H(s) = 0$ for the system with $G(s)H(s) = K/[s(s + 10)]$.

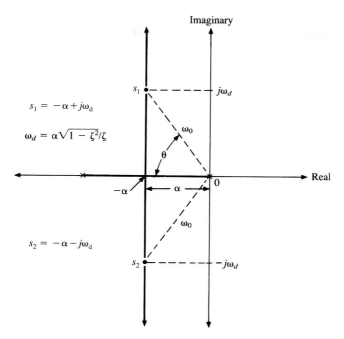

Figure 16.21   Root-locus diagram showing the relationship between the roots of the characteristic equation, the damping ratio ζ, and the damping coefficient α.

Figure 16.21 illustrates the relationship between the roots of the characteristic equation, the damping ratio (ζ), and the damping coefficient (α). The following symbols and terminology will be used in the following discussion.

α = damping coefficient, second^{-1}

ω_o = resonant frequency, radian/second

ω_d = damped resonant frequency, radian/second

ζ = damping ratio

θ = operating point angle

When the roots of the characteristic equation are imaginary, they always come in conjugate pairs given by the following equations.

$$s_1 = -\alpha + j\omega_d$$
$$s_2 = -\alpha - j\omega_d$$

Notice in Figure 16.21 that ω_o is the hypotenuse of a right triangle, and ω_d and α are the two legs of the right triangle. Thus, by the Pythagorean theorem,

$$\omega_o = \sqrt{\alpha^2 + \omega_d^2} \qquad \textbf{(16.15)}$$

The damping ratio is given by

$$\zeta = \frac{\alpha}{\omega_o}$$

Therefore

$$\theta = \cos^{-1}\zeta \qquad\qquad \textbf{(16.16)}$$

and

$$\alpha = \zeta\sqrt{\alpha^2 + \omega_d^2}$$
$$\alpha^2 = \zeta^2(\alpha^2 + \omega_d^2)$$
$$\omega_d^2 = \frac{\alpha^2(1 - \zeta^2)}{\zeta^2}$$
$$\omega_d = \alpha\frac{\sqrt{1 - \zeta^2}}{\zeta} \qquad\qquad \textbf{(16.17)}$$

Equation (16.17) can be used to determine the value of K that will result in a specified damping ratio as illustrated in Example 16.8.

Example 16.8

Find the value of gain (K) that will give a damping ratio of 0.5 in the control system in Example 16.7.

Solution

Equation (16.17) gives the required value of ω_d. From Example (16.7), $\alpha = 0.5$.

$$\omega_d = \frac{\alpha\sqrt{1 - \zeta^2}}{\zeta}$$
$$= \frac{5\sqrt{1 - 0.5^2}}{0.5}$$
$$= 8.66$$

In Example 16.7, when $(25 - K)$ is negative, the roots are imaginary and $\omega_d = \sqrt{K - 25}$.

$$\omega_d = \sqrt{K - 25} = 8.66$$
$$K - 25 = 8.66^2$$
$$K = 8.66^2 + 25 = 100$$

Thus the control system will have a damping ratio of 0.5 when the gain K is equal to 100.

The methods used in the previous examples are not adequate for the more realistic example that follows. Let us now consider the following general form of the open-loop transfer function:

$$G(s)H(s) = K\left[\frac{(s - z_1)(s - z_2)\dots(s - z_m)}{(s - p_1)(s - p_2)\dots(s - p_n)}\right] \qquad \textbf{(16.18)}$$

where $K > 0$

$z_1, z_2, \dots z_m$ are the *zeros* of $G(s)H(s)$
(zeros are values of s for which $G(s)H(s) = 0$)

$p_1, p_2, \dots p_n$ are the *poles* of $G(s)H(s)$
(poles are values of s for which $G(s)H(s) = \infty$)

Note that m is the number of zeros in $G(s)H(s)$ and n is the number of poles. We will be using those two numbers to compute certain characteristics of the root-locus plot. Consider also the following form of the closed-loop characteristic equation:

$$G(s)H(s) = -1 = 1\underline{/-180°} \qquad \textbf{(16.19)}$$

Any point s_o that makes the angle of $G(s_o)H(s_o)$ equal to an odd multiple of $\pm 180°$ will satisfy Equation (16.19). This is commonly called the *angle condition* of the closed-loop characteristic equation. All points that satisfy the following angle condition also satisfy the closed-loop characteristic equation and lie on the root locus.

Angle Condition of the Closed-loop Characteristic Equation

$$\text{Angle of } [G(s_o)H(s_o)] = \pm N(180) \qquad \textbf{(16.20)}$$

where $N = 1, 3, 5, 7, 9, \dots = \{\text{set of odd integers}\}$

If s_o is a point on the root locus, then the value of K can be computed by the following *magnitude condition*.

Magnitude Condition of the Closed-loop Characteristic Equation

$$K = \text{magnitude of } \left|\frac{(s_o - p_1)(s_o - p_2)\dots(s_o - p_n)}{(s_o - z_1)(s_o - z_2)\dots(s_o - z_m)}\right| \qquad \textbf{(16.21)}$$

A detailed study of the angle condition has resulted in the development of the following rules for construction of root locus plots.*

---

* Chester L. Nachtigal, *Instrumentation and Control Fundamentals and Applications* (New York: Wiley, 1990) pp. 630–31.

Root Locus Rules

$$G(s)H(s) = K\left[\frac{(s - z_1)(s - z_2)\ldots(s - z_m)}{(s - p_1)(s - p_2)\ldots(s - p_n)}\right] \qquad \textbf{(16.18)}$$

m = the number of zeros

n = the number of poles

Begin the root locus plot by drawing an s-plane diagram with equal scales for the real and imaginary axes. Then plot each zero and pole of $G(s)H(s)$ on the s-plane diagram. Use an X to mark the location of each pole and an 0 to mark the location of each zero. Then apply the following rules to plot the root locus of $1 + G(s)H(s)$.

1. The root locus will have n branches that begin at the n poles with the value $K = 0$.

2. The root locus will have m branches that will terminate on the m finite zeros as K approaches infinity.

3. The remaining $n-m$ branches will go to infinity along asymptotes as K approaches infinity.

4. The asymptotes meet at a point called the *centroid*, C, which is computed as follows:

$$C = \frac{\Sigma \text{ poles} - \Sigma \text{ zeros}}{(\text{number of poles}) - (\text{number of zeros})} \qquad \textbf{(16.22)}$$

$$C = \frac{(p_1 + p_2 + \ldots + p_n) - (z_1 + z_2 + \ldots + z_m)}{(\text{number of poles}) - (\text{number of zeros})}$$

The $n-m$ asymptote angles are given by:

$$\theta_N = \pm\frac{N(180°)}{n - m}; \qquad N = 1, 3, 5, 7, 9, \ldots \qquad \textbf{(16.23)}$$

For each N, 2 angles are computed. Use enough values of N to compute the required $n-m$ angles.

5. Begin at the right and number the zeros and poles that lie on the real axis without regard to whether the points are a zero or a pole. A typical numbering will appear as follows:

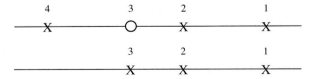

If there is an even number of points on the real axis, then the root locus will lie on the part of the real axis between each odd point and the following even point as illustrated below.

If there is an odd number of points on the real axis, then the root locus will extend to infinity from the highest numbered point as illustrated below.

6. If two zeros are connected, there must be a breakin point between them.

   If two poles are connected, there must be a breakout point between them.

   If a pole and a zero are connected, they usually are a full branch, starting at the pole with $K = 0$ and ending at the zero as $K \rightarrow \infty$.

   Use rule 6 to draw arrows and break points as shown below.

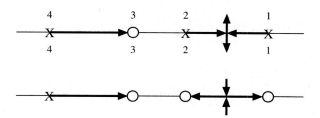

7. The break points occur where the derivative of K with respect to s is equal to zero. Computation of the break points is demonstrated in Example 16.9.

8. The points where the branches cross the imaginary axis are determined by replacing s by $j\omega$ in the characteristic equation and then solving for ω.

9. The angle condition may be used to determine the angle of departure from complex poles or the angle of arrival at complex zeros. Draw a line from each of the other zeros or poles to the zero or pole in question. Measure the angle formed by the positive real axis and the line to each of the other zeros and poles. The departure or arrival angle is equal to 180° plus the sum of all the zero angles minus the sum of all the pole angles.

10. The operating point for a given damping ratio, zeta (ζ), can be determined as follows. First, determine the operating point angle, $\theta = \cos^{-1}(\zeta)$. Then draw a line from the origin of the s-plane that forms an angle of θ with the negative real axis. This line intersects the root locus at the operating point. The length of the line from the origin to the operating point represents the natural frequency, ω_o.

11. The value of K at the operating point can be determined as follows. First, draw lines from each pole and each zero to the operating point. Measure the length of each line and determine the product of the lengths of the lines from the poles and the product of the lengths of the lines from the zeros. If there are no zeros (or poles) then use 1 as the product. The value of K at the operating point is equal to the poles product divided by the zeros product.

Example 16.9

Determine the root locus of a control system with the following open-loop transfer function:

$$G(s)H(s) = \frac{K}{s(s + 3)(s + 9)}$$

Then determine the operating point, s_o, for a damping ratio of 0.5 ($\zeta = 0.5$). Also find the gain, K, and the resonant frequency, ω_o, at the operating point.

Solution

1. The open-loop transfer function has no zeros and three poles at $s = 0$, $s = -3$, and $s = -9$. The root locus will have three branches that will all go to infinity along 3 asymptotes (Rules 1, 2, and 3).

2. The asymptotes meet at the centroid, C, determined by Equation (16.22) (Rule 4):

$$C = \frac{(\Sigma \text{ poles}) - (\Sigma \text{ zeros})}{(\text{number of poles}) - (\text{number of zeros})}$$

$$C = \frac{[0 + (-3) + (-9)] - [0]}{3 - 6} = -4$$

The asymptote angles are determined by Equation (16.23):

$$\theta_1 = \pm\frac{1(180)}{3 - 0} = \pm 60°$$

$$\theta_3 = \pm\frac{3(180)}{3 - 0} = \pm 180°$$

Therefore, the asymptote angles are $+60°$, $-60°$, and $+180°$.

3. The root locus lies on the portion of the real axis between $s = 0$ and $s = -3$, and the portion from $s = -9$ to $s = -\infty$ (Rule 5).

4. There is a breakout point on the real axis between $s = 0$ and $s = -3$ (Rule 6).

5. The breakout point occurs where the derivative of K with respect to s is equal to zero ($dK/ds = 0$). To obtain this derivative, we first solve the characteristic equation for K:

$$1 + \frac{K}{s(s + 3)(s + 9)} = 0$$

$$K = -(s^3 + 12s^2 + 27s)$$

The power rule may be used to obtain dK/ds.

$$dK/ds = -(3s^2 + 24s + 27)$$

Now set $dK/ds = 0$ to obtain the breakout point.

$$dK/ds = -(3s^2 + 24s + 27) = 0$$

Dividing both sides of the equation by -3 and then applying the quadratic formula gives us two candidates for the desired breakout point.

$$s^2 + 8s + 9 = 0$$

$$s = \frac{-8 \pm \sqrt{64 - 36}}{2}$$

$$s = -1.35425 \text{ and } -6.64575$$

The breakout point is between 0 and -3, so $s = -1.35425$ is the desired breakout point (Rule 7).

6. The points where the branches cross the imaginary axis are obtained from the characteristic equation with s replaced by $j\omega$.

$$s^3 + 12s^2 + 27s + K = 0$$
$$(j\omega)^3 + 12(j\omega)^2 + 27(j\omega) + K = 0$$
$$-j\omega^3 - 12\omega^2 + 27j\omega + K = 0$$
$$(K - 12\omega^2) + j(27\omega - \omega^3) = 0$$

Set the imaginary part equal to zero and solve for ω.

$$27\omega - \omega^3 = 0$$
$$\omega^2 = 27$$
$$\omega = \pm\sqrt{27} = \pm 3\sqrt{3}$$

Set the real part equal to zero and solve for K.

$$K = 12\omega^2 = 12(27) = 324$$

Therefore, the branches cross the imaginary axis at $\omega = \pm 3\sqrt{3}$ with $K = 324$ (Rule 8).

7. The root-locus plot is shown in Figure 16.22. The characteristic equation, $s^3 + 12s^2 + 27s + K = 0$, involves the solution of a cubic equation, which is considerably more involved than a quadratic equation. However, the roots when $K = 0$ are easy to find. They are 0, -3, and -9. The next step is to use the fact that one of the three roots will always be real. Assume that the real root is $-a$, and divide the characteristic equation by the factor $(s + a)$ to get a general equation for the remaining two roots.

$$\frac{s^3 + 12s^2 + 27s + K}{s + a} = s^2 + (12 - a)s + 27 - a(12 - a)$$

The remainder of the division is $K - 27a + 12a^2 - a^3$. We can now express the characteristic equation as follows:

$$(s + a)[s^2 + (12 - a)s + 27 - a(12 - a)] = 0 \qquad \textbf{(16.24)}$$

The value of K must be such that the remainder is 0.

$$K = 27a - 12a^2 + a^3 \qquad \textbf{(16.25)}$$

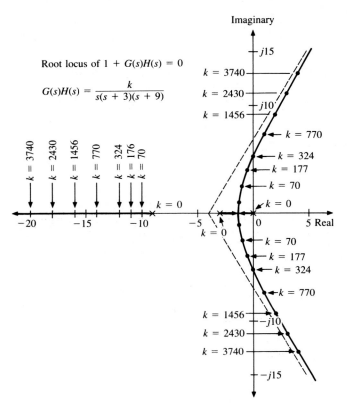

Root locus of $1 + G(s)H(s) = 0$

$$G(s)H(s) = \frac{k}{s(s + 3)(s + 9)}$$

Figure 16.22 Root locus of $1 + G(s)H(s) = 0$ for the system with $G(s)H(s) = K/[s(s + 3)(s + 9)]$.

A calculator or computer program may be used to obtain exact points on the root locus using Equations (16.24) and (16.25) and the quadratic formula. A trial and error procedure may also be employed to obtain the exact operating point. However, the following graphic analysis provided quick results with sufficient accuracy for most applications.

8. The expanded view of the root locus in Figure 16.23 shows the determination of the operating point for a damping ratio of 0.5 (Rule 10). The line of constant damping ratio makes an angle of θ as given by Equation (16.16).

$$\theta = \cos^{-1} 0.5 = 60°$$

The operating point is at the intersection of the damping ratio $(\zeta) = 0.5$ line and the root locus. From the graph, we obtain the approximate value $-1.12 + j1.95$ for the coordinates of the operating point. Using a calculator, the more exact value of the coordinates is $-1.125 + j1.9486$.

$$\text{Operating point} = -1.125 + j1.9486 = 2.25\underline{/60°}$$

The resonant frequency is equal to the length of the line from the origin to

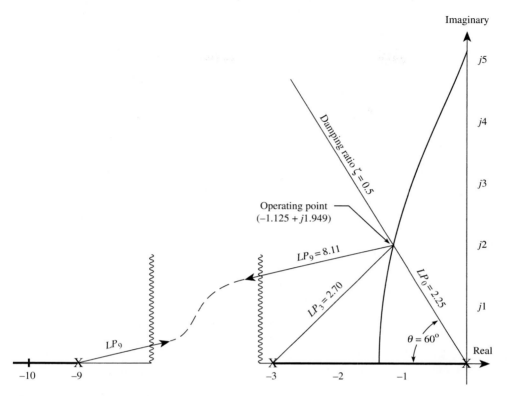

Figure 16.23   Expanded view of the operating point region of the root locus shown in Figure 16.22. At the operating point the system has a damping ratio of 0.5 and a resonant frequency of 2.25 radians/second.

the operating point. From the graph, this distance is estimated to be 2.25 (the same as the calculated value).

$$\text{Resonant frequency} = 2.25 \text{ radians/second}$$

9. The value of K at the operating point is determined by the poles product divided by the zeros product (Rule 11). From the graph we obtain the following lengths:

$$LP_0 = \text{length from the pole at } (0,0) = 2.25$$
$$LP_3 = \text{length from the pole at } (-3,0) = 2.70$$
$$LP_9 = \text{length from the pole at } (-9,0) = 8.11$$
$$\text{Poles product} = (2.25)(2.70)(8.11) = 49.3$$
$$\text{Zeros product} = 1 \text{ (There are no zeros.)}$$
$$K = 49.3/1 = 49.3$$

The more exact calculated value is $K = 49.36$.

GLOSSARY

Angle condition: A condition used to locate points on the root locus of a control system. If s_o is on the root locus, then it will satisfy the closed-loop characteristic equation (i.e., $1 + G(s_o)H(s_o) = 0$), and the angle of $G(s_o)H(s_o)$ will equal an odd multiple of $\pm 180°$. (16.10)

Centroid: The point where the asymptotes of a root locus meet. (16.10)

Characteristic equation, closed-loop: The equation $(1 + G(s)H(s) = 0)$ that defines the poles of the closed-loop transfer function and is used to construct the root-locus plot. A necessary condition of stability is that all of the roots of this equation must have negative real parts. (16.10)

Closed-loop frequency response (Bode diagram): Graphs of frequency versus the gain and phase angle change from the setpoint to the measuring transmitter output with the transmitter output connected to the error detector. (16.4)

Closed-loop transfer function: The C_m/SP transfer function of a control system with the measuring transmitter output connected to the error detector. (16.1)

Deviation ratio: The ratio of the closed-loop error magnitude over the setpoint magnitude and a measure of how accurately a control system can follow a change in the setpoint. (16.5)

Error ratio: The ratio of the closed-loop error magnitude over the open-loop error magnitude and a measure of the accuracy of a control system. (16.5)

Frequency limit: The maximum frequency of setpoint changes or disturbances that a control system can handle. A control system cannot follow setpoint changes or reduce the error caused by disturbances that are above its frequency limit. (16.5)

Gain margin: A safety factor for the stability of a closed-loop control system that specifies how much additional gain is required to make the system unstable. (16.8)

Gain margin point: A point on the Nyquist diagram, located at $-0.5 + j0$, that is used to determine if a closed-loop control system meets the gain margin criterion. (16.9)

Magnitude condition: A condition used to determine K for a point on the root locus of a control system. If s_o is on the root locus, then the magnitude of $G(s_o)H(s_o)$ will be equal to 1, and K can be determined accordingly. (16.10)

Marginally stable: A control system is marginally stable if it satisfies the stability conditions, but does not meet the gain margin, the phase margin, or both the gain and phase margin criteria. (16.8)

Nyquist stability criterion: A closed-loop control system is stable if the open-loop system is stable and the Nyquist plot does not encircle the stability point, $-1 + j0$. (16.9)

Open-loop frequency response (Bode diagram): Graphs of frequency versus the gain and phase angle change from the setpoint to the measuring transmitter output with the transmitter output disconnected from the error detector. (16.3)

Open-loop transfer function: The C_m/SP transfer function of a control system with the measuring transmitter output disconnected from the error detector. (16.1)

Overall gain: The amplitude of the output of the last of several components connected in series, divided by the amplitude of the input to the first component. The overall gain is equal to the product of the gains of the individual components. (16.2)

Overall phase angle: The phase angle of the output of the last of several components connected in series, minus the phase angle of the input to the first component. The overall phase angle is equal to the sum of the phase angles of the individual components. (16.2)

Phase margin: A safety factor for the stability of a closed-loop control system that specifies how much additional phase lag is required to make the system unstable. (16.8)

Phase margin point: A point on the Nyquist diagram, located at $1/-140°$, that is used to determine if a closed-loop control system meets the phase margin criterion. (16.9)

Root locus: The plot of all possible roots of the closed-loop characteristic equation as the controller gain (K) is varied from 0 to infinity. (16.10)

Stability conditions: Any condition that assures that a closed-loop control system will be stable.

1. A closed-loop control system is stable if the open-loop gain is less than 1 at the frequency for which the open-loop phase angle is $-180°$.
2. A closed-loop control system is stable if all the roots of the closed-loop characteristic equation, $1 + G(s)H(s) = 0$, have negative real parts. (16.8)
3. *See* Nyquist stability criterion.

Stability point: A point on the Nyquist diagram, located at $-1 + j0$, that is used to determine if a closed-loop control system meets the stability conditions defined above. (16.9)

Three zones of control: The graph of the error ratio versus frequency is divided into three distinct zones. Zone 1 covers the low frequencies; zone 2, the midrange frequencies; and zone 3, the high-range frequencies. In zone 1 the closed-loop control system reduces the error. In zone 2, closed-loop control increases the error. In zone 3, closed-loop control neither increases nor decreases the error. The frequency limit of the control system is the frequency on the border between zone 1 and zone 2. (16.5)

Unstable: A control system is unstable if it does not meet the stability conditions defined above. (16.8)

EXERCISES

In Exercises 16.1 through 16.10, you are to complete each of the following assignments for the control system given in the exercise.

Six Assignments for Exercises 16.1 to 16.10

1. Determine the open-loop transfer function.
2. Use the program "DESIGN" to make a printed copy of the Bode Data Table.
3. Construct an open-loop Bode diagram and an error ratio graph.
4. Determine the gain margin and the phase margin from the open-loop Bode diagram. State whether or not the control system meets the minimum criteria for gain margin and phase margin.

5. Determine the maximum frequency limit from the error ratio graph.
6. Construct a Nyquist diagram of the control system. State whether or not the control system meets the Nyquist criterion for stability.

16.1 A proportional controller is used to control the pressure in a first-order lag plus dead-time process. The control system has the following transfer functions:

$$G(s) = [5]$$

$$H(s) = \left(\frac{2}{1 + 0.5s}\right) e^{-0.05s}$$

Complete the six assignments listed at the beginning of the exercises.

16.2 A thermal control system has a proportional controller with a gain of 15. The first-order lag plus dead-time process has the following transfer function:

$$H(s) = \left(\frac{1}{1 + 1000s}\right) e^{-200s}$$

Complete the six assignments listed at the beginning of the exercises.

16.3 A PI controller is used to control the level in a first-order lag plus dead-time process. The control system has the following transfer functions:

$$G(s) = \frac{0.628 + 9.1s}{s}$$

$$H(s) = \left(\frac{1}{1 + 50s}\right) e^{-2s}$$

Complete the six assignments listed at the beginning of the exercises.

16.4 An armature-controlled dc motor is used in a position control system. The controller and process transfer functions are given below:

$$G(s) = \frac{3.2 + 0.48s}{1 + 0.015s}$$

$$H(s) = \frac{19}{s + 0.329s^2 + 0.00545s^3}$$

Complete the six assignments listed at the beginning of the exercises.

16.5 A solid flow control system uses a PI controller. The controller and process transfer functions are given below.

$$G(s) = \frac{1 + 36s}{18s}$$

$$H(s) = e^{-20s}$$

Complete the six assignments listed at the beginning of the exercises.

16.6 A first-order lag plus dead-time blending process has a time constant of 1000 s and a dead-time lag of 20 s. The PI controller used to control the process has a gain of 44 and an integral action rate of $\frac{1}{230}$ s^{-1}. Complete the six assignments listed at the beginning of the exercises.

16.7 A PID controller is used to control a second-order lag plus dead-time process. The transfer functions of the controller and process are given below.

$$G(s) = \frac{42.86 + 12s + 2.16s^2}{s + 0.018s^2}$$

$$H(s) = \left(\frac{0.1}{1 + 0.02s + 0.01s^2}\right)e^{-0.05s}$$

Complete the six assignments listed at the beginning of the exercises.

16.8 A pressure system consists of a measuring transmitter with a lag coefficient of 0.2 s, a critically damped control valve, and a pressure tank with a time constant of 1 s. The transfer functions of the components are given below.

$$\text{Controller TF} = G(s) = \frac{1.8 + 0.9s + 0.324s^2}{s + 0.036s^2}$$

$$\text{Measuring transmitter TF} = \frac{1}{1 + 0.2s}$$

$$\text{Control valve TF} = \frac{5}{1 + 0.2s + 0.01s^2}$$

$$\text{Process TF} = \frac{2}{1 + s}$$

Complete the six assignments listed at the beginning of the exercises.

16.9 An amplidyne position control system consists of a PD controller, an amplidyne unit, a dc motor, and a load. The transfer functions are given below.

$$\text{Controller TF} = G(s) = \frac{0.01 + 0.00067s}{1 + 0.0067s}$$

$$\text{Amplidyne TF} = \frac{8500}{1 + 0.025s + 0.0001s^2}$$

$$\text{Motor and load TF} = \frac{1}{s + 0.35s^2 + 0.015s^3}$$

Complete the six assignments listed at the beginning of the exercises.

16.10 The AJ Food Company uses a PID controller to control a jacketed kettle similar to Figure 15.6a. Heated liquid from the kettle flows through a pipe line to the bottling machine, where it is placed in 1-L bottles. A temperature sensor measures the temperature of the liquid product as it leaves the jacketed kettle. The transfer functions of the system components are given below.

$$\text{Controller TF} = G(s) = \frac{0.0075 + 3s + 600s^2}{s + 20s^2}$$

$$\text{Measuring transmitter TF} = \frac{1}{1 + 290s + 12{,}000s^2}$$

$$\text{Process TF} = \left(\frac{2}{1 + 1070s}\right)e^{-1.1s}$$

$$\text{Control valve} = \frac{4}{1 + 0.074s + 0.0021s^2}$$

Complete the six assignments listed at the beginning of the exercises.

16.11 Construct a root-locus diagram of the control system with the following open-loop transfer function.

$$G(s)H(s) = K\left(\frac{s + 12}{s + 4}\right)$$

16.12 Construct a root-locus diagram of the control system with the following open-loop transfer function.

$$G(s)H(s) = \frac{4K}{s(s + 16)}$$

Determine the value of K that will produce a damping ratio of 0.65.

16.13 Construct a root-locus diagram of the control system with the following open-loop transfer function.

$$G(s)H(s) = \frac{10K}{s^2 + 20s + 16}$$

Determine the value of K that will produce a damping ratio of 0.8.

16.14 Construct a root-locus diagram of the control system with the following open-loop transfer function.

$$G(s)H(s) = \frac{K}{s(s + 2)(s + 12)}$$

16.15 Construct a root-locus diagram of the control system with the following open-loop transfer function:

$$G(s)H(s) = \frac{K(s + 5)}{s^2}$$

16.16 Construct a root-locus diagram of the control system with the following open-loop transfer function:

$$G(s)H(s) = \frac{K(s + 5)}{s^2(s + 80)}$$

16.17 Construct a root-locus diagram of the control system with the following open-loop transfer function:

$$G(s)H(s) = \frac{K(s + 10)}{s(s + 1)(s + 50)}$$

16.18 Construct a root-locus diagram of the control system with the following open-loop transfer function:

$$G(s)H(s) = \frac{K}{s(s + 5)(s + 15)}$$

Then determine the operating point, s_o, for a damping ratio of 0.707. Also find the gain, K, and the resonant frequency, ω_o, at the operating point.

16.19 Construct a root-locus diagram of the control system with the following open-loop transfer function.

$$G(s)H(s) = \frac{K}{(s + 3)(s^2 + 8s + 32)}$$

Then determine the operating point, s_o, for a damping ratio of 0.707. Also find the gain, K, and the resonant frequency, ω_o, at the operating point.

CHAPTER 17

Controller Design

OBJECTIVES

Controller design involves the selection of the control modes to be used and the determination of the value of each mode. For continuous systems, the proportional mode is usually combined with the integral and/or derivative modes to form a PI, PD, or PID controller. The parameters for these three modes are the proportional gain (P), the integral action rate (I), and the derivative action time constant (D).

The simplest design method is to specify a PID controller and then determine the control mode parameters in the field during plant startup. This field adjustment is called "tuning the controller," and a procedure called the ultimate cycle method is used to determine the mode parameters. Some controllers are designed to perform this field adjustment automatically or semiautomatically.

A more formal design method uses Bode diagrams and a computer-aided graphic technique to design the controller. Using the Bode diagram method, the designer can determine the control mode parameters before the plant is constructed. This speeds up the plant startup considerably, although some fine tuning of the controller is usually required in the field. A major difficulty with this method is the requirement of an open-loop Bode diagram of the process. Since the plant is not yet constructed, the designer must construct a model of the process to obtain the necessary open-loop Bode diagram.

The purpose of this chapter is to give you an entry-level ability to discuss controller design, tune a controller, complete a computer-aided Bode design of a controller, and design simple compensation networks. After completing this chapter, you will be able to

1. Describe the ultimate cycle method for tuning a controller
2. Describe the process reaction method for tuning a controller
3. Discuss self-tuning adaptive controllers
4. Complete a computer-aided Bode design of a PID controller
5. Specify compensation networks that will improve the response of the control system

17.1 INTRODUCTION

Controller design consists of the selection of the control modes and control mode settings and/or compensation networks that will result in a stable system that meets the control objectives. The control objectives may specify some or all of the following characteristics.

1. The accuracy and speed of response of the measuring transmitter (refer to Chapter 6).
2. The residual error allowed by the controller after a load change (refer to Chapter 14).
3. The response of the control system to a step change in load or setpoint. Quarter amplitude decay or critical damping may be specified (refer to Chapter 2).
4. The maximum frequency limit. The control system is able to follow setpoint changes and minimize disturbances with frequencies less than the maximum limit (refer to Chapter 16).
5. The response time, the rise time, and the settling time of the control system.
6. Minimum cost. The initial cost of the hardware is only one part of this objective. Maintenance costs, operating costs, reject product costs, and environmental costs are examples of the other costs that may be a factor. This objective requires a judgment based on economic and social conditions.

Two different approaches are used to design a control system. This does not mean that all design methods use only one approach or the other. Often parts of each approach are found in the design of a control system.

One approach to controller design uses a standard P, PI, PD or PID controller. In this method, a major design decision is the selection of which control modes to include in the controller. Then startup control mode settings are chosen. The actual mode settings are determined during startup by a process called *"tuning the controller."* The tuning procedure consists of deliberately inducing a small oscillation in the system so that the period of the oscillation can be measured. (Controller adjustment formulas give the optimum control mode values based on the period of oscillation of the system.) An advantage of this method is that accurate models of the process are not required (accurate transfer functions of many industrial processes are very difficult to determine). This method is used when the cost of determining the controller adjustments during startup is less than the cost of the analysis and design required to define the control system before startup. Even the most accurate process control system design may require final adjustments in the field.

The second approach uses PID modes or compensation networks designed for a specific application. (The term "compensation network" often includes the I and D control modes under the names integration and lead-lag networks.) The second method is used when the transfer function of the process is known. (Accurate transfer functions of most electromechanical systems are relatively easy to determine.) The design procedure consists of the use of compensation networks to modify the open-loop frequency response such that a simple loop gain adjustment will result in a stable system that meets the performance objectives. This method is used when the cost of determining the controller adjustments during startup is greater than the cost of the analysis and design required to define the control system before startup.

The frequency response of most industrial processes is difficult to determine or measure. For this reason, process control design is based more on the first approach. However, frequency response methods are sometimes used to select the control modes and determine approximate mode settings. Field adjustments are almost always required to determine the exact mode settings.

The second approach has been used very successfully to design servomechanisms. The frequency response of most electromechanical components is relatively easy to determine and measure. Dead time is usually negligible. The designer has the option of using the root-locus method or the Bode method to design compensation networks which are inserted into the control loop. Field adjustments are often unnecessary, but some loop gain adjustment may be necessary to "fine tune" the system in the field.

17.2   THE ULTIMATE CYCLE METHOD

The *ultimate cycle method* uses controller adjustment formulas to determine the controller settings. The formulas require two measurements from the process: the minimum controller gain that causes the control system to oscillate (ultimate gain, G_u) and the period of the oscillation (ultimate period, P_u). To use the ultimate cycle method, the controller must be installed and the system ready to operate.

The following procedure is used to determine the ultimate gain and ultimate period. First, set the integral and derivative modes to the least effective setting. Then, start up the process, with the controller gain at a low value. Increase the gain setting until the controlled variable begins to oscillate. The last gain setting is the ultimate gain (G_u). The period of the oscillation is the ultimate period (P_u). The controller settings are determined from Table 17.1.

The original ultimate cycle method was developed by Nichols and Ziegler.* The problem with the original method is that the formulas do not guarantee an optimum response. The modified ultimate cycle method was developed to solve this problem in a very direct way. The gain setting is adjusted in a trial-and-error procedure until the desired response is obtained. The most common "desired response" is quarter amplitude decay (see Section 1.11). This criterion is illustrated in Figure 17.1. As the name suggests, quarter amplitude decay is an underdamped response in which each successive peak is one fourth as large as the preceding peak.

The modified method requires only the measurement of the ultimate period (P_u), which is used to compute the integral and derivative mode settings. After the integral and derivative modes have been set, the process is disturbed with a small change in setpoint, the response of the controlled variable is observed, and the gain is adjusted.

---

* N. B. Nichols and J. G. Ziegler, "Optimum Settings for Automatic Controllers," *ASME Transactions*, Vol. 64, No. 8, 1942, pp. 759–768.

Table 17.1   Ultimate Cycle Method

Control Modes	Original Method	Modified Method
Proportional control (P)	$P = 0.5G_u$	Adjust the gain to obtain quarter amplitude decay response to a step change in setpoint (see Figure 17.1)
Proportional plus integral control (PI)	$I = \dfrac{1.2}{P_u} \, (\text{min}^{-1})$	$I = 1/P_u \, (\text{min})$
	$P = 0.45G_u$	Adjust the gain to obtain quarter amplitude decay response to a change in setpoint
Proportional plus integral plus derivative control (PID)	$I = \dfrac{2.0}{P_u} \, (\text{min}^{-1})$	$I = \dfrac{1.5}{P_u} \, (\text{min}^{-1})$
	$D = \dfrac{P_u}{8} \, (\text{min})$	$D = \dfrac{P_u}{6} \, (\text{min})$
	$P = 0.6G_u$	Adjust the gain to obtain quarter amplitude decay response to a change in setpoint

The gain adjustment is simple—if the response is overdamped, the gain is increased; if the response is underdamped, the gain is decreased. The sequence of setpoint change, observation of the response, and adjustment of the gain is repeated until a quarter amplitude decay response is obtained. The same procedure can be used to tune a controller to obtain a critically damped response (Section 1.11). The general rule on gain adjustment is that increasing the gain will decrease the damping and vice versa.

The modified ultimate cycle method is very reliable and works on all types of processes. It is particularly effective on relatively fast processes where the waiting time is short. However, it can be very time consuming on a slow process. The author once spent most of a day tuning a process with a cycle time of about 1 hour. In a slow process, a process model can be very helpful in speeding up the tuning procedure. With a model of the process, the designer has the option of using the process reaction method (Section 17.3) or the Bode design method (Section 17.5).

Figure 17.1   The quarter amplitude decay response is a common control system design criterion.

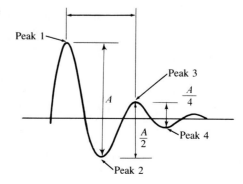

Example 17.1

A process control system is tested at startup. The derivative mode is turned off, and the integral mode is set at the lowest setting. The gain is gradually increased until the controlled variable starts to oscillate. The gain setting is 2.2 and the period of oscillation is 12 min. Use the original ultimate cycle method to determine the PID controller settings.

Solution

$$I = \frac{2.0}{P_u} = \frac{2.0}{12} = 0.167 \text{ min}^{-1}$$

$$D = \frac{P_u}{8} = \frac{12}{8} = 1.5 \text{ min}$$

$$P = 0.6G_u = (0.6)(2.2) = 1.32$$

17.3 THE PROCESS REACTION METHOD

The *process reaction method** uses controller adjustment formulas to determine the controller settings. This method assumes that the process can be approximated by a first-order lag plus dead-time model. The formulas use three parameters obtained from an open-loop step response test of the process. The test begins by operating the process on manual control (open loop) until the measured variable remains constant. Then a step change in the manipulating element is made and the response of the controlled variable is recorded. This record of the controlled variable is called the *process reaction graph*. A typical process reaction graph is shown in Figure 17.2.

The process reaction formulas require three variables from the step response test. Two variables are obtained from the process reaction graph. The third variable is (Δp), the percent change in the manipulating element output (e.g., percent change in the control valve stem position). The two variables from the reaction graph are the effective delay (L) and the slope of the tangent line (N). The controller settings are determined from the following formulas.

Process Reaction Method

$$L = \text{effective delay, minute}$$

$$N = \frac{\Delta C_m}{\Delta t} \qquad \text{percent/minute}$$

$$\Delta P = \text{change in the manipulating element, percent}$$

---

* Ibid.

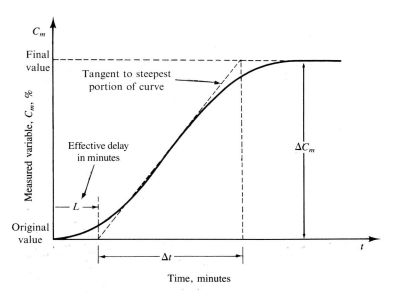

Figure 17.2 A process reaction graph is a plot of the controlled variable after a step change in the output of the manipulating element. A first-order lag plus dead-time model of the process is derived from the graph with the dead time equal to L and the first-order time constant equal to Δt.

1. Proportional control (P)

$$P = \frac{\Delta P}{NL}$$

2. Proportional plus integral control (PI)

$$P = 0.9\left(\frac{\Delta P}{NL}\right)$$

$$I = \frac{0.3}{L} \qquad \text{min}^{-1}$$

3. Proportional plus integral plus derivative control (PID)

$$P = 1.2\left(\frac{\Delta P}{NL}\right)$$

$$I = \frac{0.5}{L} \qquad \text{min}^{-1}$$

$$D = 0.5L \text{ min}$$

A major advantage of the process reaction method over the modified ultimate cycle method is that once the process model is obtained, the control modes can be determined and implemented immediately. This is particularly useful on a very slow process where considerable elapsed time is required to complete the ultimate cycle method. The disadvantage of the process reaction method is that most processes are more complex than a simple first-order lag plus dead-time model. This usually means that some final adjustment of the gain is still required to obtain the desired response (e.g., quarter amplitude decay or critical damping). However, the process reaction method does provide an initial setting of the control modes, and these initial settings are usually close to the final values. However, a realistic expectation is that some tweaking of the gain will be required to obtain the desired response.

Example 17.2

During startup, the control valve of a process control system was maintained constant until the controlled variable stopped changing and reached a steady value. Then the control valve position was changed by 10%. A response graph similar to Figure 17.2 was obtained. An analysis of the reaction graph produced the following values.

$L = 5$ min

$\Delta t = 10$ min

Initial value of C_m: 40%

Final value of C_m: 48%

Use the process reaction method to determine the settings of a three-mode controller.

Solution

$$N = \frac{\Delta C_m}{\Delta t} = \frac{48 - 40}{10} = 0.8\%/\text{min}$$

$$\Delta P = 10\%$$

$$L = 5 \text{ min}$$

$$P = 1.2\left(\frac{\Delta P}{NL}\right) = 1.2\left(\frac{(10)}{(0.8)(5)}\right) = 3$$

$$I = \frac{0.5}{L} = \frac{0.5}{5} = 0.1 \text{ min}^{-1}$$

$$D = 0.5L = (0.5)(5) = 2.5 \text{ min}$$

17.4 SELF-TUNING ADAPTIVE CONTROLLERS

A major problem in designing or tuning process control systems is that the process models are often very complex, difficult to obtain, and inaccurate. Most models assume lumped elements (e.g., resistance, capacitance, inertance, inductance, inertia, dead-time). In many processes, these elements are distributed in much the same way as they are in an electrical transmission line. Often the elements in the process change for various reasons. A change in production rate almost always changes the dead-time delays in a process. In one plant startup, the author used the modified ultimate cycle method to tune a temperature control loop. The system responded to small step changes in setpoint with almost perfect quarter amplitude decay response. The next time the author checked, the response was unstable, oscillating with an unacceptably large amplitude. The problem was caused by a reduction in the production rate. A repeat of the modified ultimate cycle method resulted in new settings that returned the system to quarter amplitude decay response. However, when the production rate was returned to the original rate, the system had an overdamped response.

The control engineer has three choices in handling the control of a process whose model changes with changes in production rate (or other reasons). The first choice is to tune the controller for the worst case and accept a sluggish response for other conditions. The second choice is to change the controller mode settings every time the process model changes. (This choice assumes we know what causes the change in the process model—this is not always the case.) The third choice is to design a controller that will change itself every time the process changes. The self-tuning adaptive controller is an implementation of the third choice.

The objective of *self-tuning adaptive controllers* is to maintain the desired control criteria when the load or production rate changes from one value to another. The adaptive controller accomplishes this objective by observing the response of the control system and automatically adjusting the PID values accordingly.

Adaptive controllers use a conventional PID algorithm within an outer loop that handles the adaptive algorithm. The outer loop is called a shell because it encloses the PID algorithm and determines the values of P, I, and D. In other words, the adaptive shell establishes the "PID environment" used by the PID algorithm. The adaptive shell is also a feedback loop. It observes the control system and uses these observations to determine the values of P, I, and D that will achieve optimum performance.

Many different techniques are used to "adapt" a controller to changes in the process. Adaptive controllers fall into one of three categories: those that use programmed adjustment of the controller gain, those that use a model of the process to determine the PID values, and those that use pattern recognition to determine the PID values. The model-based adaptive controllers use a variation of the process reaction method for computing the PID values. The pattern-recognition adaptive controllers use a variation of the ultimate cycle method for determining the PID values.

Programmed Adaptive Controllers

A *programmed adaptive controller* automatically adjusts the proportional gain (P) as a function of a number of process-related variables. The proportional gain may be programmed as a function of the controlled variable, the setpoint, the error, the controller output, or a remote input variable. The adaptive algorithm can be set for independent gain adjustment using any combination of the process-related variables. For example, the remote input could be the production rate. The controller could adjust the controller gain according to the production rate to maintain the quarter amplitude decay criteria for all production rates. The controller could also make an independent adjustment in the gain based on the value of the setpoint.

Model-Based Adaptive Controllers

A *model-based adaptive controller* uses an internal model of the process to determine the optimum PID values. The adaptive algorithm uses an identification or "self-learning" mode to determine the model of the process. In this mode, the controller introduces step changes above and below the setpoint and observes the reaction of the process to these changes. The setpoint changes and observations are repeated until sufficient data have been obtained to establish a process reaction curve similar to Figure 17.2. The process model is obtained from the process reaction curve, and the PID values are determined by formulas similar to the process reaction formulas in Section 17.3.

Once the model has been established, the adaptive algorithm uses an operating mode in which the PID values remain fixed at the values established in the identification mode. Some algorithms also provide a continuous update operating mode. In this mode, the process model is continuously updated in the same manner as it was in the identification mode. If the process changes, the continuous update mode will adjust the process model and the PID values accordingly.

Pattern-Recognition Adaptive Controllers

A *pattern-recognition adaptive controller* uses a graph of error versus time similar to Figure 17.1 to obtain the optimum PID values. The adaptive algorithm constantly examines the response of the control system to naturally occurring disturbances caused by changes in load or setpoint. Whenever the magnitude of the error exceeds a threshold value, the adaptive algorithm searches for peaks in the magnitude of the error (see peaks 1, 2, etc., in Figure 17.1). When the algorithm finds two or three peaks, it uses the time between peaks and the magnitudes of the peaks to determine the amplitude decay ratio and the period of the oscillation. Formulas similar to the ultimate cycle formulas in Section 17.2 are then used to compute the optimum PID values.

17.5 COMPUTER-AIDED PID CONTROLLER DESIGN

The design procedure presented in this section uses the open-loop Bode diagram of the control system as the major design tool. For this reason the design procedure is known as the *Bode design method*. The objective of the design procedure is to obtain a stable control system that provides accurate control for a wide range of frequencies. Stable control means that the final design satisfies the gain margin and phase margin criteria covered in Chapter 16. Control over a wide range of frequencies means that the maximum frequency limit of the control system is as high as possible. Accurate control means that the open-loop gain is as large as possible at frequencies below the $-180°$ phase-angle frequency. The design procedure consists of using the PID control modes to modify the open-loop Bode diagram in such a way that the control objectives are satisfied. The Bode design of a controller is accomplished in four steps.

STEP 1. The open-loop Bode diagram of the process is used to determine the usefulness of the derivative control mode. If the derivative mode is found to be useful, the Bode diagram is used to determine the value of the derivative action time constant, D. If the derivative mode is not useful, D is given a value of 0.

STEP 2. The open-loop Bode diagram of the process plus the derivative control mode is used to determine the usefulness of the integral control mode. If the integral mode is found to be useful, the Bode diagram is used to determine the value of the integral action rate, I. If the integral mode is not useful, I is given a value of 0.

STEP 3. The open-loop Bode diagram of the process plus the integral and derivative control modes is used to determine the proportional mode gain, P.

STEP 4. Three diagrams of the process and PID controller are used to evaluate the final design of the control system: (i) The open-loop Bode diagram is used to determine the gain margin and the phase margin of the control system. (ii) The error ratio graph is used to determine the maximum frequency limit of the control system. (iii) The closed-loop Bode diagram presents the frequency response of the closed-loop control system.

The Bode design method is a graphical method that helps the designer visualize the effect of each control mode on the frequency response of the control system. The designer uses straight-line approximations on semilog graph paper or a computer program to construct the open-loop Bode diagrams used in each step of the process. In step 4, the designer uses a Nichols chart or a computer program to translate the open-loop Bode diagram into the closed-loop Bode diagram. A major advantage of the Bode design method is the excellent visual aid provided by the graphical approach. The designer can observe exactly how each control mode reshapes the open-loop frequency response of the system.

In this section we present a BASIC computer program to facilitate the Bode design method. The program is named "Control System Design" and the file name is "DESIGN". The designer may use program "DESIGN" to complete the four-step, Bode-method design of a PID controller. Program "DESIGN" can handle a process with up to nine components (first, second, or third order) and ten dead-time delays. It uses the bottom 23 lines of the computer screen to display one of the following graphs as selected by the designer:

1. Open-loop Bode diagram
2. Closed-loop Bode diagram
3. Nyquist diagram
4. Error ratio graph

A table of "Design Decision Data" is displayed in the upper right corner of the graph. The designer uses data from this table to determine values of P, I, D, and α. The top line of the computer screen is the *Status Line*. It usually displays the current values of P, I, D, and α, but may display other information pertinent to the completion of a command.

The second line of the screen is the *Command Line*. It usually prompts the user for one of the following commands: *Dmode, Imode, Pmode, Analysis, Zoom, Unzoom,* and *Quit*. A command is given by simply pressing the first letter of the command (i.e., the capitalized letter). When a P, I, or D command is given, the Command Line prompts the designer for the appropriate control mode parameter(s). When the Analysis command is given, the Command Line prompts the designer for the choice of Closed-loop, Nyquist, Error-ratio, or Open-loop. The Zoom command provides a close-up view of the graph in the region of 0 dB gain and $-180°$ phase angle. The Unzoom command returns the graph to normal size.

Program "DESIGN" begins with $P = 1$, $I = 0$, $D = 0$, and $\alpha = 0.1$. As the design steps unfold, the designer enters new values of D, α, I, and P in that order. After each mode is entered, the program displays both the old and the new gain and phase angle graphs. The designer can observe exactly how the mode just entered has changed the open-loop gain and phase angle graphs. After viewing the changes just made, the designer presses any key to move to the next step in the design procedure.

A major feature of program "DESIGN" is the ease with which the designer can go back and change any control mode. This enables the designer to use a "what-if" analysis in a "trial-and-examine" procedure to search for the best possible control system design. When the Quit command is given, the designer has the option of printing a final design report, which includes a design summary and a table of open-loop, closed-loop, and error ratio data.

Next is a detailed explanation of each step in the Bode design procedure. Section 17.6 presents a process with three control loops and uses the program "DESIGN" to design a PID controller for each control loop.

Design Step 1: Derivative Mode

The derivative control mode is used to increase the maximum-frequency limit of the control system by moving the $-180°$ phase angle to a higher frequency. This means that the control system will be able to correct disturbances over a somewhat wider range of frequencies. The derivative mode also permits the use of a larger proportional gain setting, which increases the accuracy of the control system and reduces the residual offset.

Design step 1 begins with the determination of the usefulness of the derivative control mode. This determination is based on an estimate of how much the derivative mode will increase the $-180°$ frequency, which depends on the slope of the phase angle graph between the $-180°$ and $-270°$ frequencies. A steep slope results in less increase in the $-180°$ frequency than a shallow slope does. The ratio of the $-270°$ frequency divided by the $-180°$ frequency is used as a measure of the slope of the phase angle graph between those two frequencies. Tucker and Wills* suggest the following rule for determining whether the derivative control mode is useful.

---

* G. K. Tucker and D. M. Wills, *A Simplified Technique of Control System Engineering* (Philadelphia: Minneapolis-Honeywell Regulator Company, 1962), p. 63.

$\omega_{-180°}$ = frequency for which the phase angle is $-180°$

$\omega_{-270°}$ = frequency for which the phase angle is $-270°$

1. The derivative mode is definitely useful if $\dfrac{\omega_{-270°}}{\omega_{-180°}} > 5$.

2. The derivative mode is probably useful if $2 < \dfrac{\omega_{-270°}}{\omega_{-180°}} < 5$.

3. The derivative mode is marginally useful if $\dfrac{\omega_{-270°}}{\omega_{-180°}} < 2$.

Figure 17.3 is the Bode diagram of a derivative control mode with a derivative limiter coefficient (α) equal to 0.1. The limiter break-point frequency (ω_L) is equal to $1/\alpha$ times the derivative action break-point frequency.

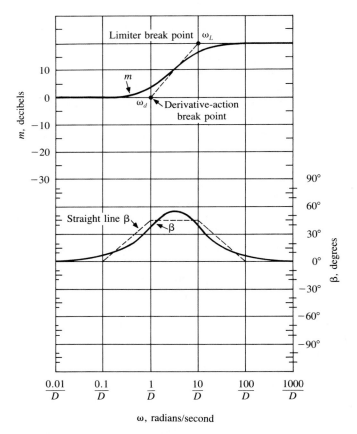

Figure 17.3   The Bode diagram of the derivative control mode displays the positive phase angle and the high-frequency amplification that characterize this mode of control.

Table 17.2 Gain and Phase Values for the Derivative Control Mode $(Ds + 1)/(0.1Ds + 1)$

ω	m (dB)	β (deg)
$0.1/D$	0	5.4
$0.2/D$	0.2	10.2
$0.5/D$	1.0	23.6
$1.0/D$	3.0	39.0
$2.0/D$	6.8	52.2
$3.16/D$	10.0	55
$5.0/D$	13.2	52.2
$10/D$	17	39
$20/D$	20	23.6
$50/D$	20	10.2
$100/D$	20	5.4

$$\omega_L = \frac{1}{\alpha D} = \frac{1}{\alpha}\,\omega_d = \frac{1}{0.1}\,\omega_d = 10\omega_d$$

Notice the positive phase angle of the derivative mode, especially between ω_d and ω_L. It is this positive phase angle that enables the derivative mode to increase the frequency at which the open-loop phase angle is $-180°$. Notice also the increase in the gain graph between ω_d and ω_L. This increase is called the *derivative amplitude*, and it is an unwanted side effect in the application of the derivative control mode. In Figure 17.3, the derivative amplitude is 20 dB, or a gain of 10. The phase-angle graph reaches a maximum value of 55° halfway between ω_d and ω_L. Other phase-angle values are given in Table 17.2.

The design of the derivative control mode consists of locating the derivative action break point such that the derivative action produces the maximum improvement in the control system. The positive phase angle of derivative action increases the phase margin. However, the gain of the derivative action reduces the gain margin, which nullifies some of the benefit of the increased phase margin. The designer must compromise between the benefit produced by the phase lead and the harm produced by the gain of the derivative mode.

In the program "DESIGN", $\alpha = 0.1$ and the first $+50°$ point on the derivative phase graph is located at the $-180°$ frequency on the open-loop Bode diagram.* This results in a derivative break frequency equal to one half the $-180°$ frequency.

$$\omega_d = 0.5\omega_{-180°} \tag{17.1}$$

The derivative action time constant is given by

$$D = \frac{1}{\omega_d} = \frac{2}{\omega_{-180°}} \tag{17.2}$$

This method tends to minimize the increase in gain without seriously reducing the increase in the $-180°$ frequency.

---

* Ibid, p. 70.

Design Step 2: Integral Mode

The integral control mode is used to increase the static accuracy of the control system by increasing the gain of the controller at low frequencies. The integral mode completely eliminates the residual offset that occurs in P and PD controllers. For stability, the integral mode gain is reduced to almost 0 dB as the frequency approaches the $-180°$ frequency and remains there for all higher frequencies. The integral mode also provides 90° of phase lag at low frequencies. This phase lag approaches 0° at about the same frequency that the gain approaches 0 dB. The high gain and 90° phase lag of the integral control mode have the potential to cause stability problems. The objective of the integral mode design is to select the integral rate, I, that provides the maximum accuracy benefit while minimizing the adverse effect on stability.

Design step 2 begins with the determination of the usefulness of the integral control mode. The integral mode is useful whenever the open-loop gain of the process does not increase as the frequency decreases toward zero. A process that displays this gain characteristic would not have an inherent integration term in its open-loop transfer function. If it did have an inherent integration term, the integral control mode would not be useful and should not be used. In fact, using the integral mode on a process that has an inherent integration term can cause serious stability problems.

The simplest way to determine the usefulness of the integral mode is to observe the process open-loop gain graph. The criteria for the usefulness of the integral mode can be stated as follows:

> If the open-loop gain of the process does not increase as the frequency approaches 0, the integral mode is useful and should be used to increase the static accuracy of the system.
>
> If the open-loop gain increases as the frequency approaches 0, the integral mode is not useful and should not be used because it would cause stability problems.

Another way to determine the usefulness of the integral mode is to observe the transfer function of each component in the process. If a component has an inherent integration term, then its transfer function will have the factor $1/s$. In other words, if $B(0) = 0$ in the transfer function of any component in the control system, the process has an integration term and the integral mode should not be used.

The criteria for the integral mode can also be stated as follows:

> If no control system component has a value of 0 for $B(0)$ in its transfer function, the integral mode is useful. If any component has a value of 0 for $B(0)$ in its transfer function, the integral mode is not useful and should not be used.*

---

* This criterion assumes that no component has a value of 0 for $A(0)$ in its transfer function. No component in this book has a value of 0 for $A(0)$, and it does not occur in normal processes, measuring transmitters, or final control elements.

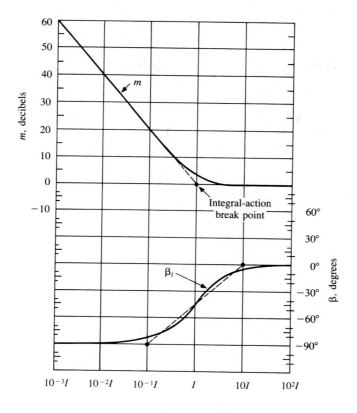

Figure 17.4   The Bode diagram of the integral control mode displays the high gain and $-90°$ phase angle at low frequencies that characterize this mode of control.

Figure 17.4 is the Bode diagram of the integral control mode. Notice the shape of the gain graph. The gain increases steadily as the frequency decreases below the integral action break point. Above the break point, the gain is 1 (0 dB). Notice that the phase angle is $-90°$ at low frequencies and changes to $0°$ at high frequencies.

The design of the integral control mode involves a compromise between the benefit of the high gain at low frequencies and the negative effect of the phase lag, also at low frequencies. The designer applies the integral mode after the derivative mode, using the open-loop Bode diagram of the process plus the derivative control mode. The design objective is to place the integral action break point such that the negative effect of the phase lag is limited to about 5% at $\omega_{-180°}$. This is accomplished by setting the integral break frequency (ω_i) equal to about one-fifth of $\omega_{-170°}$ on the open-loop Bode diagram. The integral action rate (I) is equal to the integral action break-point frequency (ω_i).

$$\omega_i = 0.2\omega_{-170°} \tag{17.3}$$

$$I = \omega_i = 0.2\omega_{-170°} \tag{17.4}$$

However, Equation (17.4) does not always produce a satisfactory result. One exception occurs with first-order lag plus dead-time processes when the first-order time constant is greater than about 10 times the dead-time delay. If Equation (17.4) is used in this situation, the combined process plus derivative plus integral phase graph dips below $-140°$ at a very low frequency as illustrated in Figure 17.5. As the frequency increases, the phase graph rises and then drops rapidly below $-180°$ as the dead time takes over. The problem becomes evident in design step 3 when the phase margin criterion is used to determine the proportional gain (P). The phase margin requires the system gain to be 0 dB at the $-140°$ frequency. The result is an extremely overdamped, sluggish system.

Figure 17.5 illustrates the problem as it occurs in the design of a process with a first-order time constant of 100 s and a dead-time lag of 2 s. In design step 1, $\omega_{-270°}/\omega_{-180°} = 2$ and we elected not to use the derivative mode. Using Equation (17.4) in step 2 of the design procedure results in an integral action rate $I = 0.136 \text{ s}^{-1}$. In step 3, the phase margin criteria force a very low proportional gain (P). In step 4 the phase margin is $40°$ and the gain margin is 51.0 dB. The extremely large gain margin is a clear indication of the problem presented by the dip in the phase graph between 0.01 and 0.1 rad/s.

The preceding design can be improved by moving the integral break point to a lower frequency. The desired break point is obtained as follows:

1. Divide $-140°$ by 1.1 to provide a 10% safety factor:

$$-140°/1.1 = -127°$$

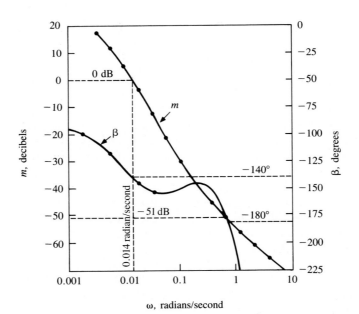

Figure 17.5 Open-loop Bode diagram of a first-order lag plus dead-time process and a PI controller (step 4). The integral mode was determined by $I = 0.2\omega_i$, which resulted in an unusually low proportional gain and a very overdampd control system.

2. Subtract the $-45°$ phase angle of the integral mode at its break point:

$$-127° - (-45°) = -82°$$

We therefore set the integral break-point frequency equal to the frequency at which the process plus derivative phase angle is $-82°$.

$$I = \omega_i = \omega_{-82°} \tag{17.5}$$

Figure 17.6 shows the result of the redesign of the control system in Figure 17.5, using Equation (17.5) instead of Equation (17.4) to determine the integral action rate (I). Notice that the dip in the phase graph has been raised above $-140°$ and $\omega_{-140°}$ is about 0.38 rad/s (compared with 0.014 rad/s in Figure 17.5). The redesigned system has a gain margin of 6 dB and a phase margin of 41°. The redesign has raised the gain by over 40 dB, and has an ideal gain and phase margin combination.

In some processes with high-order transfer functions, the $\omega_{-82°}$ value for I does not retard the integral mode break point enough to correct the excessive gain margin problem. Exercise 17.13 is an example of this type of process. In this situation, the designer may choose between a trial-and-examine procedure to determine the optimum integral mode setting and a compensation network to move the first-order

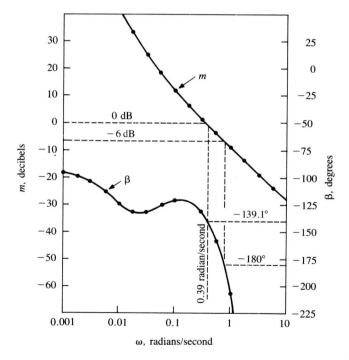

Figure 17.6   Open-loop Bode diagram of the first-order plus dead-time process from Figure 17.5 with a redesigned PI controller. The integral mode was determined by $I = \omega_{-82°}$, which resulted in a much higher proportional gain and a quarter amplitude underdamped control system.

break point to a higher frequency. The design of the compensation network is explained in Section 17.7. In either choice, program "DESIGN" is a very useful tool to carry out the design procedure.

Equations (17.4) and (17.5) are combined in the following equation for integral action rate (I).

$$I = \text{minimum of } \{\omega_{-82°} \text{ or } 0.2\omega_{-170°}\} \tag{17.6}$$

Design Step 3: Proportional Mode

In step 3, the designer uses the open-loop Bode diagram of the process plus I and D to determine the value of the proportional gain (P) that will satisfy both the gain margin and the phase margin. The gain margin is satisfied if the sum of the controller decibel gain (P_{dB}) and the Bode diagram open-loop gain at $-180°$ ($m_{-180°}$) is less than or equal to -6 dB.

$$\text{Gain margin: } P_{dB} \leq -m_{-180°} - 6 \tag{17.7}$$

The phase margin is satisfied if the sum of the controller decibel gain (P_{dB}) and the Bode diagram open-loop gain at $-140°$ ($m_{-140°}$) is less than or equal to 0 dB.

$$\text{Phase margin: } P_{dB} \leq -m_{-140°} \tag{17.8}$$

Equations (17.7) and (17.8) can be combined into the following equation for the proportional gain (P).

$$P_{dB} = \min \{-m_{-140°} \text{ or } (-m_{-180°} - 6)\} \tag{17.9}$$

The controller gain (P) is

$$P = 10^{P_{dB}/20} \tag{17.10}$$

Design Step 4: Concluding Report

The final design step uses three diagrams of the process and PID controller. The open-loop Bode diagram is used to determine the gain margin and the phase margin. The error ratio graph is used to determine the maximum frequency limit of the process (the frequency that separates zone 1 from zone 2). The closed-loop Bode diagram gives the frequency response of the control system.

PID CONTROLLER DESIGN SUMMARY

Comment: The following is a summary of the four-step procedure for the design of a PID controller using program "DESIGN".

Step 1: Derivative Mode

   Effects:

   1. Increase the maximum frequency limit.
   2. Increase the phase margin.
   3. Decrease the gain margin.

Usefulness:

1. Definitely useful if $\omega_{-270°}/\omega_{-180°} > 5$.
2. Probably useful if $2 < \omega_{-270°}/\omega_{-180°} < 5$.
3. Marginally useful if $\omega_{-270°}/\omega_{-180°} < 2$.

Design equation:

$$D = \frac{2}{\omega_{-180°}} \qquad (17.2)$$

Step 2: Integral Mode

Effects:

1. Increase the low-frequency gain.
2. Reduce the maximum frequency limit.
3. Reduce the phase margin.

Usefulness:

1. Useful if the open-loop gain of the process does not increase as the frequency approaches 0.
2. Not useful if the open-loop gain of the process increases as the frequency approaches 0.

Design equation:

$$I = \text{minimum of } \{\omega_{-82°} \text{ or } 0.2\omega_{-170°}\} \qquad (17.6)$$

Step 3: Proportional Mode

Effects:

1. Move the gain graph up or down.
2. Does not affect the phase graph.

Usefulness:

Always useful.

Design equation:

$$P_{dB} = \min \{-m_{-140°} \text{ or } (-m_{-180°} - 6)\} \qquad (17.9)$$
$$P = 10^{P_{dB}/20} \qquad (17.10)$$

Step 4: Concluding Report

Open-loop Bode diagram
Closed-loop Bode diagram
Error ratio graph
Design summary

17.6 EXAMPLE DESIGN OF A THREE-LOOP CONTROL SYSTEM

In this section, the program "DESIGN" will be used to design three controllers for the blending and heating process shown in Figure 17.7. The process blends and heats a mixture of water and syrup and has three control loops: concentration, temperature, and level. The concentration control loop manipulates the syrup input flow rate to maintain the desired concentration of syrup in the finished product. The concentration analyzer draws a sample of product from the outlet line and measures the concentration of syrup in the sample. The level control loop manipulates the water input flow rate to maintain the level of liquid in the blending tank. The temperature control system manipulates the flow rate of the heating fluid to maintain the temperature of the liquid in the blending tank.

Blending and Heating Process Specifications
1. Product
   Production rate: 2.5×10^{-4} m³/s (15 L/min)
   Composition: 90% water and 10% syrup
   Density, ρ: 1005 kg/m³
   Viscosity, μ: 0.1 Pa·s
   Specific heat, H_h: 4170 J/kg·K
2. Mixing Tank
   Diameter: 0.75 m
   Height: 1 m
   Operating level: 0.75 m
   Product temperature: 60°C
   Heating fluid temperature: 100°C
   Film coefficient $h_i = h_o$:   1349 W/m²·K
   Outlet pipe diameter: 0.0351 m

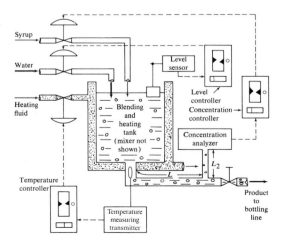

Figure 17.7 The blending and heating process has three loops to control concentration, temperature, and liquid level. Examples 17.3 to 17.5 use the program "DESIGN" to design PID controllers for these three loops.

Distance from outlet
  to temperature probe: 0.26 m
Wall thickness: 0.01 m
Wall material: steel

3. Concentration Measuring Transmitter
  Model: First-order lag plus dead time
  Time constant, ω_1:   200 s
  Dead-time delay, t_{d1}:   15 s
  Gain: 1 (% output/% input)

4. Level Measuring Transmitter
  Model: First-order lag plus dead time
  Time constant, ω_2:   2 s
  Dead-time delay, t_{d2}:   0.5 s
  Gain: 1 (% output/% input)

5. Temperature Measuring Transmitter
  Model: Overdamped second-order lag
  Time constants, τ_3:   50 s
       τ_4:   240 s
  Gain: 1 (% output/% input)

6. Control Valves
  Model: Underdamped second-order lag
  Water valve, ω_{01}:   10.2 rad/s
      ζ_1:   0.75
     Gain: 5 (% output/% input)
  Syrup valve, ω_{02}:   2.4 rad/s
      ζ_2:   0.90
     Gain: 31.25 (% output/% input)
  Heating valve, ω_{03}:   21.6 rad/s
      ζ_3:   0.8
     Gain: 8 (% output/% input)

7. Concentration Process
  Model (τ_5):   First-order lag
  Gain (G):   1 (% output/% input)

8. Level Process
  Model: Integral (nonregulating)
  FS_{in}:   0.001 m^3/s
  FS_{out}:   1 m

9. Thermal Process
  Model (τ_6, t_{d3}):   First-order lag plus dead time
  Gain (G):   1 (% output/% input)

Example 17.3

Design a PID controller for the concentration control loop in the blending and heating process, Figure 17.7.

Solution

1. Determine the concentration process time constant, τ_5

   Liquid volume, $V = $ (area)(operating height)

   $$= \frac{\pi D^2 h}{4}$$

   $$= \frac{\pi (0.75)^2 (0.75)}{4}$$

   $$= 0.3313 \text{ m}^3$$

   Liquid flow rate, $Q = 2.5 \times 10^{-4} \text{ m}^3/\text{s}$

   Time constant, $\tau_5 = \dfrac{V}{Q} = \dfrac{0.3313}{2.5 \times 10^{-4}}$

   $$= 1325 \text{ s}$$

2. Determine the syrup control valve transfer function coefficients

   $$B(1) = \frac{2\zeta_2}{\omega_{02}} = \frac{(2)(0.9)}{2.4} = 0.75$$

   $$B(2) = \frac{1}{\omega_{02}^2} = \frac{1}{2.4^2} = 0.1736$$

3. Determine the overall transfer function of the measuring transmitter, process, and control valve.

   $$\text{TF} = \left(\frac{e^{-15s}}{1 + 200s}\right)\left(\frac{1}{1 + 1325s}\right)\left(\frac{31.25}{1 + 0.75s + 0.1736s^2}\right)$$

4. The following inputs were entered in a run of program "DESIGN" to initiate the design of the concentration control system.

```
Transfer function coefficients of component number 1:
  A(0) = 1,   A(1) =    0,   A(2) = 0,   A(3) = 0
  B(0) = 1,   B(1) = 200,    B(3) = 0,   B(3) = 0

Transfer function coefficients of component number 2:
  A(0) = 1,   A(1) =     0,  A(2) = 0,   A(3) = 0
  B(0) = 1,   B(1) = 1325,   B(3) = 0,   B(3) = 0

Transfer function coefficients of component number 3:
  A(0) = 31.25,  A(1) = 0,      A(2) = 0,      A(3) = 0
  B(0) =  1,     B(1) = 0.75,   B(3) = 0.1736, B(3) = 0

Dead time delay number 1 = 15
```

Design Step 1

Figure 17.8 is the initial graphic display of program "DESIGN" after entering the above inputs. The graph shows the open-loop Bode diagram of the process and the Design Decision Data obtained from the open-loop Bode graph.

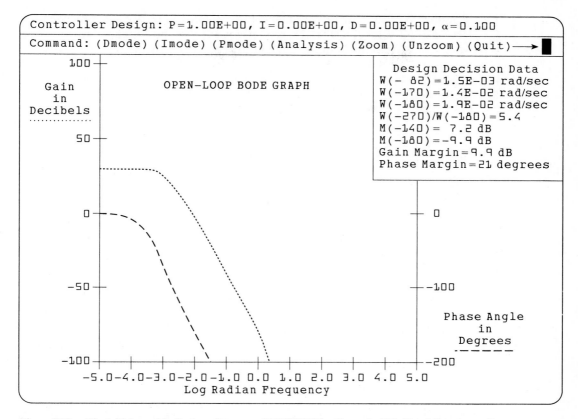

Figure 17.8   The initial graphic display of program "DESIGN" for Example 17.3. Line 2 lists the available commands. The designer initiates design step 1 by pressing the "D" key to invoke the Dmode command.

From the Design Decision Data, we see that the ratio of the $-270°$ frequency over the $-180°$ frequency is 5.4. Therefore, we conclude that the derivative mode is definitely useful. We also observe that $\omega(-180) = 0.019$ radian/second. Using Equation (17.2), we obtain the following derivative mode setting:

$$D = 2/0.019 = 105 \text{ seconds}$$

We now press the "D" key to initiate the derivative mode design. The top two lines of the screen appear as follows:

```
Derivative mode design: Dmode is definitely useful.

┌─────────────────────────────────────────────────────────┐
│  Suggested D = 1.06E+02, Current D = 0.00E+00, Enter D → ▌ │
└─────────────────────────────────────────────────────────┘
```

The first line is called the Status Line. It displays various information about the control modes. The second line is enclosed in a box that we will call the Command

Line Window. Notice that the Command Line Window includes a suggested value for D, the current value for D, and a prompt for entry of the design value of D. The suggested value of 106 for D differs slightly from the value of 105 that we computed above. This difference is due to the higher precision used by the computer to obtain the suggested value. We choose the value of 106 for D and enter that value at the D → ▮ prompt in the Command Line Window.

The Command Line Window now appears as follows:

```
Current α = 0.10      Enter desired α → ▮
```

We enter the value of 0.1 for α. As the program computes the new open-loop Bode data, it scrolls the data through the Command Line Window. When the computations are completed, the screen appears as shown in Figure 17.9.

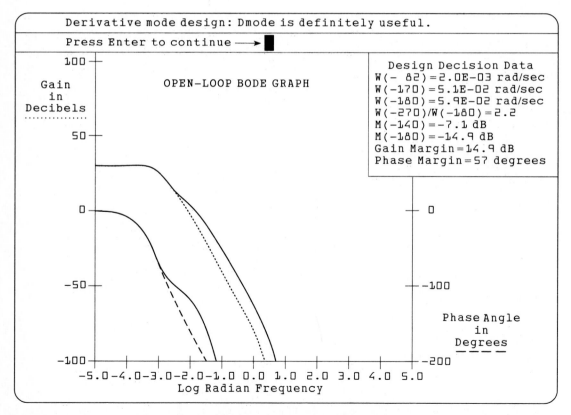

Figure 17.9 The graphic display of program "DESIGN" is on hold after completion of design step 1 for Example 17.3. The dashed lines are the original gain and angle graphs. The solid lines show how the derivative mode has changed the gain and angle graphs.

The graph in Figure 17.9 clearly shows the effect of the derivative mode on the open-loop Bode graph. The dashed lines are the original gain and phase graphs; the solid lines are the gain and phase graphs with the derivative mode included. The Design Decision Data is also updated to include the derivative mode. The screen is on hold and we conclude design step 1 by pressing the Enter key. The graph on the screen now displays the open-loop Bode graph including the derivative control mode. The Status and Command Lines appear as follows:

```
Controller Design: P = 1.00E+00, I = 0.00E+00, D = 1.06E+02, α = 0.100
```

```
Command:  (Dmode)(Imode)(Pmode)(Analysis)(Zoom)(UnZoom)(Quit) → █
```

Notice that the Status Line displays the values just entered for D and α, and the Command Line Window displays the commands shown in Figure 17.8.

Design Step 2

The graph in Figure 17.9 shows the open-loop Bode graph including the derivative mode. We observe that the gain graph levels off at the low frequency end. Therefore, we conclude that the integral mode is useful and obtain the following design data from the table in Figure 17.9:

$$\omega(-82°) = 0.002 \text{ radian/second}$$
$$\omega(-170°) = 0.051 \text{ radian/second}$$

Using Equation (17.6), we obtain the following integral mode setting:

$$I = \text{minimum } \{0.002 \text{ or } 0.2(0.051)\}$$
$$I = 0.002 \text{ 1/second}$$

We now press the "I" key to initiate the integral mode design. The Command Line Window appears as follows:

```
Suggested I = 2.00E-03, Current I = 0.00E+00, Enter I → █
```

We choose the value of 0.002 for I, and enter that value at the I → █ prompt in the Command Line Window. As before, the new open-loop Bode data scrolls through the Command Line Window. When the computations are completed, the screen appears as shown in Figure 17.10.

The graph in Figure 17.10 shows the effect of the integral mode on the open-loop Bode graph. The dashed lines are the open-loop gain and phase graphs including the derivative mode; the solid lines are the gain and phase graphs including the integral and derivative modes. The Design Decision Data is also updated to include the integral and derivative control modes. The screen is on hold and we conclude design step 2 by pressing the Enter key.

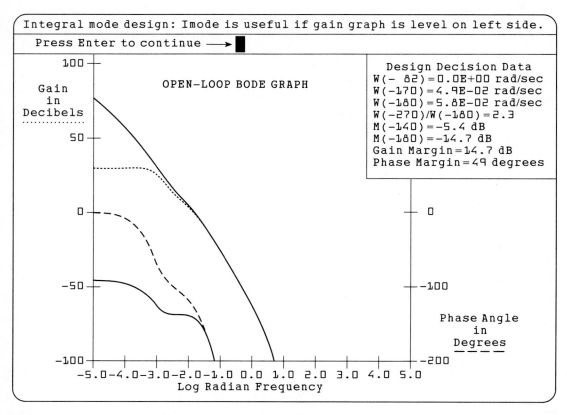

Figure 17.10   The graphic display of program "DESIGN" is on hold after completion of design step 2 for Example 17.3. The dashed lines are the gain and angle graphs before adding the integral mode. The solid lines show how the integral mode has changed the gain and angle graphs.

The graph on the screen now displays the open-loop Bode graph including the integral and derivative control modes. The Status and Command Lines appear as follows:

```
Controller Design:  P = 1.00E+00,  I = 2.00E-03,  D = 1.06E+02,  α = 0.100
```

```
Command:  (Dmode)(Imode)(Pmode)(Analysis)(Zoom)(UnZoom)(Quit) → ▮
```

Notice that the Status Line displays the value just entered for I.

Design Step 3

The graph in Figure 17.10 shows the open-loop Bode graph including the integral and derivative modes. We obtain the following design data from the table in Figure 17.10.

$$m(-140°) = -5.4 \text{ decibels}$$
$$m(-180°) = -14.7 \text{ decibels}$$

Using Equations (17.9) and (17.10), we obtain the following proportional mode setting:

$$P_{dB} = \text{minimum} \{-(-5.4) \text{ or } -(-14.7) - 6\}$$
$$P_{dB} = 5.4 \text{ decibel}$$
$$P = 10^{5.4/20} = 1.86$$

We now press the "P" key to initiate the proportional mode design. The Command Line Window appears as follows:

```
Suggested P = 1.85E+00, Current P = 1.00E+00, Enter P → ▮
```

Once again we choose the suggested value of 1.85 over the computed value of 1.86 to take advantage of the greater precision of the suggested value. We enter the value 1.85 at the P → ▮ prompt in the Command Line. The new open-loop Bode data again scrolls through the Command Line Window. When the computations are completed, the screen appears as shown in Figure 17.11.

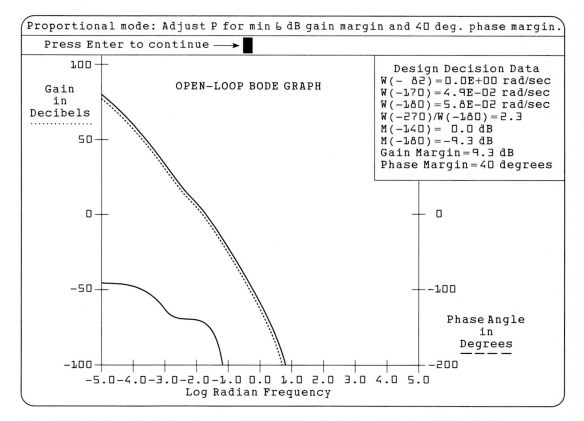

Figure 17.11   The graphic display of program "DESIGN" is on hold after completion of design step 3 for Example 17.3. The dashed lines are the gain and angle graphs before adding the proportional mode. The solid line shows how the proportional mode has changed the gain graph. Notice that the angle graph is unchanged.

The graph in Figure 17.11 shows the effect of the proportional mode on the open-loop Bode graph. The dashed lines are the open-loop gain and phase graphs including the integral and derivative modes; the solid lines are the gain and phase graphs including the proportional, integral, and derivative modes. *This is the final design open-loop Bode graph.* The Design Decision Data in Figure 17.11 is the final design data. We observe that the control system gain margin is 9.3 decibels and the phase margin is 40°. The screen is on hold, and we conclude design step 3 by pressing the Enter key.

The graph on the screen now displays the open-loop Bode graph including the PID control modes. The Status and Command Lines appear as shown below. Notice that the Status Line displays the value just entered for P.

```
Controller Design: P = 1.86E+00, I = 2.00E-03, D = 1.06E+02, α = 0.100
```

```
Command: (Dmode)(Imode)(Pmode)(Analysis)(Zoom)(UnZoom)(Quit) → ▮
```

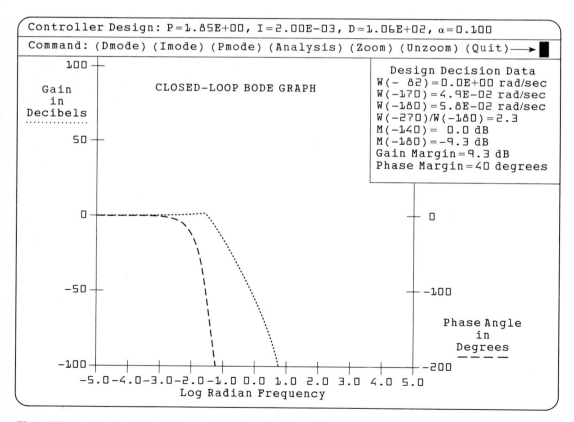

Figure 17.12 The closed-loop Bode graph for Example 17.3 as displayed by program "DESIGN" after completion of design step 3.

Design Analysis

The (Analysis) command may be used to observe the closed-loop Bode graph, the error ratio graph and the Nyquist diagram of the final design. Figure 17.12 shows the closed-loop Bode graph. Figure 17.13 shows the error-ratio graph, and Figure 17.14 shows the Nyquist diagram.

Design Summary

After entering the (Quit) command, the program presents the option to obtain a printed copy of the Bode data table and the design summary. The design summary is shown on the next page.

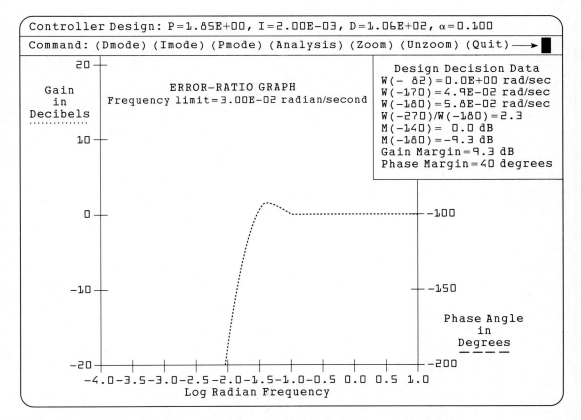

Figure 17.13   A close-up view of the error ratio graph for Example 17.3 as displayed by program "DESIGN" after completion of design step 3. The Zoom command was used to get the close-up view of the graph.

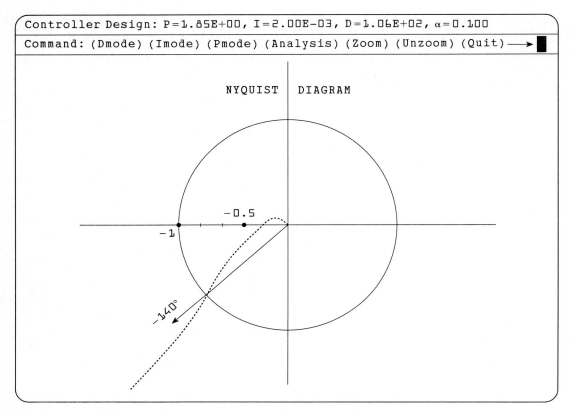

Figure 17.14 A close-up view of the Nyquist diagram for Example 17.3 as displayed by program "DESIGN" after completion of design step 3. The Zoom command was used to get the close-up view of the graph.

```
DESIGN SUMMARY
    Proportional gain:  1.85E+00
    Integral action rate:  2.00E-03 1/second
    Derivative action time constant:  1.06E+02 second
    Derivative limiter:  0.100
    M(-180):  -9.3 decibel
    ANGLE(0DB):  -140.0 degree
    Gain margin:  9.3 decibel
    Phase margin:  40 degree
    Frequency limit:  3.0E-02 radian/second
```

Example 17.4

Design a PID controller for the temperature control loop in the blending and heating process, Figure 17.7.

Solution

1. Determine the thermal process time constant, τ_6
   Thermal resistance: (from the program "THERMRES")

$$h_i = h_o = 1349 \text{ W/m}^2 \cdot \text{K}$$

$$A = \frac{\pi D^2}{4} + \pi Dh = \pi D \left(\frac{D}{4} + h \right)$$

$$= \pi(0.75) \left(\frac{0.75}{4 + 0.75} \right)$$

$$= 2.209 \text{ m}^2$$

$$x = 0.01 \text{ m}$$

$$K = 45 \text{ W/m} \cdot \text{K}$$

From "THERMRES": $R_T = 7.718\text{E} - 04 \text{ K/W}$

Thermal capacitance, $C = mH_h$

$$m = \frac{\rho h \pi D^2}{4} = \frac{(1005)(0.75)\pi(0.75)^2}{4}$$

$$= 333 \text{ kg}$$

$$C = (333)(4170) = 1.389\text{E} + 06 \text{ J/K}$$

$$\tau_6 = RC = (7.71\text{E} - 04)(1.389\text{E} + 06)$$

$$= 1070 \text{ s}$$

2. Determine the thermal dead-time lag, t_{d3}

$$\text{Outlet flow rate} = U = \frac{Q}{A}$$

$$= \frac{2.5\text{E} - 04}{0.0351^2 \pi/4}$$

$$= 0.258 \text{ m/s}$$

$$\text{Dead-time lag} = t_{d3} = \frac{\text{distance}}{\text{velocity}}$$

$$= \frac{0.26}{0.258}$$

$$= 1 \text{ s}$$

3. Determine the temperature measuring transmitter transfer function.

$$B(1) = 50 + 240 = 290$$
$$B(2) = (50)(240) = 12000$$

4. Determine the heating control valve transfer function coefficients

$$B(1) = \frac{2\zeta_3}{\omega_{03}} = \frac{(2)(0.8)}{21.6} = 0.0741$$

$$B(2) = \frac{1}{\omega_{03}^2} = \frac{1}{21.6^2} = 0.00214$$

5. Determine the overall transfer function of the measuring transmitter, process, and control valve.

$$TF = \left(\frac{1}{1 + 290s + 12,000s^2}\right)\left(\frac{e^{-s}}{1 + 1070s}\right)\left(\frac{8}{1 + 0.0741s + 0.00214s^2}\right)$$

6. The following summary was obtained from a run of the program "DESIGN".

```
DESIGN SUMMARY

    Proportional gain:  2.33E+00
    Integral action rate:  1.00E-03 1/second
    Derivative action time constant:  1.97E+02 second
    Derivative limiter:  0.100
    M(-180):  -12.7 decibel
    ANGLE(0DB):  -140.0 degree
    Gain margin:  12.7 decibel
    Phase margin:  40 degree
    Frequency limit:  1.6E-02 radian/second
```

Example 17.5

Design a PID controller for the level control loop in the blending and heating process, Figure 17.7.

Solution

1. Determine the level process integral action time constant,

$$T_i = A\frac{FS_{out}}{FS_{in}}$$

$$A = \frac{\pi D^2}{4} = \frac{\pi(0.75)^2}{4} = 0.4418$$

$$T_i = 0.4418\left(\frac{1}{0.001}\right) = 442 \text{ s}$$

2. Determine the water control valve transfer function coefficients

$$B(1) = \frac{2\zeta_1}{\omega_{01}} = \frac{(2)(0.75)}{10.2} = 0.147$$

$$B(2) = \frac{1}{\omega_{01}^2} = \frac{1}{10.2^2} = 0.00961$$

3. Determine the overall transfer function of the measuring transmitter, process, and control valve.

$$TF = \left(\frac{e^{-0.5s}}{1 + 2s}\right)\left(\frac{1}{442s}\right)\left(\frac{5}{1 + 0.147s + 0.00961s^2}\right)$$

4. The following summary was obtained from a run of the program "DESIGN".

```
DESIGN SUMMARY

    Proportional gain:  7.44E+01
    Integral action rate:  0.00E+00 1/second
    Derivative action time constant:  2.43E+00 second
    Derivative limiter:  0.100
    M(-180):  -6.0 decibel
    ANGLE(ODB):  -134.8 degree
    Gain margin:  6.0 decibel
    Phase margin:  45 degree
    Frequency limit:  1.3E-00 radian/second
```

17.7 CONTROL SYSTEM COMPENSATION

The design methods discussed in this section are used in the design of servo control systems (i.e., position, velocity, or force control of mechanical loads). The component transfer functions in a servo loop are well defined and seldom involve dead-time delays. This is quite different from process control, where transfer functions are usually not well known and dead-time delays are common.

The design of a servomechanism begins with the establishment of the performance objectives (e.g., static accuracy, response time, overshoot, stability criteria, etc.). Next the servo designer selects the system components: a servo actuator, a power supply, a drive amplifier, and a feedback transducer. The component transfer functions are then used to construct the open-loop Bode diagram or root-locus plot of the system.

The designer begins the servo controller design with an analysis of a simple loop closure. If a loop gain adjustment (proportional control mode) satisfies the performance objectives, the designer simply provides the means of establishing the necessary loop gain. This may involve adjusting the gain of one or more components in the loop, or it may involve the addition of an amplifier in the loop.

Often the simple adjustment of loop gain does not satisfy the performance objectives. The designer must then modify the open-loop frequency response of the system. This can be done in one of two ways—either change or modify components in the loop, or insert additional components into the loop. In either case, the objective is to alter the open-loop frequency response so that the performance objectives can be met with a simple gain adjustment. The second option (adding components) is known as frequency compensation or simply *compensation. The objective of compensation is to alter the open-loop frequency response so that the performance objectives can be met by a subsequent gain adjustment.* The derivative and integral modes of a PID controller are examples of compensation.

Figure 17.15 shows the Bode diagrams of the following four simple compensation networks:

1. Integration-lead network compensation

$$\frac{V_{\text{out}}}{V_{\text{in}}} = \left(\frac{1 + T_i s}{T_i s}\right) \tag{17.11}$$

2. Lead-lag network compensation

$$\frac{V_{\text{out}}}{V_{\text{in}}} = \left(\frac{1 + T_{\text{Lead}} s}{1 + T_{\text{Lag}} s}\right) \tag{17.12}$$

$$T_{\text{Lead}} > T_{\text{Lag}}$$

3. Lag network compensation

$$\frac{V_{\text{out}}}{V_{\text{in}}} = \left(\frac{1}{1 + T_{\text{Lag}} s}\right) \tag{17.13}$$

4. Lag-lead network compensation

$$\frac{V_{\text{out}}}{V_{\text{in}}} = \left(\frac{1 + T_{\text{Lead}} s}{1 + T_{\text{Lag}} s}\right) \tag{17.14}$$

$$T_{\text{Lead}} < T_{\text{Lag}}$$

The *integration-lead compensation* network is actually the same as the integral control mode in a PID controller. To verify this fact, replace T_i in Equation (17.11) by $1/I$ and rearrange the result to get the equation of the integral control mode shown below:

$$\frac{V_{\text{out}}}{V_{\text{in}}} = \left(\frac{I + s}{s}\right) \tag{17.15}$$

Integration-lead compensation is used for the same reason and in the same way as the integral control mode—to increase the static accuracy by increasing the gain at low frequencies, while preserving stability by not changing the gain at the $-180°$ frequency or above. As with the integral control mode, integration-lead compensation must not be used when the loop has an inherent integration. Serious stability problems result when this rule is not observed. The integral-action time constant, T_i, is the reciprocal of the PID integral action rate, I (i.e., $T_i = 1/I$). The rules used to

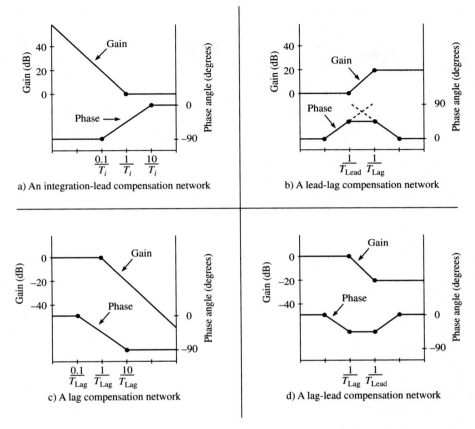

a) An integration-lead compensation network

b) A lead-lag compensation network

c) A lag compensation network

d) A lag-lead compensation network

Figure 17.15   Straight-line Bode diagrams of four compensation networks. Observe the similarity between network (a) and the integral mode, between network (b) and the derivative mode, and between network (c) and a low-pass filter.

locate I for a PID controller also apply to the determination of T_i for the integration-lead compensation network (allowing for the reciprocal relationship between T_i and I, of course).

The *lead-lag compensation* network is very similar to the derivative control mode in a PID controller. To verify this fact, take Equation (17.12) and replace T_{Lead} by D and replace T_{Lag} by αD. The result is the following equation of the derivative control mode:

$$\frac{V_{\text{out}}}{V_{\text{in}}} = \left(\frac{1 + Ds}{1 + \alpha Ds} \right) \qquad \textbf{(17.16)}$$

$$0 < \alpha < 1$$

In the lead-lag network, both the lead and the lag time constants are specified. In the derivative mode, the derivative action time constant, D, and the derivative limiter, α, are specified. In either case, the result is the specification of the two break-point frequencies: the lead network break-point, ω_{Lead}, and the lag break-point, ω_{Lag} (recall that $\omega_{\text{Lead}} = 1/T_{\text{Lead}}$ and $\omega_{\text{Lag}} = 1/T_{\text{Lag}}$). The lead-lag network is sometimes used for the same reason and in the same way as the derivative control mode. When this is done, the rules used to locate D for a PID controller also apply to the determination of T_{Lead} for the lead-lag compensation network. However, the servo designer frequently uses the lead-lag network to cancel a dominant first-order lag and replace it with a first-order lag that breaks at a higher frequency. This use of the lead-lag compensation network to cancel and replace a first-order lag is illustrated in Example 17.6.

The *lag compensation* network is actually a low-pass filter. It is used to remove undesirable high-frequency components of a signal. Often the high-frequency components are predominantly noise and are not part of the original signal.

The *lag-lead compensation* network is used to cancel dominant low-frequency lead components in much the same way that lead-lag compensation is used to cancel dominant low-frequency lag components. Note that the lead-lag and lag-lead networks have the same transfer function. The only difference is the relative size of the time constants. In the lead-lag network, the lead time constant is larger than the lag time constant ($T_{\text{Lead}} > T_{\text{Lag}}$). In the lag-lead network, the lag time constant is larger ($T_{\text{Lag}} > T_{\text{Lead}}$). The same design approach is used for both types of compensation.

In many systems, the open-loop frequency response is dominated by a few low-frequency lag elements. First-order lags are typical of temperature control systems and dc servos with short electrical time constants. Second-order lags are typical of dc servos with long electrical time constants. Faced with this type of process, a servo system designer will use lead-lag compensation to "cancel" a low-frequency lag and replace it with a higher frequency lag in the compensation network. With reasonable care in the design, lead-lag compensation can increase the dominant lag break-point frequency by a factor of 10 or more (i.e., $T_{\text{Lead}} > 10T_{\text{Lag}}$). Example 17.6 illustrates the use of integration-lead and lead-lag compensation to improve the control of a system with three predominant first-order lags.

Example 17.6

A control engineer analyzed a certain process and determined that it had three dominant first-order lags with time constants of 1000, 100, and 50 seconds. Use program "DESIGN" to complete the following five controller designs for this system. Plot the error ratio for each design on a single graph, and determine the frequency limit for each design.

a. Design a controller that uses only the proportional mode (simple gain adjustment).

b. Design a controller that uses the proportional mode plus integration-lead compensation.

c. Design a controller that uses the proportional mode plus integration-lead compensation plus lead-lag compensation that cancels the 50-second time constant and replaces it with a 5-second time constant.

d. Design a controller that uses the proportional mode plus integration-lead compensation plus lead-lag compensation that cancels the 100-second time constant and replaces it with a 10-second time constant.

e. Design a controller that uses the proportional mode plus integration-lead compensation plus two lead-lag compensation networks (one to replace the 50-second time constant with a 5-second time constant, and one to replace the 100-second time constant with a 10-second time constant).

Solution

We will use program "DESIGN" to complete the design of the five servo controllers. The error ratio graphs of the five designs are plotted in Figure 17.16. The five plots provide a very graphic way of comparing the accuracy of the five designs. We begin

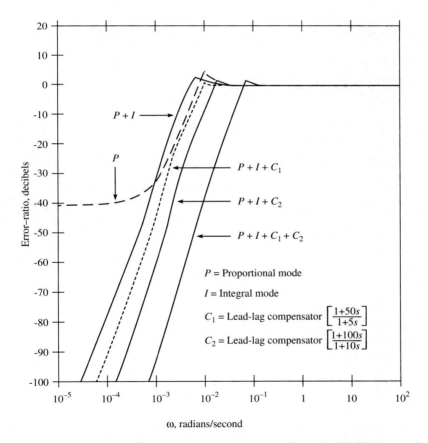

Figure 17.16   Error ratio graphs for the five controller designs in Example 17.6. The error ratio is the closed-loop error divided by the open-loop error. A low error ratio means accurate control. The lower the error ratio is, the more accurate the control is. The frequency limit is the lowest frequency at which the error ratio crosses the 0 decibel line.

by writing the transfer function of the process:

$$\text{Process TF} = \left(\frac{1}{1 + 1000s}\right)\left(\frac{1}{1 + 100s}\right)\left(\frac{1}{1 + 50s}\right)$$

a. In the first run of program "DESIGN", a Pmode value of 10.3 resulted in a gain margin of 10.8 decibels, a phase margin of 40 degrees, and a frequency limit of 0.0090 radian/second. The error-ratio graph of this design is labeled "P" in Figure 17.16.

Notice that the error ratio levels out at 40.4 dB for frequencies less than 0.0001 radian/second. This is the proportional offset error that characterizes the proportional control mode. There are two ways to reduce the proportional offset: increase the controller gain (P) or add integration-lead compensation. We could increase the gain by 4.8 dB and still have a marginally stable system (it would no longer meet the phase margin criteria). This would reduce the proportional offset, but proportional mode acting alone cannot eliminate the offset. The remaining four designs all include integration-lead compensation.

b. We will use the Imode setting in program "DESIGN" to implement the integration-lead compensation network. This is valid, since the integration-lead compensator and the integral control mode have identical transfer functions. If we wish to report T_i rather than I, we will use the fact that T_i is the reciprocal of I.

In the second run of program "DESIGN", an Imode value of 0.00207 1/seconds ($T_i = 483.1$ seconds) and a Pmode value of 4.69 resulted in a gain margin of 14.8 decibels, a phase margin of 40 degrees, and a frequency limit of 0.0055. The error-ratio graph is labeled "$P + I$" in Figure 17.16. We can make two important observations about the error ratio graph. First, notice that the error ratio continues to decrease as the frequency decreases. As the frequency approaches zero, the error ratio also approaches zero. This is the reason we use integration-lead compensation—to give high-static and low-frequency accuracy. Second, notice that the "$P + I$" graph shows a larger error ratio than the "P" graph at frequencies above 0.001 radian/second. Also, the frequency limit is reduced from 0.009 radian/second to 0.0055 radian/second. This is the cost of integration-lead compensation—it reduces the frequency limit (bandwidth) of the system and increases the intermediate frequency error.

c. In the third design, we will use the following lead-lag compensation network to cancel the 50-second time constant and replace it with a 5-second time constant:

$$\text{Compensation TF} = \left(\frac{1 + 50s}{1 + 5s}\right)$$

We can enter the compensation network in program "DESIGN" as a fourth component, giving us the following process plus compensation transfer function:

$$\text{TF} = \left(\frac{1}{1 + 1000s}\right)\left(\frac{1}{1 + 100s}\right)\left(\frac{1}{1 + 50s}\right)\left(\frac{1 + 50s}{1 + 5s}\right)$$

A simpler approach is to complete the algebraic cancellation and enter the following equivalent transfer function:

$$TF = \left(\frac{1}{1 + 1000s}\right)\left(\frac{1}{1 + 100s}\right)\left(\frac{1}{1 + 5s}\right)$$

We used the second approach in the third run of program "DESIGN". An Imode value of 0.00244 1/seconds (T_i = 409.8 seconds) and a Pmode value of 9.05 resulted in a gain margin of 25.9 decibels, a phase margin of 40 degrees, and a frequency limit of 0.0096. The error ratio graph is labeled "$P + I + C_1$" in Figure 17.16. Notice that the addition of the lead-lag compensation network regained the intermediate accuracy and bandwidth lost in design 2 and increased the low-frequency accuracy as well. The third design is more accurate than either of the first two designs.

d. In the fourth design, we will use the following lead-lag compensation network to cancel the 100-second time constant and replace it with a 10-second time constant.

$$\text{Compensation TF} = \left(\frac{1 + 100s}{1 + 10s}\right)$$

The process plus compensation transfer function can be expressed in either of the following two forms:

$$TF = \left(\frac{1}{1 + 1000s}\right)\left(\frac{1}{1 + 100s}\right)\left(\frac{1}{1 + 50s}\right)\left(\frac{1 + 100s}{1 + 10s}\right)$$

or

$$TF = \left(\frac{1}{1 + 1000s}\right)\left(\frac{1}{1 + 10s}\right)\left(\frac{1}{1 + 50s}\right)$$

We used the second form of the transfer function in the fourth run of program "DESIGN". An Imode value of 0.00303 1/seconds (T_i = 330.0 seconds) and a Pmode value of 15.4 resulted in a gain margin of 16.8 decibels, a phase margin of 40 degrees, and a frequency limit of 0.017. The error ratio graph is labeled "$P + I + C_2$" in Figure 17.16. Notice that the fourth design is more accurate than the first three designs.

e. In the fifth design, we will use the following lead-lag compensation networks to cancel both the 50-second and 100-second time constants, replacing them with 5- and 10-second time constants:

$$\text{Compensation TF} = \left(\frac{1 + 100s}{1 + 10s}\right)\left(\frac{1 + 50s}{1 + 5s}\right)$$

The process plus compensation transfer function can be expressed in either of the following two forms:

$$TF = \left(\frac{1}{1 + 1000s}\right)\left(\frac{1}{1 + 100s}\right)\left(\frac{1}{1 + 50s}\right)\left(\frac{1 + 100s}{1 + 10s}\right)\left(\frac{1 + 50s}{1 + 5s}\right)$$

or

$$TF = \left(\frac{1}{1 + 1000s}\right)\left(\frac{1}{1 + 10s}\right)\left(\frac{1}{1 + 5s}\right)$$

We used the second form of the transfer function in the fifth run of program "DESIGN". An Imode value of 0.00477 1/seconds ($T_i = 209.6$ seconds) and a Pmode value of 69.4 resulted in a gain margin of 12.5 decibels, a phase margin of 40 degrees, and a frequency limit of 0.082. The error ratio graph is labeled "$P + I + C_1 + C_2$" in Figure 17.16. Notice that the fifth design is the most accurate design and has the highest bandwidth (frequency limit).

In conclusion, we make the following observations from the results of Example 17.6:

1. If the process transfer function is dominated by two or three first-order lags, lead-lag compensation can effect significant improvements in accuracy and bandwidth.
2. The greatest improvement is achieved by lead-lag compensation networks that cancel and replace the two lags with the second- and third-largest time constants. These are the first-order lags that increase the system phase lag from 90 to 270 degrees. We ignored the largest time constant and compensated the second- and third-largest time constants because we wanted to improve the frequency response in the region near the -180 degree frequency.
3. If only one lead-lag compensation network is used, it should be applied to the second-largest time constant—the one that increases the phase lag from 90 to 180 degrees.

GLOSSARY

Bode design method: A design method that uses open-loop frequency response graphs of a control system to design control modes that alter the frequency response to obtain a stable, accurate system. (17.5)

Compensation: A design method that inserts networks into a control system to alter the open-loop frequency response of the system so that performance objectives can be met by a subsequent gain adjustment. (17.7)

Derivative amplitude: The high-frequency gain produced by the derivative control mode. (17.5)

Integration-lead compensation: Another name for the integral control mode. A network that increases static accuracy by increasing the gain at low frequencies while it preserves stability by not changing the gain at high frequencies. (17.7)

Lag compensation: Another name for a low-pass filter. Used to remove undesirable high-frequency components of a signal. (17.7)

Lag-lead compensation: A network that increases the phase lag over a range of frequencies and decreases the gain at high frequencies. Used to move the break points of first-order leads to higher frequencies. (17.7)

Lead-lag compensation: Similar to the derivative control mode. A network that decreases the phase lag over a range of frequencies and raises the gain at high frequencies. Used to increase the $-180°$ frequency. Also used to move the break points of first-order lags to higher frequencies. (17.7)

Model-based adaptive controller: A controller that uses an internal model of the process to determine the optimum control mode settings. Essentially an automated version of the process reaction method. (17.4)

Pattern-recognition adaptive controller: A controller that examines the response of the system to naturally occurring disturbances to determine the period of oscillation and the amplitude decay ratio. Formulas are then used to determine the optimum control mode settings. Essentially an automated version of the ultimate cycle method. (17.4)

Process reaction graph: The step response graph used to determine the first-order lag plus dead time model of a process. (17.3)

Process reaction method: A method of tuning an installed controller in which the step response is used to determine a first-order lag plus dead time model of the process. Formulas are then used to determine the control mode settings from the model of the process. (17.3)

Programmed adaptive controller: A controller that automatically adjusts the proportional gain (P) as a function of a number of process-related variables (e.g., controlled variable, setpoint, error, controller output, etc.). (17.4)

Self-tuning adaptive controllers: Controllers capable of changing their control mode settings to adapt to changes in the process. (17.4)

Tuning the controller: The process of determining the control mode settings of a controller during startup. (17.1)

Ultimate cycle method: A method of tuning an installed controller in which the ultimate period and ultimate gain of the system are measured and used to determine the control mode settings. (17.2)

EXERCISES

17.1 Several process control systems are tested at startup. The derivative modes are turned off, and the integral modes are set at the lowest setting. The gain of each controller is gradually increased until the control variable starts to oscillate. The gain setting and period of oscillation of each system are given below. Determine the PI and PID controller settings for each system.

System	Ultimate Gain	Ultimate Period (min)
1	0.42	20
2	6.3	6
3	0.8	2
4	1.2	18
5	2.0	0.5

17.2 During startup, the manipulating element of a process control system is maintained constant until the controlled variable levels out at 20% of the full-scale value. A 10% change is produced in the manipulating element position at time 0, and the following data are obtained.

Time (min)	C_m (%)	Time (min)	C_m (%)
0	20.0	22	35.0
2	20.0	24	36.0
4	20.0	26	37.0
6	20.0	28	37.8
8	20.3	30	38.2
10	21.8	32	38.7
12	23.6	34	39.0
14	26.0	36	39.3
16	29.0	38	39.5
18	31.5	40	39.6
20	33.4		

Construct the process reaction graph and use the process reaction method to determine the settings of a PI and a PID controller.

17.3 The maximum frequency limit (ω_{max}) of a control system is the frequency where the error ratio crosses the 0-dB line. In Example 17.3, the error ratio crosses zero between 0.032 and 0.056 rad/s, so the maximum frequency limit is between those two frequencies. Use the following linear interpolation formula to estimate a more exact value for ω_{max}.

$$\omega_{max} = \omega_1 + (\omega_2 - \omega_1)\frac{0 - ER_1}{ER_2 - ER_1}$$

where ω_{max} = maximum frequency limit, radian/second

ω_1 = frequency just before zero crossing

ω_2 = frequency just after zero crossing

ER_1 = error ratio just before zero crossing

ER_2 = error ratio just after zero crossing

17.4 A dc motor position control system has the following transfer function:

$$\frac{C_m}{V} = \frac{16.1}{s + 0.201s^2 + 3320s^3}$$

Use the program "DESIGN" to design a PID controller. Determine the maximum frequency limit from the error ratio graph.

17.5 An amplidyne position control system has the following open-loop transfer function:

$$\frac{C_m}{V} = \frac{8500}{(1 + 0.025s + 0.0001s^2)(s + 0.35s^2 + 0.015s^3)}$$

Use the program "DESIGN" to design a PID controller. Determine the maximum frequency limit from the error ratio graph.

17.6 Design a PID controller for a thermal process with the following transfer function:

$$H(s) = \left(\frac{1}{1 + 700s}\right)e^{-20s}$$

17.7 Design a PID controller to control the level in a first-order lag plus dead-time process with the following transfer function:

$$H(s) = \left(\frac{1}{1 + 25s}\right)e^{-5s}$$

17.8 Design a PID controller for a dc motor positioning system with the following transfer function:

$$H(s) = \frac{55}{s + 0.744s^2 + 0.00965s^3}$$

17.9 Design a PID controller for a solid flow control system with the following transfer function:

$$H(s) = e^{-45s}$$

17.10 A first-order lag plus dead-time blending process has a time constant of 145 s and a dead-time delay of 35 s. Design a PID controller to control the process.

17.11 Design a PID controller to control the following process:

$$H(s) = \left(\frac{0.6}{1 + 0.08s + 0.01s^2}\right)e^{-0.05s}$$

17.12 Design a PID controller for a pressure system that has the following transfer functions:

$$\text{Measuring transmitter TF} = \frac{2}{1 + 0.25s}$$

$$\text{Control valve TF} = \frac{5}{1 + 0.1s + 0.0025s^2}$$

$$\text{Process TF} = \frac{15}{1 + 5s}$$

17.13 Design a PID controller for a blending system that has the following transfer functions:

$$\text{Measuring transmitter TF} = \frac{0.5}{1 + 85s}$$

$$\text{Control valve TF} = \frac{25}{1 + 0.4s + 0.016s^2}$$

$$\text{Process TF} = \frac{12}{1 + 90s}$$

17.14 The AJ Food Company uses a PID controller to control a jacketed kettle similar to Figure 15.6a. Design a PID controller to control the temperature of the product as it leaves the heat exchanger. The transfer functions of the system components are as follows:

$$\text{Measuring transmitter TF} = \frac{2}{1 + 60s + 500s^2}$$

$$\text{Process TF} = \left(\frac{4}{1 + 750s}\right)e^{-2.5s}$$

$$\text{Control valve} = \frac{12}{1 + 0.068s + 0.0044s^2}$$

17.15 Design PID controllers for the three control loops in the blending and heating process illustrated in Figure 17.7 for the following variation in the specifications given in Section 17.6.

Production rate:	3.5×10^{-4} m^3/s
Mixing tank	
Diameter:	0.85 m
Operating level:	0.471 m
Film coefficient h_i:	90 W/m$^2 \cdot$K
Film coefficient h_o:	1800 W/m$^2 \cdot$K
Distance to temperature probe:	0.45 m
Wall thickness:	1.5 cm
Wall material:	aluminum
Concentration measuring transmitter	
Time constant τ_1	100 s
Dead-time lag t_{d1}:	25 s
Gain:	2
Level measuring transmitter	
Time constant τ_2	0.5 s
Dead-time lag t_{d2}:	0.8 s
Gain:	4
Temperature measuring transmitter	
Time constant τ_3	80 s
Time constant τ_4	220 s
Gain:	1
Water control valve	
Damping ratio ζ_1:	0.92
Resonant frequency ω_1:	18.6 rad/s
Gain:	20
Syrup control valve	
Damping ratio ζ_2:	0.84
Resonant frequency ω_2:	2.4 rad/s
Gain:	50
Heating control valve	
Damping ratio ζ_3:	0.90
Resonant frequency ω_3:	9.6 rad/s
Gain:	15

> Concentration process
>> Gain: 2
> Level process
>> FS_{in}:   0.002 m³/s
>> FS_{out}:   0.5 m
> Thermal process
>> Gain: 2

17.16 Repeat Example 17.6 for a process that has three dominant first-order lags with time constants of 200, 50, and 10 seconds.

17.17 Figure 17.5 illustrates a problem that sometimes occurs when Equation 17.4 is used to determine the integral action rate, I, in the design of a PI controller for a first-order lag plus dead time process with a time constant of 100 seconds and a dead time delay of 2 seconds. Use program "DESIGN" to complete the PI controller designs illustrated in Figures 17.5 and 17.6. Use Equation 17.4 to determine I in the first design (Figure 17.5). Use Equation 17.5 to determine I in the second design (Figure 17.6). Compare your results with Figures 17.5 and 17.6.

17.18 Repeat Exercise 17.17, but this time design a PID controller for the process. You will find that the large gain margin was not eliminated when Equation 17.5 was used to determine I. Use an iterative procedure to find a value of I that will result in a gain margin of about 6 dB. (*Hint:* Try the following values of I: 0.046, 0.069, 0.080, 0.078, and 0.079. Your final design should have a gain margin of 6.0 degrees.)

17.19 Repeat the PI design in Exercise 17.17, but this time add lead-lag compensation to replace the 100-second time constant with a 10-second time constant. Compare the error ratio graph and the frequency limit of your final design with those of the final design in Exercise 17.17. What conclusion can you make from the comparison? Of the designs in Exercises 17.17, 17.18, and 17.19, which one has the highest frequency limit?

17.20 A dc motor position control system has the following transfer function:

$$H(s) = \left(\frac{55}{s}\right)\left(\frac{1}{1 + 0.731s}\right)\left(\frac{1}{1 + 0.0132s}\right)$$

Explain why a PID controller is not a good idea for this process. Use program "DESIGN" to design a PD controller for this process. Repeat the design, but this time use a lead-lag compensation network to cancel the 0.731 time constant and replace it with a 0.0731 time constant. Then use program "DESIGN" to design a PD controller for the compensated process. Compare the two designs.

Properties of Materials

Properties of Solids

Solid	Density kilogram cubic meter	Thermal Conductivity watt meter kelvin	Heat Capacity joule kilogram kelvin
Aluminum	2,700	204	910
Asbestos	2,400	0.16	815
Asphalt	1,041	0.17	1,675
Brass	8,470	100	370
Cast iron	7,400	47	460
Copper	8,940	380	400
Glass	2,600	1	490
Gold	19,300	294	130
Graphite	2,000	5	900
Ice	900	2.25	2,000
Insulation	—	0.036	—
Lead	11,340	35	130
Nickel	8,900	60	460
Paraffin	897	0.22	2,931
Rubber	1,500	0.2	2,000
Silver	10,500	400	234
Solder (50–50)	8,842	48	168
Steel	7,800	45	500
Wood (typical oak)	740	0.2	2,400
Wood (typical pine)	440	0.15	2,800

Melting Point and Latent Heat of Fusion

Solid/Liquid	Melting Point degree Celsius	Latent Heat of Fusion kilojoule/kilogram
Aluminum	660	393
Asphalt	121	93
Ice/water	0	333
Paraffin	56	147
Solder (50–50)	216	39.5

Properties of Liquids

Liquid	Density kilogram/meter^3	Absolute Viscosity pascal second	Thermal Conductivity watt/meter kelvin	Heat Capacity joule/kg kelvin
Ethyl alcohol	800	0.0013	0.18	2,300
Gasoline	740	0.0005	0.14	2,100
Glycerine	1,260	0.83	0.29	2,400
Kerosene	800	0.0024	0.15	2,070
Mercury	13,600	0.0015	8.0	140
Oil	880	0.160	0.16	2,180
Turpentine	870	0.0015	0.13	1,720
Water	1,000	0.001	0.6	4,190

Properties of Gases[a]

Gas	Molecular Weight, M	Density kilogram/meter^3	Absolute Viscosity pascal second	Thermal Conductivity watt/meter kelvin	Heat Capacity[b] joule/kg kelvin
Hydrogen (H_2)	2.016	0.0854	8.89×10^{-6}	0.163	14200
Helium (He)	4.002	0.169	1.97×10^{-5}	0.140	
Carbon monoxide (CO)	28.0	1.19		0.023	1015
Nitrogen (N_2)	28.016	1.19	1.77×10^{-5}	0.024	1010
Air	28.8	1.22	1.81×10^{-5}	0.024	1030
Oxygen (O_2)	32.0	1.36		0.024	910
Argon (A)	39.944	1.69	2.20×10^{-5}		515
Carbon dioxide (CO_2)	44.0	1.88	1.46×10^{-5}	0.014	906

[a] At standard atmospheric conditions: 15°C and 76 cm of mercury.

[b] The heat capacity is for constant pressure.

Standard Atmospheric Conditions

1. Temperature: 288 kelvin
                          15°Celsius
                          59°Fahrenheit
2. Pressure: 1.013×10^5 pascals
                          1.013×10^6 dynes/square centimeter
                          14.7 pounds/square inch
                          76 centimeters of mercury
                          29.92 inches of mercury
                          10.336 meters of water
                          34 feet of water
3. Air density: 1.23 kilograms/cubic meter
                          1.23×10^{-3} gram/cubic centimeter
                          0.07651 pound/cubic foot

REFERENCES

The entries in the tables of properties of solids, liquids, and gases were converted to SI units from data obtained from the following sources.

Binder, R. C. *Fluid Mechanics*, 2nd edition. Englewood Cliffs, N.J.: Prentice-Hall, Inc., 1950, pp. 6, 27, 62–65.

Carmichael, Colin. *Kent's Mechanical Engineers' Handbook: Design and Production*, 12th edition. New York: John Wiley & Sons, Inc., 1950, pp. 1–04, 1–29, 2–57, 2–58, 5–78, 19–06.

Jennings, Burgess H. and Samuel R. Lewis. *Air Conditioning and Refrigeration*, 3rd edition. Scranton, Pa.: International Textbook Company, 1951, pp. 22, 99, 104–114.

Lemon, Harvey B. and Michael Ference, Jr. *Analytical Experimental Physics*. Chicago: The University of Chicago Press, 1946, pp. 37, 38, 134, 142, 177, 178, 181, 188, 219.

Salisbury, J. Kenneth. *Kent's Mechanical Engineers' Handbook: Power*, 12th edition. New York: John Wiley & Sons, Inc., 1950, pp. 2–48, 2–59, 3–04, 3–05, 3–06, 3–14, 3–15, 3–16, 3–34, 3–37, 3–58, 5–03, 6–43, 6–44.

Units and Conversion

SYSTEMS OF UNITS*

The different systems of units are best understood when applied to Newton's law of motion.

$$f = kma$$

where f = force acting on a body

m = mass of the body

a = acceleration of the body

k = constant whose value depends on the system of units

1. The SI system, $k = 1$

$$f(\text{newtons}) = m(\text{kilograms}) \cdot a(\text{meters/second}^2)$$

2. The cgs absolute system, $k = 1$

$$f(\text{dynes}) = m(\text{grams}) \cdot a(\text{centimeters/second}^2)$$

3. The fps absolute system, $k = 1$

$$f(\text{poundals}) = m(\text{pounds}) \cdot a(\text{feet/second}^2)$$

4. The engineering fps system, $k = 1/g_s = 1/32.2$

$$f(\text{pounds force}) = m(\text{pounds}) \cdot a(\text{feet/second}^2)/g_s$$

5. The engineering fss system, $k = 1$

$$f(\text{pounds force}) = m(\text{slugs}) \cdot a(\text{feet/second}^2)$$

---

* From H. B. Lemon and M. Ference, Jr., *Analytical Experimental Physics* (Chicago: The University of Chicago Press, 1946), pp. 37–38.

Conversion Factors[a]

Quantity	To Obtain:	Multiply:	By:
Area, A	meter^2	foot^2	0.09290
	foot^2	meter^2	10.764
	meter^2	inch^2	6.452×10^{-4}
	inch^2	meter^2	1550
Density, ρ	kg/meter^3	lbm/foot^3	16.018
	lbm/foot^3	kg/meter^3	0.06243
Energy (work)	joule	foot lbf	1.356
	foot lbf	joule	0.7376
	joule	Btu	1055
	Btu	joule	9.480×10^{-4}
Flow rate, Q	meter^3/second	gallon/minute	6.3088×10^{-5}
	meter^3/second	liter/minute	1.6667×10^{-5}
	gallon/minute	meter^3/second	15850.9
	gallon/minute	liter/minute	0.26418
	liter/minute	meter^3/second	60,000
	liter/minute	gallon/minute	3.7853
Force, f	newton	lbf	4.448
	lbf	newton	0.2248
Latent heat	joule/kilogram	Btu/lbm	2326
	Btu/lbm	joule/kilogram	4.2992×10^{-4}
Length, L	meter	foot	0.3048
	foot	meter	3.281
	meter	inch	0.0254
	inch	meter	39.37
Liquid resistance, R_L	$\dfrac{\text{pascal second}}{\text{meter}^3}$	psi/gpm	1.093×10^8
	psi/gpm	$\dfrac{\text{pascal second}}{\text{meter}^3}$	9.148×10^{-9}
Mass, m	kilogram	lbm	0.4536
	lbm	kilogram	2.205
Moment of inertia, I	kilogram meter^2	lbm foot^2	4.214×10^{-2}
	lbm foot^2	kilogram meter^2	23.73
	lbm foot^2	ounce inch^2	4.340×10^{-4}
	ounce inch^2	lbm foot^2	2304
	ounce inch^2	ounce inch sec^2	386.09
Power	watt	horsepower	746
	horsepower	watt	1.341×10^{-3}
	watt	Btu/hour	0.2931
	Btu/hour	watt	3.4129
Pressure, P	pascal	lbf/foot^2	47.88
	pascal	lbf/inch^2	6.895×10^3
	lbf/foot^2	pascal	0.02088
	lbf/foot^2	lbf/inch^2	144
	lbf/inch^2	pascal	1.45×10^{-4}
	lbf/inch^2	lbf/foot^2	6.9444×10^{-3}

[a] lbf = pound force, lbm = pound mass.

Conversion Factors[a] (*Continued*)

Quantity	To Obtain:	Multiply:	By:
Thermal (heat) capacity, H_h	$\dfrac{\text{joule}}{\text{kilogram kelvin}}$	Btu/lbm °F	4187
	Btu/lbm °F	$\dfrac{\text{joule}}{\text{kilogram kelvin}}$	2.388×10^{-4}
Thermal conductance, C	$\dfrac{\text{watt}}{\text{meter}^2 \text{ kelvin}}$	$\dfrac{\text{Btu}}{\text{hr ft}^2 \text{ °F}}$	5.678
	$\dfrac{\text{Btu}}{\text{hr ft}^2 \text{ °F}}$	$\dfrac{\text{watt}}{\text{meter}^2 \text{ kelvin}}$	0.17611
Thermal conductivity, K	$\dfrac{\text{watt}}{\text{meter kelvin}}$	$\dfrac{\text{Btu inch}}{\text{hr ft}^2 \text{ °F}}$	0.1441
	$\dfrac{\text{Btu inch}}{\text{hr ft}^2 \text{ °F}}$	$\dfrac{\text{watt}}{\text{meter kelvin}}$	6.938
Torque	newton meter	lbf foot	1.356
	lbf foot	newton meter	0.7376
	lbf foot	ounce inch	192
	ounce inch	lbf foot	0.005208
Viscosity, absolute, μ	pascal second	lbf second/ft²	47.88
	pascal second	lbm/ft second	1.488
	pascal second	poise	0.1
	pascal second	centipoise	0.001
	lbf second/ft²	pascal second	0.0209
	lbm/ft second	pascal second	0.672
Volume, V	liter	gallon	3.7853
	gallon	liter	0.26418
	meter³	gallon	3.7853×10^{-3}
	gallon	meter³	264.18

Digital Fundamentals

REVIEW OF BINARY ARITHMETIC

Binary Addition

The sums of all possible combinations of two binary digits are given below.

$$
\begin{array}{cccc}
0 & 0 & 1 & 1 \\
+0 & +1 & +0 & +1 \\
\hline
0 & 1 & 1 & 1\ 0 \\
\end{array}
$$

carry

The carry from $(1 + 1)$ is added to the next digit on the left.

Examples:

(1) ⟵————— carry   (1)(1)(1)(1) ⟵————— carries

$$
\begin{array}{rr}
1\ 1\ 0\ 1 & 13 \\
+\quad 1\ 0\ 0 & +\ \ 4 \\
\hline
1\ 0\ 0\ 0\ 1 & 17 \\
\end{array}
\qquad
\begin{array}{rr}
1\ 0\ 1\ 1\ 0 & 22 \\
+\ 1\ 1\ 1\ 1\ 1 & +\ 31 \\
\hline
1\ 1\ 0\ 1\ 0\ 1 & 53 \\
\end{array}
$$

Binary Subtraction

The differences of all possible combinations of two binary digits are given below.

$$
\begin{array}{cccc}
0 & 0 & 1 & 1 \\
-0 & -1 & -0 & -1 \\
\hline
0 & (-1)\ 1 & 1 & 0 \\
\end{array}
$$

borrow

The borrow from $(0 - 1)$ is subtracted from the next digit on the left.

Examples:

$$(-1) \longleftarrow \text{borrow} \qquad (-1)(-1)(-1) \longleftarrow \text{borrows}$$

1 0 1 1	11	1 0 0 0 1	17
− 1 0 1	− 5	− 1 1	− 3
1 1 0	6	0 1 1 1 0	14

Binary Multiplication

The products of all possible combinations of two binary digits are given below.

0	0	1	1
× 0	× 1	× 0	× 1
0	0	0	1

Example:

```
        1 0 1 1 0        22
      ×     1 0 1        05
        1 0 1 1 0       110
      0 0 0 0 0
    1 0 1 1 0
    1 1 0 1 1 1 0  ←——— 110 decimal
```

Binary Division

Binary division is done by the same procedure used for division of decimal numbers.

Examples:

```
110/5 = 22  R 0                 20/6 = 3   R 2
          1 0 1 1 0  R 0                    1 1   R 10
   1 0 1 ) 1 1 0 1 1 1 0           1 1 0 ) 1 0 1 0 0
          1 0 1                            1 1 0
            1 1 1                          1 0 0 0
            1 0 1                            1 1 0
              1 0 1                          0 0 1 0
              1 0 1
                1 0 1
                1 0 1
                0 0 0
```

REVIEW OF BOOLEAN ALGEBRA

The variables in Boolean algebra are called *Boolean variables* and are represented by names similar to the variable names used in ordinary algebra. Boolean variables have only two possible values, 0 or 1, and they are used to represent the inputs and outputs of the elements in a logic circuit.

Boolean algebra uses the three logic operators: AND, OR, and NOT. The three Boolean operators have the same meaning as their logic counterparts. Boolean expressions are formed by combinations of Boolean variables and Boolean operators. Examples of Boolean expressions are given below.

Boolean Expressions

$$A \qquad A + B \qquad \bar{A} \cdot B \qquad A \cdot \bar{B} + \bar{A} \cdot B$$

The AND operator has precedence over the OR operator. Thus in the expression $A \cdot B + C \cdot D$, the two AND operations are performed before the OR operation. Parentheses may be used to alter the order of operations. For example, in the expression $A \cdot (B + C) \cdot D$, the OR operation would be performed before the two AND operations.

Equality between two Boolean variables is indicated by use of the equal sign, $=$, in a Boolean equation. In Boolean algebra, the equal sign is both symmetric and transitive. By symmetric we mean that $A = B$ implies that $B = A$. By transitive we mean that $A = B$ and $B = C$ implies that $A = C$. Examples of Boolean equations are given below.

Boolean Equations

$$C = \bar{A} \cdot \bar{B} + A \cdot B$$
$$P = \bar{A} \cdot B + A \cdot \bar{B}$$
$$S = A \cdot B \cdot C + \bar{A} \cdot \bar{B} \cdot C + \bar{A} \cdot B \cdot \bar{C} + A \cdot \bar{B} \cdot \bar{C}$$

Boolean algebra is useful in both the design and analysis of logic circuits. Boolean equations provide a convenient means of representing logic circuits. Boolean algebra can be used to reduce logic equations to their simplest forms, thereby reducing to a minimum the hardware required to implement a logic function.

Rules of Boolean Algebra

Universal, Null, and Identity Rules

(1) $A + 1 = 1$ (2) $A + \bar{A} = 1$
(3) $A \cdot 0 = 0$ (4) $A \cdot \bar{A} = 0$
(5) $A \cdot A = A$ (6) $A + A = A$
(7) $A \cdot 1 = A$ (8) $A + 0 = A$

Commutative Law

(9) $A \cdot B = B \cdot A$ (10) $A + B = B + A$

Distributive Law

(11) $A \cdot (B + C) = (A \cdot B) + (A \cdot C)$
(12) $A + B \cdot C = (A + B) \cdot (A + C)$

Associative Law

(13) $(A \cdot B) \cdot C = A \cdot (B \cdot C)$
(14) $(A + B) + C = A + (B + C)$

DeMorgan's Theorem

(15) $\overline{A \cdot B} = \bar{A} + \bar{B}$

(16) $\overline{A + B} = \bar{A} \cdot \bar{B}$

Absorption

(17) $A + A \cdot B = A$

(18) $A \cdot (A + B) = A$

Double Inverse Rule

(19) $\bar{\bar{A}} = A$

Last Rule

(20) $A + \bar{A} \cdot B = A + B$

Example: Use Boolean algebra to reduce the following Boolean expression to a simpler form:

$$X = A \cdot B \cdot \bar{C} + A \cdot B \cdot C$$

Apply rule 11: $X = A \cdot B \cdot (\bar{C} + C)$

Apply rule 10: $X = A \cdot B \cdot (C + \bar{C})$

Apply rule 2: $X = A \cdot B \cdot 1$

Apply rule 7: $X = A \cdot B$

Instrumentation Symbols and Identification

The material in this section was abstracted from ANSI/ISA-S5.1-1984, "Instrumentation Symbols and Identification," copyright © Instrument Society of America, 1984, and is reprinted by permission.

1 PURPOSE

The purpose of this standard is to establish a uniform means of designating instruments and instrumentation systems used for measurement and control. To this end, a designation system that includes symbols and an identification code is presented.

2 SCOPE

2.1.2 Process equipment symbols are not part of this standard, but are included only to illustrate applications of instrumentation symbols.

2.2.1 The standard is suitable for use in the chemical, petroleum, power generation, air conditioning, metal refining, and numerous other process industries.

2.3.1 The standard is suitable for use whenever any reference to an instrument or a control system function is required for the purpose of symbolization and identification. Such references may be required for the following uses, as well as others:

Design sketches

Teaching examples

Technical papers, literature, and discussions

Instrumentation system diagrams, loop diagrams, and logic diagrams

Functional descriptions

Flow diagrams: process, mechanical, engineering, systems, piping (process), and instrumentation

Construction drawings

Specifications, purchase orders, manifests, and other lists

Identification (tagging) of instruments and control functions

Installation, operating and maintenance instructions, drawings, and records

2.3.2 The standard is intended to provide sufficient information to enable anyone reviewing any document depicting process measurement and control (who has a reasonable amount of process knowledge) to understand the means of measurement and control of the process. The detailed knowledge of a specialist in instrumentation is not a prerequisite to this understanding.

3 DEFINITIONS

Balloon: synonym for *bubble.*

Bubble: the circular symbol used to denote and identify the purpose of an *instrument* or *function.* It may contain a tag number. Synonym for *balloon.*

Computing device: a device or *function* that performs one or more calculations or logic operations, or both, and transmits one or more resultant output signals. A computing device is sometimes called a *computing relay.*

Converter: a device that receives information in one form of an instrument signal and transmits an output signal in another form. An *instrument* that changes a sensor's output to a standard signal is properly designated as a *transmitter*, not a *converter.* Typically, a temperature element (TE) may connect to a transmitter (TT), not to a converter (TY). A converter is also referred to as a *transducer;* however, "transducer" is a completely general term, and its use specifically for signal conversion is not recommended.

4 OUTLINE OF THE IDENTIFICATION SYSTEM

Comment: Each instrument or function is identified by a *tag number* that is placed inside a *balloon.*

4.1.1 Each instrument or function to be identified is designated by an alphanumeric code or tag number, as shown in Figure D.1. The loop identification part of the tag number generally is common to all instruments or functions of the loop. A suffix or prefix may be added to complete the identification. Typical identification is shown in Figure D.1.

4.2.1 The *functional identification* of an instrument or its functional equivalent consists of letters from Table D.1 and includes one first-letter (designating the measured or initiating variable) and one or more succeeding-letters (identifying the functions performed).

4.2.2 The *functional identification* of an instrument is made according to the function and not according to the construction. Thus a differential-pressure recorder used for flow measurement is identified by FR; a pressure indicator and a pressure-actuated switch connected to the output of a pneumatic level transmitter are identified by LI and LS, respectively.

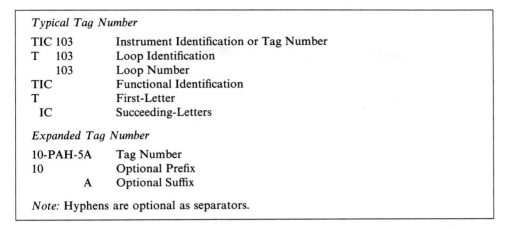

Typical Tag Number

TIC 103	Instrument Identification or Tag Number
T 103	Loop Identification
103	Loop Number
TIC	Functional Identification
T	First-Letter
IC	Succeeding-Letters

Expanded Tag Number

10-PAH-5A	Tag Number
10	Optional Prefix
A	Optional Suffix

Note: Hyphens are optional as separators.

Figure D.1   Tag numbers

4.2.3 In an instrument loop, the first-letter of the *functional identification* is selected according to the measured or initiating variable, not according to the manipulated variable. Thus a control valve varying flow according to the dictates of a level controller is an LV, not an FV.

4.2.4 The succeeding-letters of the *functional identification* designate one or more readout or passive functions and/or output functions. A modifying-letter may be used, if required, in addition to one or more other succeeding-letters. Modifying-letters may modify either a first-letter or succeeding-letters, as applicable. The TDAL contains two modifiers. The letter D changes the measured variable T into a new variable, "differential temperature." The letter L restricts the readout function A, alarm, to represent a low alarm only.

4.2.5 The sequence of *identification letters* begins with a first-letter selected according to Table D.1. Readout or passive functional letters follow in any order, and output functional letters follow these in any sequence, except that output letter C (control) precedes output letter V (valve) (e.g., PCV, a self-actuated control valve). However, modifying-letters, if used, are interposed so that they are placed immediately following the letters they modify.

4.3.1 The *loop identification* consists of a first-letter and a number. Each instrument within a loop has assigned to it the same loop number, and in the case of parallel numbering, the same first-letter. Each instrument loop has a unique loop identification. An instrument common to two or more loops should carry the identification of the loop that is considered predominant.

4.3.2 Loop numbering may be parallel or serial. *Parallel numbering* involves starting a numerical sequence for each new first-letter (e.g., TIC-100, FRC-100, LIC-100, AI-100, etc.). *Serial numbering* involves using a single sequence of numbers for a project or for large sections of a project, regardless of the first-letter of the loop identification (e.g., TIC-100, FRC-101, LIC-102, AI-103, etc.). A loop numbering sequence may begin with 1 or any other convenient number, such as 001, 301, or 1201. The number may incorporate coded information; however, simplicity is recommended.

Table D.1   Identification Letters

	First-Letter		Succeeding-Letters		
	Measured or Initiating Variable	**Modifier**	**Readout or Passive Function**	**Output Function**	**Modifier**
A	Analysis		Alarm		
B	Burner, combustion		User's choice	User's choice	User's choice
C	User's choice			Control	
D	User's choice	Differential			
E	Voltage		Sensor (primary element)		
F	Flow rate	Ratio (fraction)			
G	User's choice		Glass, viewing device		
H	Hand				High
I	Current (electrical)		Indicate		
J	Power	Scan			
K	Time, time schedule	Time rate of change		Control station	
L	Level		Light		Low
M	User's choice	Momentary			Middle, intermediate
N	User's choice		User's choice	User's choice	User's choice
O	User's choice		Orifice, restriction		
P	Pressure, vacuum		Point (test) connection		
Q	Quantity	Integrate, totalize			
R	Radiation		Record		
S	Speed, frequency	Safety		Switch	
T	Temperature			Transmit	
U	Multivariable		Multifunction	Multifunction	Multifunction
V	Vibration, mechanical analysis			Valve, damper, louver	
W	Weight, force		Well		
X	Unclassified	X axis	Unclassified	Unclassified	Unclassified
Y	Event, state or presence	Y axis		Relay, compute, convert	
Z	Position, dimension	Z axis		Driver, actuator, unclassified final control element	

4.4.1 The examples in this standard illustrate the symbols that are intended to depict instrumentation on diagrams and drawings. No inference should be drawn that the choice of any of the schemes for illustration constitutes a recommendation for the illustrated methods of measurement or control.

4.4.2 The bubble may be used to tag distinctive symbols, such as those for control valves, when such tagging is desired. In such instances, the line connecting the bubble to the instrument symbol is drawn close to, but not touching, the symbol. In other instances, the bubble serves to represent the instrument proper.

4.4.3 A distinctive symbol whose relationship to the remainder of the loop is easily apparent from a diagram need not be individually tagged on the diagram. For example, an orifice flange or a control valve that is part of a larger system need not be shown with a tag number on a diagram. Also, where there is a primary element connected to another instrument on a diagram, use of a symbol to represent the primary element on the diagram is optional.

4.4.4 A brief explanatory notation may be added adjacent to a symbol or line to clarify the function of an item.

Instrument Line Symbols

All lines are to be fine in relation to process piping lines.

1. Instrument supply* or connection to process
2. Undefined signal
3. Pneumatic signal[†]
4. Electric signal ――――――――― or
5. Hydraulic signal
6. Capillary tube
7. Electromagnetic or sonic signal (guided)[‡]
8. Electromagnetic or sonic signal (not guided)[‡]

"Or" means user's choice. Consistency is recommended.

* The following abbreviations are suggested to denote the types of power supply. These designations may also be applied to purge fluid supplies.

AS	Air supply	HS	Hydraulic supply
IA	Instrument air ⎫ options	NS	Nitrogen supply
PA	Plant air ⎭	SS	Steam supply
ES	Electric supply	WS	Water supply
GS	Gas supply		

The supply level may be added to the instrument supply line (e.g., AS-100, a 100-psig air supply; ES-24DC, a 24-volt direct-current power supply).

† The pneumatic signal symbol applies to a signal using any gas as the signal medium. If a gas other than air is used, the gas may be identified by a note on the signal symbol or otherwise.

‡ Electromagnetic phenomena include heat, radio waves, nuclear radiation, and light.

9. Internal system link (software or data link)

10. Mechanical link

The following are optional binary (on–off) symbols.

11. Pneumatic binary signal

12. Electric binary signal   or

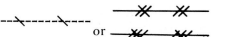

Selected Symbols

Flow rate F

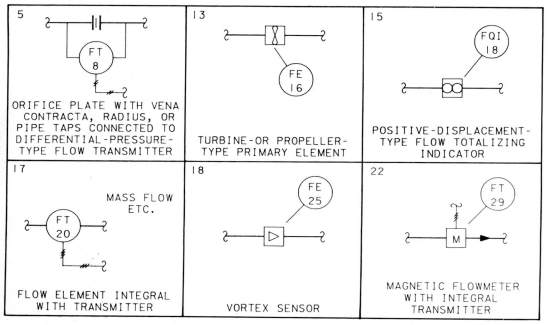

Level L

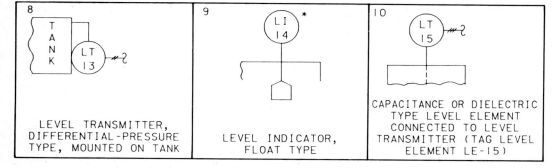

Temperature T

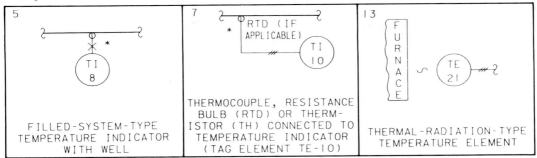

5 TI 8 *	7 RTD (IF APPLICABLE) * TI 10	13 FURNACE TE 21
FILLED-SYSTEM-TYPE TEMPERATURE INDICATOR WITH WELL	THERMOCOUPLE, RESISTANCE BULB (RTD) OR THERM-ISTOR (TH) CONNECTED TO TEMPERATURE INDICATOR (TAG ELEMENT TE-10)	THERMAL-RADIATION-TYPE TEMPERATURE ELEMENT

Speed S   Control C

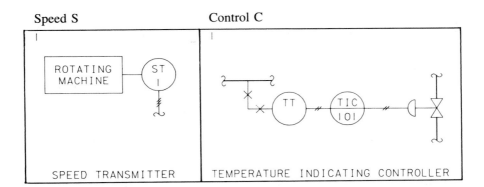

1 ROTATING MACHINE ST 1	1 TT TIC 101
SPEED TRANSMITTER	TEMPERATURE INDICATING CONTROLLER

Control C

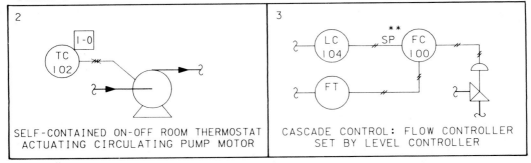

2 TC 102 1-0	3 LC 104 SP FC 100 FT
SELF-CONTAINED ON-OFF ROOM THERMOSTAT ACTUATING CIRCULATING PUMP MOTOR	CASCADE CONTROL: FLOW CONTROLLER SET BY LEVEL CONTROLLER

E

Complex Numbers

INTRODUCTION

Complex numbers were introduced to permit the extraction of the square root of negative numbers. For example, using the quadratic formula to solve the equation $x^2 - 8x + 25 = 0$ requires the extraction of the square root of the number -36. Because the square root of negative numbers is not allowed, the number -36 is written as $(-1)(36)$, the symbol j is defined as the square root of (-1), and the square root of $(-1)(36)$ is written as $j6$.* Thus the j operator permits the extraction of negative numbers, and the equation $x^2 - 8x + 25 = 0$ has the following solution:

$$x_1, x_2 = \frac{8 \pm \sqrt{(-8)^2 - 4(1)(25)}}{2(1)} = \frac{8 \pm \sqrt{-36}}{2} = 4 \pm j3$$

Thus $(4 + j3)$ and $(4 - j3)$ are the two values of x that satisfy the equation $x^2 - 8x + 25 = 0$. Values such as $(4 + j3)$ and $(4 - j3)$ belong to a set called complex numbers. Notice that complex numbers may have two components. The first component is a real number; it is called the real part of the complex number. The second component includes the j operator and a real number; the real number is called the imaginary part of the complex number (the j operator indicates the imaginary component, but it is not considered to be part of it). The complex number $(4 + j3)$ has the real part 4 and the imaginary part 3. A complex number may have both a real and an imaginary part, only a real part, or only an imaginary part. The following are examples of complex numbers:

$$(10 + j20) \qquad (10) \qquad (j20)$$

---

* In mathematics, the symbol i is used as the square root of (-1). In electrical engineering, i is used as the symbol for current, and j is used to denote $\sqrt{-1}$.

RECTANGULAR AND POLAR FORMS OF COMPLEX NUMBERS

There are two ways to designate complex numbers: the rectangular form and the polar form. In the introduction, the rectangular form was used to designate the complex numbers $(4 + j3)$ and $(4 - j3)$. The polar form of a complex number is based on Euler's identity, which is stated in the following two equations:

$$(e^{+j\theta} = \cos\theta + j\sin\theta)$$

and

$$(e^{-j\theta} = \cos\theta - j\sin\theta)$$

We will use Euler's identity and a rearrangement of the complex number $(N = 4 + j3)$ to obtain N in its polar form.

We begin by defining θ as the angle whose tangent is equal to the imaginary part of N divided by the real part of N.

$$\theta = \arctan(\tfrac{3}{4}) = 36.87° \tag{E.1}$$

Next we define c as the square root of the sum of the squares of the real and imaginary parts of N.

$$c = \sqrt{4^2 + (-3)^2} = \sqrt{25} = 5 \tag{E.2}$$

Then we multiply and divide N by $c = 5$.

$$N = 4 + j3 = 5(\tfrac{4}{5} + j\tfrac{3}{5}) = 5(0.8 + j0.6)$$

Now observe the following:

$$\tfrac{4}{5} = 0.8 = \cos(36.87°) = \cos\theta$$
$$\tfrac{3}{5} = 0.6 = \sin(36.87°) = \sin\theta$$

and

$$N = 5(\cos 36.87° + j\sin 36.87°) \tag{E.3}$$

Applying Euler's identity to Equation (E.3), we obtain the following polar form of N:

$$N = 5e^{j36.87°} \tag{E.4}$$

The symbol $\underline{/\theta}$ is frequently used in place of $e^{j\theta}$, and the polar form of N is usually written as follows:

$$N = 5\underline{/36.87°} \tag{E.5}$$

Equation (E.5) is more convenient for writing and printing the polar form, but Equation (E.4) is more useful in determining how to multiply and divide complex numbers and how to raise a complex number to a power. These operations are all based on the rules of exponents and will be covered in the sections about operations on complex numbers.

CONVERSION OF COMPLEX NUMBERS

Euler's identity is also the basis for the conversion of complex numbers from rectangular to polar form or vice versa. The conversion formulas are summarized below.

Conversion of: $N = a + jb = c\underline{/\theta}$

Rectangular to polar form:

$$c = \sqrt{a^2 + b^2} \tag{E.6}$$

$$\theta = \arctan(b/a) \tag{E.7}$$

Polar to rectangular form:

$$a = c(\cos \theta) \tag{E.8}$$

$$b = c(\sin \theta) \tag{E.9}$$

GRAPHICAL REPRESENTATION OF COMPLEX NUMBERS

Complex numbers are represented graphically by a point on a two-dimensional graph called the complex plane. Figure E.1 shows the graphical representation of the number $N = 4 + j3 = 5\underline{/36.87°}$.

Either form of a complex number may be used to locate the point on the complex plane that represents the number (N). Using the rectangular form, the real part of N ($a = 4$) is plotted on the horizontal axis, which is called the real axis. The imaginary part of N ($b = 3$) is plotted on the vertical axis, which is called the imaginary axis.

Using the polar form, the angle of N ($\theta = 36.87°$) is measured in a counterclockwise direction from the positive real axis. The magnitude of N ($c = 5$) is measured

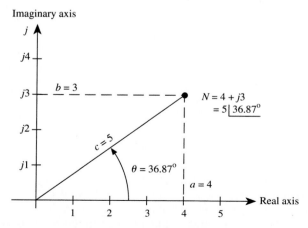

Figure E.1 Graphical representation of the complex number $N = 4 + j3$.

from the origin to the point N, which lies on the line determined by the angle θ.

Complex numbers may be in any of the four quadrants of the complex plane as shown in Figure E.2. The complex number $N = a + jb$ will be in the first quadrant if both a and b are positive; in the second quadrant if a is negative and b is positive; in the third quadrant if a and b are both negative; and in the fourth quadrant if a is positive and b is negative.

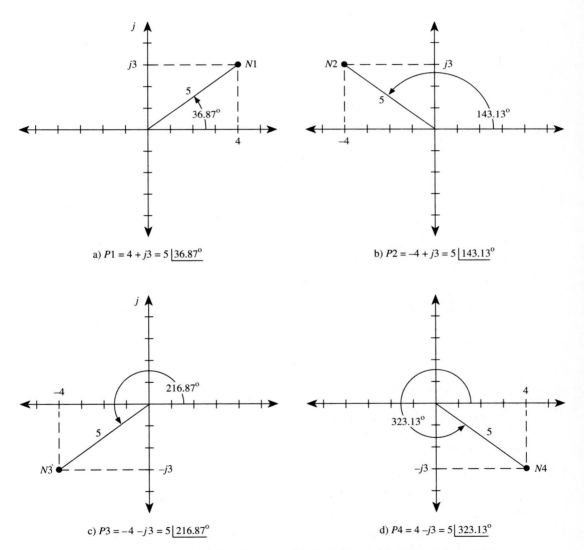

a) $P1 = 4 + j3 = 5\underline{|36.87^\circ}$

b) $P2 = -4 + j3 = 5\underline{|143.13^\circ}$

c) $P3 = -4 - j3 = 5\underline{|216.87^\circ}$

d) $P4 = 4 - j3 = 5\underline{|323.13^\circ}$

Figure E.2 The signs of the real and imaginary parts determine which quadrant the complex number will be in.

The *conjugate* of a complex number is formed by reversing the sign of the imaginary part. In Figure E.2, $P1$ and $P4$ are conjugates of each other. Also, $P2$ and $P3$ are conjugates of each other. Graphically, the conjugate of a complex number is its mirror image about the real axis.

ADDITION AND SUBTRACTION OF COMPLEX NUMBERS

Complex numbers are most conveniently added when they are expressed in the rectangular form. The sum of two complex numbers is formed by adding the real parts to obtain the real part of the sum and then adding the imaginary parts to form the imaginary part of the sum. The following example illustrates the addition of two complex numbers:

$$N1 = 5 + j12$$
$$N2 = 4 - j3$$
$$N3 = N1 + N2 = (5 + 4) + j(12 - 3) = 9 + j9$$

Subtraction of $N2$ from $N1$ is accomplished by changing the sign of both parts of $N2$ and then adding the result to $N1$.

$$N4 = N1 - N2 = (5 - 4) + j(12 + 3) = 1 + j15$$

Numbers that are in polar form must be converted to rectangular form before they can be added or subtracted. A pocket calculator with the polar to rectangular and rectangular to polar functions is the most convenient way to convert complex numbers from one form to the other.

MULTIPLICATION AND DIVISION OF COMPLEX NUMBERS

Complex numbers are most conveniently multiplied or divided when they are expressed in polar form, but the operations can also be carried out in the rectangular form.

In polar form, the product of two complex numbers is obtained by multiplying the magnitudes of the two numbers and adding their angles. The quotient of two complex numbers is obtained by dividing their magnitudes and subtracting the angle of the divisor from the angle of the dividend. The following examples illustrate multiplication and division of complex numbers:

$$N5 = 5 + j12 = 13\underline{/67.38°}$$
$$N6 = 4 + j3 = 5\underline{/36.87°}$$
$$N5 \cdot N6 = (13 \cdot 5)\underline{/67.38° + 36.87°} = 65\underline{/104.25°}$$
$$N5/N6 = (\tfrac{13}{5})\underline{/67.38° - 36.87°} = 2.6\underline{/30.51°}$$

The rules for multiplying and dividing complex numbers are based on the rules of exponents. This is more obvious if we use the exponential version of the polar form, as the following examples illustrate:

$$N5 \cdot N6 = (13e^{j67.38°}) \cdot (5e^{j36.87°}) = (13)(5)e^{j(67.38° + 36.87°)}$$
$$= 65e^{j104.25°}$$

$$N5/N6 = (13e^{j67.38°})/(5e^{j36.87°}) = (\tfrac{13}{5})e^{j(67.38° - 36.87°)}$$
$$= 2.6e^{+j30.51°}$$

The following example illustrates the multiplication of $N5$ times $N6$ in rectangular form:

$$N5 \cdot N6 = (5 + j12)(4 + j3)$$
$$= 5(4 + j3) + j12(4 + j3)$$
$$= 20 + j15 + j48 + j^2 36 \qquad \{\text{Note: } j^2 = -1\}$$
$$= -16 + j63$$
$$= 65\underline{/104.25°}$$

The first step in dividing two complex numbers in rectangular form is to multiply both the numerator and the denominator by the conjugate of the denominator. This reduces the denominator to a real number. The final step is to divide this real number into the new numerator. The following example illustrates division of $N5$ by $N6$ in rectangular form:

$$\frac{N5}{N6} = \frac{5 + j12}{4 + j3} = \frac{(5 + j12)}{(4 + j3)} \cdot \frac{(4 - j3)}{(4 - j3)}$$
$$= \frac{56 + j33}{25} = 2.24 + j1.32 = 2.6\underline{/30.51°}$$

INTEGER POWER OF A COMPLEX NUMBER

Raising the complex number N to an integer power k can be viewed as the product of k factors, each equal to N. The result is a complex number whose magnitude is equal to the magnitude of N raised to the k power and whose angle is k times the angle of N.

$$N^k = (ce^{j\theta})^k = c^k e^{jk\theta}$$

For example,

$$(3\underline{/15°})^4 = 3^4\underline{/4 \cdot 15°} = 81\underline{/60°}$$

ROOTS OF A COMPLEX NUMBER

Finding the kth root of a complex number is equivalent to solving the equation

$$R^k = ce^{j\theta} \qquad \text{(E.10)}$$

where R is the kth root of the complex number $ce^{j\theta}$. Since Equation (E.10) is an equation of degree k, there will be k roots, $R_1, R_2, \ldots, R_k$. To find the k roots, we note that complex numbers are circular numbers that repeat as the angle is increased by multiples of 2π radians (or 360°). In other words,

$$ce^{j\theta} = ce^{j(\theta + 2\pi)} = ce^{j(\theta + 4\pi)} = ce^{j(\theta + 6\pi)} = \cdots \qquad \textbf{(E.11)}$$

The roots all have the same magnitude, $|R| = c^{1/k}$. The angles of the roots are obtained by dividing k into each of the angles of the complex numbers in Equation (E.11). Enough terms must be used to obtain the k roots. Additional terms will simply repeat the first k roots.

For example, let us find the 4th root of $16\underline{/80°}$. We know there will be 4 roots, so we write the following 4 equivalent complex numbers:

$$16\underline{/80°} = 16\underline{/(80 + 360)°} = 16\underline{/(80 + 720)°} = 16\underline{/(80 + 1080)°}$$

The magnitude of the roots will be 16 raised to the $\frac{1}{4}$ power.

$$|R| = 16^{1/4} = 2$$

The angles of the roots will be:

$$\tfrac{80}{4} = 20°, \ (80 + 360)/4 = 110°,$$
$$(80 + 720)/4 = 200°, \text{ and } (80 + 1080)/4 = 290°$$

The four roots are: $2\underline{/20°}$, $2\underline{/110°}$, $2\underline{/200°}$, and $2\underline{/290°}$.

Program Listings

PROGRAM "LIQRESIS"

```
1     'Program "Liquid Resistance" - File Name "LIQRESIS"
2     '  Computes the liquid flow resistance and pressure drop in
3     '  a pipe for both laminar and turbulent flow conditions.
4     '
5     '----------------- GLOBAL VARIABLES --------------------
6     ' DIA = Inside diameter of the pipe
7     ' FF = Friction factor
8     ' FLOW$ = Type of flow (laminar, turbulent, transition)
9     ' L = Length of the pipe
10    ' KT = Turbulent flow coefficient
11    ' MU = Absolute viscosity of the fluid
12    ' NUM$ = Number format for print using string
13    ' PL = Pressure drop for laminar flow
14    ' PR = Pressure drop for laminar or turbulent flow
15    ' PR$ = Format string for printing pressure
16    ' PRU$ = Units used in PR$
17    ' PT = Pressure drop for turbulent flow
18    ' Q = Fluid flow rate
19    ' RES = Resistance for laminar or turbulent flow
20    ' RE$ = Format string for printing resistance
21    ' REU$ = Units used in RE$
22    ' REY = Reynolds number
23    ' RHO = Density of the fluid
24    ' RL = Resistance for laminar flow
25    ' RT = Resistance for turbulent flow
26    ' TP$ = Type of line (tube or pipe)
27    ' TPR$ = Format string for transition pressure
28    ' TRE$ = Format string for transition resistance
29    ' UNIT$ = Type of units (SI or English)
39    '
40    '----------------- CONVERSION FACTORS -------------------
41    GPM2CMS = 6.3088E-05 ' gallon/min to cubic meter/second
42    LPM2CMS = 1.6667E-05 ' liter/min to cubic meter/second
43    INCH2METER = .0254    ' inch to meter
44    FEET2METER = .3048    ' feet to meter
45    VISCON = 1.488164     ' lb/ft second to pascal second
```

```
46    DENSCON = 16.018      ' lb/cubic foot to kg/cubic meter
47    RESCON = 9.148E-09    ' Pa sec/cubic meter to psi/gpm
48    PRESCON = .000145     ' Pascal to psi
49    '
60    '-------------------- CONSTANTS ----------------------
61    PI = 3.14159          ' circumference/diameter
99    '
100   '================== MAIN PROGRAM ======================
110    GOSUB 1000  '  Print the input screen
120    GOSUB 2000  '  Get the input parameters
130    GOSUB 3000  '  Get the Reynolds number and type of flow
140    GOSUB 4000  '  Compute resistance and pressure drop
150    GOSUB 5000  '  Print the output screen
160    END
170   '=====================================================
180   '
1000  '============== PRINT THE INPUT SCREEN ===============
1010   CLS : KEY OFF
1020   LOCATE 2,18 : COLOR 15
1025   PRINT "LIQUID FLOW RESISTANCE AND PRESSURE DROP";
1030   LOCATE 4,10 : COLOR 7
1035   PRINT "Smooth tubing [T] or commercial pipe [P] -------->";
1040   LOCATE 5,10
1045   PRINT "SI units [S] or English units [E] -------------->";
1050   LOCATE 6,10
1055   PRINT "Flow rate ------------------------------------->";
1060   LOCATE 7,10
1065   PRINT "Inside diameter ------------------------------->";
1070   LOCATE 8,10
1075   PRINT "Length ---------------------------------------->";
1080   LOCATE 9,10
1085   PRINT "Absolute viscosity ---------------------------->";
1090   LOCATE 10,10
1095   PRINT "Fluid density --------------------------------->";
1098   RETURN
1099  '
2000  '================= GET THE INPUT DATA ==================
2010   LOCATE 4,62,1              ' Get type of line (tube or pipe)
2011   TP$ = INPUT$(1)
2012   IF TP$ = "t" THEN TP$ = "T"
2013   IF TP$ = "p" THEN TP$ = "P"
2014   IF (TP$ <> "T") AND (TP$ <> "P") THEN GOSUB 2800 : GOTO 2010
2015   IF TP$ = "T" THEN TP$ = "TUBE" ELSE TP$ = "PIPE"
2016   PRINT TP$
2017   GOSUB 2900
2018   '
2020   LOCATE 5,62                ' Get type of units (SI or English)
2021   UNIT$ = INPUT$(1)
2022   IF UNIT$ = "s" THEN UNIT$ = "S"
2023   IF UNIT$ = "e" THEN UNIT$ = "E"
2024   IF (UNIT$ <> "S") AND (UNIT$ <> "E") THEN GOSUB 2800 : GOTO 2020
2025   IF UNIT$ = "S" THEN UNIT$ = "SI" ELSE UNIT$ = "ENGLISH"
2026   PRINT UNIT$;
2027   GOSUB 2900
2028   '
2030   LOCATE 7,26 : PRINT "OF "; TP$      ' Print type of line
2031   LOCATE 8,17 : PRINT "OF "; TP$      '   on the screen
2032   IF UNIT$ = "SI" THEN GOSUB 2400 ELSE GOSUB 2500 ' Print units
2039   '
```

```
2040   ROW = 6  : GOSUB 2200      ' Get valid flow rate, Q
2045     IF UNIT$ = "SI" THEN Q = X*LPM2CMS ELSE Q = X*GPM2CMS
2050   ROW = 7  : GOSUB 2200      ' Get valid diameter, DIA
2055     IF UNIT$ = "SI" THEN DIA = X/100 ELSE DIA = X*INCH2METER
2060   ROW = 8  : GOSUB 2200      ' Get valid length, L
2065     IF UNIT$ = "SI" THEN L = X ELSE L = X*FEET2METER
2070   ROW = 9  : GOSUB 2200      ' Get valid viscosity, MU
2075     IF UNIT$ = "SI" THEN MU = X ELSE MU = X*VISCON
2080   ROW = 10 : GOSUB 2200      ' Get valid density, RHO
2085     IF UNIT$ = "SI" THEN RHO = X ELSE RHO = X*DENSCON
2090   RETURN
2095   '
2200   '---------------- GET VALID INPUT X --------------------
2210    LOCATE ROW,62 : INPUT "", N$
2220    IF VAL(N$) <> 0 THEN X = VAL(N$) ELSE GOSUB 2800 : GOTO 2210
2230    RETURN
2240    '
2400   '----------------- PRINT SI UNITS ----------------------
2410   LOCATE 6,19 : PRINT " in liters/minute "
2420   LOCATE 7,33 : PRINT " in centimeters "
2430   LOCATE 8,24 : PRINT " in meters "
2440   LOCATE 9,28 : PRINT " in Pascal seconds "
2450   LOCATE 10,23 : PRINT " in kilograms/cubic meter "
2460   RETURN
2500   '---------------- PRINT ENGLISH UNITS --------------------
2510   LOCATE 6,19 : PRINT " in gallons/minute "
2520   LOCATE 7,33 : PRINT " in inches "
2530   LOCATE 8,24 : PRINT " in feet "
2540   LOCATE 9,28 : PRINT " in pounds/foot second "
2550   LOCATE 10,23 : PRINT " in pounds/cubic foot "
2560   RETURN
2800   '--------------- PRINT ERROR MESSAGE -------------------
2810   LOCATE 25,25 : COLOR 14
2820   PRINT "INVALID INPUT, PLEASE TRY AGAIN";
2830   BEEP : COLOR 7
2840   RETURN
2900   '--------------- REMOVE ERROR MESSAGE ------------------
2910   LOCATE 25,25
2920   PRINT "                                        ";
2930   RETURN
2940   '
3000   '========= GET REYNOLDS NUMBER AND TYPE OF FLOW =========
3010    VELOCITY = 4*Q/(PI*DIA*DIA)
3020    REY = RHO*VELOCITY*DIA/MU      ' REYNOLDS NUMBER
3030    IF REY <= 2000 THEN FLOW$ = "Flow is laminar"
        ELSE IF REY >= 4000 THEN FLOW$ = "Flow is turbulent"
        ELSE FLOW$ = "Flow is in transition"
3040    RETURN
3050    '
4000   '========= COMPUTE RESISTANCE AND PRESSURE DROP =========
4010    RL = 128*MU*L/(PI*DIA^4)
4020    PL = RL*Q
4030    GOSUB 4200   ' GET FRICTION FACTOR
4040    KT = 8*RHO*FF*L/(PI*PI*DIA^5)
4050    RT = 2*KT*Q
4060    PT = KT*Q*Q
4070    IF UNIT$ = "ENGLISH" THEN GOSUB 4600 ' Convert units
4080    IF REY <= 2000 THEN RES = RL : PR = PL
4090    IF REY >= 4000 THEN RES = RT : PR = PT
```

```
4100   RETURN
4110   '
4200   '---------------- GET THE FRICTION FACTOR ---------------
4201   'Interpolates the friction factor from Moody's table of
4202   '  "Friction Factors for Pipe Flow"  ASME Transactions, 66,
4203   '  #8, Nov. 1944, pp 671.
4204   'Inputs are Reynolds number, pipe diameter in meters, and
4205   '  the type of pipe (tubing or commercial pipe)
4206   'Output is the friction factor, FF
4207   '
4210   'Set R = Reynolds number exponent
4211       R = 7 ' Initially set R to the largest allowable value
4212       IF REY < 10^7 THEN R = 6
4213       IF REY < 10^6 THEN R = 5
4214       IF REY < 10^5 THEN R = 4
4215       IF REY < 10^4 THEN R = 3
4216   '
4220   'Set S = size of the tubing of pipe
4221       S = 4 ' Initially set S to the largest allowable size
4222       IF DIA < .08 THEN S = 3
4223       IF DIA < .04 THEN S = 2
4224       IF DIA < .02 THEN S = 1
4225   '
4230   'Set T = Type of line (tube or pipe)
4235       IF TP$ = "TUBE" THEN T = 1 ELSE T = 2
4239   '
4240   'Read the Friction Table into array FT
4245   FOR TTYPE = 1 TO 2
4250       FOR SIZE = 1 TO 4
4255           FOR EXPONENT = 3 TO 8
4260               READ FT(TTYPE,SIZE,EXPONENT)
4265           NEXT EXPONENT
4270       NEXT SIZE
4275   NEXT TTYPE
4280   RESTORE
4285   '
4290   'Interpolate friction factor from friction table array FF
4295       FF1 = FT(T,S,R+1) - FT(T,S,R)
4300       FF2 = (REY - 10^R)/(10^(R+1) - 10^R)
4305       FF = FT(T,S,R) + FF1*FF2
4310   RETURN
4315   '
4400   '------------- FRICTION FACTORS FOR PIPE FLOW -----------
4401   'The first column of the table below was extrapolated
4402   'from Table 4.3 to extend the table to a Reynolds number
4403   'of 1E03.  The extrapolation is such that values of f
4404   'for a Reynolds number of 4E03 interpolate from this
4405   'table to the values in the 4E03 column in table 4.3.
4406   '
4411   DATA 0.0435, 0.030, 0.018, 0.014, 0.012, 0.012
4412   DATA 0.0435, 0.030, 0.018, 0.013, 0.011, 0.010
4413   DATA 0.0435, 0.030, 0.018, 0.012, 0.010, 0.009
4414   DATA 0.0435, 0.030, 0.018, 0.012, 0.009, 0.008
4415   '
4416   DATA 0.0440, 0.035, 0.028, 0.026, 0.026, 0.026
4417   DATA 0.0435, 0.033, 0.024, 0.023, 0.023, 0.023
4418   DATA 0.0435, 0.030, 0.022, 0.020, 0.019, 0.019
4419   DATA 0.0435, 0.030, 0.020, 0.018, 0.017, 0.017
4420   '
```

```
4600  '------------- CONVERT SI UNITS TO ENGLISH -------------
4610    RL = RL*RESCON    :    RT = RT*RESCON
4620    PL = PL*PRESCON   :    PT = PT*PRESCON
4630    RETURN
4640    '
5000  '=============== PRINT THE OUTPUT SCREEN ==================
5010    IF UNIT$ = "SI" THEN REU$= " Pascal second/cubic meter"
                          ELSE REU$ = " psi/gpm"
5020    IF UNIT$ = "SI" THEN PRU$ = " Pascal"
                          ELSE PRU$ = " psi"
5030    NUM$ = "##.###^^^^"
5040    RE$ = "Resistance =" + NUM$ + REU$
5050    PR$ = "Pressure drop =" + NUM$ + PRU$
5060    TRE$ = "Resistance =" + NUM$ + " TO" + NUM$ + REU$
5070    TPR$ = "Pressure drop =" + NUM$ + " TO" + NUM$ + PRU$
5080    LOCATE 14,10 : COLOR 14
5090    PRINT USING "Reynolds number =" + NUM$ + ", " + FLOW$; REY;
5100    IF FLOW$ = "Flow is in transition" THEN GOSUB 5400
                                            ELSE GOSUB 5200
5110    COLOR 7
5120    RETURN
5130    '
5200  '------- PRINT LAMINAR OR TURBULENT FLOW OUTPUT ---------
5210    LOCATE 15,10
5220    PRINT USING RE$; RES
5230    LOCATE 16,10
5240    PRINT USING PR$;PR
5250    RETURN
5260    '
5400  '----------- PRINT TRANSITION FLOW OUTPUT --------------
5410    LOCATE 15,10
5420    PRINT USING TRE$; RL,RT
5430    LOCATE 16,10
5440    PRINT USING TPR$; PL, PT
5450    RETURN
5460    '
```

PROGRAM "THERMRES"

```
1     'PROGRAM "Thermal Resistance" - FILE NAME "THERMRES"
2     '  Computes the thermal resistance and heat flow in a
3     '  composite wall section with surface area A.
5     '================== GLOBAL VARIABLES ==================
6     ' A     = area of the surface
7     ' C(I) = choice of film coefficient
8     ' D     = inside diameter of a pipe
9     ' H(0) = outside film coefficient
10    ' H(1) = inside film coefficient
11    ' I     = counter and subscript
12    ' K(I) = thermal conductivity of layer I
13    ' N     = number of layers in the composite wall
14    ' Q     = total heat flow
15    ' RU    = thermal resistance of one unit of area
16    ' RT    = total thermal resistance of entire surface
17    ' TD    = temperature difference
18    ' T0    = outside temperature
19    ' T1    = inside temperature
```

```
20  ' TW    = water temperature
21  ' V     = velocity
22  ' VI    = absolute viscosity
23  ' X(I)  = thickness of layer I
99  '
100 '======================= MAIN PROGRAM =====================
110  KEY OFF
120  GOSUB 200  ' First input screen
130  GOSUB 300  ' Second input screen
140  GOSUB 400  ' Compute resistance and heat flow
150  GOSUB 500  ' Print results
160  END
170 '================================================================
180  '
200 '=================== FIRST INPUT SCREEN ===================
201  P1$ = "Total surface area in square meters ---------------->"
202  P2$ = "Inside temperature in degrees Celsius --------------->"
203  P3$ = "Outside temperature in degrees Celsius ------------->"
204  P4$ = "Number of inner layers in the composite wall -------->"
205  P5$ = "Layer ##: Thickness in centimeters ----------------->"
206  P6$ = "          Thermal conductivity (watt/meter kelvin) -->"
210  CLS
215  PRINT TAB(12);"THERMAL RESISTANCE AND HEAT FLOW"
220  PRINT
225  PRINT P1$; : INPUT " ", A
230  PRINT P2$; : INPUT " ", T1
235  PRINT P3$; : INPUT " ", T0
240  PRINT P4$; : INPUT " ", N
245  FOR I = 1 TO N
250  PRINT
255  PRINT USING P5$; I; : INPUT " ", X(I)
260  PRINT P6$; : INPUT " ", K(I)
265  X(I) = X(I)/100  ' Convert to meters
270  NEXT I
275  RETURN
280  '
300 '=================== SECOND INPUT SCREEN ===================
301  F$(1) = "Natural convection, air, horizontal surface, facing up"
302  F$(2) = "Natural convection, air, horizontal surface, facing down"
303  F$(3) = "Natural convection, air, vertical surface"
304  F$(4) = "Natural convection, water"
305  F$(5) = "Natural convection, oil"
306  F$(6) = "Forced convection, air, smooth surface & inside pipe"
307  F$(7) = "Forced convection, water, straight pipes"
308  F$(8) = "Enter a value from the keyboard"
309  F1$ = "   Fluid to surface temperature difference (Celsius)"
310  F2$ = "   Water temperature (Celsius)"
311  F3$ = "   Viscosity of the oil (pascal seconds)"
312  F4$ = "   Air velocity (meters/second)"
313  F5$ = "   Water velocity (meters/second)"
314  F6$ = "   Inside diameter of pipe (centimeters)"
315  F7$ = "   Film coefficient (watt/square meter kelvin)"
316  B$ = "                                          "
319  '
320  CLS
325  LOCATE 1,12 : PRINT "THERMAL RESISTANCE AND HEAT FLOW"
330  LOCATE 3,16 : PRINT "FILM COEFFICIENT CHOICES"
335  LOCATE 5,1
340  FOR I = 1 TO 8
```

```
345   PRINT I; ".   "; F$(I)
350   NEXT I
355   LOCATE 14,1 : I = 1 ' Get inside film choice
360   INPUT "Your choice for inside film coefficient ---> ", C(I)
365   ON C(I) GOSUB 1100,1200,1300,1400,1500,1600,1700,1800
370   LOCATE 14,1: FOR I = 1 TO 4 : PRINT B$;B$ : NEXT I
375   LOCATE 14,1 : I = 0 ' Get outside film choice
380   INPUT "Your choice for outside film coefficient --> ", C(I)
385   ON C(I) GOSUB 1100,1200,1300,1400,1500,1600,1700,1800
390   RETURN
395   '
400   '=================== COMPUTE RESISTANCE =====================
410   IF H(0) <> 0 THEN RU = 1/H(0) ELSE RU = 0
420   IF H(1) <> 0 THEN RU = RU + 1/H(1)
430   FOR I = 1 TO N
440      RU = RU + X(I)/K(I)
450   NEXT I
460   RT = RU/A
470   Q = ABS(T1 - T0)/RT
480   RETURN
490   '
500   '==================== PRINT RESULTS ========================
501   P1$ = "Surface area:       ###.# square meter"
502   P2$ = "Inside temperature:  #### degrees Celsius"
503   P3$ = "Outside temperature: #### degrees Celsius"
505   P4$ = "###.#    "
506   P5$ = "###.### "
507   P6$ = "Unit resistance:  ##.##^^^^ kelvin square meter/watt"
508   P7$ = "Total resistance: ##.##^^^^ kelvin/watt"
509   P8$ = "Total heat flow:  ##.##^^^^ watts"
510   P9$ = "Heat flows from inside to outside."
511   P10$ = "Heat flows from outside to inside."
512   TD$ = "    Td: ### Celsius"
513   TW$ = "    Tw: ### Celsius"
514   VI$ = "    Viscosity: ##.##^^^^ pascal second"
515   VEL$ = "    Fluid velocity: ###.# meter/second"
516   DIA$ = "    Pipe diameter: ###.# centimeter"
517   FC$ = "    h\\  ##.##^^^^ watt/square meter kelvin"
518   M$ = "       "  ' Left margin
520   CLS
525   PRINT TAB(12);"THERMAL RESISTANCE AND HEAT FLOW"
530   PRINT
535   PRINT USING M$ + P1$; A
540   PRINT USING M$ + P2$; T1
545   PRINT USING M$ + P3$; T0
550   PRINT M$; "Inside film: "; F$(C(1))
555     ON C(1) GOSUB 710, 710, 710, 720, 730, 740, 750
560     PRINT USING M$ + FC$; "i:"; H(1)
565   PRINT M$; "Outside film, "; F$(C(0))
570     ON C(0) GOSUB 710, 710, 710, 720, 730, 740, 750
575     PRINT USING M$ + FC$; "o:"; H(0)
580   PRINT M$; "INNER LAYERS:"
585     PRINT M$; "  x (cm)    ";
590       FOR I = 1 TO N
595         PRINT USING P4$; 100*X(I);
600       NEXT I
605       PRINT
610   PRINT M$; "  K (W/mk) ";
615   FOR I = 1 TO N
```

```
620        PRINT USING P5$; K(I);
625    NEXT I
630    PRINT : PRINT
635    PRINT USING M$ + P6$; RU
640    PRINT USING M$ + P7$; RT
645    PRINT USING M$ + P8$; Q
650    IF T0 < T1 THEN PRINT M$; P9$ ELSE PRINT M$; P10$
655    RETURN
660    '
700    '================== PRINT FILM PARAMETERS ====================
710    PRINT USING M$ + TD$; TD
715    RETURN
720    PRINT USING M$ + TD$ + "," + TW$; TD, TW
725    RETURN
730    PRINT USING M$ + TD$ + "," + VI$; TD, VI
735    RETURN
740    PRINT USING M$ + VEL$; V
745    RETURN
750    PRINT USING M$ + TW$ + "," + VEL$; TW, V
760    PRINT USING M$ + DIA$; D*100
770    RETURN
780    '
1100   ' ============== FILM COEFFICIENT CHOICE #1 =============
1101   ' Natural convection, air, Horizontal surface facing up
1110      PRINT F1$; : INPUT " ---> ", TD ' Get temp. diff.
1120      H(I) = 2.5 * TD^.25
1130      RETURN
1140   '
1200   ' ============== FILM COEFFICIENT CHOICE #2 =============
1201   ' Natural convection, air, Horizontal surface facing down
1210      PRINT F1$; : INPUT " ---> ", TD ' Get temp. diff.
1220      H(I) = 1.32 * TD^.25
1230      RETURN
1240   '
1300   ' =========== FILM COEFFICIENT CHOICE #3 ============
1301   ' Natural convection, air, vertical surface
1310      PRINT F1$; : INPUT " ---> ", TD ' Get temp. diff.
1320      H(I) = 1.78 * TD^.25
1330      RETURN
1340   '
1400   ' =========== FILM COEFFICIENT CHOICE #4 ============
1401   ' Natural convection in still water.
1410      PRINT F1$; : INPUT " ---> ", TD ' Get temp. diff.
1420      PRINT F2$; : INPUT " ---> ", TW ' Get water temp.
1430      H(I) = 2.26*(TW + 34.3)*SQR(TD)
1440      RETURN
1450   '
1500   ' =========== FILM COEFFICIENT CHOICE #5 ============
1501   ' Natural convection in oil.
1510      PRINT F1$; : INPUT " ---> ", TD ' Get temp. diff.
1520      PRINT F3$; : INPUT " ---> ", VI ' Get oil viscosity
1530      H(I) = 7 * TD^.25 / VI^.4
1540      RETURN
1550   '
1600   ' =========== FILM COEFFICIENT CHOICE #6 ============
1601   ' Forced convection, air, smooth surface, inside pipe
1610      PRINT F4$; : INPUT " ---> ", V  ' Get air velocity
1620      IF V < 4.6 THEN H(I) = 4.54 + 4.1 * V
                     ELSE H(I) = 7.75 * V^.75
```

```
1630    RETURN
1640  '
1700  ' =========== FILM COEFFICIENT CHOICE #7 ============
1701  ' Forced convection, turbulent water, straight pipes
1710    PRINT F2$; : INPUT " ---> ", TW  ' Get water temperature
1720    PRINT F5$; : INPUT " ---> ", V   ' Get water velocity
1730    PRINT F6$; : INPUT " ---> ", D   ' Get pipe diameter
1740    D = D/100     ' Convert from centimeters to meters
1750    H(I) = 20.93 * (68.3 + TW) * V^.8 / D^.2
1760    RETURN
1770  '
1800  ' =========== FILM COEFFICIENT CHOICE #8 ============
1801  ' Enter film coefficient from keyboard
1810    PRINT F7$; : INPUT " ---> ", H(I)   ' Get film coefficient
1820    RETURN
```

PROGRAM "BODE"

```
1   'PROGRAM "Bode" - FILE NAME "BODE"
2   ' Uses high resolution graphics.  Runs on PC compatibles
3   ' with a color monitor.  Inputs the transfer function
4   ' polynomial coefficients, produces a Bode graph on the
5   ' screen, and makes a printed copy of the Bode data for
6   ' radian frequencies from 1.0E-6 to 5.5E+5.
7   ' TO RUN ON A MONOCHROME MONITOR, REMOVE LINE 130.
9   '
10  '=================== GLOBAL VARIABLES ===================
11  ' A = angle part of complex number (polar form)
12  ' A(0)..A(3) = coefficients of numerator polynomial
13  ' ANGLE(J) = array of phase angles of component
14  ' ANGLEA = angle of numerator
15  ' ANGLEB = angle of denominator
16  ' LOGBASE = log of frequency base, LW = EXPONENT+LOGBASE
17  ' B(0)..B(3) = coefficients of denominator polynomial
18  ' DB(J) = array of decibel gains of component
19  ' EXPONENT = frequency exponent,  W = BASE*10^EXPONENT
20  ' GAIN = gain of component at W radian/second
21  ' I = imaginary part of complex number (rectangular form)
22  ' J = FOR loop counter
23  ' K = number of elements in Bode data table
24  ' LW(J) = array of radian frequencies
25  ' M = magnitude part of complex number (polar form)
26  ' MAGA = magnitude of numerator
27  ' MAGB = magnitude of denominator
28  ' R = real part of complex number (rectangular form)
29  ' W = radian frequency, radian/second
30  '
31  '===================== CONSTANTS =======================
32   RD = 57.2958 'radian to degree conversion multiplier
39  '
40  '================= PRINT USING STRINGS ==================
41  P$ = "            TRANSFER FUNCTION COEFFICIENTS              "
42  L$ = "-------------------------------------------------------"
43  A$ = "A(0)..A(3): ##.##^^^^ ##.##^^^^ ##.##^^^^ ##.##^^^^"
44  B$ = "B(0)..B(3): ##.##^^^^ ##.##^^^^ ##.##^^^^ ##.##^^^^"
45  T$ = "                    BODE DATA TABLE                    "
```

```
46  H$ = " Frequency, W            Gain              Phase"
47  U$ = "(radian/second)        (decibel)         (degrees)"
48  N$ = "     ##.#^^^^            ####.#            ####.#"
49  M$ = "                    "
50  '
60  '========= DIMENSION OF SUBSCRIPTED VARIABLES ===========
61   DIM LW(200), DB(200), ANGLE(200)
99  '
100 '=================== MAIN PROGRAM ======================
105  KEY OFF
110  GOSUB 200  ' get transfer function coefficients
115  LOCATE 22,10 : PRINT A(0); A(1); A(2); A(3)
116  LOCATE 23,10 : PRINT B(0);B(1);B(2);B(3)
120  GOSUB 300   ' Compute gain and angle data
130  GOSUB 600   ' Put Bode diagram on screen
140  GOSUB 700   ' Process request for printed data table
150  END
160  '
200 '========= INPUT TRANSFER FUNCTION COEFFICIENTS =========
205  FOR C = 0 TO 3 : A(C) = 0 : B(C) = 0 : NEXT C ' Zero coefficients
210  GOSUB 1000 ' Draw Component Input and Display Box
215  FOR R = 0 TO 1 : FOR C = 0 TO 3 ' Write values in Display Box
220     ROW = 9 + 4 * R  :  COLUMN = 8 + 18 * C
225     IF R = 0 THEN X = A(C) ELSE X = B(C)
230     GOSUB 1900 : IF X = 0 THEN U$ = "#.#" ' Get U$
235     LOCATE ROW, COLUMN : PRINT USING U$; X
240  NEXT C : NEXT R
245  '
250 'Get the transfer function coefficients
255    FOR R = 0 TO 1 : FOR C = 0 TO 3
260       ROW = 9 + 4 * R  :  COLUMN = 8 + 18 * C
265       GOSUB 1600 ' Return with valid input in X
270       IF R = 0 THEN A(C) = X ELSE B(C) = X
275       LOCATE ROW, COLUMN : GOSUB 1900 ' Get U$
280       IF R = 0 THEN PRINT USING U$;A(C) ELSE PRINT USING U$;B(C)
285    NEXT C : NEXT R
290    LOCATE 22, 10 : INPUT"Do you wish to correct your input"; C$
291    LOCATE 22, 10 : PRINT"                                  ";
292    IF (C$ = "y") OR (C$ = "Y") GOTO 210
295  RETURN
299  '
300 '============= COMPUTE GAIN AND ANGLE DATA ==============
305  CLS : K = 0
310  FOR EXPONENT = -6 TO 5
315     FOR J = 1 TO 4
320        K = K + 1
325        READ LOGBASE : DATA 0.0, 0.25, 0.5, 0.75
330        LW(K) = EXPONENT + LOGBASE
335        W = 10^LW(K)
340        GOSUB 400 ' Compute gain and phase
345     NEXT J
350     RESTORE
355  NEXT EXPONENT
360  RETURN
365  '
400 '---------- COMPUTE AND PRINT GAIN AND PHASE -----------
405 'compute magnitude and angle parts of numerator
410  R = A(0) - A(2)*W^2     'real part of numerator
420  I = A(1)*W - A(3)*W^3    'imaginary part of numerator
```

```
425   GOSUB 500                      'convert R and I into M and A
430   MAGA = M                       'magnitude of numerator
435   ANGLEA = A                     'angle of numerator
440   'compute magnitude and angle parts of denominator
445   R = B(0) - B(2)*W^2      'real part of denominator
450   I = B(1)*W - B(3)*W^3    'imaginary part of denominator
455   GOSUB 500                      'convert R and I into M and A
460   MAGB = M                       'magnitude of denominator
465   ANGLEB = A                     'angle of denominator
470   'compute and print gain and phase
475   GAIN = MAGA/MAGB
480   DB(K) = 20*LOG(GAIN)/2.30259
485   ANGLE(K) = ANGLEA - ANGLEB
490   PRINT USING M$+N$; W,DB(K), ANGLE(K)
495   RETURN
499   '
500   '----------- RECTANGULAR TO POLAR CONVERSION ------------
510   M = SQR(R^2 + I^2)
520   IF R = 0   AND    I = 0 THEN A =   0 : RETURN
530   IF R = 0 THEN IF I > 0 THEN A =  90 : RETURN ELSE A = -90 : RETURN
540   IF I = 0 THEN IF R > 0 THEN A =   0 : RETURN ELSE A = 180 : RETURN
550   IF R < 0 THEN A = ATN(I/R)*RD + 180 : RETURN
560   A = ATN(I/R)*RD                     : RETURN
570   '
600   '=============== PUT BODE DIAGRAM ON SCREEN ==============
605   KEY OFF
610   SCREEN 2
615   CLS
620   GOSUB 2000  ' Print fixed labels
625   FMIN =    -5 : FMAX =   5  ' Set min & max frequency
630   GMIN = -100 : GMAX = 100  ' Set min & max gain
635   AMIN = -300 : AMAX = 100  ' Set min & max angle
640   CMND$ = ""
645   WHILE CMND$ <> "Q"
650   GOSUB 2200   ' Print labels on axes
655   GOSUB 3000   ' Print gain and angle graphs
660   GOSUB 4000   ' Get command and implement changes
665   WEND
670   RETURN
675   '
700   '======== PROCESS REQUEST FOR PRINTED DATA TABLE ========
710   SCREEN 0 : CLS : LOCATE 10,10
720   PRINT "Type Y if you want a printed copy of the data-->"
730   LOCATE 10,60,1
740   C$ = INPUT$(1)
750   IF (C$="Y") OR (C$="y") THEN GOSUB 800 ' Print data table
760   RETURN
770   '
800   '------------ MAKE PRINTED COPY OF DATA TABLE -----------
810   CLS
820   LPRINT M$; L$   :  LPRINT M$; P$   :  LPRINT M$; L$
830   LPRINT USING M$ + A$; A(0), A(1), A(2), A(3)
840   LPRINT USING M$ + B$; B(0), B(1), B(2), B(3)
850   LPRINT M$; L$   :  LPRINT M$; T$   :  LPRINT M$; L$
860   LPRINT M$; H$   :  LPRINT M$; U$   :  LPRINT M$; L$
870   FOR J = 1 TO K
880      LPRINT USING M$ + N$; 10^LW(J), DB(J), ANGLE(J)
890      IF J MOD 4 = 0 THEN LPRINT
900   NEXT J
```

```
910   RETURN
920   '
1000  '======== DRAW COMPONENT INPUT AND DISPLAY BOX ========
1001  H1$ = "             COMPONENT INPUT AND DISPLAY BOX"
1002  H2$ = "             Transfer Function Coefficients:"
1003  H3$ = "A(0)              A(1)              A(2)              A(3)"
1004  H4$ = "2              3"
1005  H5$ = "+         S    +         S    +         S "
1006  H6$ = "B(0)              B(1)              B(2)              B(3)"
1007  H7$ = "Press <Enter> to accept the displayed value."
1008  H8$ = "Type a new value and press <Enter> to accept the new value."
1009  H9$ = "Press the Q key to Quit and accept all displayed values.
1010  '
1015  CLS
1020  'Draw the top part of the Display Box
1021      ROW = 3 : COLUMN = 4 : WIDE = 72 : HIGH = 2
1022      UL = 201 : UR = 187 : LL = 204 : LR = 185
1023      TOP = 205 : BOTTOM = 205 : SIDE = 186
1024      GOSUB 1100  ' Draw a Box
1025  'Draw the bottom part of the Display Box
1026      ROW = 6 : HIGH = 13
1027      UL = 204 : UR = 185 : LL = 200 : LR = 188
1028      GOSUB 1100  ' Draw a Box
1029  '
1030  'Print dialog in the Display Box
1031      LOCATE  4,11 : PRINT H1$
1032      LOCATE  5,11 : PRINT H2$
1033      LOCATE  7,11 : PRINT H3$
1034      LOCATE  8,56 : PRINT H4$
1035      LOCATE  9,22 : PRINT H5$
1036      LOCATE 11, 7 : FOR I = 1 TO 67 : PRINT CHR$(196); : NEXT I
1037      LOCATE 12,56 : PRINT H4$
1038      LOCATE 13,22 : PRINT H5$
1039      LOCATE 15,11 : PRINT H6$
1040      LOCATE 17,6 : PRINT CHR$(249); H7$
1041      LOCATE 18,6 : PRINT CHR$(249); H8$
1042      LOCATE 19,6 : PRINT CHR$(249); H9$
1049  '
1050  'Draw the input boxes
1051      WIDE = 10 : HIGH = 1
1052      UL = 218 : UR = 191 : LL = 192 : LR = 217
1053      TOP = 196 : BOTTOM = 196 : SIDE = 179
1054      FOR R = 0 TO 1 : FOR C = 0 TO 3
1055          ROW = 8+4*R : COLUMN = 7+18*C
1056          GOSUB 1100 ' DRAW A BOX
1057      NEXT C : NEXT R
1058   RETURN
1099  '
1100  '--------------- Draw A Box ---------------
1105  LOCATE ROW, COLUMN : PRINT CHR$(UL);   'Upper Left corner
1110  FOR I = 1 TO WIDE : PRINT CHR$(TOP); : NEXT I 'Top
1115  PRINT CHR$(UR) 'Upper Right corner
1120  FOR I = 1 TO HIGH
1125     LOCATE ROW+I, COLUMN : PRINT CHR$(SIDE);
1130     LOCATE ROW+I, COLUMN+WIDE+1 : PRINT CHR$(SIDE)
1135  NEXT I
1140  LOCATE ROW+HIGH+1, COLUMN : PRINT CHR$(LL); 'Lower Left corner
1145  FOR I = 1 TO WIDE : PRINT CHR$(BOTTOM); : NEXT I 'Bottom
1150  PRINT CHR$(LR) 'Lower Right corner
```

```
1155   RETURN
1160   '
1600   '=============== GET VALID INPUT ===============
1601   ' Returns a valid input value in X.  The user may edit the
1602   ' value in the input box or leave the·value unchanged.  When
1603   ' the user presses <Enter> the program reads the value in
1604   ' the screen input box and checks its validity.  If the input
1605   ' is invalid, the program prints an error message and repeats
1606   ' the input routine.  If the input is valid, the value is
1607   ' returned in the variable X.
1608   '
1610   ER$ = " IS AN INVALID INPUT, PLEASE TRY AGAIN"
1615   BL$ = "                                              "
1620   '
1625   N1$ = "" : VALID = 1
1630   LOCATE ROW, COLUMN, 1, 6, 7  : M1$ = INPUT$(1)
1635   IF ASC(M1$) <> 13 THEN PRINT M1$; : INPUT "", M1$
1640   FOR I = COLUMN TO COLUMN + 9      ' Read the input box
1645       M1 = SCREEN(ROW,I)  :  M1$ = CHR$(M1)
1650       IF (M1 <> 32) AND (M1 <> 0) THEN N1$ = N1$ + M1$
1655   NEXT I
1660   X = VAL(N1$)
1665   IF (X = 0) AND (N1$ <> "0") AND (N1$ <> "0.0") THEN VALID = 0
1670   IF VALID = 0 THEN LOCATE 25,20: PRINT N1$;ER$;: BEEP: GOTO 1625
1675   '
1680   LOCATE 25,20 : PRINT BL$; ' Remove error message
1685   RETURN
1695   '
1900   '--------- GET PRINT USING STRING U$ ----------
1905   IF X < 1       THEN U$ = "#.########"
1910   IF X >= 1      THEN U$ = "  ####.###"
1915   IF X >= 1000   THEN U$ = "  ########"
1920   IF X <= .00001 THEN U$ = "##.###^^^^"
1925   IF X >= 1E+07  THEN U$ = "##.###^^^^"
1930   IF  X = 0      THEN U$ = "      #.#"
1935   RETURN
1940   '
2000   '---------------- PRINT FIXED LABELS ------------------
2010   LOCATE  1,33 : PRINT "BODE DIAGRAM"
2020   LOCATE 12, 3 : PRINT "Gain"
2030   LOCATE 13, 4 : PRINT "in"
2040   LOCATE 14, 1 : PRINT "Decibels"
2045   LOCATE 11,72 : PRINT "Phase"
2050   LOCATE 12,72 : PRINT "Angle"
2060   LOCATE 13,73 : PRINT "in"
2070   LOCATE 14,71 : PRINT "Degrees"
2080   LOCATE 25,27 : PRINT "Log Radian Frequency";
2090   '
2100   LINE(  0,  7)-(639,  7),1,,&HAAAA
2110   LINE(  0,116)-( 63,116),1,,&HAAAA
2120   LINE(559,116)-(615,116),1,,&HCCCC
2130   RETURN
2140   '
2200   '--------------- PRINT LABELS ON AXES ----------------
2210   VIEW (0,0)-(639,199)
2220   '
2230   FOR J = 0 TO 4                ' Print gain axis labels
2240       GMARK = GMAX - J*(GMAX - GMIN)/4
2250       LOCATE 3+J*5,7 : PRINT USING "####"; GMARK
```

```
2260   NEXT J
2270   '
2280   FOR J = 0 TO 4                    ' Print angle axis labels
2290      AMARK = AMAX - J*(AMAX - AMIN)/4
2300      LOCATE 3+J*5,66 : PRINT USING "####"; AMARK
2310   NEXT J
2320   '
2330   FOR J = 0 TO 10            ' Print frequency axis labels
2340      FMARK = FMIN + J*(FMAX - FMIN)/10
2350      LOCATE 24,11+J*5 : PRINT USING "##.#"; FMARK;
2360   NEXT J
2370   '
2380   RETURN
2390   '
3000   '============= PRINT GAIN AND ANGLE GRAPHS =============
3010   GMAR = (GMAX - GMIN)/40
3020   AMAR = (AMAX - AMIN)/40
3030   FMAR = (FMAX - FMIN)/40
3040   VIEW (88,14)-(510,182)
3050   WINDOW(FMIN-FMAR,GMIN-GMAR/2)-(FMAX+FMAR,GMAX+GMAR)
3060   CLS
3070   GOSUB 3100   ' Draw the axes
3080   GOSUB 3300   ' Draw the gain graph
3090   GOSUB 3400   ' Draw the angle graph
3095   RETURN
3099   '
3100   '------------------ Draw the axes -------------------
3110   LINE(FMIN,GMAX+GMAR)-(FMIN,GMIN-GMAR),1
3120   LINE(FMAX,GMAX+GMAR)-(FMAX,GMIN-GMAR),1
3130   FOR J = 0 TO 4
3140      GMARK = GMAX - J*(GMAX - GMIN)/4
3150      LINE(FMIN-FMAR, GMARK)-(FMIN+FMAR, GMARK),1,,&HAAAA
3160      LINE(FMAX-FMAR, GMARK)-(FMAX+FMAR, GMARK),1,,&HAAAA
3170   NEXT J
3180   '
3190   LINE(FMIN+FMAR,GMIN)-(FMAX+FMAR,GMIN),1,,&HAAAA
3200   FOR J = 0 TO 9
3210      FMARK = FMAX - J*(FMAX - FMIN)/10
3220      LINE(FMARK,GMIN+GMAR)-(FMARK,GMIN-GMAR/2),1
3230   NEXT J
3240   RETURN
3250   '
3300   '---------------- Draw the gain graph ----------------
3310   LINE(LW(1),DB(1))-(LW(2),DB(2)),1,,&HAAAA
3320   FOR J = 3 TO K
3330      LINE -(LW(J),DB(J)),1,,&HAAAA
3340   NEXT J
3350   RETURN
3360   '
3400   '---------------- Draw the angle graph ----------------
3410   WINDOW(FMIN-FMAR,AMIN-AMAR)-(FMAX+FMAR,AMAX+AMAR)
3420   LINE (LW(1),ANGLE(1))-(LW(2),ANGLE(2)),1,,&HCCCC
3430   FOR J = 3 TO K
3440      LINE -(LW(J),ANGLE(J)),1,,&HCCCC
3450   NEXT J
3460   RETURN
3470   '
4000   '========= GET COMMAND AND IMPLEMENT CHANGES ===========
```

```
4001  C1$ = "Command:  (Q)uit  (Z)oom  (U)nzoom  ------------> "
4002  C2$ = "Which variable:  (G)ain  (A)ngle  (F)requency --> "
4003  C3$ = "Where:  (T)op/left  (M)iddle  (B)ottom/right ---> "
4005  CMND$ = ""
4010  VIEW (3,8)-(636,15),,1
4020  C0$=C1$ : X$="Q" : Y$="Z" : Z$="U" ' First command values
4030  GOSUB 4100  ' Get valid input for first command
4040  IF CMND$ = "Z" THEN GOSUB 4200 ' Process Zoom command
4050  IF CMND$ = "U" THEN GOSUB 4400 ' Process UnZoom command
4060  RETURN
4070  '
4100  '--------------- GET VALID COMMAND ------------------
4110  VALID = 0
4120  LOCATE 2,3 : PRINT C0$;CHR$(178); : C$ = INPUT$(1)
4130  IF ASC(C$) > 96 THEN C$ = CHR$(ASC(C$)-32) ' Upcase C$
4140  IF (C$ = X$) OR (C$=Y$) OR (C$=Z$) THEN VALID = 1
4150  IF VALID = 0 THEN BEEP : GOTO 4120
4160  CMND$ = CMND$ + C$
4170  RETURN
4180  '
4200  '--------------- PROCESS ZOOM COMMAND ----------------
4210  C0$=C2$ : X$="G" : Y$="A" : Z$="F" ' Second command values
4220  GOSUB 4100  ' Get valid input for second command
4230  C0$=C3$ : X$="T" : Y$="M" : Z$="B" ' Third command values
4240  GOSUB 4100  ' Get valid input for third command
4250  IF CMND$ = "ZAT" THEN AMIN = -100 : AMAX = 100
4260  IF CMND$ = "ZAM" THEN AMIN = -200 : AMAX = 0
4270  IF CMND$ = "ZAB" THEN AMIN = -300 : AMAX = -100
4280  IF CMND$ = "ZFT" THEN FMIN = -6  : FMAX = -1
4290  IF CMND$ = "ZFM" THEN FMIN = -2  : FMAX = 3
4300  IF CMND$ = "ZFB" THEN FMIN = 0   : FMAX = 5
4310  IF CMND$ = "ZGT" THEN GMIN = 0   : GMAX = 100
4320  IF CMND$ = "ZGM" THEN GMIN = -50 : GMAX = 50
4330  IF CMND$ = "ZGB" THEN GMIN = -100 : GMAX = 0
4340  RETURN
4350  '
4400  '--------------- PROCESS UNZOOM COMMAND ----------------
4410  C0$=C2$ : X$="G" : Y$="A" : Z$="F" ' Second command values
4420  GOSUB 4100  ' Get valid input for second command
4430  IF CMND$ = "UA" THEN AMIN = -300 : AMAX = 100
4440  IF CMND$ = "UF" THEN FMIN = -5  : FMAX = 5
4450  IF CMND$ = "UG" THEN GMIN = -100 : GMAX = 100
4460  RETURN
4470  '
```

PROGRAM "DESIGN"

```
1  ' PROGRAM "Control System Design" : FILE NAME "DESIGN"
2  '  Completes the design of a PID controller for a process
3  '  that can have up to 9 components and 10 dead time lags.
4  '  Produces tables of frequency response data and summarizes
5  '  the control mode parameters, the gain margin and the
6  '  phase margin of the final design.  The frequency ranges
7  '  from 1.0E-06 to 5.6E05 radian/second.
9  '
10 '==================== GLOBAL VARIABLES ====================
11 ' A = angle part of complex number (polar form)
```

```
12  ' A(i,0)..A(i,3) = array of numerator coefficients
13  ' ANGLE(J) = array of open-loop angles
14  ' ANGLEC(J) = array of closed-loop angles
15  ' ANGLE0DB = phase angle at which gain is 0 decibel
16  ' B(i,0)..B(i,3) = array of denominator coefficients
17  ' DB(J) = array of open-loop decibel gains
18  ' DBC(J) = array of closed-loop decibel gains
19  ' DEADTIME(..) = array of dead time lags
20  ' DMODE = derivative action time constant of controller
21  ' ER(J) = array of error ratios
22  ' I = imaginary part of complex number (rectangular form)
23  ' IMODE = integral action rate of the controller
24  ' K = Number of elements in data arrays
25  ' LW(J) = array of log radian frequenies
26  ' M = magnitude part of complex number (polar form)
27  ' M140 = gain for which the phase angle is -140 degrees
28  ' M180 = gain for which the phase angle is -180 degrees
29  ' NI(J) = array of imaginary parts of open-loop response
30  ' NR(J) = array of real parts of open-loop response
31  ' NUMBEROFCOMPONENTS = number of component transfer functions
32  ' NUMBEROFDEADTIMES = number of dead time lags
33  ' R = real part of complex number (rectangular form)
34  ' W = radian frequency, radian/second
35  ' W82 = frequency at which phase angle = -82 degrees
36  ' W140 = frequency at which phase angle = -140 degrees
37  ' W170 = frequency at which phase angle = -170 degrees
38  ' W180 = frequency at which phase angle = -180 degrees
39  ' W270 = frequency at which phase angle = -270 degrees
40  '
45  '=============== GLOBAL STRING CONSTANTS ================
46   PID$ = "Controller Design:  P =##.##^^^^,  I =##.##^^^^"
         + ",   D =##.##^^^^,  " + CHR$(224) + " = #.### "
50  '===================== CONSTANTS =========================
51   RD = 57.2958  ' radian to degree conversion multiplier
52  '
80  '================== DATA FOR FREQUENCY BASE ================
81   DATA  0.0,  0.25, 0.5, 0.75
89  '
90  '========== DIMENSION OF SUBSCRIPTED VARIABLES =========
91   DIM  LW(50), DB(50), ANGLE(50), DBC(50), ANGLEC(50)
92   DIM  NR(50), NI(50), ER(50)
100  '========================= MAIN PROGRAM =========================
110    GOSUB 200                    ' get transfer functions of components
120    GOSUB 400                    ' set initial values
130    GOSUB 500                    ' compute Bode data
140    GOSUB 600                    ' find Design Decision Data
150    GOSUB 700                    ' print open-loop Bode graph
160    GOSUB 800                    ' process commands until done
170    GOSUB 900                    ' process request for printed data
180    SCREEN 0 : CLS               ' return to text mode
190    END
199  '
200  '======= INPUT COMPONENT COEFFICIENTS AND DELAYS =======
205    NUMBEROFCOMPONENTS = 0 : NUMBEROFDEADTIMES = 0
210    COMPONENT = 0 : DTNUMBER = 0 : ACTIVEBOX$ = "TFC"
215    GOSUB 1000                              ' Draw Input and Display Box
220    GOSUB 4700                              ' Get Input Box Command
225    WHILE CMND$ <> "Q"              ' Loop until Quit
230      IF CMND$ = "A" THEN GOSUB 2050 ' Add a component
235      IF CMND$ = "E" THEN GOSUB 2700 ' Edit a component
```

```
240      IF CMND$ = "N" THEN GOSUB 2800 ' Next component
245      IF CMND$ = "P" THEN GOSUB 2900 ' Previous component
250      IF CMND$ = "M" THEN GOSUB 2000 ' Move to other Box
255      GOSUB 4700                      ' Get Input Box Command
260    WEND
265    IF (COMPONENT = 0) AND (DTNUMBER = 0) THEN CLS : END
295    RETURN
299    '
300    '============== ASSIGN STRING VARIABLES ===============
301    '------------------ Active Box Status Line --------------------
302    AB$ = "Active Box: " : AB11$ = "Transfer Function Coefficents."
303    AB12$ = "  You have entered # components. "
304    AB21$ = "Dead Time Delays."
305    AB22$ = "  You have entered # dead time delays.         "
306    '------------------- Command Mode Line 1 ----------------------
307    CM1$ = "Command Mode: (Add) (Edit) (Next) (Previous) "
308    CM2$ = "(Move to other box) (Quit)---->  "
309    '-------------- TFC - Command Mode Lines 2 and 3·---------------
310    CM11$ = "Use (Add) to add a component, "
311    CM12$ = "(Edit) to change the displayed component,     "
312    CM13$ = "(Next) to display next component, "
313    CM14$ = "(Previous) to display previous component. "
314    '-------------- DTD - Command Mode Lines 2 and 3 ---------------
315    CM21$ = "Use (Add) to add a dead time delay, "
316    CM22$ = "(Edit) to change the dispalyed delay,     "
317    CM23$ = "(Next) to display next delay, "
318    CM24$ = "(Previous) to display previous delay.         "
319    '--------------- TFC - Input Mode Lines 1 - 3------------------
320    IM$ = "Input Mode: "
321    IM11$ = "Use the cursor keys to move from box to box. "
322    IM12$ = "Change coefficients #"
323    IM13$ = "as required.  When all coefficients are correct, "
324    IM14$ = "press <Enter> to#" : IM15$ = "accept each coefficient "
325    IM16$ = "from A(0) to A(3) to B(0) to B(3).         #
326    '--------------- DTD - Input Mode Lines 1 - 3------------------
327    IM21$ = "Change delay as required.   "
328    IM22$ = "Press <Enter) to accept new value.#"
329    IM23$ = "                              "
330    IM24$ = "                                   "
331    '----------------- Input and Display Box Text ------------------
332    H1$ = "             COMPONENT INPUT AND DISPLAY BOX"
333    H2$ = "         Transfer Function Coefficients of component #"
334    H3$ = "A(0)                A(1)                A(2)                A(3)"
335    H4$ = "2                   3"
336    H5$ = "+                S  +                S  +                S "
337    H6$ = "B(0)                B(1)                B(2)                B(3)"
338    H7$ = "             Dead Time Delay Number #"
360    '--------------------- PRINT USING STRINGS ---------------------
361    P$="             COMPONENT TRANSFER FUNCTIONS             "
362    L$="-----------------------------------------------------------"
363    A$ = "A(0)..A(3): ##.##^^^^ ##.##^^^^ ##.##^^^^ ##.##^^^^"
364    B$ = "B(0)..B(3): ##.##^^^^ ##.##^^^^ ##.##^^^^ ##.##^^^^"
366    E$=" Dead-time(##) = ##.##^^^^^"
367    M$ = "         "
370    P1$ = "                  BODE DATA TABLE"
371    P2$ = "-----------------------------------------------------------"
372    P3$ = "|                Open-loop    |    Closed-Loop   |         |"
373    P4$ = "|              ---------------  ---------------  | Error   |"
374    P5$ = "| Frequency  | Gain   Angle   | Gain   Angle   | Ratio   |"
375    P6$ = "| (rad/sec)  | (db)   (deg.)  | (db)   (deg.)  | (db)    |"
```

```
376   N1$ = "|  ##.#^^^^  | ####.#  ####.# | ####.#  ####.# | ####.# |"
390   RETURN
399   '
400   '=================== INITIALIZE CONTROLLER =================
410    PMODE = 1             ' proportional mode off
420    IMODE = 0             ' integral mode off
430    DMODE = 0             ' derivative mode off
440    ALPHA = .1            ' set derivative limiter
450    GOSUB 1200            ' update controller
460    FMIN =   -5 : FMAX =   5 ' set min & max frequency
470    GMIN = -100 : GMAX = 100 ' set min & max gain
480    AMIN = -200 : AMAX = 200 ' set min & max angle
490    NMIN =  -10 : NMAX =  10 ' set min & max Nyquist
495   RETURN
499   '
500   '================= COMPUTE BODE DATA ===================
505    F$ = "    ##.#^^^^      ####.#     ####.#"
510   RESTORE : K = 0
515   FOR EXPONENT = -6 TO 5
520      FOR J = 1 TO 4
525         K = K + 1
530         READ LOGBASE       ' values =  0.0, 0.25, 0.5, 0.75
535         LW(K) = EXPONENT + LOGBASE
540         W = 10^LW(K)
545         GOSUB 1300     ' compute open loop gain and angle
550         DB(K) = 20 * LOG(GAIN)/2.30259
555         ANGLE(K) = ANGLE
560         GOSUB 1500       ' Compute closed loop & Nyquist data
565         LOCATE 2,2 : PRINT USING F$;W,DB(K),ANGLE(K)
570         IF ANGLE <= -9999 GOTO 590
575      NEXT J
580      RESTORE              ' 81   DATA  0.0, 0.25, 0.5, 0.75
585   NEXT EXPONENT
590   RETURN
599   '
600   '============== FIND DESIGN DECISION DATA ==============
605   TARGET = -82  : GOSUB 2100  ' find W(-82)
610   IF FOUND = 1 THEN W82 = WTARGET    ELSE W82 = 0
615   '
620   TARGET = -140 : GOSUB 2100  ' find M(-140)
625   IF FOUND = 1 THEN M140 = DBTARGET ELSE M140 = 0
630   '
635   TARGET = -170 : GOSUB 2100  ' find W(-170)
640   IF FOUND = 1 THEN W170 = WTARGET    ELSE W170 = 0
645   '
650   TARGET = -180 : GOSUB 2100  ' find W(-180) and M(-180)
655   IF FOUND = 1 THEN W180 = WTARGET   ELSE W180 = 0
660   IF FOUND = 1 THEN M180 = DBTARGET ELSE M180 = 0
665   '
670   TARGET = -270 : GOSUB 2100  ' find W(-270)
675   IF FOUND = 1 THEN W270 = WTARGET   ELSE W270 = 10^5
680   '
685   GOSUB 2200  ' find ANGLE0DB
690   GOSUB 2300  ' find FREQLIMIT
695   RETURN
699   '
700   '============= PRINT OPEN-LOOP BODE GRAPH ===============
701   TITLE$ = " OPEN-LOOP BODE GRAPH"
710   GRAPH$ = "O"           ' set graph flag to open-loop
```

```
720    KEY OFF : SCREEN 2      ' set high resolution graphic mode
730    VIEW (0,0)-(639,199) ' set view to full screen
740    CLS
750    GOSUB 3000             ' print text on Bode graph
760    GOSUB 3200             ' print Design Decision Table
770    GOSUB 3400             ' print axes and their labels
780    GOSUB 3600             ' print dashed gain and angle graphs
790    RETURN
799    '
800    '============ PROCESS COMMANDS UNTIL FINISHED ============
810    GOSUB 4000                        ' get valid command
815    WHILE CMND$ <> "Q"                ' loop until Quit
820       IF CMND$ = "D" THEN GOSUB 4200 ' Dmode design
825       IF CMND$ = "I" THEN GOSUB 4400 ' Imode design
830       IF CMND$ = "P" THEN GOSUB 4600 ' Pmode design
835       IF CMND$ = "A" THEN GOSUB 5000 ' Analysis
840       IF CMND$ = "Z" THEN GOSUB 6000 ' Zoom
845       IF CMND$ = "U" THEN GOSUB 6200 ' Unzoom
850       GOSUB 4000                     ' get valid command
855    WEND
890    RETURN
899    '
900    '============ PROCESS REQUEST FOR PRINTED DATA ============
910    SCREEN 0 : CLS : LOCATE 10,10
920    PRINT "Type Y if you want a printed copy of the data-->"
930    LOCATE 10,60,1 : C$ = INPUT$(1)
940    IF (C$="Y") OR (C$="y") THEN GOSUB 7000 ' Print data table
950    RETURN
999    '
1000   '======== DRAW COMPONENT INPUT AND DISPLAY BOX ========
1001   CLS : KEY OFF
1005   GOSUB 300                          ' Assign string variables
1010   GOSUB 1040                         ' Draw Input Display Box
1015   IF ACTIVEBOX$ = "TFC" THEN GOSUB 1060 : GOSUB 1050
1020   IF ACTIVEBOX$ = "DTD" THEN GOSUB 1050 : GOSUB 1060
1025   GOSUB 1070                         ' Print dialog in Box
1030   GOSUB 1085                         ' Draw input boxes
1035   RETURN
1039   '
1040   '------ Draw the top part of the Display Box ------
1041   ROW = 5 : COLUMN = 4 : WIDE = 72 : HIGH = 1
1042   UL = 201 : UR = 187 : LL = 204 : LR = 185
1043   TOP = 205 : BOTTOM = 205 : SIDE = 186
1044   GOSUB 1100   ' Draw a Box
1045   RETURN
1046   '
1050   '------ Draw the middle part of the Display Box ------
1051   IF ACTIVEBOX$ = "TFC" THEN COLOR 2 ELSE COLOR 7
1052   ROW = 7 : COLUMN = 4 : WIDE = 72 : HIGH = 10
1053   UL = 204 : UR = 185 : LL = 204 : LR = 185
1054   TOP = 205 : BOTTOM = 205 : SIDE = 186
1055   GOSUB 1100   ' Draw a Box
1056   COLOR 7
1057   RETURN
1058   '
1060   '------ Draw the bottom part of the Display Box ------
1061   IF ACTIVEBOX$ = "DTD" THEN COLOR 2 ELSE COLOR 7
1062   ROW = 18 : COLUMN = 4 : WIDE = 72 : HIGH = 4
1063   UL = 204 : UR = 185 : LL = 200 : LR = 188
```

```
1064    TOP = 205 : BOTTOM = 205 : SIDE = 186
1065    GOSUB 1100  ' Draw a Box
1066    COLOR  7
1067    RETURN
1068    '
1070    '------ Print dialog in the Display Box ------
1071    LOCATE  6,11 : PRINT H1$
1072    LOCATE  8,11 : PRINT USING H2$; COMPONENT
1073    LOCATE  9,11 : PRINT H3$
1074    LOCATE 10,56 : PRINT H4$
1075    LOCATE 11,22 : PRINT H5$
1076    LOCATE 13, 7 : FOR I = 1 TO 67 : PRINT CHR$(196); : NEXT I
1077    LOCATE 14,56 : PRINT H4$
1078    LOCATE 15,22 : PRINT H5$
1079    LOCATE 17,11 : PRINT H6$
1080    LOCATE 19,11 : PRINT USING H7$; DTNUMBER
1081    RETURN
1082    '
1085    '------ Draw the input boxes ------
1086    WIDE = 10 : HIGH = 1
1087    UL = 218 : UR = 191 : LL = 192 : LR = 217
1088    TOP = 196 : BOTTOM = 196 : SIDE = 179
1089    FOR R = 0 TO 1 : FOR C = 0 TO 3
1090       ROW = 10 + 4 * R : COLUMN = 7 + 18 * C
1091       GOSUB 1100 ' Draw a box
1092    NEXT C : NEXT R
1093    ROW = 20 : COLUMN = 33
1094    GOSUB 1100 ' Draw a box
1096    RETURN
1099    '
1100    '-------------- Draw A Box ---------------
1105    LOCATE ROW, COLUMN : PRINT CHR$(UL);  'Upper Left corner
1110    FOR I = 1 TO WIDE : PRINT CHR$(TOP); : NEXT I 'Top
1115    PRINT CHR$(UR) 'Upper Right corner
1120    FOR I = 1 TO HIGH
1125       LOCATE ROW+I, COLUMN : PRINT CHR$(SIDE);
1130       LOCATE ROW+I, COLUMN+WIDE+1 : PRINT CHR$(SIDE)
1135    NEXT I
1140    LOCATE ROW+HIGH+1, COLUMN : PRINT CHR$(LL); 'Lower Left corner
1145    FOR I = 1 TO WIDE : PRINT CHR$(BOTTOM); : NEXT I 'Bottom
1150    PRINT CHR$(LR) 'Lower Right corner
1155    RETURN
1199    '
1200    '======= UPDATE CONTROLLER TRANSFER FUNCTION =======
1201    ' Update the controller transfer function with the
1202    ' current values of Pmode, Imode, Dmode, and ALPHA,
1203    ' The following controller transfer function is used:
1204    '
1205    '         V           I + (1 + I*D)*s + D*s^2
1206    '        --- = P*[---------------------------]
1207    '         E             s + ALPHA*D*s^2
1209    '
1210    N = NUMBEROFCOMPONENTS + 1
1220    A(N,0) = PMODE*IMODE
1230    A(N,1) = PMODE*(1 + IMODE*DMODE)
1240    A(N,2) = PMODE*DMODE
1250    A(N,3) = 0
1260    B(N,0) = 0
1270    B(N,1) = 1
```

```
1280  B(N,2) = ALPHA * DMODE
1290  B(N,3) = 0
1295  RETURN
1299  '
1300  '============= COMPUTE OPEN LOOP GAIN AND ANGLE =============
1305  GAIN = 1  :  ANGLE = 0
1310  FOR COMPONENT = 1 TO NUMBEROFCOMPONENTS + 1
1315      R = A(COMPONENT,0) - A(COMPONENT,2)*W^2
1320      I = A(COMPONENT,1)*W - A(COMPONENT,3)*W^3
1325      GOSUB 1800                    ' convert to polar form
1330      GAIN = GAIN * M
1335      ANGLE = ANGLE + A
1340      R = B(COMPONENT,0) - B(COMPONENT,2)*W^2
1345      I = B(COMPONENT,1)*W - B(COMPONENT,3)*W^3
1350      GOSUB 1800                    ' convert to polar form
1355      GAIN = GAIN / M
1360      ANGLE = ANGLE - A
1365  NEXT COMPONENT
1370  FOR DTNUMBER = 1 TO NUMBEROFDEADTIMES
1375      ANGLE = ANGLE - 57.2958*W*DEADTIME(DTNUMBER)
1380  NEXT DTNUMBER
1390  RETURN
1399  '
1475  LOCATE ROW, COLUMN : PRINT USING "#.#"; X
1500  '======== COMPUTE CLOSED LOOP AND NYQUIST DATA ========
1501  ' PRECONDITION: GAIN = open-loop gain, ANGLE = open-loop
1502  ' angle.  POSTCONDITION:  DBC(K) = closed-loop gain in
1503  ' decibles, ANGLEC(K) = closed-loop angle.  ER(K) = the
1504  ' error ratio, NR(K) = Nyquist real coordinate, and
1505  ' NI(K) = Nyquist imaginary coordinate.  NR(K) and NI(K)
1506  ' are limited to prevent overflow when they are plotted.
1510  R = GAIN * COS(ANGLE/RD)
1515  I = GAIN * SIN(ANGLE/RD)
1520  R = R + 1
1525  GOSUB 1800                    ' convert to polar form
1530  DBC(K) = 20*LOG(GAIN/M)/2.30259 ' closed-loop gain
1535  IF R < 0 THEN A = A - 360     ' range = (-270 < A < 90)
1540  ANGLEC(K) = ANGLE - A         ' closed-loop angle
1545  ER(K) = 1/M                   ' partial error ratio
1550  R = 2 - R
1555  I = - I
1560  GOSUB 1800                    ' convert to polar form
1565  ER(K) = 20*LOG(ER(K)/M)/2.30295 ' complete error ratio
1570  IF GAIN > 100 THEN GAIN = 100 ' limit gain for graphing
1575  NR(K) = GAIN * COS(ANGLE/RD)  ' Nyquist real coordinate
1580  NI(K) = GAIN * SIN(ANGLE/RD)  ' Nyquist imaginary coord.
1585      PRINT USING U$; X
1595  RETURN
1599  '
1600  '========== RETURN VALID INPUT IN X ==============
1601  ' Returns a valid input value in X.  The user may edit the
1602  ' value in the input box or leave the value unchanged.  When
1603  ' the user presses <Enter> the program reads the value in
1604  ' the screen input box and checks its validity.  If the input
1605  ' is invalid, the program prints an error message and repeats
1606  ' the input routine.  If the input is valid, the value is
1607  ' returned in the variable X.
1608  '
1610  ER$ = " IS AN INVALID INPUT, PLEASE TRY AGAIN"
```

```
1615  BL$ = "                                                        "
1620  '
1625  N1$ = "" : VALID = 1
1630  LOCATE ROW, COLUMN, 1, 6, 7  : M1$ = INPUT$(1)
1635  IF ASC(M1$) <> 13 THEN PRINT M1$; : INPUT "", M1$
1640  FOR I = COLUMN TO COLUMN + 9        ' Read the input box
1645      M1 = SCREEN(ROW,I) :  M1$ = CHR$(M1)
1650      IF (M1 <> 32) AND (M1 <> 0) THEN N1$ = N1$ + M1$
1655  NEXT I
1660  X = VAL(N1$)
1665  IF (X = 0) AND (N1$ <> "0") AND (N1$ <> "0.0") THEN VALID = 0
1670  IF VALID = 0 THEN LOCATE 25,20: PRINT N1$;ER$;: BEEP: GOTO 1625
1675  '
1680  LOCATE 25,20 : PRINT BL$; ' Remove error message
1685  RETURN
1695  '
1800  '========== RECTANGULAR TO POLAR CONVERSION ============
1810  M = SQR(R^2 + I^2)
1820  IF R = 0 AND     I = 0 THEN A =   0 : RETURN
1830  IF R = 0 THEN IF I > 0 THEN A =  90 : RETURN ELSE A = -90 : RETURN
1840  IF I = 0 THEN IF R > 0 THEN A =   0 : RETURN ELSE A = 180 : RETURN
1850  IF R < 0 THEN A = ATN(I/R)*RD + 180 : RETURN
1860  A = ATN(I/R)*RD                     : RETURN
1870  RETURN
1880  '
1899  '
1900  '========== RETURN FORMAT STRING IN U$ =============
1905    IF X < 1        THEN U$ = "#.########"
1910    IF X >= 1       THEN U$ = "  ####.###"
1915    IF X >= 1000    THEN U$ = "  ########"
1920    IF X <= .00001  THEN U$ = "##.###^^^^"
1925    IF X >= 1E+07   THEN U$ = "##.###^^^^"
1930    IF  X = 0       THEN U$ = "       #.#"
1935    IF FIRST = 1    THEN U$ = "#.#       "
1940    RETURN
1945    '
2000  '========== MOVE TO THE OTHER INPUT BOX ===========
2005  IF ACTIVEBOX$ = "TFC" THEN AX$ = "DTD" ELSE AX$ = "TFC"
2010  ACTIVEBOX$ = AX$
2015  IF ACTIVEBOX$ = "TFC" THEN GOSUB 1060 : GOSUB 1050
2020  IF ACTIVEBOX$ = "DTD" THEN GOSUB 1050 : GOSUB 1060
2025  RETURN
2030  '
2050  '======= ADD A COMPONENT OR DEAD TIME DELAY =========
2055  IF ACTIVEBOX$ = "TFC" THEN GOSUB 2400 ELSE GOSUB 2450
2060  RETURN
2065  '
2100  '============ FIND TARGET FREQUENCY AND GAIN ===========
2101  ' Searches the Bode data for frequencies immediately
2102  ' above and below the TARGET frequency.  If found, KEY
2103  ' is given a value of 1, else KEY has a value of 0.
2104  ' If found, the frequency of TARGET is interpolated and
2105  ' returned in WTARGET, and the decibel gain of TARGET is
2106  ' interpolated and returned in DBTARGET.
2107  '
2120  FOUND = 0 : I = 1
2125  WHILE (FOUND = 0) AND (I < K)
2130      IF (ANGLE(I+1) < TARGET) AND (ANGLE(I) >= TARGET)
              THEN FOUND = 1 ELSE I = I + 1
```

```
2135  WEND
2140  IF FOUND = 0 THEN RETURN
2145  INCLW = LW(I+1) - LW(I)
2150  INCANGLE = ANGLE(I+1) - ANGLE(I)
2155  INCDB = DB(I+1) - DB(I)
2160  INCTARGET = TARGET - ANGLE(I)
2165  WTARGET = 10^(LW(I) + INCLW * INCTARGET / INCANGLE)
2170  DBTARGET = DB(I) + INCDB * INCTARGET / INCANGLE
2175  RETURN
2199  '
2200  '================== FIND 0 DB ANGLE ==================
2201  ' Searches the Bode data for frequencies immediately
2202  ' above and below the 0 db frequency.  If found, KEY
2203  ' is given a value of 1, else KEY has a value of 0.
2204  ' If found, the angle of 0 db is interpolated and
2205  ' returned in ANGLE0DB.
2206  '
2220  FOUND = 0 : I = 1
2225  WHILE (FOUND = 0) AND (I < K)
2230     IF (DB(I+1) < 0) AND (DB(I) >= 0)
            THEN FOUND = 1 ELSE I = I + 1
2235  WEND
2240  IF FOUND = 0 THEN RETURN
2250  INCANGLE = ANGLE(I+1) - ANGLE(I)
2255  INCDB = DB(I+1) - DB(I)
2260  INC0DB =  0 - DB(I)
2265  ANGLE0DB = ANGLE(I) + INCANGLE * INC0DB / INCDB
2275  RETURN
2290  '
2300  '================= FIND FREQUENCY LIMIT =================
2301  ' Searches the Error-ratio for frequencies immediately
2302  ' above and below the 0 db error-ratio.  If found, KEY
2303  ' is given a value of 1, else KEY has a value of 0.
2304  ' If found, the frequency of 0 error-ratio is
2305  ' interpolated and returned in FREQLIMIT.
2306  '
2320  FOUND = 0 : I = 1 : FREQLIMIT = 0
2325  WHILE (FOUND = 0) AND (I < K)
2330     IF (ER(I+1) > 0) AND (ER(I) <= 0)
            THEN FOUND = 1 ELSE I = I + 1
2335  WEND
2340  IF FOUND = 0 THEN RETURN
2345  INCLW = LW(I+1) - LW(I)
2355  INCER = ER(I+1) - ER(I)
2360  INC0ER =  0 - ER(I)
2365  FREQLIMIT = 10^(LW(I) + INCLW * INC0ER / INCER)
2375  RETURN
2390  '
2400  '================= ADD A COMPONENT ===============
2405  NUMBEROFCOMPONENTS = NUMBEROFCOMPONENTS + 1
2410  COMPONENT = NUMBEROFCOMPONENTS
2415  FOR C = 0 TO 3
2420     A(COMPONENT,C) = 0 : B(COMPONENT,C) = 0
2425  NEXT C
2430  FIRST = 1 : GOSUB 2500  ' Print coefficients in Box
2435  GOSUB 2550  ' Get transfer function coefficients
2440  RETURN
2445  '
2450  '============== ADD A DEAD TIME DELAY =============
```

```
2455   NUMBEROFDEADTIMES = NUMBEROFDEADTIMES + 1
2460   DTNUMBER = NUMBEROFDEADTIMES
2465   DEADTIME(DTNUMBER) = 0
2470   FIRST = 1 : GOSUB 2600 ' Print dead time delay in Box
2475   GOSUB 2650              ' Get Dead Time Delay
2480   RETURN
2485   '
2500   '========= PRINT COEFFICIENTS IN INPUT/DISPLAY BOX ========
2505   LOCATE  8,11 : PRINT USING H2$; COMPONENT
2510   FOR R = 0 TO 1 : FOR C = 0 TO 3 ' Write values in Display Box
2515      ROW = 11 + 4 * R   :   COLUMN = 8 + 18 * C
2520      IF R = 0 THEN X = A(COMPONENT,C) ELSE X = B(COMPONENT,C)
2525      GOSUB 1900                    ' Get format string in U$
2530      LOCATE ROW, COLUMN : PRINT USING U$; X
2535   NEXT C : NEXT R
2540   RETURN
2545   '
2550   '========= GET TRANSFER FUNCTION COEFFICIENTS =========
2555   FOR R = 0 TO 1 : FOR C = 0 TO 3
2560      ROW = 11 + 4 * R   :   COLUMN = 8 + 18 * C
2565      GOSUB 1600 ' Return with valid input in X
2570      IF R = 0 THEN A(COMPONENT,C) = X ELSE B(COMPONENT,C) = X
2575      FIRST = 0 : GOSUB 1900           ' Get format string in U$
2580      LOCATE ROW, COLUMN
2585      PRINT USING U$; X
2590   NEXT C : NEXT R
2595   RETURN
2599   '
2600   '===== PRINT DEAD TIME DELAY IN INPUT/DISPLAY BOX =====
2605   LOCATE 19,11 : PRINT USING H7$; DTNUMBER
2610   X = DEADTIME(DTNUMBER)
2615   GOSUB 1900                          ' Get format string in U$
2620   ROW = 21 : COLUMN = 34
2625   LOCATE ROW, COLUMN : PRINT USING U$; X
2630   RETURN
2635   '
2650   '=============== GET DEAD TIME DELAY =================
2655   ROW = 21 : COLUMN = 34
2660   GOSUB 1600                     ' Return with valid input in X
2665   DEADTIME(DTNUMBER) = X
2670   FIRST = 0 : GOSUB 1900          ' Get format string in U$
2675   LOCATE ROW, COLUMN : PRINT USING U$; X
2680   RETURN
2685   '
2700   '=============== EDIT A COMPONENT OR DEAD TIME ============
2705   IF ACTIVEBOX$ = "TFC" THEN GOSUB 2550 ELSE GOSUB 2650
2710   RETURN
2715   '
2800   '========= NEXT COMPONENT OR DEAD TIME DELAY ========
2810   FIRST = 0 : AX$ = ACTIVEBOX$
2820   X = NUMBEROFCOMPONENTS : Y = NUMBEROFDEADTIMES
2830   IF AX$ = "TFC" THEN COMPONENT = COMPONENT MOD X + 1
2840   IF AX$ = "DTD" THEN DTNUMBER = DTNUMBER MOD Y + 1
2850   IF AX$ = "TFC" THEN GOSUB 2500 ' Print in Box
2860   IF AX$ = "DTD" THEN GOSUB 2600 ' Print in Box
2870   RETURN
2900   '====== PREVIOUS COMPONENT OR DEAD TIME DELAY ========
2910   FIRST = 0 : AX$ = ACTIVEBOX$
2920   X = NUMBEROFCOMPONENTS : Y = NUMBEROFDEADTIMES
```

```
2930   IF AX$ = "TFC" THEN COMPONENT = (COMPONENT+X-2) MOD X + 1
2940   IF AX$ = "DTD" THEN DTNUMBER = (DTNUMBER+Y-2) MOD Y + 1
2950   IF AX$ = "TFC" THEN GOSUB 2500 ' Print in Box
2960   IF AX$ = "DTD" THEN GOSUB 2600 ' Print in Box
2970   RETURN
3000   '============== PRINT TEXT ON BODE GRAPH ================
3010   LOCATE 1,2 : PRINT USING PID$; PMODE,IMODE,DMODE,ALPHA
3020   LOCATE 4,24: PRINT TITLE$;
3025   '
3030   LOCATE 4,3 : PRINT "Gain";
3035   LOCATE 5,4 : PRINT "in";
3040   LOCATE 6,1 : PRINT "Decibels";
3045   LINE (0,51)-(63,51),1,,&HAAAA
3050   '
3055   LOCATE 25,24 : PRINT "LOG RADIAN FREQUENCY";
3060   '
3065   LOCATE 18,68 : PRINT "Phase";
3070   LOCATE 19,68 : PRINT "Angle";
3075   LOCATE 20,69 : PRINT "in";
3080   LOCATE 21,68 : PRINT "Degrees";
3085   LINE (536,171)-(592,171),1,,&HCCCC
3090   RETURN
3099   '
3200   '============== PRINT DESIGN DECISION TABLE =============
3201   LINE1$ = "  Design Decision Data   "
3202   LINE2$ = "W(- 82)=##.#^^^^ rad/sec"
3203   LINE3$ = "W(-170)=##.#^^^^ rad/sec"
3204   LINE4$ = "W(-180)=##.#^^^^ rad/sec"
3205   LINE5$ = "W(-270)/W(-180)=###.#    "
3206   LINE6$ = "M(-140)=###.# db         "
3207   LINE7$ = "M(-180)=###.# db         "
3208   LINE8$ = "Gain Margin=###.# db     "
3209   LINE9$ = "Phase Margin=### degrees"
3210   '
3215   LINE (435,16)-(435,88),1,,&HAAAA
3220   LINE (436,88)-(639,88),1,,&HAAAA
3225   LINE (639,16)-(639,88),1,,&HAAAA
3230   '
3235   LOCATE  3,56 : PRINT LINE1$;
3240   LOCATE  4,56 : PRINT USING LINE2$; W82;
3245   LOCATE  5,56 : PRINT USING LINE3$; W170;
3250   LOCATE  6,56 : PRINT USING LINE4$; W180;
3255   LOCATE  7,56 : PRINT USING LINE5$; W270/W180
3260   LOCATE  8,56 : PRINT USING LINE6$; M140;
3265   LOCATE  9,56 : PRINT USING LINE7$; M180;
3270   LOCATE 10,56 : PRINT USING LINE8$; -M180;
3275   LOCATE 11,56 : PRINT USING LINE9$; 180+ANGLE0DB;
3280   RETURN
3299   '
3400   '========== PRINT AXES AND LABELS ==========
3410   ' Print gain axis, hash marks, and labels
3415   LINE (75,16)-(75,179),1,,&HAAAA
3420   FOR J = 0 TO 4
3425      GMARK = GMAX - J*(GMAX - GMIN)/4
3430      LOCATE 3+J*5,4 : PRINT USING "####"; GMARK;
3445      LINE (68, 19+J*40)-(82,19+J*40),1,,&HAAAA
3450   NEXT J
3455   '
3460   ' Print angle axis, hash marks, and labels
```

```
3465  LINE (475,88)-(475,179),1,,&HAAAA
3470  FOR J = 0 TO 2
3475     AMARK = AMAX - (J+2)*(AMAX - AMIN)/4
3480     LOCATE 13+J*5,62 : PRINT USING "####"; AMARK;
3485     LINE (468,99+J*40)-(482,99+J*40),1,,&HAAAA
3490  NEXT J
3495  '
3500  ' Print frequency axis, hash marks and labels
3505  LINE (75,179)-(475,179),1,,&HAAAA
3510  FOR J = 0 TO 10
3515     FMARK = FMIN + J*(FMAX - FMIN)/10
3520     LOCATE 24,9+J*5 : PRINT USING "##.#"; FMARK;
3525     LINE (75+J*40,176)-(75+J*40,182),1
3530  NEXT J
3535  RETURN
3599  '
3600  '============= DRAW GAIN AND ANGLE GRAPHS ==============
3610  VIEW (75,19)-(475,179)              ' set view for graph
3615  '
3620  WINDOW (FMIN,GMIN)-(FMAX,GMAX)        ' set graph window
3625  LINE (LW(1),DB(1))-(LW(2),DB(2)),1,,&HAAAA
3630  FOR J = 3 TO K
3635     LINE -(LW(J),DB(J)),1,,&HAAAA      ' draw angle graph
3640  NEXT J
3645  '
3650  WINDOW (FMIN,AMIN)-(FMAX,AMAX)        ' set graph window
3655  LINE (LW(1),ANGLE(1))-(LW(2),ANGLE(2)),1,,&HCCCC
3660  FOR J = 1 TO K
3665     LINE -(LW(J),ANGLE(J)),1,,&HCCCC ' draw angle graph
3670  NEXT J
3675  '
3680  WINDOW : RETURN
3699  '
3800  '====== DRAW SOLID LINE GAIN AND ANGLE GRAPHS ======
3810  VIEW (75,19)-(475,179)              ' set view for graph
3815  '
3820  WINDOW (FMIN,GMIN)-(FMAX,GMAX)  ' set graph window
3825  LINE (LW(1),DB(1))-(LW(2),DB(2)),1
3830  FOR J = 3 TO K
3835     LINE -(LW(J),DB(J)),1           ' draw gain graph
3840  NEXT J
3845  '
3850  WINDOW (FMIN,AMIN)-(FMAX,AMAX)  ' set graph window
3855  LINE (LW(1),ANGLE(1))-(LW(2),ANGLE(2)),1
3860  FOR J = 1 TO K
3865     LINE -(LW(J),ANGLE(J)),1        ' draw angle graph
3870  NEXT J
3875  WINDOW
3880  RETURN
3899  '
4000  '=============== GET VALID COMMAND ================
4002  LINE2$ = "Command: (Dmode) (Imode) (Pmode) (Analysis) "
             + "(Zoom) (Unzoom) (Quit)----> "
4005  VIEW (2,8)-(637,15),,1
4010  VALID = 0
4020  LOCATE 2,3  : PRINT LINE2$;
4030  LOCATE 2,75 : PRINT CHR$(219); : C$ = INPUT$(1)
4040  IF ASC(C$) > 96 THEN C$ = CHR$(ASC(C$)-32) ' Upcase C$
4050  IF (C$ = "D") OR (C$ = "I") OR (C$ = "P") THEN VALID = 1
```

```
4060   IF (C$ = "A") OR (C$ = "Z") OR (C$ = "U") THEN VALID = 1
4070   IF (C$ = "Q") THEN VALID = 1
4080   IF VALID = 0 THEN BEEP : GOTO 4010    ' Try again
4090   CMND$ = C$
4095   RETURN
4099   '
4200   '=========== DESIGN DERIVATIVE MODE ============
4201   GOSUB  700  ' print open-loop Bode graph
4202   LINE1$ = "Derivative mode design:  Dmode is "
4203   DU$ = "definitely useful.                        "
4204   PU$ = "probably useful.                          "
4205   MU$ = "marginally useful.                        "
4206   LINE2A$ = "Suggested D =##.##^^^^ second,  Current "
              + "D =##.##^^^^,  Enter D--> "
4207   LINE2B$ = "Current " + CHR$(224) + " =##.##        Enter"
              + " desired " + CHR$(224) + "--> "
4210   LOCATE 1,2 : PRINT LINE1$;
4215   IF W270/W180 > 5 THEN PRINT DU$;
       ELSE IF W270/W180 > 2 THEN PRINT PU$; ELSE PRINT MU$;
4220   DREC = 2/W180
4225   GOSUB 4360  ' clear and put border around view window
4230   LOCATE 2,2: PRINT USING LINE2A$; DREC, DMODE;
4235        INPUT "", DMODE
4240   GOSUB 4360  ' clear and put border around view window
4245   LOCATE 2, 2 : PRINT USING LINE2B$; ALPHA;: INPUT"", ALPHA
4250   GOSUB 4300  ' implement control mode design
4255   RETURN
4299   '
4300   '========== IMPLEMENT CONTROL MODE DESIGN ============
4305   GOSUB 1200   ' update controller transfer function
4310   GOSUB 4360   ' clear and put border around view window
4315   GOSUB  500   ' compute Bode data
4320   GOSUB  600   ' find Design Decision data
4325   GOSUB 3200   ' update Design Decision Table
4330   GOSUB 3800   ' plot gain and phase - solid lines
4335   GOSUB 4360   ' clear and put border around view window
4340   GOSUB 4380   ' hold the screen
4345   GOSUB 700    ' print open-loop Bode graph
4350   RETURN
4355   '
4360   '===== CLEAR AND PUT BORDER AROUND VIEW WINDOW ========
4365   VIEW (2,8)-(637,15),,1 : CLS : RETURN
4370   '
4380   '================== HOLD THE SCREEN ==================
4385   HOLD$ = "Press Enter to continue----> " + CHR$(219)
4390   LOCATE 2,2 : PRINT HOLD$; : C$ = INPUT$(1) : RETURN
4399   '
4400   '=============== DESIGN INTEGRAL MODE ================
4401   GOSUB  700   ' print open-loop Bode graph
4402   LINE1$ = "Integral mode design:  Imode is useful if "
              + "gain graph is level on left side.  "
4403   LINE2$ = "Suggested I =##.##^^^^ 1/second,  Current I "
              + "=##.##^^^^,  Enter I--> "
4410   '
4420   IF W82 < .2*W170 THEN IREC = W82 ELSE IREC = .2*W170
4430   LOCATE 1,2 : PRINT LINE1$;
4440   GOSUB 4360   ' clear and put border around view window
4450   LOCATE 2,2 : PRINT USING LINE2$; IREC, IMODE;
4460        INPUT "", IMODE
```

```
4470   GOSUB 4300  ' implement control mode design
4490   RETURN
4499   '
4600   '============= DESIGN PROPORTIONAL MODE ==============
4601   GOSUB  700  ' print open-loop Bode graph
4602   LINE1$ = "Proportional mode:  Adjust P for min 6db gain"
           + "margin and 40 deg. phase margin."
4603   LINE2$ = "Suggested P =##.##^^^^,  Current P ="
           + "##.##^^^^,  Enter P--> "
4605   '
4610   MGM = -M180 - 6  :  MPM = -M140
4620   IF MGM < MPM THEN PDB = MGM ELSE PDB = MPM
4630   PREC = PMODE*10^(PDB/20)
4640   LOCATE 1,2 : PRINT LINE1$;
4650   GOSUB 4360  ' clear and put border around view window
4660   LOCATE 2,2 : PRINT USING LINE2$; PREC, PMODE;
4670       INPUT "", PMODE
4680   GOSUB 4300  ' implement control mode design
4690   RETURN
4699   '
4700   '============= GET INPUT BOX COMMAND ===============
4705   LOCATE 1,2 : COLOR 2 : PRINT AB$; : AX$ = ACTIVEBOX$
4710   IF AX$="TFC" THEN PRINT AB11$;:PRINT USING AB12$;NUMBEROFCOMPONENTS
4715   IF AX$="DTD" THEN PRINT AB21$;:PRINT USING AB22$;NUMBEROFDEADTIMES
4720   VALID = 0
4725   LOCATE 2,2 : COLOR 12 : PRINT CM1$; CM2$ : COLOR 7
4730   LOCATE 3,4 : IF AX$ = "TFC" THEN PRINT CM11$; CM12$
4735   LOCATE 3,4 : IF AX$ = "DTD" THEN PRINT CM21$; CM22$
4740   LOCATE 4,4 : IF AX$ = "TFC" THEN PRINT CM13$; CM14$
4745   LOCATE 4,4 : IF AX$ = "DTD" THEN PRINT CM23$; CM24$
4750   LOCATE 2,79,1,0,7 : CMND$ = INPUT$(1)
4755   IF ASC(CMND$) > 96 THEN CMND$ = CHR$(ASC(CMND$)-32) 'Upcase CMND$
4760   IF (CMND$ = "A") OR (CMND$ = "E") THEN VALID = 1
4765   IF (CMND$ = "N") OR (CMND$ = "P") THEN VALID = 1
4770   IF (CMND$ = "M") OR (CMND$ = "Q") THEN VALID = 1
4775   IF VALID = 0 THEN BEEP : GOTO 4720    ' Try again
4780   RETURN
4799   '
5000   '==================== ANALYSIS =====================
5001   LINE1$ = "Analysis options: Closed-loop Bode, Nyquist"
           + " diagram, Error-ratio diagram, or Open-loop Bode"
5002   LINE2$ = "Command: (Closed-loop) (Nyquist) (Error-ratio)"
           + " (Open-loop) ----> "
5005   '
5010   LOCATE 1,2 : PRINT LINE1$;
5012   GOSUB 4360  ' clear and put border around view window
5015   LOCATE 2,2 : PRINT LINE2$;
5020   VALID = 0
5025   LOCATE 2,66 : PRINT CHR$(219); : C$ = INPUT$(1)
5030   IF ASC(C$) > 96 THEN C$ = CHR$(ASC(C$)-32) ' Upcase C$
5035   IF (C$="C") OR (C$="E") OR (C$="N") OR (C$="O") THEN VALID = 1
5040   IF VALID = 0 THEN BEEP : GOTO 5025    ' Try again
5045   IF C$ = "C" THEN GOSUB 5100   ' Closed-loop Bode
5050   IF C$ = "E" THEN GOSUB 5300   ' Error-ratio
5055   IF C$ = "N" THEN GOSUB 5500   ' Nyquist diagram
5060   IF C$ = "O" THEN GOSUB  700   ' Nyquist diagram
5090   RETURN
5099   '
```

```
5100 '================ CLOSED-LOOP BODE GRAPH ==================
5101 TITLE$ = "CLOSED-LOOP BODE GRAPH"
5105 GRAPH$ = "C"            ' set graph flag to closed-loop
5110 SCREEN 2                ' set high resolution graphic mode
5115 VIEW (0,0)-(639,199) ' set view to full screen
5120 CLS
5125 GOSUB 3000             ' print text on Bode graph
5130 GOSUB 3200             ' print Design Decision Table
5135 GOSUB 3400             ' print axes and their labels
5140 GOSUB 5200             ' print closed-loop graphs
5145 RETURN
5199 '
5200 '======= PRINT CLOSED-LOOP GAIN AND ANGLE GRAPHS =======
5210 VIEW (75,19)-(475,179)
5215 '
5220 WINDOW (FMIN,GMIN)-(FMAX,GMAX)        ' set graph window
5225 LINE (LW(1),DBC(1))-(LW(2),DBC(2)),1,,&HAAAA
5230 FOR J = 3 TO K
5235    LINE -(LW(J),DBC(J)),1,,&HAAAA      ' draw gain graph
5240 NEXT J
5245 '
5250 WINDOW (FMIN,AMIN)-(FMAX,AMAX)        ' set graph window
5255 LINE (LW(1),ANGLEC(1))-(LW(2),ANGLEC(2)),1,,&HCCCC
5260 FOR J = 1 TO K
5265    LINE -(LW(J),ANGLEC(J)),1,,&HCCCC ' draw angle graph
5270 NEXT J
5275 '
5280 WINDOW : RETURN
5299 '
5300 '================== ERROR-RATIO GRAPH ==================
5301 TITLE$ = "  ERROR-RATIO GRAPH      "
5305 GRAPH$ = "E"           ' set graph flag to error-ratio
5310 SCREEN 2               ' set high resolution graphic mode
5315 VIEW (0,0)-(639,199) ' set view to full screen
5320 CLS
5325 GOSUB 3000             ' print text on Bode graph
5330 GOSUB 3200             ' print Design Decision Table
5335 GOSUB 3400             ' print axes and their labels
5340 GOSUB 5400             ' print error-ratio graph
5345 RETURN
5399 '
5400 '============== PRINT ERROR-RATIO GRAPH ===============
5401 FL$ = "Frequency limit = ##.##^^^^ radian/second"
5410 VIEW (75,19)-(475,179)          ' set view for graph
5415 '
5420 WINDOW (FMIN,GMIN)-(FMAX,GMAX)  ' set graph window
5425 LINE (LW(1),ER(1))-(LW(2),ER(2)),1,,&HAAAA
5430 FOR J = 3 TO K
5435   LINE -(LW(J),ER(J)),1,,&HAAAA ' draw error-ratio graph
5440 NEXT J
5445 '
5450 LOCATE 5,13
5455 PRINT USING FL$; FREQLIMIT
5460 WINDOW : RETURN
5499 '
5500 '==================== NYQUIST DIAGRAM ==================
5501 TITLE$ = "    NYQUIST DIAGRAM     "
5505 GRAPH$ = "N"                        ' graph flag = Nyquist
```

```
5510  SCREEN 2                            ' set hi-res graph mode
5515  VIEW (0,0)-(639,199)                ' set view to full screen
5520  CLS
5525  LOCATE 4,24: PRINT TITLE$;
5530  LOCATE 1,2 : PRINT USING PID$; PMODE,IMODE,DMODE,ALPHA
5535  '
5540  VIEW (75,19)-(475,179)              ' set view for graph
5545  WINDOW (NMIN,NMIN)-(NMAX,NMAX)      ' set graph window
5550  LINE (0,NMAX)-(0,NMIN),1,,&HAAAA    ' draw y axis
5555  LINE (NMIN,0)-(NMAX,0),1,,&HAAAA    ' draw x axis
5560  LINE (0,0)-(-1,-.8391)              ' draw -140 degree line
5565  CIRCLE (0,0),1,,,,5/12              ' draw unit circle
5570  '
5575  LINE (NR(1),NI(1))-(NR(2),NI(2)),1,,&HAAAA
5580  FOR J = 1 TO K
5585      LINE -(NR(J),NI(J)),1,,&HAAAA   ' draw Nyquist graph
5590  NEXT J
5595  WINDOW : RETURN
5599  '
6000  '======================= ZOOM =====================
6010  F0 = CINT(LOG(W180)/LOG(10))
6015  FMIN = F0 - 3 : FMAX = F0 + 2       ' set min & max freq.
6020  GMIN = -20    : GMAX = 20           ' set min & max gain
6025  AMIN = -200   : AMAX = 0            ' set min & max angle
6030  NMIN =   -2   : NMAX =   2          ' set min & max Nyquist
6035  IF GRAPH$ = "O" THEN GOSUB 700      ' print open loop graph
6040  IF GRAPH$ = "C" THEN GOSUB 5100     ' print closed-loop graph
6045  IF GRAPH$ = "E" THEN GOSUB 5300     ' print error-ratio graph
6050  IF GRAPH$ = "N" THEN GOSUB 5500     ' print Nyquist diagram
6055  RETURN
6099  '
6200  '======================= UNZOOM =====================
6210  FMIN =    -5 : FMAX =    5          ' set min & max freq.
6215  GMIN = -100 : GMAX = 100            ' set min & max gain
6220  AMIN = -200 : AMAX = 200            ' set min & max angle
6225  NMIN = -10  : NMAX = 10             ' set min & max Nyquist
6230  IF GRAPH$ = "O" THEN GOSUB 700      ' print open loop graph
6235  IF GRAPH$ = "C" THEN GOSUB 5100     ' print closed-loop graph
6240  IF GRAPH$ = "E" THEN GOSUB 5300     ' print error-ratio graph
6245  IF GRAPH$ = "N" THEN GOSUB 5500     ' print Nyquist diagram
6250  RETURN
6299  '
7000  '================== PRINT DATA TABLE ===================
7010  CLS : LOCATE 10,10,1
7020  PRINT "Turn on the printer and press ENTER when ready.";
7030  C$ = INPUT$(1)
7040  GOSUB 7200     ' print component transfer functions
7050  GOSUB 7400     ' print dead time lags
7060  GOSUB 7800     ' print data table
7070  GOSUB 8000     ' print design summary
7080  RETURN
7099  '
7200  '============ PRINT COMPONENT TRANSFER FUNCTIONS =========
7210  LPRINT M$; L$  :  LPRINT M$; P$  :  LPRINT M$; L$
7220  FOR J = 1 TO NUMBEROFCOMPONENTS
7230      LPRINT USING M$ + A$; A(J,0), A(J,1), A(J,2), A(J,3)
7240      LPRINT USING M$ + B$; B(J,0), B(J,1), B(J,2), B(J,3)
7250      LPRINT M$ + L$
7260  NEXT J
```

```
7270  LPRINT
7280  RETURN
7299  '
7400  '================== PRINT DEAD-TIME LAGS ==================
7410  FOR J = 1 TO NUMBEROFDEADTIMES
7420      LPRINT USING M$ + E$;J,DEADTIME(J)
7430  NEXT J
7440  RETURN
7499  '
7800  '==================== PRINT DATA TABLE ====================
7810  LPRINT : LPRINT
7820  LPRINT M$+H1$: LPRINT M$+H2$: LPRINT M$+H3$: LPRINT M$+H4$
7830  LPRINT M$+H5$ : LPRINT M$+H6$ : LPRINT M$+H2$
7840  FOR J = 1 TO K
7850    W = 10^LW(J)
7860    IF (ANGLE(J)<-999) OR (ANGLEC(J)<-999) GOTO 7890
7870    LPRINT USING M$+N1$;W,DB(J),ANGLE(J),DBC(J),ANGLEC(J),ER(J)
7880  NEXT J
7890  RETURN
7899  '
8000  '================ PRINT DESIGN SUMMARY ==================
8001  H1$="DESIGN SUMMARY"
8002  H2$="     Proportional gain:                ##.##^^^^"
8003  H3$="     Integral action rate:            ##.##^^^^ 1/sec."
8004  H4$="     Derivative action time constant: ##.##^^^^ second"
8005  H5$="     Derivative limiter: ##.###"
8006  H6$="     M(-180):          ###.# decible"
8007  H7$="     ANGLE(0DB):      ####.# degree"
8008  H8$="     Gain margin:      ###.# decibel"
8009  H9$="     Phase margin:     ####  degree"
8010  H0$="     Frequency limit:##.#^^^^ radian/second"
8011  LPRINT : LPRINT
8015  LPRINT M$ + H1$
8020  LPRINT USING M$ + H2$; PMODE
8025  LPRINT USING M$ + H3$; IMODE
8030  LPRINT USING M$ + H4$; DMODE
8035  LPRINT USING M$ + H5$; ALPHA
8040  LPRINT USING M$ + H6$; M180
8045  LPRINT USING M$ + H7$; ANGLE0DB
8050  LPRINT USING M$ + H8$; -M180
8055  LPRINT USING M$ + H9$; 180 + ANGLE0DB
8060  LPRINT USING M$ + H0$; FREQLIMIT
8065  RETURN
8099  '
```

References

Arthur, K. *Transducer Measurements*. Beaverton, Ore.: Tektronix, Inc. 1970.

Bateson, R. N. *Motomatic Servo Control Course Manual*. Hopkins, Minn.: Electro-Craft Corporation, 1971.

Binder, R. C. *Fluid Mechanics*. Englewood Cliffs, N.J.: Prentice-Hall, Inc., 1949.

Bogart, T. F., Jr. *Laplace Transforms and Control Systems Theory for Technology*. New York: John Wiley & Sons, Inc., 1982.

Bower, J. L. and P. M. Schultheiss. *Introduction to the Design of Servomechanisms*. New York: John Wiley & Sons, Inc., 1964.

Bryan, G. T. *Control Systems for Technicians*. Hart Publishing Company, Inc., 1967.

Buckley, P. S. *Techniques of Process Control*. New York: John Wiley & Sons, Inc., 1964.

Chemical Engineering, Vol. 76, No. 12, June 2, 1969.

Diefenderfer, A. J. *Principles of Electronic Instrumentation*. Philadelphia: W. B. Saunders Company, 1972.

Eckman, D. P. *Industrial Instrumentation*. New York: John Wiley & Sons, Inc., 1950.

Erickson, W. H. and N. H. Bryant. *Electrical Engineering Theory and Practice*. New York: John Wiley & Sons, Inc., 1952.

Floyd, Thomas L. *Digital Fundamentals*. New York: Macmillan Publishing Company, 1990.

Fundamentals of Industrial Instrumentation. Philadelphia: Honeywell, Inc., 1957.

Harriott, P. *Process Control*. New York: McGraw-Hill Book Company, 1964.

Harrison, H. L. and J. G. Bollinger. *Introduction to Automatic Controls*. Scranton, Pa.: International Textbook Company, 1969.

Herman, S. L. and W. N. Alerich. *Industrial Motor Control*. Albany, N.Y.: Delmar Publishers, Inc., 1985.

Hordeski, M. F. *Design of Microprocessor, Sensor and Control Systems*. Englewood Cliffs, N.J.: Reston Publishing Company, Inc., 1985.

Jennings, B. H. and S. R. Lewis. *Air Conditioning and Refrigeration*. Scranton, Pa.: International Textbook Company, 1949.

Johnson, C. D. *Process Control Instrumentation Technology.* New York: John Wiley & Sons, Inc., 1982.

Jones, C. T. and L. A. Bryan. *Programmable Controllers: Concepts and Applications.* Atlanta: International Programmable Controls, Inc., 1983.

Kafrissen, E. and M. Stephans. *Industrial Robots and Robotics.* Englewood Cliffs, N.J.: Reston Publishing Company, Inc., 1984.

Lesa, A. and R. Zaks. *Microprocessor Interfacing Techniques.* Oakland, Calif.: SYBEX Inc., 1978.

Liptak, B. G. and K. Venczel. *Instrument Engineers' Handbook Process Measurement.* Radnor, Pa.: Chilton Book Company, 1982.

Maloney, T. J. *Industrial Solid-State Electronics Devices and Systems.* Englewood Cliffs, N.J.: Prentice-Hall, Inc., 1979.

McGlynn, D. R. *Microprocessors: Technology, Architecture, and Applications.* New York: John Wiley & Sons, Inc., 1976.

McNamara, J. E. *Technical Aspects of Data Communication.* Maynard, Mass.: Digital Equipment Corporation, 1977.

Morris, N. M. *Control Engineering.* Maidenhead, Berkshire, England: McGraw-Hill Book Company (U.K.) Ltd., 1968.

Nachtigal, Chester L. *Instrumentation and Control.* New York: John Wiley & Sons, Inc., 1990.

Norton, H. N. *Sensor and Analyzer Handbook.* Englewood Cliffs, N. J.: Prentice-Hall, Inc., 1982.

Olesten, Nils O. *Numerical Control.* New York: John Wiley & Sons, Inc., 1970.

Principles of Automatic Process Control. Research Triangle Park, N.C.: Instrument Society of America, 1968 (text and film).

Principles of Frequency Response. Research Triangle Park, N.C.: Instrument Society of America, 1958 (text and film).

Seyer, M. D. *RS-232 Made Easy: Connecting Computers, Printers, Terminals, and Modems.* Englewood Cliffs, N.J.: Prentice-Hall, Inc., 1984.

Shinskey, F. G. *Process Control Systems.* New York: McGraw-Hill Book Company, 1967.

Stein, D. H. *Introduction to Digital Data Communications.* Albany, N.Y.: Delmar Publishers, Inc., 1985.

Tucker, G. K. and D. M. Wills. *A Simplified Technique of Control System Engineering.* Philadelphia: Minneapolis-Honeywell Regulator Company, 1962.

Weathers, Tom, Jr. and Claud C. Hunter. *Automotive Computers and Control Systems.* Englewood Cliffs, N.J.: Prentice-Hall, Inc., 1984.

Webb, John W. *Programmable Logic Controllers*, 2nd edition. New York: Macmillan Publishing Company, 1992.

Wilson, H. S. and L. M. Zoss, *Control Theory Notebook.* Research Triangle Park, N.C.: Instrument Society of America. Reprint from *ISA Journal.*

Zeines, B. *Automatic Control Systems.* Englewood Cliffs, N.J.: Prentice-Hall, Inc., 1972.

Ziegler, J. G. and N. B. Nichols. "Optimum Settings for Automatic Controllers," *ASME Transactions*, Vol. 64, No. 8, 1942, pp. 759–768.

Answers to Selected Exercises

Chapter 1

1.3 Size and timing. The transfer function.

1.4 Gain = 0.492 gpm/psi, Phase difference = 22°, Transfer function = $0.492\underline{/22°}$.

1.8 Open-loop advantages: inexpensive, simple, trouble-free, no instability.
Disadvantages: cannot compensate for disturbances, must be calibrated.

1.12 Component 1: Dead band and saturation
Component 2: Hysteresis
Component 3: Nonlinearity
Component 4: Hysteresis, dead band, and saturation.

1.13

Component	Maximum Output Difference
1	24 at input = 4 to 20
2	40 at input = 12.5
3	Not applicable
4	63 at input = 12.5

1.15 Critical damping.

Chapter 2

2.2

Time	Temperature	Quantization Error
11:30	71.1°F	0.1°F
11:31	71.2°F	0.2°F
11:32	71.4°F	0.4°F
11:33	71.6°F	0.6°F
11:34	71.8°F	0.8°F
11:35	71.9°F	0.9°F
...	...	...
11:42	73.1°F	0.1°F

2.3 (Possible answers)

Regulator system: automobile cruise control
Follow-up system: automobile power steering

2.7 (a) $k_f = 6.32 \times 10^{-6}$

(b) $k = 2.13 \times 10^{-8}$

(c) $w = 2.17 \times 10^{-3}$ kg/s

2.9 Event-driven and time-driven

2.13 NC: Numerical control is a system that uses predetermined instructions to control a sequence of manufacturing operations.

CNC: Computerized numerical control uses a dedicated computer to accept the input of instructions and to perform the control functions required to produce a part.

DNC: Direct numerical control is a system in which a number of numerical control machines are connected to a central computer for real-time access to a common database of part programs and machine programs.

2.16 *Distributed control:* The controllers are close to their measuring transmitters and actuators. Advantage: Produces good control due to short loops. Disadvantage: Requires a communication network to keep the operator informed.

Centralized control: The controllers are located in a central control room. Advantage: The operator can see all controllers at once and is able to monitor the process very closely. Disadvantage: Long transmission lines are required for each control loop, which is expensive and degrades control.

2.17 Process control: 2.14, 2.16, 2.17, 2.19, 2.20, 2.22
Servomechanism: 2.15, 2.18
Sequential control: 2.21

2.19 (b) Regulator

(d) Regulator

(f) Time-driven sequential

(h) Follow-up

(j) Event-driven sequential

Chapter 3

3.1

	Binary	Octal	Decimal	Hexadecimal
(a)	1011	13	11	B
(c)	1000 1010	212	138	8A

3.2

	Octal	Binary	Decimal	Hexadecimal
(a)	17	1111	15	F
(c)	251	1010 1001	169	A9

3.3

	Decimal	Binary	Octal	Hexadecimal
(a)	47	10 1111	57	2F
(c)	132	1000 0100	204	84

3.4

	Hexadecimal	Binary	Octal	Decimal
(a)	2A	10 1010	52	42
(c)	3A6	11 1010 0110	1646	934

3.6 Largest positive number: $0111\ 1111_2 = 127$ (decimal)
Negative number with the largest magnitude:

$$1000\ 0000_2 = -128 \text{ (decimal, 2's complement)}$$

$$1000\ 0000_2 = -127 \text{ (decimal, 1's complement)}$$

$$1111\ 1111_2 = -127 \text{ (decimal, sign + magnitude)}$$

3.9 The negated-input OR has the same truth table as the NAND.
Conclusion: NOT A OR NOT B = NOT (A AND B)

3.11

Decimal	Gray	Decimal	Gray
16	11000	24	10100
17	11001	25	10101
18	11011	26	10111
...	...	...	...
23	11100	31	10000

3.13 Decimal inputs to bit 0 OR element: 1, 3, 6, 8
Decimal inputs to bit 1 OR element: 2, 3, 7, 8
Decimal inputs to bit 2 OR element: 4, 9
Decimal inputs to bit 3 OR element: 5, 6, 7, 8, 9

3.16 $Z = \bar{A} \cdot \bar{B} \cdot C + \bar{A} \cdot B \cdot C + A \cdot B \cdot \bar{C}$
$\quad = \bar{A} \cdot C + A \cdot B \cdot \bar{C}$

3.18 $Y = \bar{A} \cdot \bar{B} + A \cdot B$

Chapter 4

4.1 $R = 310\ \Omega$

4.3 $C = 7\ \mu\text{F}$

4.5 $t_d = 74\ \mu\text{s}$

4.7 Reynolds number = 9.57, flow is laminar
Resistance = 5.69 psi/gpm
Pressure drop = 11.4 psi

4.9 Reynolds number = 12,527, flow is turbulent
Resistance = 1.68×10^7 Pa·s/m³
Pressure drop = 1.48 kPa

4.11 $I_L = 5.45 \times 10^8$ Pa·s²/m³

4.13 $p = 3.6$ kPa
$p_1 = 106.6$ kPa
$R_g = 171$ kPa·s/kg

4.15 (a) $C_g = 1.50 \times 10^{-5}$ kg/Pa
(c) $C_g = 1.83 \times 10^{-5}$ kg/Pa

4.17 $R_u = 3.7 \times 10^{-2}$ K·m²/W
$R_T = 7.1 \times 10^{-3}$ K/W
$Q = 2.8$ kW
Heat flows from inside (oil) to outside (water)

4.19 $C_T = 3.31 \times 10^6$ J/K

4.21 $R_m = 1.5$ N·s/m
$F_c = 9$ N
$F = 9 + 1.5\ V$

4.23 Section A: $F_I = 8.61$ N

Section B: $F_I = -12.5$ N

4.25 (a) $V_o = 26.82$ m/s, $m = 1,179.4$ kg, $D_b = 48.77$ m

(b) $a = -7.376$ m/s^2, $f = -8,699$ N

(c) $f = 1,955.6$ lbf

(d) $a = -24.2$ f/s^2, $f = 1,954$ lbf

(e) The results are almost identical. The small difference can be attributed to the approximation used for g_c.

Chapter 5

5.1 $579 \dfrac{dh}{dt} + h = 220q_{in}$

5.3 $0.103 \dfrac{de_{out}}{dt} + e_{out} = e_{in}$

5.5 $0.0182p_{in} = 0.561 \dfrac{d^2x}{dt^2} + 9.6 \dfrac{dx}{dt} + 60,400x$

5.7 (a) $f(t) = 6.7t$

(c) $f(t) = 45e^{-72t}$

(e) $f(t) = 650te^{-8t}$

(g) $f(t) = 82(1 - e^{-t/5})$

(i) $f(t) = 28 \cos \omega t$

5.9 $\dfrac{F_o(s)}{F_i(s)} = e^{-245s}$

5.11 $\dfrac{X(s)}{I(s)} = \dfrac{0.3}{1 + 0.02s + 0.0001s^2}$

5.13 $\dfrac{H(s)}{Q(s)} = \dfrac{0.5}{s}$

5.15 $\dfrac{I(s)}{E(s)} = \dfrac{1.636 + 3.6s + 0.0288s^2}{s}$

5.17 Exercise 5.10: $\dfrac{I}{\Theta} = \dfrac{5.7}{1 + 8.6s}$

Exercise 5.12: $\dfrac{\Theta}{X} = \dfrac{125}{a + 26s + 25s^2}$

Exercise 5.15: $\dfrac{I}{E} = \dfrac{1.636 + 3.6s + 0.0288s^2}{s}$

Chapter 6

6.1 (a) Mean = 15.52 psi, $S_\chi = 0.06790$ psi

(b) Measured accuracy = $+0.61$ psi, -0 psi

Bias = 0.52 psi

Repeatability = 0.19 psi

(c) Range of measurements: 15.32 to 15.72 psi

6.3 Dead band = 0.05 psi

= 0.56% of lower input value

= 0.42% of span

6.5 *Static* characteristics describe the accuracy of a measuring instrument at room temperature with the measured variable constant or changing very slowly.

Dynamic characteristics describe the performance of a measuring instrument when the measured variable is changing rapidly.

6.7 Ideal position = 64°
$60.8° \leq$ position $\leq 67.2°$

6.9 Resolution = 0.025 V/turn = 0.125%

6.11 Sensitivity = 0.285 V/N at 20%
Sensitivity = 0.315 V/N at 80%

6.13 10 to 90% rise time = 10.7 s
Overshoot = 20%
2% settling time = 52 s

6.15 Time constant = 89.3 s

6.17 Transfer function $= \dfrac{1}{1 + 193s + 2633s^2}$

Chapter 7

7.1 $+V_{\text{sat}} = 14.40$ V, $v_2 - v_1 = 7.20 \ \mu$V
$-V_{\text{sat}} = -12.8$ V, $v_2 - v_1 = -6.40 \ \mu$V

7.3 CMMR $= 3.125 \times 10^6$, CMR $= 129.9$ dB

7.5 Inverting amplifier with $R_f = 181.8$ kΩ

7.7 The summing amplifier in Figure 7.6a with the following parameter values:

$$v_a = 1 \text{ V (arbitrarily chosen)}$$
$$v_b = v_{\text{in}}$$
$$R_a = R_f = 181.8 \text{ k}\Omega$$
$$R_b = 1 \text{ k}\Omega$$

7.9 $R_a = 1$ kΩ

7.11 (a) Single-stage low-pass filter
$\tau = 0.1$ s
$C = 10 \ \mu$F (arbitrarily chosen)
$R = 10$ kΩ

$$\text{TF} = \frac{1}{1 + 0.1s}$$

(b) Two-stage low-pass filter
$\tau = 0.01$
$C = 1 \ \mu$F (arbitrarily chosen)
$R = 10$ kΩ

$$\text{TF} = \frac{1}{1 + 0.02s + 0.0001s^2}$$

Comment: The two-stage low-pass filter has a steeper slope than the single-stage filter. Consequently, the two-stage filter has less attenuation below 1000 rad/s and more attenuation above 1000 rad/s than the single-stage filter.

7.12 $K = 40$

$$\text{TF} = \frac{1 + 0.002s + 1 \times 10^{-6}s^2}{1 + 0.040025s + 1 \times 10^{-6}s^2}$$

7.15 $v_s = 0.3096$ V, $R_2 = 294$ Ω,
$R_3 = 119.8$ Ω, $R_4 = 294$ Ω,
$\alpha = 0.2895$, $R_a = 1$ kΩ,
$R_f = 59.1$ kΩ, $R_c = 1$ kΩ,
$R_1 = 3.11$ kΩ

T (°C)	R_s (Ω)	ε	$v_b - v_a$	v_1	v_{out}
0	100.0	−0.1653	−0.0357	2.110	1.00
25	109.8				
50	119.8	0	0	0	3.11
75	129.6				
100	139.3				5.00

7.18 (a) Analog output of 0101 = 1.56 V
    (c) Analog output of 1000 = 2.50 V
7.19 (a) Number of steps = 15
       Quantization error = ±0.333 V
       Percent accuracy = ±3.33%
    (c) Number of steps = 1023
       Quantization error = ±0.00489 V
       Percent accuracy = ±0.0489

Chapter 8

8.2 Transmission time = 97.5 ms
8.4 Logical address space = 2^{28} = 268,435,454 bytes
Physical address space = 2^{18} = 262,144 bytes
8.9 $BPS_{max} = 1200$
8.12 A *baseband* network has a single channel that transmits an unmodulated signal.
A *carrierband* network has a single channel that transmits a carrier modulated signal.
A *broadband* network has many channels, each with its own carrier frequency, that can transmit messages simultaneously on all its channels.
8.15 *Polling:* The nodes on the network take turns being invited to transmit. Only one frame of a message can be transmitted with each invitation. If a message consists of several frames, the frames will be separated from each other by one polling cycle.
Contention: Each node competes for access to the network, and some method is used to resolve collisions when two or more nodes try to initiate a transmission at the same time.

Chapter 9

9.1 $N = 500$ turns
9.3 (a) $\theta = 75°$, $e_{out} = 1.60 \sin 377t$ V
    (c) $\theta = 150°$, $e_{out} = -5.37 \sin 377t$ V
9.4 (a) Amplitude = 3.1, $\theta = 60°$
    (c) Amplitude = 5.5, $\theta = 27.5°$
9.5 (a) $\theta + \theta_d = 0°$, $e_{out} = 22.5 \sin 2513t$ V
    (c) $\theta + \theta_d = 65°$, $e_{out} = 9.51 \sin 2513t$ V
9.8 Number of positions = 65,536
9.9 Displacement per pulse = 1.571×10^{-4} m
$N_T = 40,107$ pulses

9.11 Selection: IP-3 (IP-4 or IP-5 could also be used, but IP-3 is smaller and less expensive)

9.13 $K_E = 0.01728$ V/rpm

Calibration equation: $E = 0.01728S$

9.15 (a) $a_{max} = \pm 66.7$ m/s^2

(b) $f_0 = 26.0$ Hz

(c) $b = 2.35$ N·s/m

(d) $f_{max} = 10.4$ Hz

9.17 $A = 4.17$ in^2

9.19 (a) 1,629 pulses

(b) 14 bits minimum

Chapter 10

10.1 (1) vapor   (2) gas   (3) liquid   (4) mercury

10.3 $R = 129.6$ Ω

The value of R predicted by the equation is the same as the value given in Table 10.1 (within the accuracy given in Table 10.1).

10.5 $E = 0.05175T + 1.05 \times 10^{-5}T^2$

$E(50) = 2.61$ mV

10.7

Controller Setting (%)	Flow Rate (gal/min)
25	0.0754
75	0.1266

10.9 (b) $N = 32,060$, $V = 1.09$ m^3

10.11 Range = 0 to 200 kPa

10.13 $f_{min} = -132.4$ N

$f_{max} = 63.8$ N

10.15 (a) Terminal-based linearity = 1.33%

(b) $i_{lin} = 1.031i_x - 0.31$ mA, 10 mA $\leq i_x \leq$ 19.7 mA

$i_{lin} = 1.010i_x + 0.101$ mA, 19.7 mA $\leq i_x \leq$ 29.6 mA

$i_{lin} = 0.990i_x + 0.693$ mA, 29.6 mA $\leq i_x \leq$ 39.7 mA

$i_{lin} = 0.971i_x + 1.456$ mA, 39.7 mA $\leq i_x \leq$ 50 mA

Chapter 11

11.3 (1) Increase the anode voltage above the forward breakdown voltage.

(2) Apply a positive voltage pulse to the gate input of the SCR.

(3) Apply light to the gate/cathode junction.

(4) Rapidly increase the anode-to-cathode voltage, V_{AC}.

11.5 The major difference between the triac and the SCR is that the triac can conduct in both directions, whereas the SCR can conduct in only one direction. The triac is equivalent to two SCRs connected in parallel but oriented in opposite directions.

11.7 Voltage gain = -83.0

Current gain = -15.14

11.9 The 6-in. cylinder is selected. The required flow rate is 2.44 gal/min, and the working pressure is 796 psi. The 6-in. cylinder fits in the available space.

11.12 $C_v = 156$; the required valve size is 4 in.

11.14 $C_v = 4.37$; the required value size is 1 in.

11.15 $Q_1 = 19.553$ kW

Chapter 12

12.2 Your paragraph should explain the following facts.
(1) A force is exerted on a current-carrying conductor in a transverse magnetic field.
(2) A voltage is induced in a conductor moving through a transverse magnetic field.

12.4 If the commutator of a dc generator is replaced by slip rings, an ac voltage will replace the dc voltage at the brushes. The dc generator becomes an ac alternator.

12.6 Velocity TF $= \dfrac{\Omega}{V_c} = \dfrac{4.252}{1 + 0.06378s}$

Position TF $= \dfrac{\Theta}{V_c} = \dfrac{4.252}{s + 0.06378s^2}$

12.9 $e = 48.08$ V
$\omega_2 = 308.83$ rad/s
$S_2 = 2949$ rpm

12.15 $R_1 = 8$ kΩ
$R_2 = 2$ kΩ
$R_3 = 0.08$ Ω

12.17 $\dfrac{\Omega}{\text{SP}} = \dfrac{KK_T}{(rb + KK_TK_m + K_EK_T) + rb(\tau_m + \tau_e)s + rb\tau_m\tau_e s^2}$

Chapter 13

13.3 1M = CRM
= NOT MASTER-STOP AND (START-MOTORS OR 1M) AND NOT OL

13.5 CR1 = NOT CR4 AND (START OR CR1) AND LS5
CR2 = CR1 AND NOT CR3 AND LS4 AND NOT LS2
CR3 = CR1 AND (LS2 OR CR3)
CR4 = NOT LS6 AND (CR3 AND LS1 OR CR4)
TD1 = CR4 AND NOT DELAY
Sol.A = CR2
Sol.B = CR3
Sol.C = CR1
Sol.D = TD1

13.11

Timing Diagram						
Step: 1	2	3	4	5	6	7
CR1 × × × × ×						
CR2	× × × × ×					
CR3		× × × × ×				
CR4			× × × × ×			
CR5				× × × × ×		
CR6					× × × × ×	
CR7						× × × × ×

Chapter 14

14.1 *Two-position:* The output has one of two possible values, depending on the sign of the error signal.

Floating: The output is either increasing, stationary, or decreasing, depending on the size and sign of the error.

Proportional: The output is proportional to the error.

Integral: The output is proportional to the integral of the error (or the output changes at a rate that is proportional to the error).

Derivative: The output is proportional to the derivative (or rate of change) of the error.

14.3 Floating control mode

14.5 (b) $e = 20\%$

14.7 (a) At $t = 15$ s, $v = 42.8\%$

(b) At $t = 30$ s, $v = 40.75\%$

14.9 (a) Noninteracting: $\dfrac{V}{E} = \dfrac{4 + 28s + 14s^2}{7s + 0.35s^2}$

(b) Interacting: $\dfrac{V}{E} = \dfrac{4 + 30s + 14s^2}{7s + 0.35s^2}$

14.11 Two-position mode

14.13 PI mode

The derivative mode amplifies the noise spikes that often occur in a flow signal.

14.15 PID mode

14.17 $R_i = 185$ kΩ

$R_1 = 28.9$ kΩ

14.19 $R_d = 320$ kΩ

$R_1 = 35.6$ kΩ

$R_i = 231$ kΩ

$C_i = 303$ μF

Chapter 15

15.1 $h^*(t_1) = \dfrac{1}{T_i} \displaystyle\int_{t_0}^{t_1} (q^*_{in} - q^*_{out}) \, dt + h^*(t_0)$

T_i = Integral action time constant = 636 s

TF = $H^*/Q^*_{in} = 1/(636s)$

$h^*(t_0 + 50) = 1.74$ m

15.3 (a) $C_L = 2.5 \times 10^{-4}$ m^3/Pa

(b) $R_L = 6.68 \times 10^7$ Pa·s/m^3

(c) $\tau = 16{,}700$ s

(d) $G = 0.605$

(e) $16{,}700 \dfrac{dh^*}{dt} + h^* = 0.605q^*_{in}$

(f) $\dfrac{H^*(s)}{Q^*_{in}(s)} = \dfrac{0.605}{1 + 16{,}700s}$

15.5 (a) Amplitude = 0.25 V, phase = 39.3°

(b) Amplitude = 0.354 V, phase = 0°

(c) Amplitude = 0.15 V, phase = −39.3°

15.8 (a) $C_g = 6.85 \times 10^{-6}$ kg/Pa

(b) $\tau = 1.23$ s

(c) $G = 4.50$

(d) $1.23 \dfrac{dp^*}{dt} + p^* = 4.5w^*$

(e) $\dfrac{P^*(s)}{W^*(s)} = \dfrac{4.5}{1 + 1.23s}$

15.10 (a) $0.0173 \dfrac{d^2h}{dt^2} + 0.029 \dfrac{dh}{dt} + h = 6.9 \times 10^{-4} f$

(b) $\dfrac{H(s)}{F(s)} = \dfrac{6.9 \times 10^{-4}}{1 + 0.029s + 0.0173s^2}$

(c) $\omega_o = 7.61$ rad/s

(d) $\zeta = 0.110$

(e) The process is underdamped.

15.11 (a) Amplitude $= 0.01$ m, phase $= -1.2°$

(b) Amplitude $= 0.05$ m, phase $= -90°$

(c) Amplitude $= 0.0001$ m, phase $= -179°$

15.13 $360,000 \dfrac{d^2h^*}{dt^2} + 1500 \dfrac{dh^*}{dt} + h^* = 0.561q^*{}_{in}$

$\dfrac{H^*(s)}{Q^*{}_{in}(s)} = \dfrac{0.561}{1 + 1500s + 360,000s^2}$

15.15 $\tau_m = 1.6$ s, $\tau_e = 0.0625$ s

$0.00472 \dfrac{d^2\omega}{dt^2} + 0.0785 \dfrac{d\omega}{dt} + \omega = 4.33E$

$\dfrac{\Omega(s)}{E(s)} = \dfrac{4.33}{1 + 0.0785s + 0.00472s^2}$

$\omega_o = 14.6$ rad/s

$\zeta = 0.57$, underdamped

15.17 $f_o(t) = f_i(t - 18.6)$

$\dfrac{F_o(s)}{F_i(s)} = e^{-18.6s}$

15.19 $t_d = 6.91$ s, $\tau = 1220$ s

$\dfrac{H(s)}{M(s)} = \left(\dfrac{1}{1 + 1220s}\right) e^{-6.91s}$

Chapter 16

16.1 $TF = \left(\dfrac{10}{1 + 0.5s}\right) e^{-0.05s}$

Gain margin $= 4.2$ dB, phase margin $= 37°$

The system does not satisfy the minimum criteria for gain margin or phase margin.

$\omega_{max} = 23.5$ rad/s

The system satisfies the Nyquist criterion for stability.

16.3 $TF = \left(\dfrac{0.628 + 9.1s}{s}\right)\left(\dfrac{1}{1 + 50s}\right) e^{-2s}$

Gain margin $= 11.9$ dB, phase margin $= 53°$

The system satisfies the minimum criteria for gain margin and phase margin.

$\omega_{max} = 0.402$ rad/s

The system satisfies the Nyquist criterion for stability.

16.5 $TF = \left(\dfrac{1 + 36s}{18s}\right) e^{-20s}$

No gain margin, no phase margin.

The system is unstable and does not satisfy the minimum criteria for gain margin or phase margin.

The maximum frequency limit has no meaning when the system is unstable.

The system does not satisfy the Nyquist criterion for stability.

16.7 $TF = \left(\dfrac{42.86 + 12s + 2.16s^2}{s + 0.018s^2}\right)\left(\dfrac{0.1}{1 + 0.02s + 0.01s^2}\right)e^{-0.05s}$

No gain margin, no phase margin.

The system is unstable and does not satisfy the minimum criteria for gain margin or phase margin.

The maximum frequency limit has no meaning when the system is unstable.

The system does not satisfy the Nyquist criterion for stability.

16.9 $TF = \left(\dfrac{0.01 + 0.00067s}{1 + 0.0067s}\right)\left(\dfrac{8500}{1 + 0.025s + 0.0001s^2}\right)\left(\dfrac{1}{s + 0.35s^2 + 0.015s^3}\right)$

No gain margin, no phase margin.

The system is unstable and does not satisfy the minimum criteria for gain margin or phase margin.

The maximum frequency limit has no meaning when the system is unstable.

The system does not satisfy the Nyquist criterion for stability.

16.11 $s = -\left(\dfrac{12K + 4}{K + 1}\right)$

The loci of s are on the real axis from -4 to -12.

16.13 $s1 = -10 + \sqrt{84 - 10K}$, $s2 = -10 - \sqrt{84 - 10K}$

The loci of s are on the real axis from -0.835 to -19.165 and the vertical line from $s = -10 - j\infty$ to $s = -10 + j\infty$.

A value of $K = 14.03$ results in a damping ratio of $\zeta = 0.8$.

16.15 The root locus has 2 branches. One branch terminates at the zero at $s = -5$. The other branch goes to infinity along a single asymptote. (R1,R2,R3)

The asymptote passes through the centroid, $c = -5$. (R4)

The asymptote angle, $\theta = 180°$. (R4)

The root locus lies on the portion of the real axis between $s = -5$ and $s = -\infty$, and at the point $s = 0$. (R5)

There is a breakaway point on the real axis at $s = 0$, and a breakin point at $s = -10$. (R6,R7)

Characteristic equation: $s^2 + Ks + 5K = 0$

Selected points on the root locus:

K	Characteristic Equation	s_1	s_2
0	$s^2 = 0$	0	0
5	$s^2 + 5s + 25 = 0$	$-2.5 + j4.33$	$-2.5 - j4.33$
10	$s^2 + 10s + 50 = 0$	$-5.0 + j5.0$	$-5.0 - j5.0$
15	$s^2 + 15s + 75 = 0$	$-7.5 + j4.33$	$-7.5 - j4.33$
20	$s^2 + 20s + 100 = 0$	-10	-10
30	$s^2 + 30s + 150 = 0$	-6.34	-23.66

16.18 The root locus has 3 branches that start at the three poles and go to infinity along the asymptotes. (R1,R2,R3)

The asymptote passes through the centroid, $c = -6.67$. (R4)

The asymptote angle, $\theta_1 = 60°$, $\theta_2 = -60°$, $\theta_3 = 180°$. (R4)

The root locus lies on the portion of the real axis between $s = 0$ and $s = -5$, and between $s = -15$ and $s = -\infty$. (R5)

There is a breakaway point on the real axis at $s = -2.26$, with $K = 78.89$. (R6,R7)

The branches cross the imaginary axis at $s = \pm j8.66$ with $K = 1500$. (R8)

Characteristic equation: $s^3 + 20s^2 + 75s + K = 0$

Selected points on the root locus:

K	s_1	s_2	s_3
78.89	-15.49	-2.26	-2.26
136.51	-15.80	$-2.1 + j2.06$	$-2.1 - j2.06$
138.62	-15.811	$-2.094 + j2.094$	$-2.094 - j2.094$
176.0	-16.0	$-2.0 + j2.65$	$-2.0 - j2.65$
408.0	-17.0	$-1.5 + j4.66$	$-1.5 - j4.66$
702.0	-18.0	$-1.0 + j6.16$	$-1.0 - j6.16$
1,064	-19.0	$-0.5 + j7.47$	$-0.5 - j7.47$
1,500	-20.0	$+ j8.66$	$- j8.66$

Operating point angle, $\theta = \arctan 0.707 = 45°$.

Operating point, $s_{op} = -2.094 \pm j2.094$.

Gain at the operating point, $K_{op} = 138.6$.

Natural frequency, $\omega_o = 2.96$ r/s.

Chapter 17

17.1 System 2: $G_u = 6.3$, $P_u = 6$ min

PI mode: $P = 2.8$, $I = 0.20$ min^{-1}

PID mode: $P = 3.8$, $I = 0.33$ min^{-1}

$D = 0.75$ min

System 4: $G_u = 1.2$, $P_u = 18$ min

PI mode: $P = 0.54$, $I = 0.067$ min^{-1}

PID mode: $P = 0.72$, $I = 0.11$ min^{-1}

$D = 2.25$ min

17.3 $\omega_{max} = 0.035$ rad/s

17.5 Design Summary

Control modes used: PD

Proportional gain = 0.00111

Integral action rate = 0 (not used)

Derivative action time constant = 0.31 s

Gain margin = 8.5 dB

Phase margin = 40.0°

Frequency limit = 9.7 rad/s

17.7 Design Summary

Control modes used: PI

Proportional gain = 3.41

Integral action rate = 0.0606

Derivative action time constant = 0 (not used)

Gain margin = 6.6 dB

Phase margin $= 40°$
Frequency limit $= 0.17$ rad/s

17.9 Design Summary
Control modes used: PI
Proportional gain $= 0.490$
Integral action rate $= 0.0128$ s^{-1}
Derivative action time constant $= 0$ (not used)
Gain margin $= 6$ dB
Phase margin $= 101°$
Frequency limit $= 0.044$ rad/s

17.11 Design Summary
Control modes used: PID
Proportional gain $= 1.23$
Integral action rate $= 4.63$ s^{-1}
Derivative action time constant $= 0.128$ s
Gain margin $= 6$ dB
Phase margin $= 46°$
Frequency limit $= 16$ rad/s

17.13 Retard factor, R.F. $= 0.25$
Design Summary
Control modes used: PID
Proportional gain $= 4.13$
Integral action rate $= 0.0025$ s^{-1}
Derivative action time constant $= 8.35$ s
Gain margin $= 13.0$ dB
Phase margin $= 40.0°$
Frequency limit $= 0.84$ rad/s

17.15 Design Summary: Temperature Control Loop
Control modes used: PID
Proportional gain $= 2.35$
Integral action rate $= 0.000745$ s^{-1}
Derivative action time constant $= 263$ s
Gain margin $= 12.5$ dB
Phase margin $= 40.0°$
Frequency limit $= 0.012$ rad/s

17.17 First Design Summary
Proportional gain $= 0.183$
Integral action rate $= 0.136$ s^{-1}
Derivative action time constant $= 0$
Gain margin $= 51.0$ dB
Phase margin $= 40°$
Frequency limit $= 0.017$ rad/s
Second Design Summary
Proportional gain $= 36.5$
Integral action rate $= 0.044$ s^{-1}
Derivative action time constant $= 0$
Gain margin $= 6.0$ dB
Phase margin $= 40°$
Frequency limit $= 0.46$ rad/s

17.19 Design Summary

Proportional gain = 3.50

Integral action rate = 0.147 s^{-1}

Derivative action time constant = 0

Gain margin = 6.2 dB

Phase margin = $40°$

Frequency limit = 0.46 rad/s

The gain margin, phase margin, frequency limit, and error ratio graphs are almost identical to those obtained in Exercise 17.17, Design 2.

The designs with the highest frequency limit are the two in Exercise 17.18 with $I = 0.046 \text{ s}^{-1}$ and $I = 0.069 \text{ s}^{-1}$.

Index

Error ratio, 626, 636, 672
Ethernet LAN, 330
Event-driven operation, 45, 68
Event-driven process, 508–523, 532
Expert system control, 537

Fanning equation, 125, 154
Fault, 532
Feedback, 7, 10, 29
Feedback control, operations, 11
Feedforward control, 537, 570, 575
Field poles, 454, 458, 497
Filled thermal system (FTS), 381, 404
Film coefficient, 141
Final controlling element, 15
First-order lag plus dead-time process, 617
First-order lag process, 587–599, 618
    blending, 591–592
    electrical, 589
    gas pressure, 591–592
    liquid level, 589
    thermal, 590–591
Flash or parallel ADC, 288, 292
Flow rate measurement, 388–393
Flowchart, 505, 532
Follow-up system, 36, 68
Force measurement, 368–372
Force sensor, pneumatic, 371
Force sensor, strain gage, 368
Frequency domain, 171, 192
Frequency limit, 626, 672
Frequency response, 8, 160, 186–191, 192,
    219, 626
Frequency response, how obtained, 189
Frequency response graph, 582, 618
Frequency shift keying (FSK), 327
Frequency spectrum, 292
Full-duplex, 320, 335
Full-step operation, 483, 497
Function generator, 250, 292
Functional Laplace transforms, 173, 192

Gage factor, 369, 373
Gage pressure, 394, 404
Gain, 7, 29, 186
Gain margin, 627, 651, 672
Gain margin point, 655, 672
Gateway, 335
Gray code, 84
Hagen-Poiseuille law, 125, 154
Half-duplex, 319, 335
Half-step operation, 485, 497
HDLC protocol, 333
Headend repeater, 335
Hexadecimal numbering system, 75, 109
High-pass filter, 254, 292
Hydraulic motor, 428–430, 448
Hysteresis, 20, 29, 207, 224, 358

I/O interface, 532
I/O scan time, 532
IEEE-488 data bus, 324
IEEE-583 interface, 325
Incremental encoder, 353, 373
Inductance, defined, 117
    electrical, 121, 154
Inertance, defined, 117
    liquid flow, 133, 154
Inertia, 154
    defined, 117
    mechanical, 151
Instability, 29
Instrumentation amplifier (IA), 248, 292
Instrumentation symbols and identification,
    16, 735–741
Integral, 192
Integral action time constant, 584, 619
Integral mode (I), 14, 29
Integral process, 583–587, 619
Integration-lead compensation, 712, 718
Integrator, 245, 292
Integro-differential equations, 168, 192
Intel 8080, 303–304
Intel 8085, 303–304
Intel 8086/8088/80X86, 308–310
Interacting second-order process, 603, 619
Inverse Laplace transform, 180, 192
Inverse Laplace transformation, 160, 172
Inverter, 497
Inverter, pulse-width-modulated (PWM),
    497
Inverter, variable-voltage (VVI), 497
Inverting amplifier, 241, 292
Isolation, 232
Isolation amplifiers, 254, 292

Karnaugh map, 97, 109

Ladder diagram, 505, 508, 532
Lag compensation, 714, 718
Lag-lead compensation, 714, 718
Laminar flow, 125, 154
Laplace transform, 160, 170, 192
Latent head of fusion, table of, 726
Lead-lag compensation, 713, 719
Linear differential equation, 192
Linear operating region, 235, 292
Linearity, 19, 207, 224
    independent, 209, 224
    least-squares, 209, 224
    terminal-based, 209, 225
    zero-based, 209, 225
Linearization, 232, 251, 259, 292
Liquid level measurement, 398–403
Liquid tank, nonregulating, 162
Liquid tank, self-regulating, 161
Load, 29